MICROBIOLOGY RESEARCH ADVANCES

MUSHROOMS

TYPES, PROPERTIES AND NUTRITION

MICROBIOLOGY RESEARCH ADVANCES

Additional books in this series can be found on Nova's website
under the Series tab.

Additional E-books in this series can be found on Nova's website
under the E-book tab.

FOOD SCIENCE AND TECHNOLOGY

Additional books in this series can be found on Nova's website
under the Series tab.

Additional E-books in this series can be found on Nova's website
under the E-book tab.

MICROBIOLOGY RESEARCH ADVANCES

MUSHROOMS

TYPES, PROPERTIES AND NUTRITION

SOPHIE ANDRES
AND
NIKO BAUMANN
EDITORS

Nova Science Publishers, Inc.
New York

For permission to use material from this book please contact us:
Telephone 631-231-7269; Fax 631-231-8175
Web Site: http://www.novapublishers.com

Library of Congress Cataloging-in-Publication Data

Mushrooms : types, properties and nutrition / editors: Sophie Andres and Niko Baumann.
p. cm.
Includes bibliographical references and index.
ISBN 978-1-61470-110-1 (hardcover : alk. paper) 1. Mushrooms. 2. Mushrooms--Nutrition. I. Andres, Sophie. II. Baumann, Niko.
QK617.M87 2011
579.6--dc23
2011019895

Published by Nova Science Publishers, Inc. † New York

Contents

PREFACE

This book presents topical research from across the globe in the study of the types, properties and nutrition of mushrooms. Topics discussed include edible mushroom role in weight regulation; disease control in the mushroom industry; sustainable use of microbial endophytes; hemolytic lectins of higher mushrooms; radioactivity of European wild growing edible mushrooms; co-cultivation of pleurotus species with yeasts; biologically active mushroom proteins and activation and stabilization of mushroom tyrosinase by addition of polyethylene glycols.

Chapter 1 – The advantage of submerged mushroom culture for the production of mushroom derivatives is clear: mycelia formed by growing pure strains in submerged culture under controlled conditions not only have a consistent composition, but are also safer. One of the first macrofungi to be cultivated at large scale in submerged culture was the basidiomycete *Lentinula edodes* (Berk.) Pegler, otherwise known as shiitake.

The topics on submerged cultivation related to biosynthesis of extracellular polysaccharides and other medicinal substances are covered. Attention is paid to the extracellular proteins production (oxidative enzymes, ribonucleases, enzymatic activity in dual cultures of *Lentinula*) and morphogenesis in liquid culture.

Then we turn our attention to the results that our group has obtained working with *L. edodes* over the last ten years.

The formation of pigmented mycelial film in submerged culture was assayed to find correlations with both the medium composition and the age of inoculums. The quantum chemical methods and QSAR were applied to test our supposition that a differential character of interaction between the amino acids under study and Ca^{2+}, Mn^{2+} cations should be related not to the distinct electron structures of zwitter ions, but most likely to their differing hydrophobicity.

The growth parameters and the dynamics of lectin activity of *L. edodes* on liquid media enriched with selenium in the organic form have been studied. Diacetophenonylselenide (DAPS-25) exerts the appreciably positive effect upon the vital processes of *L. edodes*.

The effort to explain the metal cations effect on the lectin activity of *L. edodes* from the viewpoint of quantum chemical considerations has been attempted, starting from the dynamics of change in the extracellular lectins activity in the presence of double-charged cations of iron, cobalt, nickel in the composition of liquid synthetic culture medium.

The dependence of periods before fruiting of *L. edodes* upon the presence of divalent manganese in the liquid media, when growing the seeding mycelium by submerged cultivation, has been explored. Provided that Mn^{2+} enters the composition of liquid medium, so the considerable decrease in the duration of periods before both the brown mycelial film formation in liquid culture and the fruit bodies' appearance on using the intermediate seeding culture thus obtained, has been observed.

The feasibility has been examined and the parameters selected for the co-cultivation of *L. edodes* and bacteria from *Azospirillum* genus. The co-culture of *Lentunus* sp. and *Azospirillum* sp. has been never studied previously. An explicit advantage of co-culture with respect to its capability of suppressing the microfloral contamination has been noted.

The characterization of the group of extracellular indolic compounds of *L. edodes* in relation to the submerged cultivation conditions has been presented.

Based on the aforesaid, several novel methods for growing the seeding mycelium of shiitake has been proposed and confirmed by national patents.

Chapter 2 - Mushrooms are a class of "macrofungi" which in recent years have gained popularity among general consumers and the scientific community. Edible mushrooms have been shown to possess high nutritive value, and are being considered as "nutraceuticals" due to their potential health benefits. *In vitro* and *in vivo* studies suggest that bioactive compounds in mushrooms have immunomodulatory, anti-inflammatory, anti-oxidative, lipid lowering and anti-tumor effects. Some research findings suggest that edible mushroom consumption is associated with prevention of cardiovascular diseases. Many of these health benefits can be attributed to the nutrient profile of the mushrooms, which include protein, B-vitamins, minerals, along with low fat content. In addition, bioactive compounds like polyphenols and flavonoids present in mushrooms may contribute to health benefits.

Weight management is a growing concern for all segments of the U.S. population. Moreover, recent statistics provide compelling evidence that overweight and obesity are on the rise not only among adults, but also in children younger than twelve years of age. This growing trend in overweight and obesity imposes major health, economic and social burdens worldwide. While genetic, biologic and social factors are responsible to a great extent, many individuals engage in dietary behaviors that put them at risk of weight gain.

On the intake side of the energy balance equation, high intake of energy-dense foods is largely responsible for the problem of positive energy balance, and the resulting epidemic of overweight and obesity in the U.S. Energy regulation is complex, but there is evidence that humans have limited ability to regulate food intake in response to changes in energy density. One obvious method of preventing passive overconsumption when consuming energy dense foods is substitution of low energy dense foods. However, palatability differences, access, cost, and habit may prevent people from choosing such foods.

Edible mushrooms are very low in calories and energy density. They have a nutrient profile similar to that of many foods recommended in weight loss/maintenance diets. However many "diet" foods have low palatability, which may result in non-adherence to these foods by those seeking to lose weight and keep it off. In contrast, edible mushrooms are generally regarded by adults as highly palatable, making them a potentially good substitute for high energy density foods in the prevention and treatment of obesity. In support of this hypothesis, one study we conducted in humans on the effect of mushroom on satiety and palatability showed promising results, suggesting a potential role in improving weight regulation. In addition, mushrooms may have the added benefit of reducing oxidative stress and inflammation, conditions that are associated with obesity and a number of chronic diseases. Future research is needed on the effect of mushroom intake on body weight regulation, and associated health benefits.

Chapter 3 – Button mushroom (Agaricus bisporus), oyster mushroom (*Pleurotus* sp.) and shiitake (*Lentinus edodes*) are the most commonly cultivated basidiomycetes worldwide. The production of fruiting bodies is severely afflicted by fungal, bacterial, and viral pathogens that can cause diseases which have an effect on yield and quality. Major *A. bisporus* fungal pathogens are *Mycogone perniciosa*, *Lecanicillium fungicola*, and *Cladobotryum* with two species mycophillum and dendroides, causal agents of dry bubble, wet bubble, and cobweb disease, respectively. Various *Trichoderma* species are the causal organisms of green mould, affecting all three edible mushrooms. The usual method of controlling of diseases on farms worldwide is based on the use of fungicides. However, development of pathogen resistance to fungicides after frequent application and host sensitivity to fungicides are serious problems. For improvement of crop protection and reduction of production costs, the effects of some new fungicides are being tested. The strains of edible mushrooms that were recently commonly cultivated seem to be more tolerant to fungicides in vitro than the earlier commercial strains. Resistance to benzimidazole fungicides has developed and the number of available fungicides is decreasing. Since studies of fungicides' efficacy on cultivated mushrooms by agrochemical companies are very rare, only few fungicides are officially recommended in the mushroom industry: prochloraz in EU countries, and chlorothalonil, thiabendazol and tiophanate-methyl in U.S. and Canada. Decreased sensitivity of *L. fungicola* to prochloraz was noted. Also, inefficiency of this fungicide was recorded in experimental growing room, at a level of spotting symptoms of cobweb disease. However, with regard to resistance development, harm to the environment and human health, special attention should be focused on a good programme of hygiene. The introduction of new fungicides of biological origin creates new possibilities for crop protection from fungal pathogens.

Chapter 4 – Mushroom culture is a biotechnological process that recycles ligninocellulosic wastes and can be produced on natural materials from agriculture, woodland, animal husbandry, and manufacturing industries and the spent substrates can be used in different ways (soil conditioner, compost, cattle feed). Edible mushrooms are highly appreciated and have a commercial potential in many countries. They have a chemical composition, which is attractive from the nutritional point of view for protein. Mushroom yields and biological efficiencyvary according to biological factors, environmental conditions and growing substrates.

Mushroom cultivation is a well-established and profitable agribusiness carried out worldwide on a large or small scale. Cultivation on sugarcane substrates is an option to be considered due to low cost as substrate, physic-chemical structure, and their positive economical, social and environmental factors for the sugarcane cultivation areas in a sugarcane diversification concept in comparison with other agricultural by-products. However, successful technology transfer programs to increase yields and reduce production costs of edible mushrooms are limited. The sustainable model for production of edible mushrooms represents a strategy that allows large-scale, small-scale, and domestic cultivation to promote regional development.

This article discusses the development of an agro-industry of fresh and dried edible mushrooms with sugarcane production; major fundamental technologies involved: Spawn technology, Mushroom production technology, and processing technology and how their economic and nutritional potentials will increase the productivity of mushrooms on sugarcane substrates with better flavor, appearance, texture, nutritional qualities at low cost.

Chapter 5 – In the past two decades, a great deal of information on the potential role of endophytes in nature has been collected. Endophytes, microorganisms that colonize internal host tissues of living plants, are a relatively unstudied and dependable source of bioactive and chemically novel compounds with potential for exploitation in a wide variety of medical and pharmacological areas. The mechanism through which endophytes exist and respond to their surroundings must be better understood. In most cases their relationship with the host plant is symbiotic and probably mutualistic. Many are capable of synthesizing bio-active compounds that can be used by the plant for defense against pathogenic fungi and bacteria. Some of these compounds have proven useful for novel drug discovery. The opportunity to find new and interesting endophytes among myriad of plants is great. Endophytes represent a huge diversity of microbial adaptations that have developed in special and sequestered environments, and their diversity and specialized habituation make them an exciting field of study in the search for new medicines. The hunt for new drugs is particularly important in view of the fact that so many diseases are developing immunity to some of the current treatments. This review will concentrate on what has been discovered, and what is still unknown about endophytes that synthesize chemicals with bioactive properties.

Chapter 6 – Tens of wild growing mushroom species are widely consumed as a delicacy in a part of Europe. Knowledge of their nutritional value has so far been fragmentary mainly due to the very limited information on the bioavailability of their constituents. Dry matter content varies usually between 80 and 140 g kg^{-1}. Usual medians of crude protein, lipid and ash content are about 25, 3 and 8 g per 100 g of dry matter, respectively. Various carbohydrates form the rest. However, great variations occur. Energy is low, about 150 kJ per 100 g of fresh mushrooms. The proportion of essential amino acids seems to be nutritionally favorable, while the content of n-3 fatty acids is negligible. Chitin, glycogen, mannitol and trehalose are typical carbohydrate constituents. Potassium is the highly prevailing element within minerals. Relatively high proportion of fiber, health-promoting β-glucans, compounds with antioxidation activity and flavor constituents are the topics provoking an increasing interest of both researchers and consumers. Nevertheless, several popular species accumulate high levels of cadmium, mercury and lead if growing on heavily polluted soils.

Chapter 7 – Traditionally mushrooms considered in daily diet especially in the orient as they are rich in nutrition and have abundant therapeutic property. Nowadays large scale mushroom cultivation takes attention of many pharmaceutical companies around the world due to their rich and variety of medicinal compounds. Extensive research in isolation, characterization and production of these biological components is conducted in last decades. Among them numerous studies dedicates to polysaccharide production as they exhibit immunomodulator and anti-tumor activities. Glucan with different active unit linkage such as (1→3), (1→6)-β-glucan and (1→3)-α- glucans constitute mushroom polysaccharides which perform immunomodultor activity as they are biological response modifiers (BRMs). Beta-glucan obtained from the *Pleurotus ostreatus (*Oyster mushroom*)* is known in the scientific literature as pleuran. It is a structural cell wall compound. It stimulates the body′s defense system against infections (bacterial, viral, yeast, and parasitic) and modulates the blood producing activity of bone marrow. In this chapter we are going to focus on the structure of pleuran, followed by its production strategies. Furthermore the therapeutic application of pleuran will be demonstrated and further researched will be suggested.

Chapter 8 – The *Agaricus subrefescens*Peck [*A. blazei* (Murrill) ss. Heinemann] mushroom has been widely studied in various parts of the world, due to its medical and pharmacological properties.As substances of interest, the cell wall of the fungi contains polysaccharides called β-glucans, which have a structural function. During the 1980s, the *A. subrufescens* mushroom was imported to Japan due to its alleged health effects and is widely used today in Oriental countries both as an edible mushroom, considered a functional food, and as a natural therapy in the form of a medicinal extract, used mostly for the prevention and treatment of cancer. In accordance with Brazilian tradition, it could be useful against a variety of diseases, such as diabetes, atherosclerosis, hepatitis, hypercholesterolemia and heart disease, among others. The mushroom is commercialized in several countries as a nutraceutical product, which is a novel class of dietary supplements, including partially refined extract or dried biomass from the mushroom made into a capsule or tablet. In general, the compositional analysis of *A. subrufescens* mushrooms is water (84-87%), protein (30-33%), fat (0.82-1.3%), fiber (5.6-6.8%) and minerals (5.9-7.1%), with β-glucan contents between 4.4 and 6.9g 100 g^{-1} of mushroom. Active metabolites can be isolated from basidiomes; pure culture mycelia, culture filtrate, and currently numerous attempts are in progress to obtain active metabolites from the mycelia through submerged fermentation culture due the cheaper preparations. The first separation of active anticancer compounds purified from the sodium hydroxide extract of the fruit body occurred in 1989. Following this achievement, several controversies have arisen, mainly regarding the nutraceutical quality of the extracts obtained and in relation to the tests conducted, emphasizing that the chemical composition and the stage of maturity of the mushrooms, as well as the cultivation practices used in growing them, are crucial to obtaining quality basidiomes with high concentrations of β-glucans. Thus, in this book chapter, the problems arising from the variability of the content of β-glucans of mushrooms due to different practices and growing techniques adopted for the *A. subrufescens*production are discussed. To achieve this, five strains (ABL 99/28, ABL 99/30, ABL 03/44, ABL 04/49 and ABL 06/59), cultivated in three composts (Massai straw + sugar cane bagasse, oat straw + sugar cane bagasse and Aruana straw + sugar cane bagasse), with four casing layers (soil + charcoal, soil + coconut fiber, soil + peat moss and soil + composted pine bark) in four growing environments (climatized chamber, greenhouse with a transparent plastic film, greenhouse with a milky-white plastic film and greenhouse with a Duplalon® plastic film)were studied, with the aim of analyzing the variability of β-glucan content in the harvested mushroom and the agronomic performance of the growing practices. The ABL 04/49 strain cultivated with Massai straw and sugar cane bagasse showed the highest concentrations of β-glucans, but not a best yield, which was achieved by the ABL 99/30 strain cultivated with oat straw + sugar cane bagasse and obtained good β-glucan content and high yield. For agronomic performance, soil + composted pine bark was highlighted, influenced by the cultivation environment, to achieve high yield. Finally, following the presentation of a series of results, the creation of a growing protocol is suggested, which seeks to adopt cultivation practices that enhancethe presence of β-glucans in the mushrooms.

Chapter 9 – The results of study of 4 specific lectins isolated from higher *Amanita* genus mushrooms and possessing hemolytic activity are presented. Two of them – *A. phalloides and A. virosa*, belong to *Amanitaceae* family, while two others - *Mycena pura* and *Laetiporus sulphureus*, belong to *Tricholomataceae* and *Corticiaceae* families, correspondingly. The purification methods used for their isolation are described, and the comparative analysis of physico-chemical and biological properties of these lectins has been done. It was shown that according to the peculiarities of the action towards erythrocytes, these lectins can be divided into two groups. While in *Mycena pura* and *Laetiporus sulphureus* lectins, the hemagglutinating activity is higher than the hemolytic activity, in *A. virosa* and *A. phalloides* lectins, situation is opposite and the hemolytic activity is higher than the hemagglutinating activity. The hemolytic activity of the lectins under study was inhibited by the polyethylene glycol of high molecular mass. At the same time, the presence of polyethylene glycol did not influence the hemagglutinating ability of the lectins. This allowed studying their carbohydrate specificity, and it was found that the lectins of *Amanita* genus mushrooms considerably differed in this specificity. The hemolytic lectins also possessed antimicrobial action and inhibited growth of Gram-positive microorganisms and *Proteus*. Besides, they demonstrated cytotoxic action towards cultured mammalian leukemia cells. In general, these mushroom lectins resemble cytolytic lectins of lower invertebrates. Their potential biological role is considered.

Chapter 10 – Wild growing mushrooms are widely consumed as a delicacy in several European countries, at levels up to several kg per year per capita. Activity concentrations of the natural isotope ^{40}K are usually 800-1,500 Bq kg^{-1} dry matter (DM). Other natural radionuclides with leading ^{210}Pb and ^{210}Po are of lower importance. Activities of ^{137}Cs from nuclear weapons testing below 1,000 Bq kg^{-1} DM were commonly reported until 1986. The situation changed dramatically after the accident of Chernobyl nuclear power station in 1986. Activities up to over 100,000 Bq kg^{-1} DM of ^{137}Cs and to a lesser extent of ^{134}Cs were observed in some edible species in the following years. Commonly, mycorrhizal species accumulate radiocesium more than species with saprotrophic or parasitic nutritional strategy. *Xerocomus badius, X. chrysenteron, Suillus variegatus, Rozites caperata, Laccaria amethystina* and *Hydnum repandum* belong among the radiocesium highly accumulating and widely consumed species. Activity concentrations have been affected by several environmental factors, such as rate of soil contamination by the Chernobyl fallout, the depth from which mycelium takes nutrients and time since the accident. Most of the ^{137}Cs in forest soils appear to be available for uptake by mushrooms until now. A considerable consumption of accumulating species collected from the sites heavily contaminated in 1986 can be still of a health concern. The contamination can be reduced by soaking or cooking of dried or frozen mushroom slices. Until now, meat of wild boars eating some mushroom species from heavily contaminated areas can highly surpass statutory limit for ^{137}Cs.

Chapter 11 – White button mushroom (*Agaricus bisporus*) is the most commonly cultivated mushroom species which comprises about 32% of world mushroom production. It is a good source of proteins, amino acids, carbohydrates, vitamins (mainly vitamin D and vitamin B-complex) and minerals, such as iron, calcium, phosphorous and potassium. *Agaricus bisporus* mushroom is valued for its white appearance and browning of its cap is an indicator of poor quality. Browning and bruising of the mushroom surface lead to reduced shelf-life and lower financial returns to producers, therefore there is a need for objective evaluation of mushroom quality to ensure that only high quality produce reaches the market. Traditional methods have been widely used to evaluate mushroom quality and shelf life based on several quality parameters such as visual appearance, size, colour, maturity stage, development stage, microbial growth and weight loss. However, today's mushroom industry demands non-destructive, rapid and accurate on-line sensor technologies to monitor and control mushroom quality and safety. In this chapter, some traditional quality grading methods as well as application of novel techniques for mushroom quality and shelf life evaluation are discussed.

Chapter 12 – *Basidiomycetes* mushrooms comprise a diversity of gill fungi that occur in most terrestrial ecosystems worldwide. However, data about their diversity and biological applications in tropical ecosystems remain almost unknown. Some *Basidiomycetes* species from Brazilian tropical rainforests have been described as sources of primary and secondary bioactive compounds with antimicrobial, anti-inflammatory, immunomodulatory, antiparasitic, and antitumoral metabolites and as nutraceutical foods. In addition, some *Basidiomycetes* have demonstrated potential as producers of interesting prototype molecules for the development of drugs useful in medicine. The purpose of the present chapter will be to summarize the current status of the diversity, biotechnological use, and ability to produce different bioactive compounds of *Basidiomycetes* mushrooms from Brazilian tropical rainforests.

Chapter 13 – Fungi of the genus *Pleurotus* are one of the most popular edible fungi in the world. Most of the cultivaion substrates and wood too is very poor in nitrogen, so one of the greatest problems for all wood decay fungi including the *Pleurotus* species is to provide themselves nitrogen nutrition. To satisfy their nitrogen requirement *Pleurotus* fungi can parasitize on the bacteria, yeasts, and algae cells in nature and during the cultivation process.

Some of the widely cultivated *Pleurotus* species (*P. citrinopileatus*, *P. djamor*, *P. eryngii*, *P. ostreatus*, and *P. pulmonarius*) were observed in mixed cultures with 8 species (10 strains) of yeasts from different taxonomic groups (ascomycetous, i.e. *Debaryomyces hansenii, Hanseniaspora uvarum*, *Kluyveromyces marxianus*, *Metschnikowia pulcherrima*, *Saccharomyces cerevisiae*, and basidiomycetous, i.e. *Cryptococcus albidus, Cystofilobasidium capitatum, Rhodotorula minuta*) on agar media. To provide co-cultivation on agar media and substrate, water suspension of alive or dead yeast cells or the suspension supernatant was used.

The morphological and physiological peculiarities of *Pleurotus* mycelium were described on agar medium in mixed cultures with the suspension of alive/dead yeast cells. Some specialized mycelial structures were found only in the mixed cultures with yeasts, and weren't found in the pure culture, e.g. short nipple-like appendages and coralloid hyphae, which provide contacts between mycelium and yeast cells.

The trophic preferendum was defined for each of five *Pleurotus* species, and then *Pleurotus* fungi were cultivated on substrate in association with the yeast species from the range of trophic preferendum. On substrate all five *Pleurotus* species demonstrated the same reaction on the yeast cells presence. Mycelium growth speed was higher on substrate with yeast suspension. Biomass of the fruiting bodies was higher for different species, and the highest increase was shown for *P. eryngii*, twofold in comparison with the biomass of the fruiting bodies which grew on the substrate without yeasts. Analyses of amino acids and volatile compounds of *P. citrinopileatus* fruiting bodies, cultivated with and without yeast *S. cerevisiae,* demonstated high food value and wide spectrum of odor.

Chapter 14 – Mushrooms are functional foods largely consumed in Asia and Europe. The consumption of mushroom in Brazil is very low, in part due to the lack of information about the nutritional value of the edible mushrooms cultivated in the country. The objective of this chapter was to investigate the antioxidant properties and to quantify the phenolic contents of hydroalcoholic extracts of the most commonly mushrooms consumed in Brazil: *Agaricus bisporus* (white button mushroom), *Lentinula edodes* (shiitake), *Pleurotus ostreatus* (shimeji) and *Pleurotus eryngii* (eryngii). Among four edible mushrooms studied, *A. bisporus* presented the highest phenolic contents (19.97±2.01 mg/g dry extract), followed by *P. ostreatus* (16.05±1.10 mg/g), *P. eryngii* (12.74±0.98 mg/g) and *L. edodes* (10.00±0.76 mg/g). The antioxidant activities were evaluated using four methods. A strong correlation was found between the phenolic contents and the antioxidant activity determined by the scavenging activity of DPPH radicals, reducing power and lipid peroxidation inhibition. On the other hand, a weak correlation was found between the phenolic content and the chelating activity against Fe^{2+}. The results obtained allow the authors to conclude that the mushrooms produced in Brazil present antioxidant properties, and that the most popular of them, *A. bisporus*, appears as an excellent option to enrich the diet with antioxidant phenolic compounds.

Chapter 15 – The objective of this article is to review mushroom proteins with biological activity. They include antifungal proteins, antibacterial proteins, ribosome inactivating proteins, ribonucleases, ubiquitin- like proteins, lectins, proteases, hemolysins, laccases, xylanases, acid phosphatases, L-amino acid oxidases, α-mannosidases, deoxyribonucleases, carboxylesterases, polysaccharopeptides and polysaccharoproteins, amatoxins, phallotoxins and virotoxins;

Chapter 16 – Effect of polyethylene glycol (PEG) and its molecular weight on the activity and stability of mushroom tyrosinase was studied systematically. The kinetic parameters (such as K_m, V_{max}, optimal pH and temperature, and activation energy) and the thermostability of the enzyme at different temperatures were determined and compared in the aqueous solution with and without addition of PEGs. Polyethylene glycol showed the ability of both activating and stabilizing the enzyme (e.g., a 1.6-fold increase in V_{max}/K_m and a 1.8-fold increase in $t_{1/2}$ were achieved in the presence of PEG 10,000), and such PEG-induced activation and stabilization effect was molecular weight dependent. Both our activity and stability data suggest that PEG can affect the enzyme activity and stability basically via its interaction with the enzyme so as to modify the enzyme's active site and conformation or steric structure.

Chapter 17 – Athletes are known to be prone to illnesses such as infection and asthma, which is due to exercise-related skewness of the immune response, and injury-related inflammation in the muscle-skeleton system. The Brazilian medicinal mushroom, *Agaricus blazei* Murill (AbM), has strong immunomodulatory properties and is shown to protect against infections, allergy and asthma in animal models. An AbM-based extract (AndoSan^{TM}) is also found to have anti-inflammatory effect both in healthy individuals and in patients with inflammatory bowel diseases. Hence, it is suggested that AbM mushroom extract may be beneficial for athletes during strenuous exercise.

Chapter 18 – *Pleurotus* mushrooms are popular edible mushrooms all over the world. Studies on this genus have shown that these mushrooms possess anticancer, antimicrobial, antiviral, antioxidant, antinociceptive, antidiabetic activities and cholesterol lowering properties. Proteins, enzymes, polysaccharides such as glucans, and glycoproteins isolated from *Pleurotus* mushrooms are responsible for the said activities. Proteins such as ubiquitins, hemolysins and lectins have shown antiviral activity. It is noteworthy, that anti-HIV activity was observed in all 3 classes of aforementioned proteins. Lignolytic enzymes such as laccases present in *Pleurotus* have helped in converting agricultural waste to animal feed. Glucans are known to possess anticancer activity, antiviral activity and cholesterol lowering properties. Water insoluble glucans present in *Pleurotus* have been converted to water soluble carboxymethylated and sulfated glucans having antitumor and antiviral activities respectively.

A few secondary plant metabolites are identified from *Pleurotus* mushrooms such as pleurotin which has promising anticancer properties and lovastain with cholesterol lowering properties. An oxidized form of the abundant fungal steroid ergosterol has shown antifungal activity while ergosterol itself was implicated in bringing about inhibition of lipid peroxidation. The total phenolic content is also positively correlated to the antioxidant activity. Carbon-8 compounds arising from the hydroperoxidation of linoleic acid present in the fruiting body, is identified as a compound responsible for the production of the pleasant aroma in the mushroom. A mixture containing such volatile compounds mimicking the concentration in the fruiting bodies has shown antibacterial activity.

Chapter 19 – Mushrooms produce a number of molecules with medicinal properties, including polysaccharides, polysaccharide-peptide complexes, lectins, and antifungal proteins have been isolated from several mushroom species, but the one from *Coriolus versicolor* designated as PSP is the most extensively studied.PSP induces apoptosis in a variety of tumor cells by regulating the expression of pro- and anti-apoptotic genes. It inhibits angiogenesis, regulates the production of cytokines, counteracts the action of immunosuppressive agents, and potentiates the action of antitumor drugs.PSP is devoid of toxicity on female reproductive physiology and embryonic development. Polysaccharide-peptide complexes from other mushroom species also manifest antitumor and immunoregulatory activities.

In: Mushrooms: Types, Properties and Nutrition
Editors: S. Andres and N. Baumann
ISBN 978-1-61470-110-1

Chapter 1

SHIITAKE MUSHROOM SUBMERGED CULTURE AND MORPHOGENESIS: REGULARITIES, COMPUTATIONS, PRACTICAL USE

***Olga M. Tsivileva*[1*], *Alexei N. Pankratov*[2+] and *Valentina E. Nikitina*[1]**

[1]Laboratory of Microbiology, Institute of Biochemistry and Physiology of Plants and Microorganisms, Russian Academy of Sciences, 13 Entuziastov Ave., Saratov 410049, Russia

[2]Division of Analytical Chemistry and Chemical Ecology, Institute of Chemistry, N. G. Chernyshevskii Saratov State University, 83 Astrakhanskaya Street, Saratov 410012, Russia

ABSTRACT

The advantage of submerged mushroom culture for the production of mushroom derivatives is clear: mycelia formed by growing pure strains in submerged culture under controlled conditions not only have a consistent composition, but are also safer. One of the first macrofungi to be cultivated at large scale in submerged culture was the basidiomycete *Lentinula edodes* (Berk.) Pegler, otherwise known as shiitake.

The topics on submerged cultivation related to biosynthesis of extracellular polysaccharides and other medicinal substances are covered. Attention is paid to the extracellular proteins production (oxidative enzymes, ribonucleases, enzymatic activity in dual cultures of *Lentinula*) and morphogenesis in liquid culture.

Then we turn our attention to the results that our group has obtained working with *L. edodes* over the last ten years.

The formation of pigmented mycelial film in submerged culture was assayed to find correlations with both the medium composition and the age of inoculums. The quantum chemical methods and QSAR were applied to test our supposition that a differential character of interaction between the amino acids under study and Ca^{2+}, Mn^{2+} cations

* Corresponding author: Olga M. Tsivileva. E-mail: tsivileva@ibppm.sgu.ru

+ E-mail: PankratovAN@info.sgu.ru

should be related not to the distinct electron structures of zwitter ions, but most likely to their differing hydrophobicity.

The growth parameters and the dynamics of lectin activity of *L. edodes* on liquid media enriched with selenium in the organic form have been studied. Diacetophenonylselenide (DAPS-25) exerts the appreciably positive effect upon the vital processes of *L. edodes*.

The effort to explain the metal cations effect on the lectin activity of *L. edodes* from the viewpoint of quantum chemical considerations has been attempted, starting from the dynamics of change in the extracellular lectins activity in the presence of double-charged cations of iron, cobalt, nickel in the composition of liquid synthetic culture medium.

The dependence of periods before fruiting of *L. edodes* upon the presence of divalent manganese in the liquid media, when growing the seeding mycelium by submerged cultivation, has been explored. Provided that Mn^{2+} enters the composition of liquid medium, so the considerable decrease in the duration of periods before both the brown mycelial film formation in liquid culture and the fruit bodies appearance on using the intermediate seeding culture thus obtained, has been observed.

The feasibility has been examined and the parameters selected for the co-cultivation of *L. edodes* and bacteria from *Azospirillum* genus. The co-culture of *Lentunus* sp. and *Azospirillum* sp. has been never studied previously. An explicit advantage of co-culture with respect to its capability of suppressing the microfloral contamination has been noted.

The characterization of the group of extracellular indolic compounds of *L. edodes* in relation to the submerged cultivation conditions has been presented.

Based on the aforesaid, several novel methods for growing the seeding mycelium of shiitake have been proposed and confirmed by national patents.

1. INTRODUCTION

The production volume of mushrooms has increased dramatically over the last several decades [Van Griensven, 2001]. Products from both mushroom and mushroom derivatives (mushroom nutriceuticals/dietary supplements) should have a positive global impact on long-term food nutrition, health care, environmental conservation and regeneration, economic and social changes [Chang, 2001]. Mushrooms do not merely constitute a highly nutritious source of food. More recently, attention has focused on a second area of exploitation, following the discovery that many of these fungi produce a range of metabolites of great interest to the pharmaceutical and food (*e.g.* flavour compounds) industries [Jong & Birmingham, 1993; Buswell *et al.*, 1996]. As a result, the present-day mushroom industry is based on two main components: the application of traditional, although modernized, techniques for the production of fruiting bodies (*i.e.* mushrooms themselves) and the application of modern biotechnological techniques to produce mushroom derivatives such as nutriceuticals and dietary supplements. Products from both mushroom crops and mushroom derivatives should have a positive global impact on long-term food nutrition, health care, environmental conservation and regeneration, and economic and social change [Chang, 2001].

Whenever mushroom derivatives are the desired products, the disadvantages that traditional fruiting body production suffers from are well known: relatively long cultivation times and the possibility of variations in product quality due to unavoidable variations in growing conditions. The advantage of submerged mushroom culture for the production of mushroom derivatives is clear: mycelia formed by growing pure cultures in submerged conditions are of constant composition, and submerged culture is the best technique for obtaining consistent and safe mushroom products [Wasser *et al.*, 2000; Reshetnikov *et al.*, 2001].

Morphology, life cycles and genetics of the basidiomycete *Lentinula edodes* (Berk.) Pegler [syn.: *Lentinus edodes* (Berk.) Sing.] (shiitake) have come under scrutiny of science only in the last three decades and laid the scientific foundations of current methods of the selection and cultivation of this mushroom [Przybylowicz & Donoghue, 1991]. Nevertheless, essential gaps in studying the biochemical aspects of *L. edodes* development and cultivation, especially in liquid media, exist to the present day. This paper does not cover traditional techniques for fruiting body production. Readers interested in this topic should see *e.g.* [Harris, 1986; Kirchhoff & Lelley, 1991; Leonard &Volk, 1992; Tan & Moore, 1992; Worrall & Yang, 1992; Levanon *et al.*, 1993a, 1993b; Donoghue & Denison, 1995; Mizuno, 1995; Royse, 1995; Vinciguerra, 1995; Smith, 1997; Kalberer, 2000; Royse & Sanchez-Vazquez, 2001; Philippoussis *et al.*, 2003].

2. Biosynthesis of Extracellular Polysaccharides and Other Medicinal Substances

The medicinal properties of substances occurring in higher basidiomycetes have become a subject of thorough investigations [Wasser & Weis, 1999]. This is especially true for the shiitake mushroom, the medicinal and therapeutic values of which have been recognized for many decades [Bisen *et al.*, 2010]. Mushrooms comprise a vast and yet largely untapped source of powerful new pharmaceutical products, particularly (and most importantly for modern medicine) representing a source of a large number of polysaccharides with antitumor and immunostimulating properties [Wasser, 2002]. Their practical application depends not only on their unique properties but also on biotechnological availability. Isolation and purification of polysaccharides from mushroom material (*i.e.* intracellular polysaccharides from mycelium and extracellular ones from culture broth) is relatively simple and straightforward, and can be carried out with minimal effort [Mizuno, 1996, 1999].

The biochemical mechanisms that mediate the biological activity of polysaccharides are still not clearly understood. What is obvious is that the (1→3)-β-glucan backbone is essential, and that the most active polymers have degrees of branching (DB) between 0.20 and 0.33 (*i.e.* a branch every three to five backbone residues). (1→3)-β-Glucans that have β-glucopyranosyl units attached by (1→6) linkages as single unit branches enhance the immune system systemically [Bohn & BeMiller, 1995]. This enhancement results in antitumor, antibacterial, antiviral, anticoagulatory and wound healing activities.

The currently available information about which polysaccharide properties are essential for immunocompetence is not necessarily contradictive, although the conditions that have been used for evaluating these properties are not uniform. As a result, some data suggest that triple helical structures formed from high molecular mass polymers are possibly important for immunopotentiating activity, while other data suggest that the activity is independent of any specifically ordered structure. Interpretation of these results is complicated by the fact that there are still other data which indicate that it is the distribution of the branch units along the backbone chain that is the main determinant of the activity. Futhermore, there are data that indicate both that β-glucopyranosyl units are required for immunopotentiating activity and that the specific nature of the substituent is unimportant. Finally, there are also data that indicate both that the more water-soluble polymers are more active (up to a certain DB) and that some insoluble aggregates are more stimulatory than the soluble polymers. The best conclusion at this time is that the immunopotentiating activity of (1→3)-β-glucans depends on a helical conformation and on the presence of hydrophilic groups located on the outside surface of the helix.

Polysaccharides from mushrooms do not attack cancer cells directly, but produce their antitumor effects by activating the immune responses of the host, stimulating natural killer cells, T cells, B cells, and macrophage-dependent immune system responses [Borchers *et al.*, 1999]. The immunomodulating action of mushroom polysaccharides is especially valuable as a means of prophylaxis (a mild and non-invasive form of treatment), prevention of metastatic tumors, and as a co-treatment with chemotherapy.

The commercial polysaccharide "lentinan" from the shiitake mushroom offers the most clinical evidence for antitumor activity [Wasser, 2002]. In fact, lentinan is one of the first three major drugs developed from medicinal mushrooms, all three being β-glucans. Lentinan is produced in submerged, agitated liquid cultures, giving biomass concentrations between 7 and 8 g/L, with the lentinan yield being greatest at pH=5 [Harvey *et al.*, 2001]. Polysaccharides with similar infrared spectra to those obtained from the fruiting bodies of *L. edodes* have been extracted from biomass and from the culture liquid itself. There appears to be some difference in the molecular composition of the intra- and extracellular lentinan.

Lentinan produced extracellularly is much more effective in stimulating antibody production in mice than that extracted either from fruiting bodies or from the biomass produced by fermentation methods. However, more recently, lentinan has been extracted from *L. edodes* mycelium *via* a new cost-effective procedure that resulted in high purity (88%) and yield [Ng & Yap, 2002]. Unlike previous studies, in which lentinan was given parenterally, in this study it was administered orally. It was shown that the antitumor efficacy was still expressed through this route of administration.

More and more pharmacologically active metabolites from mushrooms are being isolated and tested. For example, edible mushrooms including *L. edodes* have been reported to have anti-hypertensive effects [Kim *et al.*, 2002]. As another example, the mycelium-free broth of *L. edodes* grown in submerged liquid culture was bacteriostatic against *Streptococcus pyogenes*, *Staphylococcus aureus* and *Bacillus megaterium*, but had no antifungal activity against *Candida albicans* [Hatvani, 2001]. The substance responsible for the activity was heat-stable, could be extracted with chloroform and had a molecular mass under 10000. These characteristics suggested that the component might be lenthionine, an antibacterial and antifungal sulphur-containing compound. The culture fluid was less toxic to human tissue culture cells than to microbes. The antibacterial activity and the toxicity to human cells did not reside in the same component.

3. Aflatoxin-Inhibiting Effect of *Lentinula edodes*

L. edodes filtrates and mycelia exhibit aflatoxin-inhibiting effects [Fanelli *et al.*, 1984]. In other words, they act as an external stimulus affecting the antioxidant status in the toxin-producing fungus and this leads to inhibition of aflatoxin production. For example, culture filtrates of *L. edodes* added to potato dextrose broth inoculated with a toxigenic strain of *Aspergillus parasiticus* Speare inhibited aflatoxin production, with filtrates from 30-day-old static cultures and 15-day-old shaken cultures having the highest inhibitory activity. In a similar study, when mycelia of *L. edodes* were incubated on wheat seeds for 20 and 30 days and then *A. parasiticus* was subsequently inoculated onto the seeds, growth of *A. parasiticus* and aflatoxin production were inhibited. This strategy could potentially be used to control infection of grains by *A. parasiticus*, thereby preventing aflatoxin production.

Some works shed light on the grounds for these effects by elucidating the factors that stimulate aflatoxin production in the producing organism. The addition of some halogenated alkanes (bromotrichloromethane, carbon tetrachloride and chloroform) to cultures of *Aspergillus parasiticus* and *Aspergillus flavus* stimulates aflatoxin biosynthesis significantly [Fanelli *et al.*, 2000]. The oxygen requirement and antioxidant status of a toxigenic strain of *A. parasiticus* was compared with that of a nontoxicogenic strain in the trophophase and idiophase of growth [Jayashree & Subramanyam, 2000]. In comparison with the nontoxigenic strain, for which the oxygen requirements were relatively unaltered at various growth phases, the toxigenic strain exhibited greater oxygen requirements at trophophase, coinciding with the onset of aflatoxin production. The activities of antioxidant enzymes such as xanthine oxidase, superoxide dismutase, and glutathione peroxidase and the mycelial contents of thiobarbituric acid-reactive substances, as well as of reduced glutathione, were all enhanced during the progression of the toxigenic strain from the trophophase to the idiophase. The combined results suggest that aflatoxin production by the toxigenic strain may be a consequence of increased oxidative stress, leading to enhanced lipid peroxidation and free radical generation.

The mechanism underlying the aflatoxin-inhibiting effect of the *Lentinula edodes* culture filtrates was studied by analyzing the anti-oxidant activity and β-glucan content of the filtrates [Reverberi *et al.*, 2005]. Lyophilized filtrates from submerged culture of *L. edodes* stimulated *A. parasiticus* to produce anti-oxidant enzymes (superoxide dismutase, catalase and glutathione peroxidase), although aflatoxin inhibition was better correlated with the β-glucan content than with the anti-oxidant activity of the filtrates. RT-PCR analyses of culture-filtrate-treated mycelia of *A. parasiticus* showed a delay in the activation of aflR and norA, genes of the aflatoxin cluster, and a synchronous activation of hsf2-like factor, a homologue of a yeast transcription factor involved in oxidative stress responses. The delay of the transcription of the genes of the aflatoxin cluster led to a marked reduction of aflatoxin production. This research suggests that new perspectives for setting suitable strategies against aflatoxins and *L. edodes* should be considered a promising tool.

4. Production of Extracellular Oxidases

One of the most economically-viable processes for the bioconversion of many types of lignocellulosic wastes is represented by edible mushroom cultivation, *L. edodes* being among the most important commercially cultivated mushrooms which exhibit high ability to utilize different lignocellulosics as growth substrate [Buswell *et al.*, 1996]. The studies concerning the factors affecting the ligninolytic activities of important white-rot fungi, including *L. edodes*, under the conditions of submerged cultivation are consequently of great interest.

The mushroom species produces a variety of extracellular oxidases and hydrolases, the most abundant of which is laccase, a polyphenoloxidase [Royse *et al.*, 1985]. The activities of three recognized lignin-modifying extracellular oxido-reductases were examined in [Buswell *et al.*, 1996], namely lignin peroxidase (LiP), manganese-dependent peroxidase (MnP) and laccase. *L. edodes* was grown on a defined medium with glucose as the sole carbon source. As Buswell *et al.* reported previously [Buswell *et al.*, 1995], MnP production by *L. edodes* was dependent upon nutrient nitrogen levels in the culture medium and, under low-nitrogen conditions (2.6 mM-N), enzyme activity was detected after 5 weeks and reached a maximum of 300 U/ml after 12 weeks. No MnP was recorded in high-nitrogen cultures (26 mM-N) over the 13 week experimental period. Lignin degradation by *L. edodes* is Mn dependent [Leatham, 1986], and the concentration of Mn in the culture medium also influences both the titre and the time of appearance of MnP in low-nitrogen cultures of *L. edodes* [Buswell *et al.*, 1995]. Highest enzyme titres were recorded in cultures containing 1.1 ppm added Mn. Since manganese supplementation at concentrations over the range 0.3-15 ppm only marginally affected fungal growth, the effect of Mn on MnP production was not due to increased biomass production. This response to manganese may result from Mn induction of MnP expression. No LiP activity was detected in the extracellular medium following growth of *L. edodes* under the growth conditions used in the study [Buswell *et al.*, 1996].

In the case of laccase, the position was reversed. Low nutrient nitrogen conditions are not required for laccase biosynthesis by the mushroom. Unlike the multiple forms of MnP and laccase reported in some other lignolytic fungi, native PAGE of crude *L. edodes* culture supernatants revealed only one MnP protein and one laccase protein. Activity staining with ABTS in the presence of H_2O_2 revealed a single band with MnP activity only in gels of low-nitrogen culture supernatants.

A laccase (EC 1.10.3.2) was isolated from the culture filtrate of *L. edodes* [Nagai *et al.*, 2002]. The enzyme was purified to a homogeneous preparation using hydrophobic, anion-exchange, and size-exclusion chromatographies. SDS-PAGE analysis showed the purified laccase, Lcc 1, to be a monomeric protein of 72.2 kDa. The enzyme had an isoelectric point of around pH 3.0. The optimum pH for enzyme activity was around 4.0, and it was most active at 40 degrees C and stable up to 35 degrees C. The enzyme contained 23.8% carbohydrate and some copper atoms. The enzyme oxidized 2,2'-azino-bis(3-ethylbenzothiazoline-6-sulfonic acid) diammonium salt, p-phenylendiamine, pyrogallol, guaiacol, 2,6-dimethoxyphenol, catechol, and ferulic acid, but not veratryl alcohol, tyrosine, and β-(3,4-dihydroxyphenyl) alanine. The N-terminal amino acid sequence of Lcc 1 showed close homology to the N-terminal sequences determined for laccases from *Phlebia radiata*, *Trametes villosa*, and *Trametes versicolor*, but only low similarity was observed to a previously reported laccase from *L. edodes*. Lcc 1 was effective in the decolorization of chemically different dyes - Remazole Brilliant Blue R, Bromophenol Blue, methyl red, and Naphtol Blue Black - without any mediators, but the decolorization of two dyes - red poly(vinylamine)sulfonate-anthrapyridone dye and Reactive Orange 16 - did require some redox mediators.

L. edodes grew very poorly in agitated submerged culture on crystalline cellulose (Avicel). This poor growth is reflected by the absence of detectable cellulolytic activity (endoglucanase, exoglucanase and β-glucosidase) in culture supernatants [Buswell *et al.*, 1996], in contrast to the solid state cultures with red oak wood (*Quercus rubra*) as substrate, where all three activities are detectable [Mishra & Leatham, 1990].

5. Changes in Activity of Extracellular Enzymes in Dual Cultures of *Lentinula*

The main problem that arises during the cultivation of *Lentinula* is that the logs on which the cultivation is performed are contaminated by competing micro-organisms, especially *Trichoderma* spp. The aim of the study [Hatvani *et al.*, 2002] was to examine the changes in activity of extracellular enzymes in dual cultures of *Trichoderma* spp. and *L. edodes*. *Trichoderma* enzymes important for the degradation of fungal cell walls (N-acetyl-β-glucosaminidase and laminarinase) were shown to be induced by inactive *L. edodes* mycelia in liquid culture. The changes that occurred in the extracellular enzyme activities of *L. edodes* and mycoparasitic *Trichoderma* spp. (*T. aureoviride*, *T. harzianum* and *T. viride*) were examined during antagonistic interactions on solid medium. The extracellular enzyme patterns of both partners proved to be altered. *Trichoderma* spp. was induced to produce N-acetyl- β-glucosaminidase and laminarinase in the presence of active *L. edodes* mycelia, similarly as observed in liquid culture. The activities of both laccase and manganese peroxidase of *L. edodes* decreased after physical contact with active *Trichoderma* mycelia, possibly in consequence of the beginning of degradation of *L. edodes* by the *Trichoderma* enzymes. However, besides a decrease in manganese peroxidase activity, an enhancement of *L. edodes* laccase activity was observed on solid media containing crude culture fluids from *Trichoderma* liquid cultures. The metabolites responsible for these effects proved to be heat stable. Consequently, induction and inhibition of several extracellular enzymes of both partners were shown in dual cultures of *L. edodes* and *Trichoderma* strains, indicating the important role of these enzymes in the antagonistic interaction between the two species. As the main problem during the large-scale cultivation of *L. edodes* is the contamination of the growth substrate by *Trichoderma* mycelia, the particular knowledge of the mechanism of this competition might be relevant.

L. edodes was cultivated in liquid media containing malt and yeast extract [Savoie *et al.*, 1998]. Extracellular laccase activity, measured in the culture fluids, was 5-18 times higher in cultures incubated for 29 days than in cultures incubated for 24 days. The addition to water-soluble lignin derivatives or *Trichoderma* sp. in cultures of *L. edodes* incubated for 11 days increased laccase activity 3- to 20 fold. The higher response was obtained with live mycelium of *Trichoderma* sp., but cell-free culture fluids of *Trichoderma* sp. in pure cultures were also effective. *Trichoderma* sp. induced changes in the laccase isoenzyme pattern as a result of the alteration of laccases secreted by *L. edodes* and not the induction of new isoforms.

6. Decrease in the Phytotoxicity of Wastes

Olive-mill wastewater (OMW) exhibits highly phytotoxic properties, mainly due to phenols. A valuable option for OMW disposal is its agricultural use provided that phytotoxic effects are removed. The investigation [D'Annibale *et al.*, 2004] was aimed at evaluating the efficacy of the lignin-degrading fungus *L. edodes* in achieving OMW detoxification. Germinability experiments on durum wheat showed that OMW phytotoxicity was significantly reduced by *L. edodes* cultures. Germinability on undiluted and twofold diluted OMW from fungal cultures was 34±5% and 57±6%, respectively, while on related incubation controls it was almost completely suppressed. These results suggest that fungal cultures of *L. edodes* would decrease the phytotoxicity of this waste.

Effect of OMW on the edible medicinal mushroom *L. edodes* growth and lignin degrading enzymes was studied [Zjalic *et al.*, 2002].

Immobilized laccase proved to be efficiently stable in removing OMW phenolics. Immobilization of *L. edodes* laccase on Eupergit C increased pH, thermal and proteolytic stability with slight modifications in laccase oxidation efficiency [D'Annibale *et al.*, 2000].

β-1-3-Glucan synthase activity and its induction by OMW was studied in ten fungal strains, including *L. edodes* [Reverberi *et al.*, 2004]. A microtiter-based enzymatic assay on β-1-3-glucan synthase activity was carried out on all mycelia growth both on the control medium and on OMW. Among the fungi assayed, *L. edodes* β-1-3-glucan synthase was highly enhanced in OMW. The main components of OMW, i.e. phenols and lipids, were added separately to the control medium, to highlight the mechanism of *L. edodes* β-1-3-glucan synthase induction. A Southern blot analysis and PCR with degenerated primers were carried out to detect the presence of fks1-like genes in these Basidiomycetes. The sequences obtained from the ten Basidiomycota were remarkably similar to fks1 from *Filobasidiella neoformans*. Spectrofluorimetric and RT-PCR analyses of β-1-3-glucan synthase were performed on the mycelia of *L. edodes*. In this fungus, a strong stimulation of β-1-3-glucan synthase mRNA and protein was recorded in the presence of OMW and phenols.

A high-field NMR study of the polysaccharidic fraction extracted from *L. edodes* mycelium grown on OMW is reported [Tomati *et al.*, 2004]. Diffusion-ordered NMR spectroscopy was applied to the polysaccharidic fraction. The results showed the presence of two polysaccharides of different sizes, whose structures were revealed using one- and two-dimensional NMR techniques. These two polysaccharides were identified as xylan and lentinan.

7. Biosynthesis of Ribonucleases

L. edodes cultivated in potato dextrose medium produced five RNases in the culture filtrate [Kobayasi *et al.*, 1998]. The two major RNases (RNase Le37 and RNase Le45) were highly purified and their molecular masses, base specificities, N-terminal amino acid sequences, and amino acid compositions were analyzed and compared to RNase Le2 isolated from the fruit bodies of the same mushroom. RNase Le37 and RNase Le45 are base non-specific and adenylic acid preferential RNases like RNase Le2 and their N-terminal sequences are very similar to RNase Le2, but they are glycoproteins and their amino acid compositions are significantly different from that of RNase Le2. In addition to these enzymes, a guanylic acid-specific RNase with a molecular mass 13 kDa was partially purified. Since RNase Le2, which has very similar N-terminal sequence to RNase Le37 and RNase Le45, was not excreted from the mycelia, the analysis of the structures of these two excreted RNase may shade a light on the mechanism of excretion of RNases in this organism.

Somewhat later the production of RNases Le2, Le37, and Le45 by *L. edodes* was described in [Inokuchi *et al.*, 2000; Kobayashi *et al.*, 2000]. The primary structures of the former two RNases, having molecular masses about 24 and 37 kDa, respectively, have been elucidated to be members of the RNase T2 family. RNase Le37 was some 30 amino acid residues longer at the C-terminal end than RNase Le2. The C-terminal region of RNase Le37 was rich in O-glycosylated serine and threonine. In fungal glucoamylases and chitinases, which hydrolyze raw-starch and chitin, respectively, have structures resembling the structure of the C-terminal of RNase Le37. In [Kobayasi *et al.*, 1998], the primary structure of RNase Le45 was estimated using the following experimental evidence. (i) The partial amino acid sequence of RNase Le45 determined that up to about 60% of total protein was identical with that of RNase Le37. (ii) The amino acid composition of RNase Le45 was identical to that of RNase Le37. (iii) The elution profiles on HPLC of lysylendopeptidase and *Staphylococcus aureus* V8 protease digests of RCM-RNase Le45 (reduced and S-carboxymethylated RNase Le45) were very similar to those of RNase Le37, except for the absence of C-terminus peptide which contained O-glycosylated peptides. However, RNase Le45 contained about 70 residues of mannose and 4 residues of hexosamine. These values were more than twice those of RNase Le37. (iv) RNase Le45 was immunologically indistinguishable from RNase Le37. (v) After treatment with both glycosidase EndoH and alpha-mannosidase, RNases Le37 and Le45 gave complex bands by slab-gel electrophoresis. However, one of the major bands with the highest mobility from RNase Le45 and Le37 showed the molecular mass of 29 kDa in common, which is slightly larger than that of RNase Le2 containing no carbohydrate. These results indicated that RNase Le45 is an enzyme which is a heavily glycosylated species of RNase Le37.

8. Morphogenesis in Liquid Culture

8.1. *Lentinula* Fruiting in Liquid Media

Although the time required for fruiting-body formation in *Lentinula* is usually too long to make it practical to obtain fruiting bodies on agar medium, even for experimental studies, there are some reports of fruiting on agar medium, but fruiting in liquid medium is of more frequent occurrence [Miles & Chang, 1987]. G.F. Leatham [Leatham, 1983] reported that he obtained fruiting of a dikaryotic stock of *L. edodes* on a chemically-defined medium within 45 days after inoculation and that sometimes fruiting occurred as early as 27 days. The purpose of the further study was to examine the fruiting ability of different dikaryotic stocks of *L. edodes* on this chemically-defined liquid medium and also on its modifications [Miles & Chang, 1987]. While the results reported in the latter work are only the first steps in elucidating the conditions for *L. edodes* fruiting in liquid media, the results nevertheless should prove helpful in future experimentation on the genetics of fruiting of this species. The differences in the time of primordia formation and of fruiting among the stocks, in the absence of substantial differences in growth, are indicative of genetic control of the fruiting process. While the authors of [Miles & Chang, 1987] do not gain evidence concerning the number or mode of operation of genes controlling fruiting in *L. edodes*, it is evident that all dikaryotic stocks of this species do not behave similarly in regard to fruiting, and that fruiting is genetically controlled as well as being under environmental regulation.

8.2. Peculiarities of Chemical Composition of *Lentinula edodes* in Submerged Culture

Comparison of *L. edodes* with true xylotrophs and humus saprotrophs cultivated under the conditions used in commercial fungal production [Feofilova *et al.*, 1998] revealed significant differences. Firstly, the overall lipid content is different: submerged mycelium contains up to 20% (m/m, dry weight basis) lipids, while fruiting bodies contain only 3–4%. Secondly, there are significant differences in composition. The lipids of the fruiting bodies of *L. edodes* contain high levels of $C_{16:0}$ fatty acids and $C_{16:1}$ is present; the neutral lipids of the fruiting body have a high monoglyceride and free fatty acid content and a low triglyceride level; and the main *L. edodes* phospholipids are phosphatidylethanolamine and cardiolipin. In contrast, the mycelium produced in submerged culture has triglycerides and free fatty acids as its main neutral lipids and the main constituents of its phospholipids are phosphatidylcholine and cardiolipin.

The composition of submerged mycelium of *L. edodes* grown in laboratory fermenters was studied by the authors of [Lobanok *et al.*, 2003]. The mycelium contained 23–24% (m/m) proteins, 8–9% lipids, up to 1.8% phenolic substances, and a significant amount of inorganic substances, including calcium and iron. The fungus produced up to 5.0% (m/m) intracellular and 3.5–4.0 g/l extracellular polysaccharides.

The total carbohydrate content in the submerged mycelium of *L. edodes* (strains 182 and 198) is about 47–48% [Shcherba & Babitskaya, 2004]. Free carbohydrates of the cell cytosol amounted to 15 and 19%, respectively, of the total cell dry weight, while structural carbohydrates amounted to 33 and 28%, respectively. Qualitative composition of carbohydrates was studied, which allowed for attribution of structural polysaccharides to hetero- and homoglycans with β- and α-glycoside bonds.

Submerged cultivation of *L. edodes* is accompanied by the production of a number of valuable polysaccharides. Thus, in [Harvey *et al.*, 2001] a process has been established for the mycelial cultivation of *L. edodes* in submerged, agitated liquid cultures. A suitable medium has been developed to yield relatively high biomass concentrations, and pH favored production of lentinan proposed. Polysaccharides with similar infrared spectra to those obtained from the fruiting bodies of *L. edodes* have been extracted from biomass and from the culture liquid itself. There appears to be some difference in the molecular composition of the intra- and extracellular lentinan.

8.3. Pigmented Mycelial Film

Pigmented mycelial film has proved an attractive and interesting topic, from the aspect of both research and production. This brown mycelial structure may be considered a step of morphogenesis peculiar to shiitake, normally followed by the formation of primordia and then fruiting bodies. The work [Aleksandrova *et al.*, 1998] deals with the development of *L. edodes* and describes the fine structure of a brown mycelial film appearing on the surface of mature mycelium and representing a dense network of intertwined pigmented hyphae with thick cell walls. Cultivation conditions of *L. edodes* ensuring a high yield of fruiting bodies (up to 7 fruiting bodies per 0.5 l of medium) and submerged mycelium (up to 32 g/l of medium) are described. Research on the capacity of *L. edodes* to grow on various media revealed a number of properties distinguishing shiitake from true xylotrophs.

We report further on studies involving this pigmented mycelial film in the next section.

9. Studies Carried Out by Our Group

Now we turn our attention to the results that our group has obtained working with *L. edodes* over the last ten years.

9.1. Biotechnological Potential of Mushroom Lectins

Many authors define lectins as proteins of non-immunoglobulin nature capable of specific recognition and reversible binding to carbohydrate moieties of complex carbohydrates without altering covalent structure of any of the recognized glycosyl ligands [Kocourek & Horejsi, 1983]. The ability of lectins to affect cell-division cycles and to implement various functions in living organisms is widely used in experimental biology, cytology, genetics, and oncology [Kilpatrick, 2000].

Lectin biosynthesis has been discovered in many microorganisms including fungi. Lectins have been isolated from fruit bodies and mycelium of many macromycete species. Use of fungal lectins in biological and medical research is possible due to their biological activity, mitogenic [She *et al.*, 1998; Wang *et al.*, 2004] and cytostatic, against a series of normal cells such as keratinocytes [Parslew *et al.*, 1999], iris fibroblasts [Batterbury *et al.*, 2002], and retinal pigment epithelial cells [Kent *et al.*, 2003]. Recently works on the antitumor activity of fungal lectins have appeared [Wang *et al.*, 2000; Liua *et al.* 2001; Mahajan *et al.*, 2002; Ngai & Ng, 2004; Wasser & Weis, 1999; Wasser, 2002]. Although researchers often put emphasis upon biological significance of the polysaccharides found in fungal mycelia, it is highly probable that the antitumor activity is associated with lectins, for which polysaccharides may be ligands. The presence of such biomolecules with lectin sites is reported in a recent work [Miyazaki *et al.*, 2007].

The application of lectins makes it possible to discriminate in situ between extracellular polymers produced by different organisms [Kamper *et al.*, 2004]. The utility of immobilized lectins in isolating and characterizing glycan structures of naturally occurring glycoproteins allows the lectins to contribute to the problem of xenotransplantation, since the glycoproteins containing known sugar epitope (linear trisaccharide Galα1,3Galβ1,4GlcNAc) is responsible for hyperacute rejection of tissues and organs from lower mammals to humans [Winter *et al.*, 2002; Loganathan *et al.*, 2003]. The strict specificity of some lectins for α2,6-linked sialic acid renders these lectins a valuable tool for glycobiological studies in biomedical and cancer research [Mo *et al.*, 2000; Ueda *et al.*, 2002].

The data strongly suggest that proteins (among those, lectins and hemolysins being good candidates) are responsible for most of the insecticidal activity of mushroom fruitbodies and may be a source of genes available for plant protection against insects [Wang *et al.*, 2002]. Some lectins have been found to be toxic to insects and possess a high insecticidal activity compared to lectin isolated from Fabaceae (*Lathyrus ochrus*) or from the snowdrop (*Galanthus nivalis*) [Trigueros *et al.,* 2003].

9.2. Detection of Shiitake Lectins on Submerged Cultivation

The vast majority of studies to locate mushroom lectins have focused on the basidiome, but *in vitro* mycelium cultures of certain species have shown that lectins are not produced solely by fruit bodies. At a start point of our Lab's research in 2000, data on the lectins from *L. edodes* (shiitake) were very limited and were presented in two works [Jeune *et al.*, 1990; Wang *et al.*, 1999] on lectin isolation from *L. edodes* fruit bodies. The literature offers no information on the preparations of extracellular lectins of xylotrophic basidiomycetes including *L. edodes*.

Our earlier work dealt with finding and studying the lectin activity of culture liquid and submerged mycelium of *L. edodes*; establishing the carbohydrate-binding specificity of the hemagglutinins revealed; and exploring the lectin activity in relation to selected physicochemical factors of cultivation [Tsivileva *et al.*, 2000, 2001]. We studied the hemagglutinating activity of the submerged mycelium and the culture liquid of several strains of *L. edodes*. The activity of the culture liquids proved to be much higher than that of mycelia. The data obtained suggest a considerable amount of extracellular agglutinins released into the culture medium. For correct comparison between the activities of different strains cultured during equal periods of time, the experiments on determining growth rates as a function of growth duration were carried out. The relevant values let one to infer from that the differences in respect to hemagglutination are markedly over the growth rates variations from one strain to another on cultivation in this medium.

We also determined the carbohydrate specificity of fungal agglutinating factors by a hemagglutination-inhibition assay. Three carbohydrates incorporating galactose and glucose residues in their structures (galactose, lactose, maltose) appeared to be the best inhibitors in our study.

The hemagglutinating activity was investigated as a function of the inoculum size, cultivation temperature, and culture age. The seeding mycelium dose-dependence of lectin activity occurred to pass through maximum, i.e. there was a highest activity that did not enhance with the increase of inoculum quantity. The temperature effect in respect to lectin activity confirmed once again that the latter is not a purely growth-associated product: poor mycelia development at temperatures lower or higher compared to optimal (26°C) was accompanied by the enhanced lectin activity, in the cases of both culture liquid and submerged mycelium.

Our earlier studies showed that lectins are synthesized during the active growth phase, with the lectin activity remaining practically constant during the stationary phase. We studied the dependence of activity of extracellular lectins on the sources of carbon and nitrogen in medium and on the C:N ratio, in the latter case using D-glucose as the carbon source [Tsivileva *et al.* 2005a]. The hemagglutination titer (T) in the growth phase varied from 4 to 4096. The maximal activity was observed on day 3-7 of culture, with an initial C:N ratio of 17:1; the minimal activity was seen in the absence of asparagine (Asn) in the glucose-containing culture medium. An increase in lectin activity was observed at initial pH 8-9; no lectin activity was detected at pH 2.0 and 2.5 on days 9 and 12, respectively. At initial pH 3.0, the hemagglutination titer decreased 32 times compared to the initial value on day 12. The use of buffer-containing media (pH 7.0) did not lead to an increase in lectin activity in the culture liquid. Addition of 10 mM phosphate buffer containing 0.15 M NaCl decreased lectin activity [Tsivileva *et al.*, 2005a].

9.3. Shiitake's Pigmented Mycelial Film Formation

As a result of studying the agglutinating activity of different morphogenetic structures of *L. edodes* F-249, including mycelium, brown mycelial film (MF), primordia, and fruiting bodies, MF was found to possess the maximum hemagglutinating activity [Tsivileva *et al.*, 2001], which can be explained by the possible involvement of agglutinins in the formation of MF as a structure composed of glued hyphae.

We established the relationship between the process of MF formation and the extracellular lectin activity in the presence of divalent metal cations (M^{2+}) and natural amino acids, during cultivation of *L. edodes* F-249 in liquid medium [Tsivileva *et al.*, 2004, 2005c].

The involvement of *L. edodes* lectins in MF formation was confirmed by a number of experiments we performed. Firstly, we elucidated the relationship between the lectin activity of extracts from mycelia grown on several agar media and the fruiting of these mycelia. The higher the mycelial hemagglutination titer was, the faster the MF formed, and the greater lectin activity of this MF therewith [Tsivileva *et al.*, 2001]. Brown films with maximum lectin activity gave subsequent fruiting. Secondly, lectin activity of *L. edodes* was examined at different morphogenesis steps. The hemagglutination titers increased as the morphology changed from mycelium to MF, and then decreased as the MF produced the primordia and the fruiting body. These results observed on different substrates [Tsivileva *et al.*, 2001].

We added divalent cations in an attempt to improve the MF formation on liquid medium. The growth duration was 28 days. MF formed in the presence of divalent calcium (concentration interval: 2 to 10 mM) or manganese (concentration interval: 0.5 to 2 mM) cations in the liquid culture medium. With these cations the MF of *L. edodes* F-249 was formed within 3 to 9 days after inoculation, depending on the cation concentration, whereas on the same growth medium, but in the absence of cations, MF only formed at two months. MF formation is characterized by a 8 to 128-fold decrease in hemagglutination titer of the culture liquid. Thus, at the manganese(II) content of 2 mM, within the period of 5 to 11 days of growth, the hemagglutination titer diminished from 512 to 4. This correlates with the high lectin activity that was obtained in extracts from the MF.

Among the natural amino acids under study as nitrogen sources, and nine double-charged metal cations as inorganic additives, the simultaneous use of Asn and either Ca^{2+} or Mn^{2+} stimulated MF formation, without regard to the age of the seeding culture. In the absence of amino acid source in culture liquid, the cations did not stimulate MF formation. The positive influence of Asn was most pronounced among the natural amino acids. The synergistic effect of Asn and calcium(II) or manganese(II) ions is understandable. But to judge on the role of Asn here, it seems reasonable to make two suppositions: i) the distinct properties of Asn as the component of liquid nutrient medium are reduced to chemical binding in solution. If it is so, the electronic structure of Asn molecule should stand out even among the structurally close amino acids, which do not display the above stimulating effect according to our assays; ii) carbohydrate binding site of the lectin probably contains an Asn residue. It is known that carbohydrates interact with lectins, among others, through complex networks of hydrogen bonds. Therewith the amide hydrogen and carbonyl oxygen of Asn in the combining sites of lectins are commonly involved in such protein-carbohydrate interactions [Liener *et al.*, 1986]. To get a better insight in this problem, the quantum chemical calculations were used.

Glutamine (Gln) is the nearest structural analogue of Asn, having an amide group ($CONH_2$) on its side chain. In spite of the fact that Gln molecule differs from the Asn molecule by a single methylene unit, the addition of Gln to the medium does not have any detectable influence on the formation of MF. It is probable that the different chemical behaviour of these two structurally analogous compounds is related to different charges on probable reactive centres. The processes under consideration occur in aqueous solutions, in which the amino acids exist in the form of zwitterions. By means of the restricted Hartree-Fock (RHF) formalism in the 6-31G(d) basis [Hehre *et al.*, 1986], we carried out *ab initio* computations of the electron structure of the Asn and Gln zwitterions. The charge characteristics of probable reactive centres for both amino acids are very close to each other. To elucidate other factors affecting the chemical behaviour of the two amino acids in their interaction with metal cations used software from the *HyperChem* package to compute the QSAR quantities of the Asn and Gln zwitterions. It appers that the distinct reactivities of the two zwitterions are, to some extent, due to their differential hydrophobicity. Asn, as a less hydrophobic reactant, possesses a more firmly attached hydrate shell and, consequently, in comparison with Gln, bonds a metal cation less rigidly, *i.e.* reversibly. Another reason for the differences between Asn and Gln is the spatial factor. Starting from criteria such as the van der Waals surface and, especially, the volume of a molecule, one could search for fragments of appropriate topology in the structure of the lectin on the basis of future data of X-ray structural analysis.

9.4. The Effect of Metal Cations on the Lectin Activity

The following series was established with respect to decreasing positive effect on the activity of shiitake extracellular lectins: Mg>Ca>Cu>Fe>Mn>Zn~Sn>Co>Ni [Tsivileva *et al.*, 2005c]. These metals differ greatly in respect to their places in the Periodic System of elements and to their properties. Therefore, when dealing with such multi-factor property as the lectin activity, the correct comparisons should be done within the series of metal cations similar to each other by their most important characteristics. Consequently, we have regarded here the triad of iron, cobalt, nickel. All these elements are positioned in the fourth period and occupy the neighbouring places in the Periodic System (order numbers are 26, 27 and 28, respectively). So, it would be interesting to follow the individual peculiarities of each cation with respect to its influence on the lectin activity. A probable mechanism of the latter influence could be the formation of mixed-ligand complexes with hydroxy groups of both lectins and glycoconjugates, thus assisting the reversible interaction among those.

Most likely the differences between cations are determined by the thermodynamics of complexation reactions, by the strength of metal-oxygen bond. For examining this supposition, we have carried out the quantum chemical *ab initio* study of the thermodynamics of reactions between hexaaqua complexes of iron(II), cobalt(II), nickel(II) and the model system ethylene glycol, a typical chelating reagent and a simple analog of carbohydrates [Tsivileva *et al.*, 2006b]. As a result of quantum chemical computations, we obtained the series of the cations capability of complexing with ethylene glycol in descending order Fe^{2+} > Co^{2+} >> Ni^{2+}. Thus, otherwise, the energies of chelating reactions of metal-hexaaqua complexes with the model ligand ethylene glycol agree with the series of these cations' decreasing effect on the lectin activity: $Fe^{2+} > Co^{2+} > Ni^{2+}$.

9.5. Lectin Activity and Organo-Selenium Compounds

The lectin activity and growth characteristics of *L. edodes* occurred to be dependent upon the presence of selenium-containing component DAPS-25 (1,5-diphenyl-3-selenopentanedione-1,5) in liquid and agar media [Tsivileva *et al.*, 2005b]. Addition of DAPS-25 stimulated the accumulation of biomass of *L. edodes* in both submerged cultivation and on agar medium. In the case of submerged cultivation, the stimulation was greatest for media that supported fast growth. In the case of growth on agar media, the stimulation was greatest for media that supported slow growth. The stimulation of the lectin activity of both culture liquid and mycelial extracts of *L. edodes* by DAPS-25 was greatest in the case of synthetic medium. In the absence of DAPS-25 this medium exhibits a high activity of extracellular lectins but a relatively low lectin activity of mycelial extracts.

Along with DAPS-25, several other compounds of the 1,5-di(4-R-phenyl)-3-selenopentanediones-1,5 series interaction with the basidiomycete *L. edodes* lectins were explored involving both computations and experiment [Pankratov *et al.*, 2011b]. The role of spatial and electron structure, hydrophobic properties and concentration of organoselenium compounds on their interaction with fungal metabolites - extracellular lectins of shiitake mushroom has been considered.

By the hybrid method of density functional theory at the B3LYP/6-31G(*d,p*) theory level, spatial and electronic structure of the 1,5-diphenyl-3-selenopentanedione-1,5 (preparation DAPS-25), 1,5-di(4-methoxyphenyl)-3-selenopentanedione-1,5 and 1,5-di(4-ethoxyphenyl)-3-selenopentanedione-1,5 molecules has been studied. The above molecules have been stated to be substantially similar to each other by their electronic and spatial characteristics.

By means of the QSAR properties evaluation by the atomic-additive schemes, it has been shown that the molecules of the preparation DAPS-25, its dimethoxy- and diethoxy-substituted are close to each other by the hydrophilic-lipophilic balance, whereas di-*n*-octoxy derivative DAPS-25 is explicitly hydrophobic.

The hemagglutinating activity of lectins in the presence of the preparation DAPS-25 and its alkyloxy-substituted increases, therewith the most effective addition is 1,5-di(4-ethoxyphenyl)-3-selenopentanedione-1,5. Apparently, the greater effectiveness of the said substance compared to DAPS -25 is caused by the formation of hydrogen bonds with a participation of unshared electron pairs of oxygen atoms from the ethoxy groups and mobile hydrogen atoms from the OH groups of glycoconjugates on erythrocytes surface. The positive effect of 1,5-di(4-*n*-octoxyphenyl)-3-selenopentanedione-1,5 is not so prominent, since the enlarged alkyl chain shields the aromatic fragments of organoselenium molecule participating in the binding with lectin [Pankratov *et al.*, 2011b].

9.6. Isolation and Characterization of *Lentinula edodes* Extracellular Lectins

Our latest research has been directed to isolation, purification, and characterization of preparations of *L. edodes* F-249 extracellular lectins. The culture of *L. edodes* F-249 occured to synthesize two extracellular lectins. We have isolated those from fungal culture liquid and purified to electrophoretic homogeneity [Tsivileva *et al.*, 2008]. The lectins differ in physicochemical properties, composition, carbohydrate specificity, and ability for agglutination of erythrocytes of various types. Extracellular lectin *L*1 from *L. edodes* is a glycoprotein of mono-subunit structure with molecular weight of 43 kDa. *L*1 is comprised of 10.5 ± 1.0% (w/w) carbohydrates represented by glucose. Extracellular lectin *L*2 is a proteoglycan of mono-subunit structure with molecular weight of 37 kDa. *L*2 is comprised of 90.3 ± 1.0% (w/w) carbohydrates represented by glucose (73% of the total mass of the carbohydrate moiety of the lectin molecule) and galactose (27% of the total mass of the carbohydrate part of the lectin molecule). The content of Asn in *L*2 is high, i.e. 42% (w/w) of total amino acids. This fact along with the composition of the carbohydrate part of the molecule (glucose + galactose) allows one to assign *L*2 to *N*-asparagine-bound proteins. Both lectins are specific to *D*-galactose and lactose at an equal for *L*1 and *L*2 minimal inhibiting concentration of these carbohydrates (2.08 mM Gal and 8.33 mM Lac). Other carbohydrates to which the lectins show affinity are different for the two lectins: *L*-rhamnose (4.16 mM) for *L*1 and *L*-arabinose (4.16 mM) and mannitol (8.33 mM) for *L*2. The purified extracellular lectins of *L. edodes* are highly selective at recognition of definite structures on the surface of trypsinized rabbit erythrocytes and do not react with the erythrocytes of other animals and humans.

In our opinion, all the aforesaid testifies to an apparent lack of information regarding the demonstration of lectin activity in fungal cultures, the preparation of purified samples of extracellular fungal lectins, and the study of their properties and supposed functions.

9.7. Application of Mn(II) for Obtaining the Seeding Mycelium of *Lentinula edodes*

We revealed the relationship between the period before fruiting and the presence of Mn^{2+} cations in the medium during submerged cultivation of seeding mycelium of *L. edodes* [Tsivileva *et al.*, 2006a]. The presence of Mn(II) in the medium significantly decreases not only the interval before the formation of the brown mycelial film in liquid culture but also the time required for the the appearance of fruiting bodies. These results led to the proposal of a method of growing seed mycelium of *L. edodes* [Nikitina & Tsivileva, 2004] to be used for mushrooms production.

9.8. Co-Culture of *Lentinula edodes* with Bacteria

From a viewpoint of biotechnological potentialities, of interest are selected factors of biological origin that promote the growth of *Lentinula*. We obtained the rather promising results on co-cultivation of *L. edodes* with *Azospirillum brasilense*, a nitrogen-fixing soil bacterium that promotes plant growth. Bacteria from the genus *Azospirillum* are capable of producing substances of phytohormonal nature and possess antifungal and antibacterial activities [Red'kina, 1990]. The vast majority of studies of the azospirilla growth-stimulating ability have focused on the higher plants, but algae have been also refered to [Gonzalez & Bashan, 2000]. In spite of the fact that the azospirilla are the most widespread nitrogen-fixing microorganisms associated with mycorrhizal fungi in nature [Li & Strzelczyk, 2000], the co-culture of *Lentinula* sp. and *Azospirillum* sp. was not investigated until now.

The assays on stimulating the growth of *L. edodes* in its double culture with *Azospirillum brasilense* allowed us to conclude on the explicit advantages of these double culture (over pure culture of *L. edodes*) with respect to the enhanced suppression of contaminating microflora [Nikitina *et al.*, 2006]. The parameters of growing *L. edodes* F-249 with *A. brasilense* Sp7 have been selected. One of the most favourable nutrient medium for growing both cultures appeared to be diluted barley wort; temperature of cultivation was 28-29°C. Under these conditions, the mycelium dry weight increased by about 60% compared to pure culture of mushroom. Several synthetic glucose- and asparagine-containing media occurred less appropriate, with the corresponding value of not greater then 30%. The optimal age of liquid seeding co-culture has been revealed to be 14 days: this is the best time, during a *L. edodes* culture, for the addition of an inoculum of *A. brasilense*. When inoculating the solid substrate with the double culture, the period of substrate colonization by mycelium became shorter [Nikitina, 2005].

The effect of co-cultivation of *A. brasilense* and *L. edodes* on the changes in carbohydrate composition of mycelium as compared with the mushroom monoculture during morphogenesis has been investigated. The increase in mannitol level in the co-culture of *L. edodes* F-249 and *A. brasilense* Sp7 has been observed, that commonly precedes the shiitake fruiting. This phenomenon could serve as one of the aspects of *A. brasilense* Sp7 positive influence on the growth of *L. edodes* F-249 we revealed before.

9.9. Shiitake Mushroom as a Non-Phyto Origin of Indolic Phytohormons

An enormous body of knowledge demonstrates that auxin indolyl-3-acetic acid (IAA) can be synthesized from tryptophan (Trp) by plants and bacteria. Several major findings have emerged. They are that: 1) plants utilize multiple pathways for the homeostasis of IAA; 2) these pathways are differentially regulated; and 3) numerous environmental factors including light, temperature and biotic or abiotic stress regulate the activity of these pathways. Only few experiments are done with fungi, therewith the published works on phytohormonic potentialities of xylotrophic mushrooms are virtually absent.

The characterization of the group of extracellular indolic compounds of basidiomycete *L. edodes* in relation to the submerged cultivation conditions has been presented. The *L. edodes* F-249 culture has been stated to synthesize the indolic-nature compounds when being grown in submerged culture. The group of extracellular indolic compounds of shiitake includes, in different proportions, the following components: L-tryptophane (Trp), β-indolyl-3-acetic acid (IAA), β-indolyl-3-acetaldehyde (IAAld), β-indolyl-3-acetamide (IAAm), β-indolyl-3-pyruvic acid (IPyA), tryptamine (TAM), 5-hydroxy-β-indolyl-3-acetic acid (5-OH-IAA) [Tsivileva *et al.*, 2011].

The time-course production of indolic derivatives in culture liquid of shiitake has been studied comparatively to reveal correlations with the growth rates. It has been established that shiitake is capable of producing IAA rather actively in the culture-age dependent manner. Maximal IAA content we marked in experiments was 10.6 mg/l provided that the initial TAM level in the medium was 1 mg/l. The highest levels of TAM (22.8 mg/l), IPyA (13.9 mg/l), IAAld (27.9 mg/l), 5-OH-IAA (9.2 mg/l) have been observed at different additives of the indole-group substances, and varied with the fungal culture age. The IAAm level changed only slightly as compared to reference probe. Starting from the synthetic Trp-free medium, the concentrations of extracellular Trp became from 13.8 mg/l (7th day) to 23.9 mg/l (21st day).

On the basis of literature data on physiologically active concentration values for auxins in plants, we use the concentration range of 10^{-1}-10^{-8} g/l of additives to explore the effects of IAA and its precursors upon the shiitake submerged culture. The positive influence of exogenic IAA within its concentration range of $2{\cdot}10^{-7}$ to $2{\cdot}10^{-4}$ g/l upon the *L. edodes* biomass accumulation has been found, a minimal growth-inhibiting phytohormone concentration being about $5{\cdot}10^{-4}$ g/l on mineral medium. When inducing the IAA biosynthesis by its exogenous micro-additives ($1{\cdot}10^{-5}$ to $1{\cdot}10^{-8}$ g/l), the increase in phytohormone level in the medium (up to 4,000-fold) accompanied by the anthranilic acid appearance (up to 1.5 mg/l) have been revealed under these experimental conditions exclusively.

The studies of biosynthetic routes for IAA realized by the mushroom culture under question have been attempted in order to conclude whether the above route is Trp-dependent, or Trp-independent pathway becomes also involved when the mushroom grows in the presence of exogenous synthetic analogs of the IAA precursors. The experimental evidences in favor of co-existence of two alternative pathways for IAA production by *L. edodes* have been obtained. Those routes are: Trp-dependent (mainly via TAM) and Trp-independent, the latter being realized in the presence of exogenous indole within the concentration range from $1{\cdot}10^{-3}$ down to $1{\cdot}10^{-4}$ g/l or the IAA micro-additives.

The induction of generative developmental stage by indolic derivative has been revealed for shiitake. It has been stated that among the compounds - IAA precursors, solely IAAm at a concentration of about 10^{-4} g/l in *L. edodes* culture liquid exerts the explicitly marked stimulating effect on the occurrence of morphological structure - brown mycelial film [Tsivileva *et al.*, 2011].

The indole participation in growth and metabolic processes in shiitake has been elucidated. The effect of exogenous indole in respect to growth characteristics of the basidiomycete *Lentinula edodes* submerged culture has been revealed to manifest itself in the mycelial biomass diminution at the indole concentration of 10^{-4}-10^{-1} g/l. The production of indolyl-3-acetic and indolyl-3-pyruvic acids was stimulated by indole. The set of mushroom's extracellular compounds of indolic nature on the indole-enriched media included also tryptamine, tryptophane, indolyl-3-acetamide. The biosynthesis of these compounds was suppressed by exogenous indole to a variable extent depending on the addition's concentration and fungal culture age. The pathway of indole biochemical transformation under these conditions was observed in the form of oxindole occurrence [Pankratov *et al.*, 2011a].

10. Conclusions and Perspectives

Fungi are a group of living organisms very interesting in theoretical and practical respect; however, extracellular fungal lectins have not been described. Hence, our first preliminary studies are relevant as a contribution to description of a new group of glycoproteins of xylotrophic basidiomycetes, extracellular lectins, physicochemical properties of which are important for understanding their physiological role.

In trying to evaluate the future needs with respect to *L. edodes* biotechnology, one should obviously take into account the important applications of submerged cultivation process, first of all for the production of the biologically active substances (polysaccharides, glycoproteins, selected low-molecular compounds), as well as for the production of mycelial biomass itself. The latter is valuable not only as food and fodder supplement, but also as the intermediate product for obtaining fruiting bodies. To study the *Lentinula* cultivation processes in labs, optimize and then fit the assays to a bioreactor scale is the general contemporary biotechnology challenge.

References

Aleksandrova, E. A., Zav'yalova, L. A, Tereshina, V. M., Garibova, L. V., & Feofilova, E. P. (1998). Obtaining of fruiting bodies and submerged mycelium of *Lentinus edodes* (Berk.) Sing [*Lentinula edodes* (Berk.) Pegler]. *Microbiology, 67*, 535-539.

Batterbury, M., Tebbs, C. A., Rhodes, J. M., & Grierson, I. (2002). *Agaricus bisporus* (edible mushroom) lectin inhibits ocular fibroblast proliferation and collagen lattice contraction. *Exp. Eye Res., 74*, 361-370.

Bisen, P. S., Baghel, R. K., Sanodiya, B. S., Thakur, G. S., & Prasad, G. B. K. S. (2010). *Lentinus edodes*: A Macrofungus with Pharmacological Activities. *Current Medicinal Chemistry, 17,* 2419-2430.

Bohn, J. A., & BeMiller, J. N. (1995). (1→3)-β-Glucans as biological response modifiers: A review of structure-functional activity relationships. *Carbohydr. Polym., 28*, 3-14.

Borchers, A. T., Stern, J. S., Hackman, R. M., Keen, C. L., & Gershwin, E. M. (1999). Mushrooms, tumors, and immunity. *Proc. Soc. Exp. Biol. Med., 221*, 281-293.

Buswell, J. A., Cai, Y. J., & Chang, S. T. (1995). Effect of nutrient nitrogen on manganese peroxidase and laccase production by *Lentinula* (*Lentinus*) *edodes*. *FEMS Microbiol. Lett., 128*, 81-88.

Buswell, J. A., Cai, Y. J., Chang, S. T., Peberdy, J. F., Fu, S. Y., & Yu, H.-S. (1996). Lignocellulolytic enzyme profiles of edible mushroom fungi. *World J. Microbiol. Biotechnol., 12*, 537-542.

Chang, S.-T. (2001). A 40-year journey though bioconversion of lignocellulosic wastes to mushrooms and dietary supplements. *Int. J. Med. Mush., 3,* 80.

D'Annibale, A., Casa, R., Pieruccetti, F., Ricci, M., & Marabottini, R. (2004). *Lentinula edodes* removes phenols from olive-mill wastewater: impact on durum wheat (Triticum durum Desf.) germinability. *Chemosphere, 54*, 887-894.

D'Annibale, A., Stazi, S. R., Vinciguerra, V., & Sermanni, G. G. (2000). Oxirane-immobilized *Lentinula edodes* laccase: stability and phenolics removal efficiency in olive mill wastewater. *J. Biotechnol., 77*, 265-273.

Donoghue, J. D., & Denison, W. C. (1995). Shiitake cultivation: Gas phase during incubation influences productivity. *Mycologia, 87*, 239-244.

Fanelli, C., Fabbri, A. A., Finotti, E., Fasella, P., & Passi, S. (1984). Free radicals and aflatoxin biosynthesis. *Cell. Mol. Life Sci., 40*, 191-193.

Fanelli, C., Tasca, V., Ricelli, A., Reverberi, M., Zjalic, S., Finotti, E., & Fabbri, A. A. (2000). Inhibiting effect of medicinal mushroom *Lentinus edodes* (Berk.) Sing (*Agaricomycetideae*) on aflatoxin production by *Aspergillus parasiticus* Speare. *Int. J. Med. Mush., 2*, 229–236.

Feofilova, E. P., Gornova, I. B., Memorskaya, A. S., & Garibova, L. V. (1998). Lipid composition of *Lentinus edodes* (Berk.) Sing [*Lentinula edodes* (Berk.) Pegler] fruiting bodies and submerged mycelium. *Microbiology, 67*, 540-544.

Gonzalez, L. E., & Bashan, Y. (2000). Increased growth of the microalga *Chlorella vulgaris* when coimmobilized and cocultured in alginate beads with the plant-growth-promoting bacterium *Azospirillum brasilense*. *Appl. Environ. Microbiol., 66*, 1527-1531.

Harris, B. (1986). *Growing Shiitake Commercially: A Practical Manual for Production of Japanese Forest Mushrooms*. Madison, WI USA: Science Tech Publishers.

Harvey, L., McNeil, B., Kristiansen, B., Smith, J. E., & Kheng, T. K. (2001). Production of lentinan by submerged cultivation of *Lentinus edodes* (Berk.) Sing. *Int. J. Med. Mush., 3*, 161.

Hatvani, N. (2001). Antibacterial effect of the culture fluid of *Lentinus edodes* mycelium grown in submerged liquid culture. *Int. J. Antimicrob. Agents, 17*, 71-74.

Hatvani, N., Kredics, L., Antal, Z., & Mécs, I. (2002). Changes in activity of extracellular enzymes in dual cultures of *Lentinula edodes* and mycoparasitic *Trichoderma* strains. *J. Appl. Microbiol., 92*, 415-423.

Hehre, W. J., Radom, L., Schleyer, PvR., & Pople, J. A. (1986). *Ab Initio Molecular Orbital Theory*. New York: Wiley.

Inokuchi, N., Kobayashi, H., Hara, J., Itagaki, T., Koyama, T., Iwama, M., Ohgi, K., & Irie, M. (2000). Amino acid sequence of an unique ribonuclease with a C-terminus rich in O-glycosylated serine and threonine from culture medium of *Lentinus edodes*. *Biosci. Biotechnol. Biochem., 64*, 44-51.

Jayashree, T., & Subramanyam, C. (2000). Oxidative stress as a prerequisite for aflatoxin production by *Aspergillus parasiticus*. *Free Radic. Biol. Med., 29*, 981-985.

Jeune, K. H., Moon, I. J., Kim, M. K., & Chung, S. R. (1990) Studies on lectins from Korean higher fungi; IV. A mitogenic lectin from the mushroom *Lentinus edodes*. *Planta Med., 56*, 592.

Jong, S. C., & Birmingham, J. M. (1993). Mushrooms as a Source of Natural Flavor and Aroma Compounds. In S.T. Chang, J.A. Buswell, & S.W. Chiu (Eds.), *Mushroom Biology and Mushroom Products* (pp. 345-366). Hong Kong: Chinese University Press.

Kalberer, P. P. (2000). Influence of urea and ammonium chloride on crop yield and fruit body size of shiitake (*Lentinula edodes*). *Mushroom Sci., 15*, 361-366.

Kamper, M., Vetterkind, S., Berker, R., & Hoppert, M. (2004). Methods for in situ detection and characterization of extracellular polymers in biofilms by electron microscopy. *J. Microbiol. Methods, 57*, 55-64.

Kent, D., Sheridan, C., Tomkinson, H. A., White, S., Hiscott, P., & Grierson, I. (2003). Edible mushroom (*Agaricus bisporus*) lectin modulated human retinal pigment epithelial cell behaviour *in vitro*. *Exp. Eye Res., 76*, 213-219.

Kilpatrick, D. C. (2000). *Handbook of animal lectins: properties and biomedical applications*. London New York: Wiley.

Kim, J. M., Ra, K. S., Noh, D. O., & Suh, H. J. (2002). Optimization of submerged culture conditions for the production of angiotensin converting enzyme inhibitor from *Flammulina velutipes*. *J. Ind. Microbiol. Biotechnol., 29*, 292-295.

Kirchhoff, B., & Lelley, J. (1991). Investigations of Shiitake (*Lentinus edodes* (Berk.) Sing.) bag-log cultivation to increase the yield in Germany. *Mushroom Sci., 13*, 509-516.

Kobayasi, H., Imanaka, M., Inokuchi, N., Koyama, T., & Irie, M. (1998). Characterization of ribonucleases from culture medium of *Lentinus edodes*. *Biosci. Biotechnol. Biochem., 62*, 1604-1608.

Kobayashi, H., Hara, J., Itagaki, T., Inokuchi, N., Koyama, T., Sanda, A., Iwama, M., Ohgi, K., & Irie, M. (2000). Relationship of two ribonucleases with molecular masses of 45 kDa and 37 kDa from the culture medium of *Lentinus edodes*. *Biol. Pharm. Bull., 23*, 800-804.

Kocourek, J., & Horejsi, V. (1983). A Note on the Recent Discussion on Definition of the Term "Lectin". In T.C. Bog-Hansen, & G.A. Spengler (Eds.), *Lectins - Biology, Biochemistry, Clinical Biochemistry* (Vol. 3, pp. 3-6). Berlin: Walter de Gruyter.

Leatham, G. F. (1983). A chemically defined medium for the fruiting of *Lentinus edodes*. *Mycologia, 75*, 905-908.

Leatham, G. F. (1986). The ligninolytic activities of *Lentinus edodes* and *Phanerochaete chrysosporium*. *Appl. Microbiol. Biotechnol., 24*, 51-58.

Leonard, T. J., & Volk, T. J. (1992). Production of Specialty Mushrooms in North America: Shiitake and Morels. In G.F. Leatham (Ed.), *Frontiers in Industrial Mycology* (pp. 1-23). New York, USA: Chapman & Hall.

Levanon, D., Rothschild, N., Danai, O., & Masaphy, S. (1993a). Strain selection for cultivation of Shiitake mushrooms (*Lentinus edodes*) on straw. *Bioresour. Technol., 45*, 9-12.

Levanon, D., Rochschild, N., Danai, O. & Masaphy, S. (1993b). Bulk treatment of substrate for the cultivation of Shiitake mushrooms (*Lentinus edodes*) on straw. *Bioresour. Technol., 45*, 63-64.

Li, C. Y., & Strzelczyk, E. (2000). Belowground microbial processes underpin forest productivity. *Phyton, 40*, 129-134.

Liener, I. E., Sharon, N., & Goldstein, I. J. (Eds.) (1986). *The Lectins: Properties, Functions, and Applications in Biology and Medicine*. Orlando: Academic Press.

Liua, W., Ho, J. C. K., & Ng, T. (2001). Suppression of cell cycle progression by a fungal lectin: activation of cyclin-dependent kinase inhibitors. *Biochemical Pharmacology, 61*, 33-37.

Lobanok, A. G., Babitskaya, V. G., Plenina, L. V., Puchkova, T. A., & Osadchaya, O.V. (2003). Composition and biological activity of submerged mycelium of the xylotrophic basidiomycete *Lentinus edodes*. *Appl. Biochem. Microbiol., 39*, 60-64.

Loganathan, D., Winter, H. C., Judd, W. J., Petryniak, J., & Goldstein, I. J. (2003). Immobilized *Marasmius oreades* agglutinin: use for binding and isolation of glycoproteins containing the xenotransplantation or human type B epitopes. *Glycobiology, 13*, 955-960.

Mahajan, R. G., Patil, S. I., Mohan, D. R., & Shastry, P. (2002). *Pleurotus eous* mushroom lectin (PEL) with mixed carbohydrate inhibition and antiproliferative activity on tumor cell lines. *J. Biochem. Mol. Biol. Biophys., 6*, 341-345.

Miles, P. G., & Chang, S. T. (1987). Fruiting of *Lentinus edodes* (Shiitake) in liquid media. *MIRCEN Journal, 3*, 103-112.

Mishra, C., & Leatham, G. F. (1990). Recovery and fractionation of the extracellular degradative enzymes from *Lentinula edodes* cultures cultivated on a solid lignocellulosic medium. *J. Ferment. Engineer., 69*, 8-15.

Miyazaki, Y., Kaneko, S., Sunagawa, M., Shishido, K., Yamazaki, T., Nakamura, M., & Babasaki, K. (2007). The fruiting-specific Le.flp1 gene, encoding a novel fungal fasciclin-like protein, of the basidiomycetous mushroom *Lentinula edodes*. *Curr. Genet., 51*, 367-375.

Mizuno, T. (1996). Development of antitumor polysaccharides from mushroom fungi. Foods Food Ingred. *J. Jpn., 167*, 69-85.

Mizuno, T. (1999). The extraction and development of antitumor-active polysaccharides from medicinal mushrooms in Japan. *Int. J. Med. Mush., 1,* 9-29.

Mo, H., Winter, H. C., & Goldstein, I. J. (2000). Purification and Characterization of a Neu5Acα2-6Galβ1-4Glc/GlcNAc-specific Lectin from the Fruiting Body of the Polypore Mushroom *Polyporus squamosus*. *J. Biol. Chem., 275*, 10623-10629.

Nagai, M., Sato, T., Watanabe, H., Saito, K., Kawata, M., & Enei, H. (2002). Purification and characterization of an extracellular laccase from the edible mushroom *Lentinula edodes*, and decolorization of chemically different dyes. *Appl. Microbiol. Biotechnol., 60*, 327-335.

Ng, M. L., & Yap, A. T. (2002). Inhibition of human colon carcinoma development by lentinan from shiitake mushrooms (*Lentinus edodes*). *J. Altern. Complem. Med., 8,* 581-589.

Ngai, P. H., & Ng, T. B. (2004). A mushroom (*Ganoderma capense*) lectin with spectacular thermostability, protein mitogenic activity on splenocytes, and antiproliferative activity toward tumor cells. *Biochem. Biophys. Res. Commun., 314*, 988-993.

Nikitina, V. E. (2005). Method of obtaining the seeding mycelium of edible fungi. *RF Patent*, 2249614.

Nikitina, V. E., & Tsivileva, O. M. (2004). A method of growing the edible fungus *Lentinus edodes*. *RF Patent,* 2239983.

Nikitina, V. E., Tsivileva, O. M., Pankratov, A. N., & Bychkov, N. A. (2007). *Lentinula edodes* Biotechnology - From Lentinan to Lectins. *Food Technol. Biotechnol., 45,* 228-235.

Nikitina, V. E., Tsivileva, O. M., & Loshchinina, E. A. (2006). Interrelations of xylotrophic basidiomycetes and soil nitrogen-fixing bacteria from the genus *Azospirillum.* In Y.V. Sergeev (Ed.), *Advances in Medical Mycology* (Vol. VII, pp 293-294). Moscow: All-Russian National Academy of Mycology.

Pankratov, A. N., Loshchinina, E. A., Tsivileva, O. M., Makarov, O. E., Yurasov, N. A., & Nikitina, V. E. (2011a). Elucidating the indole participation in growth and metabolic processes in mycelial fungus. *Proceedings of Saratov University. Ser. Chemistry. Biology. Ecology, 11*, iss. 2. *In press.*

Pankratov, A. N., Tsivileva, O. M., Drevko, B. I., & Nikitina, V. E. (2011b). Compounds of the 1,5-di(4-R-phenyl)-3-selenopentanediones-1,5 series interaction with the Basidiomycete *Lentinula edodes* lectins: Computations and Experiment. *J. Biomol. Struct. Dyn., 28*, 969-974.

Parslew, R, Jones, K. T., Rhodes, J. M., & Sharpe, G. R. (1999). The antiproliferative effect of lectin from the edible mushroom (*Agaricus bisporus*) on human keratinocytes: preliminary studies on its use in psoriasis. *Brit. J. Dermat., 140,* 56-60.

Philippoussis, A. N., Diamantopoulou, P. A., & Zervakis, G. I. (2003). Correlation of the properties of several lignocellulosic substrates to the crop performance of the shiitake mushroom *Lentinula edodes*. *World J. Micribiol. Biotechnol., 19*, 551-557.

Przybylowicz, P., & Donoghue, J. (1991). *Shiitake Growers Handbook: the Art and Science of Mushroom Cultivation*. Dubuque: Kendall / Hunt Publ. Co.

Red'kina, T. V. (1990). Fungistatic activity of bacteria of the genus *Azospirillum*. *Agrokemia es Talajtan (Agrochemistry and Soil Science), 39*, 465-468.

Reshetnikov, S. V., Wasser, S. P., & Tan, K. K. (2001). Higher Basidiomycota as a source of antitumor and immunostimulating polysaccharides. *Int. J. Med. Mushrooms, 3*, 361-394.

Reverberi, M., Di Mario, F., & Tomati, U. (2004). β-Glucan synthase induction in mushrooms grown on olive mill wastewaters. *Appl. Microbiol. Biotechnol., 66*, 217-225.

Reverberi, M., Fabbri, A. A., Zjalic, S., Ricelli, A., Punelli, F., & Fanelli, C. (2005). Antioxidant enzymes stimulation in *Aspergillus parasiticus* by *Lentinula edodes* inhibits aflatoxin production. *Appl. Microbiol. Biotechnol., 69*, 207-215.

Royse, D. J. (1995). Specialty mushrooms. *Mushroom News, 43,* 4-19.

Royse, D. J., Schisler, L. C., & Diehle, D. A. (1985). Shiitake mushrooms. Consumption, production and cultivation. *Interdisciplinary Science Review, 10*, 329–335.

Royse, D. J., & Sanchez-Vazquez, J. E. (2001). Influence of substrate wood-chip particle size on shiitake (*Lentinula edodes*) yield. *Bioresource Technol., 76,* 229-233.

Savoie, J.- M., Mata, G., & Billette, C. (1998). Extracellular laccase production during hyphal interactions between *Trichoderma* sp. and Shiitake, *Lentinula edodes*. *Appl. Microbiol. Biotechnol., 49,* 589-593.

Shcherba, V. V., & Babitskaya, V. G. (2004). The carbohydrates of submerged mycelium of xylotrophic basidiomycetes. *Appl. Biochem. Microbiol., 40,* 551-554.

She, Q.-B., Ng, T. B., & Liu, W. K. (1998). A novel lectin with potent immunomodulatory activity isolated from both fruiting bodies and cultured mycelia of the edible mushroom *Volvariella volvacea*. *Biochem. Biophys. Res. Commun., 247,* 106-111.

Smith, C. (1997). Commercial production of shiitake substrates. *Mushroom World, 3*, 4-12.

Tan, Y. H., & Moore, D. (1992). Convenient and effective methods for *in vitro* cultivation of mycelium and fruiting bodies of *Lentinus edodes*. *Mycol. Res., 96*, 1077-1084.

Tomati, U., Belardinelli, M., Galli, E., Iori, V., Capitani, D., Mannina, L., Viel, S., & Segre, A. (2004). NMR characterization of the polysaccharidic fraction from *Lentinula edodes* grown on olive mill waste waters. *Carbohydr. Res., 339*, 1129-1134.

Trigueros, V., Lougarre, A., Ali-Ahmed, D., Rahbe, Y., Guillot, J., Chavant, L., Fournier, D., & Paquereau, L. (2003). *Xerocomus chrysenteron* lectin: identification of a new pesticidal protein. *Biochimica et Biophysica Acta (BBA) - General Subjects, 1621*, 292-298.

Tsivileva, O. M., Loshchinina, E. A., Makarov, O. E., & Nikitina, V. E. (2011). Polyalternate pathways of auxin production by higher fungus *Lentinus edodes* (Berk.) Sing in the prersence of low concentations of the indole group compounds. *Appl. Biochem. Microbiol. In press*.

Tsivileva, O. M., Nikitina, V. E., & Garibova, L. V. (2005a). Effect of culture medium composition on the activity of extracellular lectins of *Lentinus edodes*. *Appl. Biochem. Microbiol., 41*, 174-176.

Tsivileva, O. M., Nikitina, V. E., & Garibova, L. V. (2006a). Application of Mn(II) for obtaining the seeding mycelium of *Lentinus edodes*. *Mycol.&Phytopathol., 40*, 142-145.

Tsivileva, O. M., Nikitina, V. E., Garibova, L. V., & Ignatov, V. V. (2001). Lectin activity of *Lentinus edodes*. *Int. Microbiol., 4*, 41-45.

Tsivileva, O. M., Nikitina, V. E., Garibova, L. V., Zav'yalova, L. A., & Ignatov, V. V. (2000). Hemagglutinating activity of the fungus *Lentinus edodes* (Berk.) Sing [*Lentinula edodes* (Berk.) Pegler]. *Microbiology, 69*, 30-35.

Tsivileva, O. M., Nikitina, V. E., & Loshchinina, E. A. (2008). Isolation and Characterization of *Lentinus edodes* (Berk.) Singer Extracellular Lectins. *Biochemistry (Moscow), 73*, 1154-1161.

Tsivileva, O. M., Nikitina, V. E., Pankratov, A. N., Drevko, B. I., Loshchinina, E. A., & Garibova, L. V. (2005b). Effect of a selenium-containing preparation DAPS-25 on growth and lectin activity of *Lentinus edodes*. *Biotechnology in Russia, 2*, 70-79.

Tsivileva, O. M., Pankratov, A. N., Loshchinina, E. A., & Nikitina, V. E. (2006b). Experimental and theoretical studies of the effect of ferrum triad metal cations on the activity of extracellular lectins of *Lentinus edodes*. *Herald of Moscow University, Series 2: Chemistry, 47*, 91-96.

Tsivileva, O. M., Pankratov, A. N., Nikitina, V. E., & Garibova, L. V. (2004). Relationship between the molecular structure of nitrogen source and the activity of the extracellular lectins of *Lentinus edodes* (Berk.) Sing [*Lentinula edodes* (Berk.) Pegler] upon submerged cultivation. *Microbiology, 73*, 410-413.

Tsivileva, O. M., Pankratov, A. N., Nikitina, V. E., & Garibova, L. V. (2005c). Effect of media components on the mycelial film formation in submerged culture of *Lentinus edodes* (Shiitake). *Food Technol. Biotechnol., 43*, 227-234.

Ueda, H., Matsumoto, H., Takahashi, N., & Ogawa, H. (2002). *Psathyrella velutina* mushroom lectin exhibits high affinity toward sialoglycoproteins possessing terminal N-acetylneuraminic acid α2,3-linked to penultimate galactose residues of trisialyl N-glycans. Comparison with other sialic acid-specific lectins. *J. Biol. Chem., 277*, 24916-24925.

Van Griensven, L. J. L. D. (2001). *Agaricus bisporus* (J. Lge) Imbach and its close relatives: their present status, use, and future in commerce and research. *Int. J. Med. Mush., 3*, 81.

Vinciguerra, V., D'Annibale, A., Delle Monache, G., & Sermanni, G. G. (1995). Correlated effects during the bioconversion of waste olive waters by *Lentinus edodes*. *Bioresour. Technol. 51*, 221-226.

Wang, H. X., Ng. T. B., & Ooi, V. E. C. (1999). Studies on purification of a lectin from fruiting bodies of the edible shiitake mushroom *Lentinus edodes*. *Int. J. Biochem. Cell Biol., 31*, 595-599.

Wang, H., Gao, J., & Ng, T. B. (2000). A new lectin with highly potent antihepatoma and antisarcoma activities from the oyster mushroom *Pleurotus ostreatus*. *Biochem. Biophys. Res. Commun., 275*, 810-816.

Wang, M., Trigueros, V., Paquereau, L., Chavant, L., & Fournier, D. (2002). Proteins as active compounds involved in insecticidal activity of mushroom fruitbodies. *J. Econ. Entomol., 95*, 603-607.

Wang, P. H., Hsu, C. I., Tang, S. C., Huang, Y. L., Lin, J. Y., & Ko, J. L. (2004). Fungal immunomodulatory protein from *Flammulina velutipes* induces interferon-gamma production through p38 mitogen-activated protein kinase signaling pathway. *J. Agric. Food. Chem., 52*, 2721-2725.

Wasser, S. P. (2002). Medicinal mushrooms as a source of antitumor and immunomodulating polysaccharides. *Appl. Microbiol. Biotechnol., 60*, 258-274.

Wasser, S. P., & Weis, A. L. (1999). Therapeutic effects of substances occurring in higher Basidiomycetes mushrooms: a modern perspective. *Crit. Rev. Immunol., 19*, 65-96.

Wasser, S. P., Nevo, E., Sokolov, D., Reshetnikov, S. V., & Timor-Tismenetsky, M. (2000). Dietary supplements from medicinal mushrooms: diversity of types and variety of regulations. *Int. J. Med. Mushrooms, 2,* 1-19.

Winter, H. C., Mostafapour, K., & Goldstein, I. J. (2002). The mushroom *Marasmius oreades* lectin is a blood group type B agglutinin that recognizes the Galα1,3Gal and Galα1,3Galβ1,4GlcNAc porcine xenotransplantation epitopes with high affinity. *J. Biol. Chem., 277*, 14996-15001.

Worrall, J. J., & Yang, C. S. (1992). Shiitake and oyster mushroom production on apple pomace and sawdust. *Hort. Sci., 27*, 1131-1133.

Zjalic, S., Fabbri, A. A., Ricelli, A., Reverberi, M., Galli, E., & Fanelli, C. (2002). Effect of olive oil mill waste waters on the edible medicinal mushroom *Lentinus edodes* (Berk.) Sing. growth and lignin degrading enzymes. *Int. J. Med. Mush., 4,* 85–93.

In: Mushrooms: Types, Properties and Nutrition
Editors: S. Andres and N. Baumann ISBN 978-1-61470-110-1

Chapter 2

EDIBLE MUSHROOMS: POTENTIAL ROLE IN WEIGHT REGULATION

Kavita H. Poddar, Mary Jo Feeney and Lawrence J. Cheskin
Johns Hopkins Bloomberg School of Public Health,
Baltimore, Maryland, USA

ABSTRACT

Mushrooms are a class of "macrofungi" which in recent years have gained popularity among general consumers and the scientific community. Edible mushrooms have been shown to possess high nutritive value, and are being considered as "nutraceuticals" due to their potential health benefits. *In vitro* and *in vivo* studies suggest that bioactive compounds in mushrooms have immunomodulatory, anti-inflammatory, anti-oxidative, lipid lowering and anti-tumor effects. Some research findings suggest that edible mushroom consumption is associated with prevention of cardiovascular diseases. Many of these health benefits can be attributed to the nutrient profile of the mushrooms, which include protein, B-vitamins, minerals, along with low fat content. In addition, bioactive compounds like polyphenols and flavonoids present in mushrooms may contribute to health benefits.

Weight management is a growing concern for all segments of the U.S. population. Moreover, recent statistics provide compelling evidence that overweight and obesity are on the rise not only among adults, but also in children younger than twelve years of age. This growing trend in overweight and obesity imposes major health, economic and social burdens worldwide. While genetic, biologic and social factors are responsible to a great extent, many individuals engage in dietary behaviors that put them at risk of weight gain.

On the intake side of the energy balance equation, high intake of energy-dense foods is largely responsible for the problem of positive energy balance, and the resulting epidemic of overweight and obesity in the U.S. Energy regulation is complex, but there is evidence that humans have limited ability to regulate food intake in response to changes in energy density. One obvious method of preventing passive overconsumption when consuming energy dense foods is substitution of low energy dense foods. However, palatability differences, access, cost, and habit may prevent people from choosing such foods.

Edible mushrooms are very low in calories and energy density. They have a nutrient profile similar to that of many foods recommended in weight loss/maintenance diets. However many "diet" foods have low palatability, which may result in non-adherence to these foods by those seeking to lose weight and keep it off. In contrast, edible mushrooms are generally regarded by adults as highly palatable, making them potentially a good substitute for high energy density foods in the prevention and treatment of obesity. In support of this hypothesis, one study we conducted in humans on the effect of mushroom on satiety and palatability showed promising results, suggesting a potential role in improving weight regulation. In addition, mushrooms may have the added benefit of reducing oxidative stress and inflammation, conditions that are associated with obesity and a number of chronic diseases. Future research is needed on the effect of mushroom intake on body weight regulation, and associated health benefits.

INTRODUCTION

Mushrooms are "macrofungi" with a distinct fruiting body which grows above or below the ground. Thousands of mushroom species are known to exist worldwide. Of these, a tenth are considered edible [1]. In recent years with technological advancements and innovative methods of cultivation, huge quantities of these "macrofungi" are being produced for consumption by human's [1]. Production of edible mushrooms has increased by more than 50% in the last decade, with the most common species found in the West being *Agaricus bisporus,* which include white button, brown crimini and portabella mushrooms and *Lentinus edods,* known as shiitake mushrooms [1, 2]. Mushrooms have been used by humans as medicine for traditional therapies since ancient times, with origins in the East [3-5]. Of the thousands of species of mushrooms, 700 species of the Basidiomycetes class of mushroom have been found to possess significant pharmacological benefits [5]. Besides possessing medicinal properties, mushrooms are gaining popularity as "nutraceuticals" and "functional foods" due to presence of certain nutrients like polyphenols and other bioactive components which may exert health benefits [2, 6, 7-10]. Thus, mushroom extracts are being commercialized as dietary supplements in different forms, like capsules and tablets, with a worldwide market value of billions of US dollars annually [11]. In addition to these benefits, edible mushrooms have been used as food since the ancient times, and to this day form a important part of the diet of many people worldwide because of their taste, aroma, flavor and nutritional value; some are considered a delicacy in parts of the world [1, 10, 12]. Edible mushrooms also possess high nutritive value, and are considered good sources of carbohydrates, fiber, vitamins and minerals; they are also low in energy, sodium and fat content and contain some protein [6, 12-15]. Edible mushrooms may confer a plethora of health benefits, including cardioprotection, antitumor activity, hypolipidemic properties, antihypertensive activity, anti inflammatory properties, and immunomodulatory effects [2, 6, 7-10]. Many of these conditions are associated with weight gain and obesity [16-18]. Most of the health benefits associated with edible mushrooms is attributable to the presence of bioactive compounds and nutrients [2, 6-10]. While several studies have looked at the benefits of edible mushrooms in health using mushroom extracts, the purpose of this review is to assess the potential role of whole mushrooms in the prevention of obesity and associated factors, including oxidative stress and inflammation.

NUTRITIVE VALUE OF EDIBLE MUSHROOMS

Edible mushrooms are good sources of protein, carbohydrates, fiber, B vitamins, minerals and certain bioactive compounds [6, 12-15]. The amount of protein in mushrooms varies from 17 to 80g/100g dry weight, depending on the species. The highest protein level of 80.93g/100g is found in *Agaricus bisporus* (button mushrooms) [6]. As the water content of mushrooms is high, typically 80-90%, the per-serving level of protein is more modest, but still significant. In addition to being a source of protein in general, some mushroom species are high in essential amino acids, making mushroom protein quality comparable to animal proteins [13, 14]. Notably, edible mushrooms are also very low in fat, ranging from <1% (*Agaricus bisporus* - button mushrooms*)* to 5% in other species [6, 14, 15]. The types of fat in edible mushrooms are predominantly polyunsaturated and monounsaturated fatty acids, with very low amounts of saturated fatty acids [6, 12-15]. In addition, edible mushrooms contain relatively large amounts of carbohydrates (8% - [*Agaricus bisporus*] to 71% in other species) [6, 12, 15]. One study found dietary fiber content ranging from 1.5g/100g of dry matter (*A. bisporus*-white strain of mushroom) to 3.3g/100g of dry matter (*L. edodes* strain of mushroom) [14]. Major types of carbohydrates and fiber present in mushrooms are polysaccharides such as glucans, monosaccharides, and disaccharides, and chitin [6] and can range from 4.5g/100g (*A. bisporu*-white button mushrooms) to 5.8g/100g (*L. edodes*) of fresh weight [14]. In other species, fiber content ranged from 17.5g/100g to 24.7g/100g of dry weight [15]. Energy content of mushrooms is low varying from 27-30 kcal/100g [14] to 300kcal/100g of dry weight [6]. The nutrient profile of different mushroom species varies widely due to mushroom type, conditions in which they are grown, cooking process, harvesting time, different techniques used in managing them, and handling conditions [6, 14, 19-21]. In addition to the nutrients already covered, mushrooms contain bioactive compounds like phenols, which are the major antioxidant present, ranging from 0.88 mg/g to 8.94 mg/g, followed by flavonoids (0.67 mg/g to 3.40mg/g), ascorbic acid (0.03 mg/g to 0.87 mg/g) and carotenoids in varying amounts [6]. Most mushrooms have high moisture content (>80%) [15], which varies depending on environmental factors like humidity, heavy watering of the crop and storage conditions, which may reduce the dry matter and thus nutrient content [14]. Besides macronutrients, mushrooms contain significant amounts of B complex vitamins, some vitamin D, and minerals, making them a rich source of a variety of micronutrients [9, 22-26].

In summary, the nutrient content of mushrooms varies, depending on the maturity levels of the edible fruiting body, the environmental conditions in which it grows and is stored, the species type, and cooking conditions [6, 12-15]. While varying in nutrient profile, in general mushrooms are rich sources of important nutrients, and have low caloric value, suggesting potential value in weight management. Nutrient values for mushrooms are also available online at the USDA, in their National Nutrient Database http://www.ars.usda.gov/main/site_main.htm?modecode=12-35-45-00).

Overweight and Obesity: Why Is It a Big Concern?

Obesity Trends in the U.S.

Despite differences in the incidence of overweight and obesity across race, gender and age groups, recent statistics provide irrefutable evidence that weight management is a growing concern for the U.S. population [27, 28]. According to the most recent National Health and Nutrition Examination Survey (NHANES 2007-2008), the incidence of overweight [body mass index (BMI; kg/m^2) >25] and obesity [body mass index (BMI; kg/m^2) >30] in the U.S. adult population (age >20) increased at a lower rate between 1999-2000 and 2007-2008 (4.7% in men and 2.1% in women) in comparison to 7% in men and women in previous years (1976-1980, 1988-1994 and 1999-2000) [27]. Though these data suggest that current rise in overweight and obesity is slower than before, more than two-thirds of the U.S. adult population is overweight/obese [27, 28]. Moreover, it has been projected that if this rising trend in overweight and obesity continues, more than three quarters of U.S. adults will be obese by 2030, and hundred percent will be obese by 2048 [29]. Disturbing is the fact that excess weight gain in adulthood often begins earlier in life [30] and overweight children and adolescents are more likely to become obese adults [31, 32]. According to NHANES 2007-2008, more than 10% of U.S. children and adolescents aged 2- 19 were at or above the 97th percentile BMI for age [33].

Consequences of Excess Body Fat

The high and increasing prevalence of overweight and obesity imposes major health, economic and social burdens on afflicted individuals, as well as the world.

Health

Overweight and obesity increases the risk of major health problems, which include cardiovascular diseases, hypertension, type-2 diabetes mellitus, metabolic syndrome, dyslipidemia and certain forms of cancers [16-18]. Moreover, those who have a BMI of higher than 35 kg/m^2 are 20 times more likely to develop diabetes and 2 times more likely to develop heart diseases, hypertension, stroke and gallstones [17]. The risk of developing diabetes, hypertension, stroke, gallstones and cardiovascular diseases goes up with the severity of the overweight condition, and those who are severely obese are likely to have more than one co-morbid condition [17]. Still worse is the fact that diseases such as type-2 diabetes and hyperlipidemias, once thought to arise almost exclusively during adulthood, are now frequently occurring in children and adolescents [34-36]. Studies of predictors of health problems in young adults suggest that those who are overweight or obese during childhood and adolescence are at higher risk of developing major health issues in young adulthood, thus underscoring the need to implement weight loss strategies in an early stage of the lifespan [34-36]. In addition to specific health issues, overweight and obesity reduces overall life expectancy in individuals who are > 25 $kg/m^{2,}$ and this reduction can be dramatic for those who have a BMI of >35 kg/km^2 [37].

Psychosocial and Economic

Obese individuals tend to face discrimination and social stigma in the U.S.; often reporting being called names and ridiculed in public settings, especially as children [38]. Moreover, adults who are obese report being discriminated against in employment settings [38] and are more likely to report having low self esteem, and higher rates of anxiety and depression [39, 40]. Younger obese individuals are less likely to marry, and have lower rates of education [41]. Besides health and psychosocial consequences, the projected US healthcare cost attributable to overweight and obesity will run into the hundreds of billions of dollars by 2030, and will account for 16-18% rise in the nation's healthcare spending [29]. This rise in healthcare spending can largely be attributed to obesity-related health conditions which are increasing with the rise in obesity rates [42], imposing a major economic burden as more than half of these expenses are paid for with public funds, including Medicare and Medicaid [43]. In addition to the economic burden that obesity imposes on the government, obese individuals and employers with obese employees also incur costs due to loss of personal opportunities, loss in workplace productivity due to the health conditions of obese employees and high health insurance premiums [43, 44].

Causes

Genetic and Environmental

Obesity is a result of complex interplay between genetic, biologic and environmental factors. Genetic predisposition can increase individual susceptibility to weight gain [45]. While rare genetic mutations which affect eating behaviors resulting in obesity account for less than 1% of the rise in rates of obesity; common genetic variations in individuals can influence taste perceptions and meal size, in part explaining why certain people may choose meals that put them at a higher risk of developing obesity [45]. Although some genes have been identified which may be linked to higher percent body fat in humans, the exact mechanism of action remains unclear [45]. In the U.S., environmental factors, such sedentary lifestyles and easy access to energy-dense food are likely the major driver of the recent obesity epidemic [46, 47]. Current urban environments promote easy access to energy-dense foods, decreased levels of physical activity, limited availability of recreational facilities, food advertising and marketing which adversely affect our choices and promote unhealthful lifestyles [46, 47]. More and more research is being directed to understanding the association between the built environment and obesity so that effective interventions can be developed [46, 47]. While changing the environment to promote healthy lifestyles has the potential to be successful, it is imperative to develop sustainable dietary intervention approaches as well.

Dietary Behaviors

Emerging data on U.S. dietary behaviors indicate that many individuals engage in behaviors that put them at risk of developing obesity [48-53]. In general, consumption of red meat high in fat, high calorie sugar-sweetened beverages, refined carbohydrates, and excess sodium in processed foods has gone up substantially, and varies by age, ethnicity and socioeconomic status [48-54]. A recent review on standard American diets found that mean total calorie consumption has gone up by more than 700 kcal/day per adult in the last five decades and is over 2,600 kcal/day currently [48]. Contribution of calories from refined carbohydrates (9.5% of total calories) and added sugars (9%) were highest [48]. Red meat consumption in 2008 went up by 7 lbs/person since the 1950's [48] and constituted 58% (50-90g/day) of the total meat (100-150g/day) consumed [49]. Taste seems to be driving this increase [50]. Besides meat, sugar-sweetened beverage (SSB) intake has also gone up [51-54] and accounts for more than 500 kcal/day for young adults (19-39 years of age) and close to 400 kcal/day in children aged 2-6 [52]. According to NHANES III (1988-1994 and 1999-2004) data, adults consuming sugar-sweetened beverages rose from 58 to 63%, with an increase in calories from SSB averaging 50 kcal/day and 6 oz per day [53]. Although SSB consumption varies by age, gender, and SES, [51-54], intake is high in the U.S., and is associated with high total energy intake [55]. Research shows that red meat intake accounts for high intake of solid fats (~7%), with SSBs accounting for more than 50% of the added sugar in American diets [56] all of which are associated with high energy density and probably the development of obesity [57].

Rationale for Reducing Dietary Energy Density to Prevent Weight Gain and Obesity

Weight gain and subsequent development of obesity can be attributed to positive energy balance: energy expenditure less than energy intake [58]. In the U.S., decreased physical activity, along with increased intake of energy-dense foods is well recognized as one of the main environmental factors associated with the rise in rates of obesity [59]. Energy density of the diet is key element in regulating body-weight, as it influences appetite control signals (hunger and satiety) [58]. Dietary fat in particular is associated with high energy density, passive overconsumption of foods and subsequent weight gain due to its relatively weak effect on satiation for a set quantity of calories [58, 60, 61]. Epidemiologic evidence suggests that overconsumption of energy-dense foods are indeed positively associated with high fat intake in the diet, and higher BMI, waist circumference, hip circumference and body weight among individuals susceptible to weight gain [62-69]. Given that high energy density of foods is one of the main causes of the current epidemic of obesity, one obvious dietary intervention method to curb obesity trends would be to substitute high energy density foods with low energy density foods. Studies have shown that intake of fruits and vegetables is associated with lower dietary energy density and lower body weight and BMI in the U.S. [63, 64, 66, 67] Fruits and vegetables also have the benefit of providing high nutrient density [68, 69]. Several experimental trials, wherein energy density of meals were manipulated by substitution of energy dense foods for low energy dense foods, have shown a direct relationship between energy density and energy intake, with low satiety [70-73]. Moreover, high energy intake is associated with high fat in the diet [73, 74]. In spite of the high nutrient quality and benefits of low energy density foods, Americans tend to be less inclined towards consuming these foods. Easy accessibility to high energy dense foods which are palatable and cost less are factors pushing us towards obesity [51, 54, 57, 60, 75].

Reducing Energy Density with Edible Mushrooms

Edible mushrooms are valuable sources of nutrients and bioactive components [6, 12-15] and possess desirable features like flavor, aroma, taste [1, 10] and satiety [72]. Notable of mushrooms is their low energy density [6, 14]; an important element in weight regulation as several studies show that high energy dense food intake is associated with high BMI and body weight [62-67]. Weight gain due to high intake of energy-dense foods like fats has been attributed to passive overconsumption because of their low satiety value and high palatability [60]. In the U.S, processed meat and most red meats that are high in fat form a significant source of saturated fat intake in the diet [56, 76].The consumption of red and processed meats has risen significantly [49-50]. Substituting red and processed meats, which have a high content of saturated fat and cholesterol [76-78], with edible mushrooms can be an effective strategy to reduce calories in the diet and potentially lose weight. Because mushrooms are a source of proteins and essential amino acids [6, 13, 14] they can substitute for some of the high quality protein found in meats [77, 78]. Also, edible mushrooms contain several vitamins and minerals that are essential for optimal health [9, 22-26]. Dietary fiber, found in high quantities in whole grains, fruits and vegetables, provides high satiety and low energy intake [71, 79, 80]. Studies show that with high dietary fiber intake, by promoting satiation overall energy intake is 7-15% lower [71, 80]. However, palatability is often rated lower on typical high=fiber diets [57, 60] again making mushrooms a good substitute [6, 12-15]. Thus, several studies have shown that low energy density foods like fruits, vegetables and whole grains lower energy intake due to increased satiety of the diet [67, 69-71] which translates into lower BMI [62-68]. Possible beneficial effects of edible mushrooms in general health are only beginning to get noticed, and data is lacking on their usefulness as a low energy density food substitute, To date only one study has been published to assess the effect of mushrooms on energy content of test meals and subsequent energy intake [72] The promising results of this study suggests that mushroom ingestion may be helpful in regulating body weight.

Rationale for Reducing Meat Intake in the Diet

Meat consumption in the US has increased dramatically [48-50] with a consequent increase in saturated fat intake and dietary cholesterol intake since meat is one of the major contributors of these nutrients in the American diet [56, 77-78] which are also one of the major dietary contributors to the obesity epidemic [74-75]. According to NHANES 2001-2002 data, red meat constituted of a large percentage (almost 40%) of the meat and bean subgroup, a major contributor of solid fats in the diet [56]. Similarly, the Bogalusa Heart Study [77] concluded that those who consumed more meat (>75th percentile) had higher intakes of energy from total fat (41%), saturated fat (13%), protein (18%), dietary cholesterol (372 mg/day), and less carbohydrates (40%) compared with those whose meat consumption was less <25th percentile, where the energy from total fat (32%), saturated fat (11%), protein (11%) was lower, and they consumed less dietary cholesterol (264 mg/day) and more carbohydrates (55%) [77]. These data collectively indicate that red meat forms a large part of the American diet, and is a significant contributor to total fat intake. Reducing intake of red meat can help decrease fat, cholesterol, and calories significantly without compromising on protein and important vitamins and minerals [77]. Nicklas, et al [77] reported that those who consumed less meat had significantly lower intake of protein, heme iron and phosphorus, yet adequately met the RDA for these nutrients and 66% of the RDA for vitamin B12, zinc and niacin. Edible mushrooms provide protein, including essential amino acids, niacin, vitamin B12 [9, 22-26] making them a good substitute for meat. Moreover substituting one meat-containing meal per week with edible mushrooms can create a significant energy deficit (~385 kcals) which is equivalent to 5 lbs of weight loss [72]. Several studies have shown that high red meat consumption may be associated with the development of cardiovascular disease risk factors like dyslipidemias, hypertension, diabetes and certain types of cancers [81-89]. One study reported no association between meat consumption and development of cardiovascular risk factors [77], and two reviews on meat consumption and associated health risks reported weak or no association [90, 91] suggesting some inconsistencies in the evidence on the effects of meat intake on health risks. It is possible that high sodium from processed meat intake [85, 86] and visible fat in meat [91] may contribute to the observed health risks associated with consumption of red meat. Since overall red meat consumption, in particular processed meat in the US population is higher [48-50, 77] than recommended levels, it is imperative that dietary strategies are developed to enable consumers to eat primarily lean meats, and in moderate amounts.

Effect of Substituting Mushrooms for Meat on Energy Intake

To date, only one study has assessed the effect of substituting mushrooms for meat on energy density of the meal and subsequent energy intake. Cheskin, *et al* [72] conducted a short randomized clinical trial (crossover design) to assess the effect of substituting low energy density mushrooms for high energy density ground beef on total energy intake, palatability, appetite, satiation and satiety in 54 men and women. The subjects of this study were classified as normal weight (42.6%), overweight (33.3%) or obese (24.1%). Four different recipes were developed for lunch which included either meat (90% lean ground beef) or white button mushrooms as the main ingredient. Similar volumes of each lunch entrée (either meat or mushroom base) were served to the study subjects who came to the research center to have their meals. The study subjects were asked to report to the research center for 8 test meals on four consecutive weekdays for 2 consecutive weeks (with a 3-day washout period between the two weeks of either meat or mushroom entrée). The subjects were randomized to consume all four meat or all four mushroom entrees the first week to control for possible order effects. Subjects could have water ad libitum, and were served only one helping of the entrée. Meals were measured before and after consumption to determine the actual amount of food consumed. Subjects were trained to complete a food dairy and were asked to record their food intake for the times that they consumed food off-site, and a trained research assistant examined the records for thoroughness. Visual analog scales were used to measure palatability, satiety, satiation and appetite using 100-mm visual analog scales.

Results

Table 1 shows the macronutrient composition of the lunch entrées with meat and mushroom versions.

Tables and Figures: Authors data [72].

Table 1. Macronutrient composition of test lunches

	Lasagna		Savory Napoleon		Sloppy Joe		Chili	
	Meat	**Mushroom**	**Meat**	**Mushroom**	**Meat**	**Mushroom**	**Meat**	**Mushroom**
Energy (kcal)	1026	451	679	359	822	248	604	300
Carbohydrate (g)	33	41	43	49	40	48	38	41
Protein (g)	88	30	29	10	71	14	33	14
Fat (g)	57	19	45	17	42	4	31	5

On an average, subjects consumed more than double the total energy (730± 7.9 kcal vs. 310 ± 5.8 kcal), fat (41.1±0.4g *vs.*10.2±0.2g), and protein (51.4 ± 0.6g *vs.* 15.2 ± 0.3g) in the meat *vs.* mushroom test meals (Figures 1-3). Interestingly, mean 24 hour energy, fat and protein intake was 2013.5 ± 70.0 kcal *vs.* 1639.5 ± 64.8 kcal, 88.4 ± 3.4 g vs. 60.3 ± 3.2g and 105.0 ± 3.9 g vs. 68.7 ± 3.0g respectively for meat vs. mushroom intake days (Figures 4-6). Overall, 24 hour energy, protein and fat intake was higher by 374.0 ± 51.8 kcal, 28.2 ± 2.3g, 36.2 ± 2.3 g in the meat condition vs. the mushroom condition (Figures 4-6).

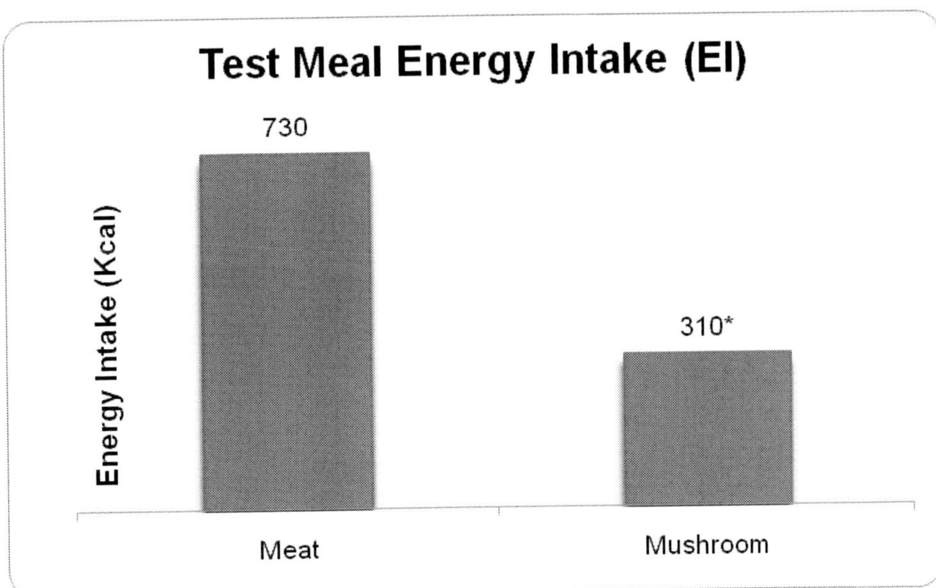

Figure 1. Mean (± SE) kcal of food consumed at meals for meat and mushroom meal conditions. Subjects consumed significantly more energy during meat meals than they did during mushroom meals (730 ± 7.9 kcal vs. 310 ± 5.8 kcal, *$p < 0.0001$).

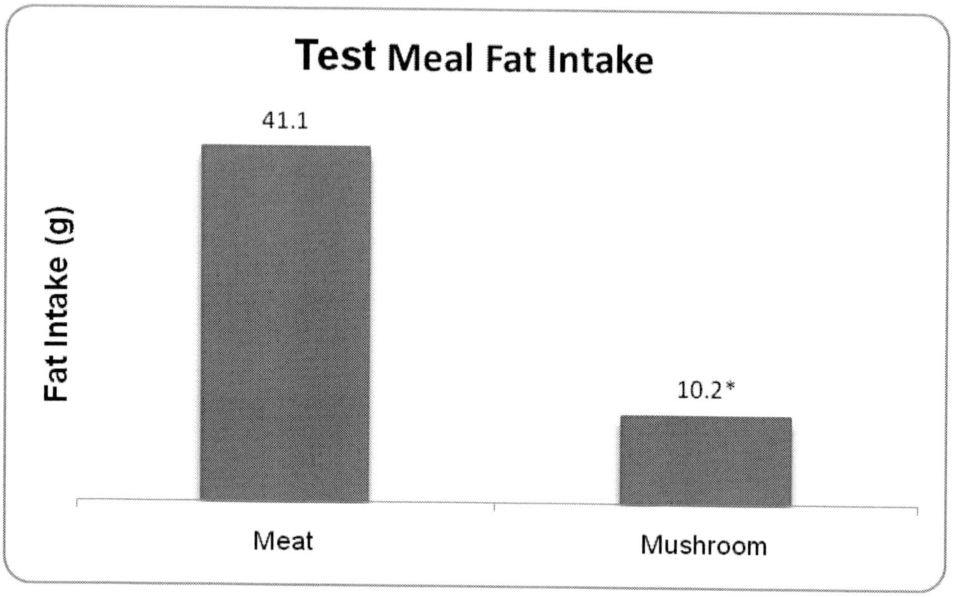

Figure 2. Mean (± SE) grams of fat consumed at meals for meat and mushroom meal conditions. Subjects consumed significantly more grams of fat during meal meals than they did during mushroom meals (41.1 ± 0.4 g vs. 10.2 ± 0.2 g, *$p < 0.0001$).

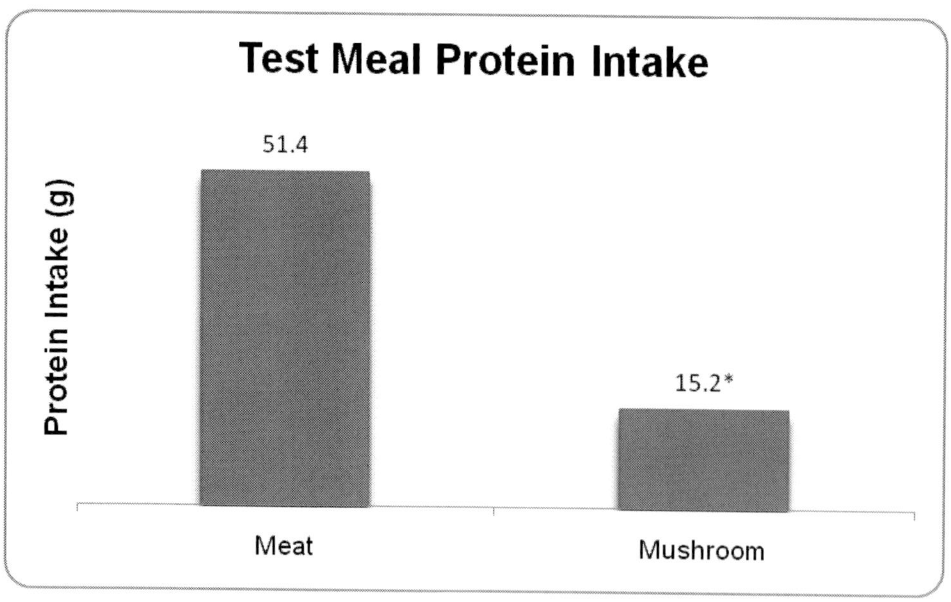

Figure 3. Mean (± SE) grams of protein consumed at meals for meat and mushroom meal conditions. Subjects consumed significantly more grams of protein during meal meals than they did during mushroom meals (51.4 ± 0.6 g vs. 15.2 ± 0.3 g, *$p < 0.0001$).

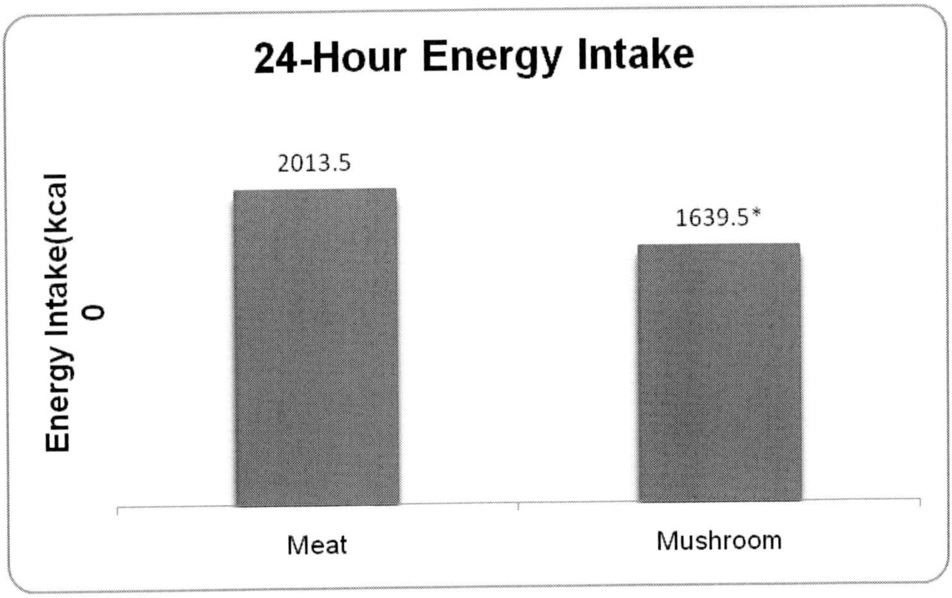

Figure 4. Mean (± SE) kcal of food consumed over 24-hours, including test meals, for meat and mushroom meal conditions. Subjects consumed significantly more energy on days when meat meals were consumed compared with days when mushroom meals were consumed (2013.5 ± 70.0 kcal vs. 1639.5 ± 64.8 kcal, $^*p < 0.0001$).

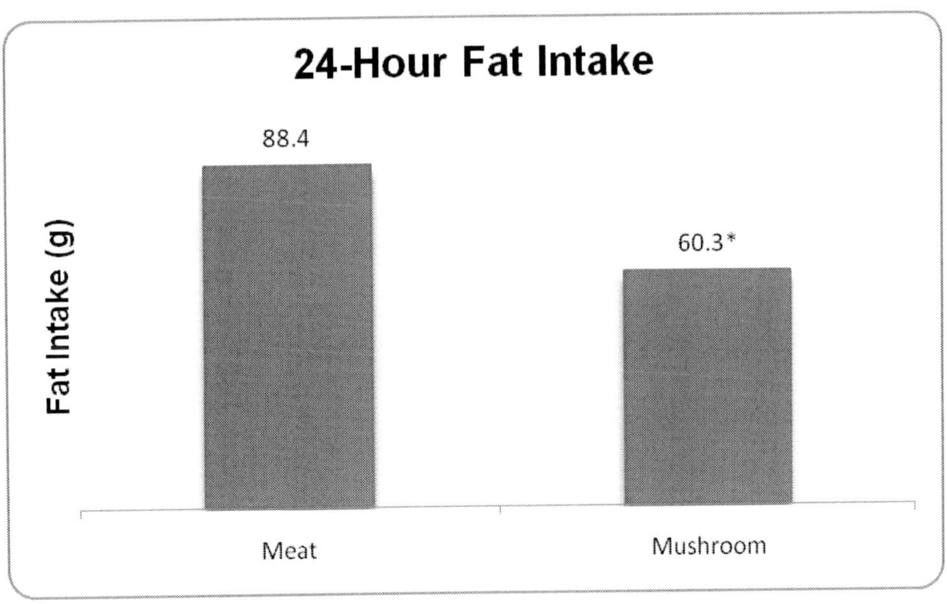

Figure 5. Mean (± SE) grams of fat consumed over 24-hours, including test meals, for meat and mushroom conditions. Subjects consumed significantly more grams of fat on days when meat meals were consumed compared with days when mushroom meals were consumed (88.4 ± 3.4 g vs. 60.3 ± 3.2 g, $^*p < 0.0001$).

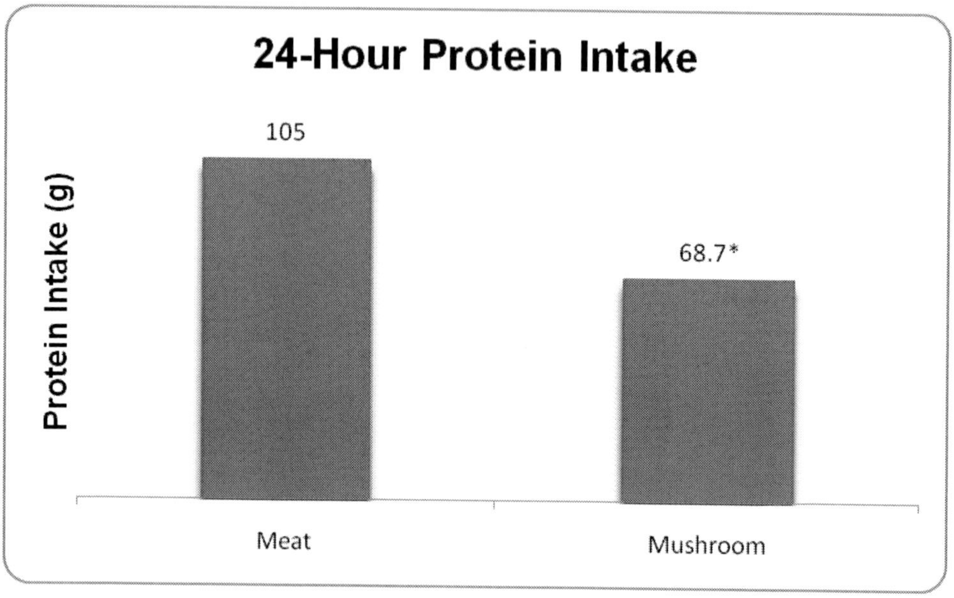

Figure 6. Mean (± SE) grams of protein consumed over 24-hours, including test meals, for meat and mushroom conditions. Subjects consumed significantly more grams of protein on days when meat meals were consumed compared with days when mushroom meals were consumed (105.0 ± 3.9 g vs. 68.7 ± 3.0 g, $^{*}p < 0.0001$).

The percentage of test meals (volume) that was consumed did not differ between meal conditions, with 93.2% of the meat meals being consumed, as compared to 91.3% of the mushroom-containing meals. No differences were observed in palatability, satiation, satiety and appetite between test meal conditions (Table 2).

Table 2. Portions consumed, subjective ratings and physical activity [a b]

	Test Meal Condition	
	Meat	Mushroom
Portion Test Meal Consumed (%)	93.2 (0.01)	91.3 (0.01)
Palatability (n=54)	13.4 (0.4)	13.0 (0.4)
Satiation (post-meal fullness) (n=47)	7.4 (0.3)	6.8 (0.3)
Satiety (general fullness) (n=47)	6.0 (0.3)	5.8 (0.3)
Appetite (n=47)	4.4 (0.3)	4.3 (0.3)
Physical Activity (kcal) (n=54)	186.60 (36)	192.2 (39)

[a] Mean (SE)

[b] No subscript denotes mean values are not statistically significantly different at $p < 0.05$ level

Subjects compensated partially for energy content (11%) and consumed fewer calories (average 1484.9 ± 210.7 kcal) over 4 days in the meat test days. The mean protein (g) content of the mushroom meals was 15 g compared to the 51(g) in the meat meals. Though protein intake was significantly lower in the mushroom meals, the study demonstrated that over a period of 24 hours, the mean protein intake with the mushroom meals was 69 g along with 1,640 calories which represents about 17% of calories from protein. This intake is within the

AMDR (Acceptable Macronutrient Distribution Range) of 10-35% as established by the IOM [92]. Dietary protein, the most thermogenic macronutrient, may suppress food intake by delaying return of hunger and by affecting satiety and energy intake *via* post-ingestive and post-absorptive signals [93]. While high protein intake is a popular approach for weight loss, Berti *et al* [94] concluded that the satiating properties of different beef preparations (roast beef, beef cooked in a vegetable broth, and canned beef described as boiled beef with jelly) did not depend on the protein content. The canned beef in jelly, characterized by high water content, was the most satiating preparation despite the lowest protein content. The authors concluded that physical characteristics, mainly energy density, seemed to be more effective determinants of satiety, and small portions of low-energy dense protein foods seemed useful in modulating satiety sensations. The authors also concluded that high protein intakes are not the only way to control food intake. These findings support those of Cheskin, *et al* (72), where despite the protein differences (See figure 3 and 6) in the meat versus mushroom meals, no differences in appetitive measures were observed, suggesting that energy density differences may override differences in the satiety hierarchy of macronutrients studied in isolation. Overall, the data demonstrates that mushrooms are satiating, help lower energy density, lower caloric intake yet meet the AMDR for protein. The results of this study showed promising effects of substituting mushrooms for meat in the diet of adults to reduce overall energy intake. Notable was the finding that there were no difference in the palatability, satiety, satiation and appetite when meat was substituted with mushrooms, suggesting its utility as a low calorie substitute since palatability exerts a strong influence on energy intake [51, 54, 57, 60]. Another study found that young children tend to adjust their energy intake more effectively in response to energy density [95]. However as they get older, energy intake is not effectively adjusted in response to energy density [95]. As fat content of a diet is positively associated with palatability and despite a sizeable reduction in fat by substituting mushrooms for lean ground beef, there was no significant difference in palatability between the meal conditions. The authors concluded that substitution of fewer than 10 meals with low energy dense foods like white button mushrooms with meat would be required to produce energy deficit equivalent to 1 lb of body fat. Substitutions of one food for high energy dense alternatives is a less dramatic way of making dietary changes for those who want to lose weight than wholesale dietary changes, and might be sustainable for a longer period than more aggressive dietary manipulations.

As a follow-up to the study by Cheskin, *et al* [72], we are now examining the effect of white button mushrooms as a substitute for meat intake on the body composition changes in obese individuals. The primary aim of this study is to assess weight loss and weight maintenance efficacy with white button mushrooms as a substitute for high calorie food like meat. The study has enrolled omnivorous, mushroom-eating adult male and female participants, who are overweight or obese (BMI 25-40), seeking weight loss. These participants are randomized to the intervention ("mushroom diet") or the control group ("standard diet that contains meat"). Those in the intervention group are asked to substitute white button mushrooms for beef in at least three meals per week over the course of 1 year. Data collected includes total percent body fat, waist circumference, and body weight. All participants are prescribed a 500kcal/d energy-deficit diet designed to achieve ~20 lbs weight loss over the first 6 months of intervention. Participants in the control group are advised on various methods to improve their diet, not including the mushroom substitution for the first six months. For the following 6 months, participants are advised on different ways to maintain their body composition changes while on the mushroom or meat diet.

Health Benefits and Edible Mushroom Intake

Mushrooms are gaining popularity in recent years because of their beneficial health effects. They can be considered functional foods or nutraceuticals, with most of the functional properties being imparted by bioactive compounds [2, 6, 7-10]. Extracts from different mushrooms species show beneficial effects in the treatment and/or prevention of certain types of cancers, and cardiovascular risk factors like dyslipidemias, inflammation, oxidative stress and hypertension. However, most of these effects have been shown in *in vitro* or *in vivo*, with very limited data coming from clinical trials [2, 7-10]. Most data comes from studies using extracts from different species of mushrooms, rather than using whole mushrooms in longer-term clinical trials. Though since ancient times people from the East have considered mushrooms (mostly extracts) as medicine to treat a variety of medical issues including viral infections [10], there is a lack of convincing scientific evidence in support of health benefits associated with mushroom as a part of healthful diet.

Anti-Inflammatory and Anti Oxidative Stress Properties of Edible Mushrooms

Obesity, Inflammation and Oxidative Stress

Obesity is associated with the development of several medical conditions [16-18] which can be attributed to low-grade inflammation and oxidative stress induced by lipid overload, resulting in high fat levels in the tissues of the body [96-100]. Chronic intake of excess calories *via* high intake of macronutrients in the diet promotes oxidative stress and inflammation through activation of several metabolic pathways [101-104] which lead to metabolic dysfunction and a plethora of diseases [96-100]. Chronic high calorie intake causes increased levels of fat deposition (triglycerides in adipose tissues), with consequent hypertrophy (increased fat cell size) and hyperplasia (increased number of fat cells) of adipose tissues. This alters the normal function of fat cells, which induces stress on the body tissues due to release of several inflammatory factors by hypertrophic fat cells [96-98]. This stress causes functional abnormalities of several tissues and activation of signaling pathways that give rise to the development of metabolic diseases. Macrophages in hypertrophic adipocytes play an important role in inducing the proinflammatory environment, along with other immune cells. High numbers of macrophages are present in hypertrophic adipocytes, which release a number of proinflammatory factors, including tumor necrosis factor-α (TNF – α), C- reactive protein (CRP), interleukin-6 (IL-6) and inducible nitric oxide synthase (iNOS), resulting in activation of proinflammatory responses [96-98]. Moreover, these adipocytes express monocyte chemoattractant protein-1 (MCP-1) and m-colony stimulating factor (M-CSF-1) that recruit more macrophages to the fat tissues, leading to higher levels of inflammation due to infiltration and release of proinflammatory cytokines [96-98]. In addition to inflammation, excess levels of lipid stores cause endoplasmic reticulum (ER) stress and disrupt the functioning of this organelle which plays a very important role in the synthesis of new proteins, lipid droplets and cholesterol regulation [96-98]. This ER stress induces activation of additional proinflammatory pathways that promote insulin resistance, which results in high levels of circulating fat and glucose concentration in blood, thereby causing further ER stress and a vicious cycle of worsening insulin resistance [96-98]. High levels of circulating free fatty acids, inflammatory factors and saturated fatty acids results in activation of inflammatory pathways that involve c-Jun N-terminal kinases (JNK) and nuclear factor kappa B (NF-KB), leading to the development of insulin resistance in additional tissues other than fat (like muscle and liver), and may reduce pancreatic β-cell mass, thereby diminishing insulin production and sensitivity [96-98]. Furthermore, the presence of chronic inflammation can induce oxidative stress due to production of reactive free radicals by increasing the expression and activity of pro-oxidant enzymes like myeloperoxidase and NADPH oxidase [96]. Presence of excess free fatty acids as seen in obesity causes mitochondrial uncoupling as the mitochondria cannot keep up with the excess fatty acids that they have to process. This uncoupling causes release of reactive oxygen species (ROS), which result in further development of insulin resistance and lipid accumulation in tissues like liver and muscles, along with diminished activity of the pancreatic β-cells [96, 97]. Moreover, some of the proinflammatory factors released by hypertrophic fat cells, like TNF-α, CRP and IL-6 stimulate ROS production [97].Certain adipokines released by the fat tissue in obesity affect energy utilization by the body resulting in fatty liver, atherosclerosis and type 2 diabetes, in addition to insulin resistance [97]. The oxidative stress induced in obesity is associated with the development of metabolic syndrome, and subsequent development and progression of cardiovascular diseases as a result of increased lipid peroxidation, coupled with impaired antioxidant defense mechanisms [99, 100]. Excess free radicals damage cellular lipids, proteins and DNA, initiating a chain of events resulting in further damage and disease progression. Oxidized LDL is an important biomarker of oxidative stress; high levels of circulating oxidized LDL have been associated with metabolic syndrome and cardiovascular risk factors [99, 100]. Increased levels of oxidized LDL in the aorta are associated with plaque formation and subsequent development of atherosclerosis [99]. Thus, complex mechanisms are involved in obesity and the development of associated health effects. Interplay between inflammation and oxidative stress initiates a vicious cycle of disruption in several signaling pathways and gene expression, which lead to the development of several metabolic diseases. Research shows that even modest changes in energy intake (favorable or unfavorable) exert a strong influence on insulin resistance, C-reactive protein, oxidized LDL, free fatty acids, and IL-6 levels, which are associated with oxidative stress and inflammation [105-110]. Moreover, individuals consuming diets high in fat exhibit increased levels of markers associated with oxidative stress and inflammatory markers [107, 109, 110], suggesting that not only chronic high energy intake, but high fat intake may also be associated with inflammation and oxidative stress.

Effect of Edible Mushrooms on Inflammation and Oxidative Stress

Dietary intake of antioxidants such as those found in fruits and vegetables have been implicated in health and well being [111-113]. The nutrient profile of mushrooms though variable, is similar to that of fruits and vegetables and contains significant amounts of plant phenols, tocopherols, ascorbic acid, carotenoids, flavonoids and certain types of polysaccharides [5-10, 19-23]. Extensive research shows that these compounds from mushrooms possess beneficial properties and have effects similar to those in fruits and vegetables in preventing oxidative stress and inflammation [5-10, 19-23]. However the effects vary depending on the species and the conditions in which they are cultivated, both of which lead to significant differences in the amount of bioactive compounds present [5-10, 19-23].

Anti-Inflammatory Properties

The role of mushrooms in modulating inflammatory responses has been studied extensively with both *in vitro* and *in vivo* models using extracts from different species [7, 9, 114]. One study reported significant anti-inflammatory effect of *Pleutorus Florida* Eger mushroom extract in mice, comparable in magnitude to the anti-inflammatory medication diclofenac. The mushroom extract also reduced platelet aggregation by 88% to 95% [114]. Kim, *et al*, [115] examined the anti-inflammatory effect of the mycelia extract from six strains of *Phellinus Linteus* mushroom species in mice. The authors determined that n-BuOH sub fraction had highest anti-inflammatory potency as determined by inhibition of croton-oil induced ear edema and angiogenesis [115]. Martin [116] examined the effect of *Agaricus bisporus* (white button and crimini), *Lentinula edodes* (shiitake), *Pleurotus ostreatus* (oyster), and *Grifola frondosa* (maitake) extracts on the pro-inflammatory cytokine-induced up regulation of adhesion molecule expression, such as VCAM-1, ICAM-1 and E-selectin, which have been linked to cardiovascular disease. The results were that white button mushrooms reduced the expression of these molecules, and all the other mushroom species reduced binding of monocytes to aortic endothelium cells in a pro-inflammatory environment, suggesting a beneficial role of these mushrooms in preventing cardiovascular diseases which is associated with inflammatory processes [116]. Besides these anti inflammatory effects, some mushroom species (white button, crimini, maitake, oyster and shiitake extracts) are known to modulate immune function by stimulating TNF – α, IFN-γ, IL-1 β production in macrophages and T-cells, with little effect on IL-10 activation. This modest proinflammatory environment induced by these species is important in killing tumorogenic cells and clearing infectious organisms [7]. One study showed that white button mushroom (*Agaricus bisporus*) extracts boost immunity by increasing the production of TNF – α, and interferon gamma (IFN-γ), and enhancing the natural killer cell activity in the splenocytes of C57BL/6 mice. [117]. Kim GY *et al* [118] reported the anti-inflammatory effect of acidic polysaccharide (PL) extract from the *Phellinus linteus* mushroom species on lipopolysaccharide (LPS)-induced cytokine production using *in vitro* and *in vivo* models. The study showed that PL treatment resulted in reduced expression of LPS-stimulated TNF – α, IL-12 and IL-1β in an *in vitro* model, and reduced the production of LPS- stimulated IFN- γ, IL-2 and TNF – α in an *in vivo* model. Moreover, the polysaccharide resulted in LPS-stimulated macrophage and lymphocyte death [118]. Several other studies in *in vitro* and *in vivo* models have shown similar anti-inflammatory effects of mushroom extracts, including Cordyceps pruinosa [119] and Inonotus obliquus [120, 121]. The primary mechanism of action in reducing inflammation appears to be via regulating the production of IL-1β, TNF – α, nitric oxide, prostaglandin E2 by suppressing gene expression of these proteins and cyclooxegenase-2 (COX-2), and downregulating the nuclear factor kappa-β pathway [119-121], one of the primary pathways inducings inflammation in obesity [96-98]. Thus, certain species of mushrooms and compounds may result in the stimulation of macrophages, which release cytokines that are important in killing tumor cells, thus modulating immunity. In other cases, mushroom extracts may inhibit production of cytokines by macrophages, thus imparting anti-inflammatory properties, and may have relevance in the clinical management of inflammatory diseases.

Anti-Oxidative Stress Properties

Several compounds from different species of mushrooms have been identified that exhibit anti-oxidative properties- by inhibiting lipid peroxidation, and preventing damage by free radical scavenging [122-124]. One study by Zhang, *et al* [122] found that the fatty acid fraction and three additional compounds from the mycelia of the edible mushroom *Grifola Frondosa* reduced liposomal peroxidation by 42% to 79%, with the highest inhibitory activity seen in the fatty acid fraction (which includes palmitic, oleic and linoleic acids). Similarly, an extract from the fruiting body of the edible mushroom *Agrocybe aegerita* inhibited liposomal peroxidation by 75% with the fatty acid fraction, 45% by palmitic acid, and 43% by ergosterol [123]. Vaz, *et al* [124] examined the antioxidant activity of extracts from two species of Portuguese wild mushrooms (*Clitocybe alexandri* and *Lepista inversa)* in an *in vitro* model which included human tumor cells. The authors evaluated the free radical scavenging property, reducing power and inhibition of lipid peroxidation to measure the antioxidant activity, and concluded that the phenolic extract of the *Clitocybe alexandri* mushroom species had a strong inhibitory effect on the growth of the tumor cells; the polysaccharide extract of the *Lepista inversa* mushroom species had strong antioxidative properties [124]. These studies indicate that phenols and other bioactive compounds in various mushrooms have the potential to eliminate excess reactive oxygen species produced in the tissues under certain stressful conditions like high calorie intake [97-99]. Moreover one study by Xu, *et al* [125] examined the effect of polysaccharides from the *L. edodes* mushroom species on expression of VCAM-1 mRNA in thoracic aorta cells in rats. The results revealed that the polysaccharides from the mushroom species had increased the activity of antioxidant enzymes in the cell, and decreased the VCAM-1 mRNA expression, thereby reducing the oxidative stress that was induced by high fat diet [125]. Similar results have been reported by Martin [116], who examined the effect of *Agaricus bisporus* (white button and crimini), *Lentinula edodes* (shiitake), *Pleurotus ostreatus* (oyster), and *Grifola frondosa* (maitake) on the regulation of adhesion molecules which cause atherosclerosis. The expression of vascular cell adhesion molecule 1 (VCAM-1), intercellular adhesion molecule 1 (ICAM-1) and endothelial-leukocyte adhesion molecule 1 (ELAM-1 or E-selectin) had significantly reduced in the human endothelial aortic cell lines with the mushroom treatment [116]. Development of atherosclerosis in a proinflammatory environment is associated with increased oxidative stress [95, 96], which can modulate expression of adhesion molecules. One study found significant reduction in the development of atherosclerotic plaques by the edible mushroom *Agaricus sylvaticus* (6.6+/-2.9%) *vs.* 47.0+/-14.0% in control in the aorta of rabbits which were induced by high cholesterol diets indicating that it has may have anti oxidative stress property [126]. Several studies with different mushroom extracts have shown hypocholesterolemic effect and lower incidence of atherosclerotic plaque development [127-129] suggesting possible anti oxidative stress mechanism. Moreover expression of adhesion molecules are associated with oxidative stress [130, 131] which form one of the underlying causes of disease progression and studies show that oxidized LDL can induce the expression of certain adhesion molecules which are associated with atherosclerotic events [132, 133]. Thus, inhibition of adhesion molecule expression by some mushroom extracts (116, 125) exhibit anti oxidative property. One study showed that extracts from *Agrocybe aegerita* mushroom species indeed inhibited human LDL oxidation in *in vitro* model [134] corroborating the usefulness of some mushroom species as antioxidants [134, 135]. *Pleurotus ostreatus,* an oyster mushroom, has been shown to increase the activity of several antioxidative enzymes, including superoxide dismutase (SOD), glutathione peroxidase (Gpx) and glutathione S-transferase in different tissues of rats [136, 137] showing its potential as a powerful antioxidant.The weight loss study that substitutes mushrooms for meat described earlier is also assessing the effects of mushroom consumption on levels of oxidative stress and inflammation. as assessed by levels of IL-6, C-reactive protein, and oxidized LDL in blood.

Conclusion

Mushrooms are foods with a rich nutrient profile, and show promise for preventing and assisting in the treatment of various health conditions. They have been considered medicinal foods in the East for centuries. Thousands of species of edible mushrooms exist, which have varying concentrations of different nutrients depending on the species and cultivation techniques. But all species possess a high quality of protein in some amounts, carbohydrates and fats. Besides these macronutrients, mushrooms possess several B-complex vitamins, minerals, and polyphenols, which have anti-oxidative, anti-inflammatory and immune-modulating effects. Obesity is a growing epidemic in the United States which is associated with a number of health complications. Though various factors are associated with the development of obesity, increased intake and passive overconsumption of high calorie foods is one of the major contributors. One strategy to address this factor is to consume lower energy density foods, substituting them for high energy density foods. Increasing dietary fruits and vegetables has been shown to lower the energy density of meals and increase satiety. One limiting factor with increasing fruit and vegetable intake is lesser palatability. Mushrooms can be a good low calorie substitute, with high nutrient profile and good palatability for most consumers. One study found that substituting red meat with mushrooms significantly reduced total calorie and fat intake, suggesting that it has the potential to reduce overall energy consumption in those who are seeking to lose weight. Moreover, several studies have shown that mushrooms have the potential to reduce inflammation and oxidative stress. Obesity is associated with low-grade inflammation and oxidative stress.

Collectively, data on edible mushrooms suggest that it is a high quality food with a rich nutrient profile, including several bioactive compounds with various health benefits. It may be an important ingredient in the diet of those seeking to lose weight, as it may help to reduce overall energy intake. However, data on the beneficial effects of mushrooms in reducing inflammation and oxidative stress have been obtained from studies that have used extracts from different mushroom species in *in vitro* and *in vivo* models. It remains to be seen whether these results can be duplicated in obese individuals with whole mushroom intake. Further research is needed to examine the role of edible mushrooms for weight control and associated health benefits in humans.

REFERENCES

[1] Miles PG, Chang ST. Mushroom Biology: Concise basics and current developments. Singapore: *World Scientific*; 1997.

[2] Guillamón E, García-Lafuente A, Lozano M, D'Arrigo M, Rostagno MA, Villares A, Martínez JA. Edible mushrooms: role in the prevention of cardiovascular diseases. *Fitoterapia*. 2010;81:715-23.

[3] Firenzuoli F, Gori L, Lombardo G. The Medicinal Mushroom Agaricus blazei Murrill: Review of Literature and Pharmaco-Toxicological Problems. *Evid Based Complement Alternat Med*. 2008;5:3-15.

[4] Wasser SP. Current findings, future trends, and unsolved problems in studies of medicinal mushrooms. *Appl Microbiol Biotechnol*. 2011;89:1323-32.

[5] Wasser SP, Weis AL. Therapeutic effects of substances occurring in higher Basidiomycetes mushrooms: a modern perspective. *Crit Rev Immunol*. 1999;19:65-96.

[6] Barros L, Cruz T, Baptista P, Estevinho LM, Ferreira IC. Wild and commercial mushrooms as source of nutrients and nutraceuticals. *Food Chem Toxicol*. 2008;46:2742-7.

[7] Yu S, Weaver V, Martin K, Cantorna MT. The effects of whole mushrooms during inflammation. *BMC Immunol*. 2009:20;10:12.

[8] Borchers AT, Krishnamurthy A, Keen CL, Meyers FJ, Gershwin ME. The immunobiology of mushrooms. *Exp Biol Med*. 2008 ;233:259-76.

[9] Lull C, Wichers HJ, Savelkoul HF. Antiinflammatory and immunomodulating properties of fungal metabolites. *Mediators Inflamm*. ;2005:63-80.

[10] Chang R. Functional properties of edible mshrooms. *Nutr Rev*. 1996;54:S91-3.

[11] Wasser S, Nevo E, Sokolov D, Reshetnikov S, Timor-Tismenetsky M. Dietary supplements from medicinal mushrooms: diversity of types and variety of regulations. *Int J Med Mushr* 2000, 2:1-19.

[12] Bano Z, Rajarathnam S. Pleurotus mushrooms. Part II. Chemical composition, nutritional value, post-harvest physiology, preservation, and role as human food. *Crit Rev Food Sci Nutr*. 1988;27:87-158.

[13] Oyctayo FL, Akindahunsi AA, Oyetayo VO. Chemical profile and amino acids composition of edible mushrooms Pleurotus sajor-caju. *Nutr Health*. 2007;18:383-9.

[14] Mattila P, Salo-Väänänen P, Könkö K, Aro H, Jalava T. Basic composition and amino acid contents of mushrooms cultivated in Finland. *J Agric Food Chem*. 2002 23;50:6419-22.

[15] Kansci G, Mossebo DC, Selatsa AB, Fotso M. Nutrient content of some mushroom species of the genus Termitomyces consumed in Cameroon. Nahrung. 2003;47:213-6.

[16] Pi-Sunyer X. The medical risks of obesity. *Postgrad Med*. 2009;121:21-33.

[17] Field AE, Coakley EH, Must A, Spadano JL, Laird N, Dietz WH, Rimm E, Colditz GA. Impact of overweight on the risk of developing common chronic diseases during a 10-year period. *Arch Intern Med.* 2001, 9;161:1581-6.

[18] Wilson PW, D'Agostino RB, Sullivan L, Parise H, Kannel WB. Overweight and obesity as determinants of cardiovascular risk: the Framingham experience. *Arch Intern Med.* 2002,9;162:1867-72.

[19] Barros L, Baptista P, Correia DM, Morais JS, Ferreira IC. Effects of conservation treatment and cooking on the chemical composition and antioxidant activity of Portuguese wild edible mushrooms. *J Agric Food Chem.* 2007,13;55:4781-8.

[20] Barros L, Baptista P, Estevinho LM, Ferreira IC. Effect of fruiting body maturity stage on chemical composition and antimicrobial activity of Lactarius sp. mushrooms. *J Agric Food Chem.* 2007,17;55:8766-71.

[21] Barros L, Baptista P, Ferreira IC. Effect of Lactarius piperatus fruiting body maturity stage on antioxidant activity measured by several biochemical assays. *Food Chem Toxicol.* 2007;45:1731-7.

[22] Ferreira IC, Barros L, Abreu RM. Antioxidants in wild mushrooms. *Curr Med Chem.* 2009;16:1543-60.

[23] Balsano C, Alisi A. Antioxidant effects of natural bioactive compounds. Curr Pharm Des. 2009;15:3063-73.

[24] Koyyalamudi SR, Jeong SC, Cho KY, Pang G. Vitamin B12 is the active corrinoid produced in cultivated white button mushrooms (Agaricus bisporus). J Agric Food Chem. 2009;57:6327-33.

[25] Györfi J, Geösel A, Vetter J. Mineral composition of different strains of edible medicinal mushroom Agaricus subrufescens peck. *J Med Food.* 2010;13:1510-4.

[26] Mattila P, Könkö K, Eurola M, Pihlava JM, Astola J, Vahteristo L, Hietaniemi V, Kumpulainen J, Valtonen M, Piironen V. Contents of vitamins, mineral elements, and some phenolic compounds in cultivated mushrooms. *J Agric Food Chem.* 2001;49:2343-8.

[27] Flegal KM, Carroll MD, Ogden CL, Curtin LR. Prevalence and trends in obesity among US adults, 1999-2008. JAMA. 2010; 20:235-41

[28] Ogden CL, Carroll MD, Curtin LR, McDowell MA, Tabak CJ, Flegal KM. Prevalence of overweight and obesity in the United States, 1999-2004. *JAMA.* 2006;295:1549-1555.

[29] Wang Y, Beydoun MA, Liang L, Caballero B, Kumanyika SK. Will All Americans Become Overweight or Obese? Estimating the Progression and Cost of the US Obesity Epidemic. *Obesity.* 2008;16:2323-30.

[30] Gordon-Larsen P, Adair LS, Nelson MC, Popkin BM. Five-year obesity incidence in the transition period between adolescence and adulthood: the National Longitudinal Study of Adolescent Health. *Am J Clin Nutr.* 2004;80:569-575.

[31] Deshmukh-Taskar P, Nicklas TA, Morales M, Yang SJ, Zakeri I, Berenson GS. Tracking of overweight status from childhood to young adulthood: the Bogalusa Heart Study. *Eur J Clin Nutr.* 2006;60:48-57.

[32] Kvaavik E, Tell GS, Klepp KI. Predictors and tracking of body mass index from adolescence into adulthood: follow-up of 18 to 20 years in the Oslo Youth Study. *Arch Pediatr Adolesc Med.* 2003;157:1212-1218.

[33] Ogden CL, Carroll MD, Curtin LR, Lamb MM, Flegal KM. Prevalence of high body mass index in US children and adolescents, 2007-2008. *JAMA*. 20;303:242-9.
[34] Janssen I, Katzmarzyk PT, Srinivasan SR, Chen W, Malina RM, Bouchard C, Berenson GS. Utility of childhood BMI in the prediction of adulthood disease: comparison of national and international references. *Obes Res*. 2005;13:1106-15.
[35] Srinivasan SR, Myers L, Berenson GS. Predictability of childhood adiposity and insulin for developing insulin resistance syndrome (syndrome X) in young adulthood: the Bogalusa Heart Study. *Diabetes*. 2002;51:204-9.
[36] Nguyen QM, Srinivasan SR, Xu JH, Chen W, Berenson GS. Changes in risk variables of metabolic syndrome since childhood in pre-diabetic and type 2 diabetic subjects: the Bogalusa Heart Study. *Diabetes Care*. 2008;31:2044-9.
[37] Finkelstein EA, Brown DS, Wrage LA, Allaire BT, Hoerger TJ. Individual and aggregate years-of-life-lost associated with overweight and obesity. *Obesity*. 2010;18:333-9.
[38] Puhl RM, Andreyeva T, Brownell KD. Perceptions of weight discrimination: prevalence and comparison to race and gender discrimination in America. *Int J Obes*. 2008 ;32:992-1000.
[39] Adam TC, Epel ES. Stress, eating and the reward system. *Physiol Behav*. 2007, 24;91:449-58.
[40] Kottke TE, Wu LA, Hoffman RS. Economic and psychological implications of the obesity epidemic. Mayo Clin Proc. 2003;78:92-4.
[41] Haas JS, Lee LB, Kaplan CP, Sonneborn D, Phillips KA, Liang SY. The association of race, socioeconomic status, and health insurance status with the prevalence of overweight among children and adolescents. *Am J Public Health*. 2003;93:2105-10.
[42] Avenell A, Broom J, Brown TJ, Poobalan A, Aucott L, Stearns SC, Smith WC, Jung RT, Campbell MK, Grant AM. Systematic review of the long-term effects and economic consequences of treatments for obesity and implications for health improvement. *Health Technol Assess*. 2004;8:1-182.
[43] Runge CF. Economic consequences of the obese. *Diabetes*. 2007;56:2668-72.
[44] Finkelstein EA, Fiebelkorn IC, Wang G. National medical spending attributable to overweight and obesity: how much, and who's paying? *Health Aff (Millwood)*. 2003;W3-219-226.
[45] Grimm ER, Steinle NI. Genetics of eating behavior: established and emerging concepts. *Nutr Rev*. 2011;69:52-60.
[46] Lake A, Townshend T. Obesogenic environments: exploring the built and food environments. *J R Soc Promot Health*. 2006;126:262-7.
[47] Booth KM, Pinkston MM, Poston WS. Obesity and the built environment. *J Am Diet Assoc*. 2005;105:S110-7.
[48] Grotto D, Zied E. The Standard American Diet and its relationship to the health status of Americans. *Nutr Clin Pract*. 2010;25:603-12.
[49] Daniel CR, Cross AJ, Koebnick C, Sinha R. Trends in meat consumption in the USA. *Public Health Nutr*. 2010 Nov 12:1-9.
[50] Wang Y, Beydoun MA, Caballero B, Gary TL, Lawrence R. Trends and correlates in meat consumption patterns in the US adult population. *Public Health Nutr*. 2010 Sep;13(9):1333-45.

[51] Kuczmarski MF, Mason MA, Schwenk EA, Evans MK, Zonderman AB. Beverage Consumption Patterns of a Low-Income Population. *Top Clin Nutr.* 2010;25:191-201.
[52] Popkin BM. Patterns of beverage use across the lifecycle. *Physiol Behav.* 2010, 26;100:4-9.
[53] Bleich SN, Wang YC, Wang Y, Gortmaker SL. Increasing consumption of sugar-sweetened beverages among US adults: 1988-1994 to 1999-2004. *Am J Clin Nutr.* 2009;89:372-81.
[54] Thompson FE, McNeel TS, Dowling EC, Midthune D, Morrissette M, Zeruto CA. Interrelationships of added sugars intake, socioeconomic status, and race/ethnicity in adults in the United States: National Health Interview Survey, 2005. *J Am Diet Assoc.* 2009;109:1376-83.
[55] Dennis EA, Flack KD, Davy BM. Beverage consumption and adult weight management: A review. *Eat Behav.* 2009;10:237-46.
[56] Bachman JL, Reedy J, Subar AF, Krebs-Smith SM. Sources of food group intakes among the US population, 2001-2002. J Am Diet Assoc. 2008;108:804-14.
[57] Drewnowski A. Obesity and the food environment: dietary energy density and diet costs. *Am J Prev Med.* 2004;27:154-62.
[58] Prentice A, Jebb S. Energy intake/physical activity interactions in the homeostasis of body weight regulation. *Nutr Rev.* 2004;62:S98-104.
[59] Ershow AG. Environmental influences on development of type 2 diabetes and obesity: challenges in personalizing prevention and management. *J Diabetes Sci Technol.* 2009;3:727-34.
[60] Westerterp KR. Perception, passive overfeeding and energy metabolism. *Physiol Behav.* 2006 Aug 30;89(1):62-5.
[61] Blundell JE, Lawton CL, Cotton JR, Macdiarmid JI. Control of human appetite: implications for the intake of dietary fat. *Annu Rev Nutr.* 1996;16:285-319.
[62] Vergnaud AC, Estaquio C, Czernichow S, Péneau S, Hercberg S, Galan P, Bertrais S. Energy density and 6-year anthropometric changes in a middle-aged adult cohort. *Br J Nutr.* 2009;102:302-9.
[63] Savage JS, Marini M, Birch LL. Dietary energy density predicts women's weight change over 6 y. *Am J Clin Nutr.* 2008 Sep;88(3):677-84.
[64] Bes-Rastrollo M, van Dam RM, Martinez-Gonzalez MA, Li TY, Sampson LL, Hu FB. Prospective study of dietary energy density and weight gain in women. *Am J Clin Nutr.* 2008 Sep;88(3):769-77.
[65] Howarth NC, Murphy SP, Wilkens LR, Hankin JH, Kolonel LN. Dietary energy density is associated with overweight status among 5 ethnic groups in the multiethnic cohort study. *J Nutr.* 2006 ;136:2243-8.
[66] Ello-Martin JA, Roe LS, Ledikwe JH, Beach AM, Rolls BJ. Dietary energy density in the treatment of obesity: a year-long trial comparing 2 weight-loss diets. *Am J Clin Nutr.* 2007;85:1465-77.
[67] Ledikwe JH, Blanck HM, Kettel Khan L, Serdula MK, Seymour JD, Tohill BC, Rolls BJ. Dietary energy density is associated with energy intake and weight status in US adults. *Am J Clin Nutr.* 2006 ;83:1362-8.
[68] Kant AK, Graubard BI. Energy density of diets reported by American adults: association with food group intake, nutrient intake, and body weight. *Int J Obes (Lond).* 2005 Aug;29(8):950-6.

[69] Ledikwe JH, Blanck HM, Khan LK, Serdula MK, Seymour JD, Tohill BC, Rolls BJ. Low-energy-density diets are associated with high diet quality in adults in the United States. *J Am Diet Assoc.* 2006;106:1172-80.
[70] Rolls BJ, Roe LS, Meengs JS. Portion size can be used strategically to increase vegetable consumption in adults. *Am J Clin Nutr.* 2010;91:913-22.
[71] Rolls BJ, Roe LS, Meengs JS. Salad and satiety: energy density and portion size of a first-course salad affect energy intake at lunch. *J Am Diet Assoc.* 2004 Oct;104(10):1570-6.
[72] Cheskin LJ, Davis LM, Lipsky LM, Mitola AH, Lycan T, Mitchell V, Mickle B, Adkins E. Lack of energy compensation over 4 days when white button mushrooms are substituted for beef. *Appetite.* 2008;51:50-7.
[73] Rolls BJ, Roe LS, Meengs JS. Reductions in portion size and energy density of foods are additive and lead to sustained decreases in energy intake. *Am J Clin Nutr.* 2006;83:11-7.
[74] Donahoo W, Wyatt HR, Kriehn J, Stuht J, Dong F, Hosokawa P, Grunwald GK, Johnson SL, Peters JC, Hill JO. Dietary fat increases energy intake across the range of typical consumption in the United States. *Obesity*, 2008;16:64-9.
[75] Drewnowski A. The real contribution of added sugars and fats to obesity. Epidemiol Rev. 2007;29:160-71.
[76] Zevenbergen H, de Bree A, Zeelenberg M, Laitinen K, van Duijn G, Flöter E. Foods with a high fat quality are essential for healthy diets. *Ann Nutr Metab.* 2009;54:15-24.
[77] Nicklas TA, Farris RP, Myers L, Berenson GS. Impact of meat consumption on nutritional quality and cardiovascular risk factors in young adults: the Bogalusa Heart Study. *J Am Diet Assoc.* 1995;95:887-92.
[78] Zanovec M, O'Neil CE, Keast DR, Fulgoni VL 3rd, Nicklas TA. Lean beef contributes significant amounts of key nutrients to the diets of US adults: National Health and Nutrition Examination Survey 1999-2004. *Nutr Res.* 2010;30:375-81.
[79] Perrigue MM, Monsivais P, Drewnowski A. Added soluble fiber enhances the satiating power of low-energy-density liquid yogurts. *J Am Diet Assoc.* 2009;109:1862-8.
[80] Flood-Obbagy JE, Rolls BJ. The effect of fruit in different forms on energy intake and satiety at a meal. *Appetite.* 2009;52:416-22.
[81] López EP, Rice C, Weddle DO, Rahill GJ. The relationship among cardiovascular risk factors, diet patterns, alcohol consumption, and ethnicity among women aged 50 years and older. *J Am Diet Assoc.* 2008 ;108:248-56.
[82] Deshmukh-Taskar PR, O'Neil CE, Nicklas TA, Yang SJ, Liu Y, Gustat J, Berenson GS. Dietary patterns associated with metabolic syndrome, sociodemographic and lifestyle factors in young adults: the Bogalusa Heart Study. *Public Health Nutr.* 2009;12:2493-503
[83] McAfee AJ, McSorley EM, Cuskelly GJ, Moss BW, Wallace JM, Bonham MP, Fearon AM. Red meat consumption: an overview of the risks and benefits. *Meat Sci.* 2010 ;84:1-13.
[84] Steffen LM, Kroenke CH, Yu X, Pereira MA, Slattery ML, Van Horn L, Gross MD, Jacobs DR Jr. Associations of plant food, dairy product, and meat intakes with 15-y incidence of elevated blood pressure in young black and white adults: the Coronary Artery Risk Development in Young Adults (CARDIA) Study. *Am J Clin Nutr.* 2005;82:1169-77.

[85] Männistö S, Kontto J, Kataja-Tuomola M, Albanes D, Virtamo J. High processed meat consumption is a risk factor of type 2 diabetes in the Alpha-Tocopherol, Beta-Carotene Cancer Prevention study. *Br J Nutr.* 2010;103:1817-22.

[86] Vang A, Singh PN, Lee JW, Haddad EH, Brinegar CH. Meats, processed meats, obesity, weight gain and occurrence of diabetes among adults: findings from Adventist Health Studies. *Ann Nutr Metab.* 2008;52:96-104.

[87] Cross AJ, Leitzmann MF, Gail MH, Hollenbeck AR, Schatzkin A, Sinha R. A prospective study of red and processed meat intake in relation to cancer risk. *PLoS Med.* 2007;4(12):e325.

[88] Rodriguez C, Sinha R, Calle EE. Meat consumption and risk of colorectal cancer. *JAMA.* 2005 12;293:172-82.

[89] Steinbrecher A, Erber E, Grandinetti A, Kolonel L, Maskarinec G. Meat consumption and risk of type 2 diabetes: the Multiethnic Cohort. *Public Health Nutr.* 2010:1-7.

[90] McAfee AJ, McSorley EM, Cuskelly GJ, Moss BW, Wallace JM, Bonham MP, Fearon AM. Red meat consumption: an overview of the risks and benefits. *Meat Sci.* 2010;84:1-13.

[91] Li D, Siriamornpun S, Wahlqvist ML, Mann NJ, Sinclair AJ. Lean meat and heart health. *Asia Pac J Clin Nutr.* 2005;14:113-9.

[92] Manore MM. Exercise and the Institute of Medicine recommendations for nutrition. *Curr Sports Med Rep.* 2005;4(4):193-8.

[93] Westerterp-Plantenga MS, Nieuwenhuizen A, Tomé D, Soenen S, Westerterp KR. Dietary protein, weight loss, and weight maintenance. *Annu Rev Nutr.* 2009;29:21-41.

[94] Berti C, Riso P, Porrini M. Satiating properties of meat-preparations: role of protein content and energy density. *J Am Coll Nutr.* 2008;27:244-52.

[95] Cecil JE, Palmer CN, Wrieden W, Murrie I, Bolton-Smith C, Watt P, Wallis DJ, Hetherington MM. Energy intakes of children after preloads: adjustment, not compensation. *Am J Clin Nutr.* 2005;82:302-8.

[96] Iyer A, Fairlie DP, Prins JB, Hammock BD, Brown L. Inflammatory lipid mediators in adipocyte function and obesity. *Nat Rev Endocrinol. 2010;6:71-82.*

[97] de Ferranti S, Mozaffarian D. The perfect storm: obesity, adipocyte dysfunction, and metabolic consequences. *Clin Chem.* 2008;54:945-55.

[98] Lionetti L, Mollica MP, Lombardi A, Cavaliere G, Gifuni G, Barletta A. From chronic overnutrition to insulin resistance: the role of fat-storing capacity and inflammation. *Nutr Metab Cardiovasc Dis.* 2009;19:146-52.

[99] Holvoet P. Relations between metabolic syndrome, oxidative stress and inflammation and cardiovascular disease. *Verh K Acad Geneeskd Belg.* 2008;70(3):193-219.

[100] Grattagliano I, Palmieri VO, Portincasa P, Moschetta A, Palasciano G. Oxidative stress-induced risk factors associated with the metabolic syndrome: a unifying hypothesis. *J Nutr Biochem.* 2008;19(8):491-504.

[101] Mohanty P, Hamouda W, Garg R, Aljada A, Ghanim H, Dandona P. Glucose challenge stimulates reactive oxygen species (ROS) generation by leucocytes. *J Clin Endocrinol Metab.* 2000;85(8):2970-3.

[102] Mohanty P, Ghanim H, Hamouda W, Aljada A, Garg R, Dandona P. Both lipid and protein intakes stimulate increased generation of reactive oxygen species by polymorphonuclear leukocytes and mononuclear cells. *Am J Clin Nutr.* 2002;75(4):767-72.

[103] Dhindsa S, Tripathy D, Mohanty P, Ghanim H, Syed T, Aljada A, Dandona P. Differential effects of glucose and alcohol on reactive oxygen species generation and intranuclear nuclear factor-kappaB in mononuclear cells. *Metabolism*. 2004;53(3):330-4.

[104] Tripathy D, Mohanty P, Dhindsa S, Syed T, Ghanim H, Aljada A, Dandona P. Elevation of free fatty acids induces inflammation and impairs vascular reactivity in healthy subjects. *Diabetes*. 2003;52(12):2882-7.

[105] Brinkworth GD, Noakes M, Buckley JD, Keogh JB, Clifton PM. Long-term effects of a very-low-carbohydrate weight loss diet compared with an isocaloric low-fat diet after 12 mo. *Am J Clin Nutr*. 2009;90(1):23-32.

[106] Koren MS, Purnell JQ, Breen PA, Matthys CC, Callahan HS, Weigle DS. Plasma C-reactive protein concentration is not affected by isocaloric dietary fat reduction. *Nutrition*. 2006;22(4):444-8.

[107] Hernandez TL, Sutherland JP, Wolfe P, Allian-Sauer M, Capell WH, Talley ND, Wyatt HR, Foster GD, Hill JO, Eckel RH. Lack of suppression of circulating free fatty acids and hypercholesterolemia during weight loss on a high-fat, low-carbohydrate diet. *Am J Clin Nutr*. 2010;91(3):578-85.

[108] Wood RJ, Volek JS, Davis SR, Dell'Ova C, Fernandez ML. Effects of a carbohydrate-restricted diet on emerging plasma markers for cardiovascular disease. *Nutr Metab (Lond)*. 2006;3:19.

[109] Rankin JW, Turpyn AD. Low carbohydrate, high fat diet increases C-reactive protein during weight loss. *J Am Coll Nutr*. 2007;26(2):163-9.

[110] Peairs AT, Rankin JW. Inflammatory response to a high-fat, low-carbohydrate weight loss diet: effect of antioxidants. *Obesity*. 2008;16(7):1573-8.

[111] Liu RH. Potential synergy of phytochemicals in cancer prevention: mechanism of action. *J Nutr*. 2004 ;134:3479S-3485S.

[112] Aviram M, Kaplan M, Rosenblat M, Fuhrman B. Dietary antioxidants and paraoxonases against LDL oxidation and atherosclerosis development. *Handb Exp Pharmacol*. 2005;(170):263-300.

[113] González-Gallego J, García-Mediavilla MV, Sánchez-Campos S, Tuñón MJ. Fruit polyphenols, immunity and inflammation. *Br J Nutr*. 2010;104:S15-27.

[114] Jose N, Ajith TA, Janardhanan KK. Methanol extract of the oyster mushroom, Pleurotus florida, inhibits inflammation and platelet aggregation. *Phytother Res*. 2004;18(1):43-6.

[115] Kim SH, Song YS, Kim SK, Kim BC, Lim CJ, Park EH. Anti-inflammatory and related pharmacological activities of the n-BuOH subfraction of mushroom Phellinus linteus. *J Ethnopharmacol*. 2004;93(1):141-6.

[116] Martin KR. Both common and specialty mushrooms inhibit adhesion molecule expression and in vitro binding of monocytes to human aortic endothelial cells in a pro-inflammatory environment. *Nutr J*. 2010;9:29.

[117] Wu D, Pae M, Ren Z, Guo Z, Smith D, Meydani SN. Dietary supplementation with white button mushroom enhances natural killer cell activity in C57BL/6 mice. *J Nutr*. 2007;137(6):1472-7.

[118] Kim GY, Roh SI, Park SK, Ahn SC, Oh YH, Lee JD, Park YM. Alleviation of experimental septic shock in mice by acidic polysaccharide isolated from the medicinal mushroom Phellinus linteus. *Biol Pharm Bull*. 2003;26(10):1418-23.

[119] Kim KM, Kwon YG, Chung HT, Yun YG, Pae HO, Han JA, Ha KS, Kim TW, Kim YM. Methanol extract of Cordyceps pruinosa inhibits in vitro and in vivo inflammatory mediators by suppressing NF-kappaB activation. *Toxicol Appl Pharmacol.* 2003;190(1):1-8.

[120] Kim HG, Yoon DH, Kim CH, Shrestha B, Chang WC, Lim SY, Lee WH, Han SG, Lee JO, Lim MH, Kim GY, Choi S, Song WO, Sung JM, Hwang KC, Kim TW. Ethanol extract of Inonotus obliquus inhibits lipopolysaccharide-induced inflammation in RAW 264.7 macrophage cells. *J Med Food.* 2007;10(1):80-9.

[121] Park YM, Won JH, Kim YH, Choi JW, Park HJ, Lee KT. In vivo and in vitro anti-inflammatory and anti-nociceptive effects of the methanol extract of Inonotus obliquus. *J Ethnopharmacol.* 2005;101:120-8.

[122] Zhang Y, Mills GL, Nair MG. Cyclooxygenase inhibitory and antioxidant compounds from the mycelia of the edible mushroom Grifola frondosa. *J Agric Food Chem.* 2002;50:7581-5.

[123] Zhang Y, Mills GL, Nair MG. Cyclooxygenase inhibitory and antioxidant compounds from the fruiting body of an edible mushroom, Agrocybe aegerita. *Phytomedicine.* 2003;10:386-90.

[124] Vaz JA, Heleno SA, Martins A, Almeida GM, Vasconcelos MH, Ferreira IC. Wild mushrooms Clitocybe alexandri and Lepista inversa: in vitro antioxidant activity and growth inhibition of human tumour cell lines. *Food Chem Toxicol.* 2010;48:2881-4.

[125] Xu C, HaiYan Z, JianHong Z, Jing G. The pharmacological effect of polysaccharides from Lentinus edodes on the oxidative status and expression of VCAM-1 mRNA of thoracic aorta endothelial cell in high-fat-diet rats. *Carbohydr Polym.* 2008;74:445-50.

[126] Percario S, Odorizzi VF, Souza DR, Pinhel MA, Gennari JL, Gennari MS, Godoy MF. Edible mushroom Agaricus sylvaticus can prevent the onset of atheroma plaques in hipercholesterolemic rabbits. *Cell Mol Biol (Noisy-le-grand).* 2008;54:1055-61.

[127] Jeong SC, Jeong YT, Yang BK, Islam R, Koyyalamudi SR, Pang G, Cho KY, Song CH. White button mushroom (Agaricus bisporus) lowers blood glucose and cholesterol levels in diabetic and hypercholesterolemic rats. *Nutr Res.* 2010;30(1):49-56.

[128] Kabir Y, Yamaguchi M, Kimura S. Effect of shiitake (Lentinus edodes) and maitake (Grifola frondosa) mushrooms on blood pressure and plasma lipids of spontaneously hypertensive rats. *J Nutr Sci Vitaminol (Tokyo).* 1987;33:341-6.

[129] Bobek P, Galbavý S. Hypocholesterolemic and antiatherogenic effect of oyster mushroom (Pleurotus ostreatus) in rabbits. *Nahrung.* 1999;43(5):339-42.

[130] Marui N, Offermann MK, Swerlick R, Kunsch C, Rosen CA, Ahmad M, Alexander RW, Medford RM. Vascular cell adhesion molecule-1 (VCAM-1) gene transcription and expression are regulated through an antioxidant-sensitive mechanism in human vascular endothelial cells. *J Clin Invest.* 1993;92(4):1866-74.

[131] Chen JW, Chen YH, Lin FY, Chen YL, Lin SJ. Ginkgo biloba extract inhibits tumor necrosis factor-alpha-induced reactive oxygen species generation, transcription factor activation, and cell adhesion molecule expression in human aortic endothelial cells. *Arterioscler Thromb Vasc Biol.* 2003;23(9):1559-66.

[132] Cominacini L, Garbin U, Pasini AF, Davoli A, Campagnola M, Contessi GB, Pastorino AM, Lo Cascio V. Antioxidants inhibit the expression of intercellular cell adhesion molecule-1 and vascular cell adhesion molecule-1 induced by oxidized LDL on human umbilical vein endothelial cells. *Free Radic Biol Med.* 1997;22(1-2):117-27.

[133] Cominacini L, Garbin U, Pasini AF, Davoli A, Campagnola M, Rigoni A, Tosetti L, Lo Cascio V. The expression of adhesion molecules on endothelial cells is inhibited by troglitazone through its antioxidant activity. *Cell Adhes Commun.* 1999;7(3):223-31.

[134] Lo, K.M. and Cheung, P.C.K. (2005). Antioxidant activities of extracts from the fruiting bodies of Agrocybe aegerita var. alba. *Food Chem* 89, 533-539.

[135] Cheung, L.M. and Cheung, P.C.K. (2005). Mushroom extracts with antioxidant activity against lipid oxidation. *Food Chem* 89, 403-409.

[136] Jayakumar T, Thomas PA, Ramesh E, Geraldine P. An extract of the pleurotus ostreatus mushroom bolsters the glutathione redox system in various organs of aged rats. *J Med Food.* 2010;13(4):771-8.

[137] Jayakumar T, Sakthivel M, Thomas PA, Geraldine P. Pleurotus ostreatus, an oyster mushroom, decreases the oxidative stress induced by carbon tetrachloride in rat kidneys, heart and brain. *Chem Biol Interact.* 2008;176(2-3):108-20.

In: Mushrooms: Types, Properties and Nutrition
Editors: S. Andres and N. Baumann
ISBN 978-1-61470-110-1

Chapter 3

CULTIVATED MUSHROOMS: DISEASE CONTROL IN MUSHROOM INDUSTRY

Ivana Potočnik *
Institute of Pesticides and Environmental Protection,
Banatska 31B, P.O. Box 163, 11080, Belgrade-Zemun, Serbia

ABSTRACT

Button mushroom (Agaricus bisporus), oyster mushroom (*Pleurotus* sp.) and shiitake (*Lentinus edodes*) are the most commonly cultivated basidiomycetes worldwide. The production of fruiting bodies is severely afflicted by fungal, bacterial, and viral pathogens that can cause diseases which have an effect on yield and quality. Major *A. bisporus* fungal pathogens are *Mycogone perniciosa*, *Lecanicillium fungicola*, and *Cladobotryum* with two species mycophillum and dendroides, causal agents of dry bubble, wet bubble, and cobweb disease, respectively. Various *Trichoderma* species are the causal organisms of green mould, affecting all three edible mushrooms. The usual method of controlling of diseases on farms worldwide is based on the use of fungicides. However, development of pathogen resistance to fungicides after frequent application and host sensitivity to fungicides are serious problems. For improvement of crop protection and reduction of production costs, the effects of some new fungicides are being tested. The strains of edible mushrooms that were recently commonly cultivated seem to be more tolerant to fungicides in vitro than the earlier commercial strains. Resistance to benzimidazole fungicides has developed and the number of available fungicides is decreasing. Since studies of fungicides efficacy on cultivated mushrooms by agrochemical companies are very rare, only few fungicides are officially recommended in mushroom industry: prochloraz in EU countries, and chlorothalonil, thiabendazol and tiophanate-methyl in US and Canada. Decreased sensitivity of *L. fungicola* to prochloraz was noted. Also, inefficiency of this fungicide was recorded in experimental growing room, at a level of spotting symptoms of cobweb disease. However, with regard to resistance development, harm to the environment and human health, special attention should be focused on good programme of hygiene. The introduction of new fungicides of biological origin creates new possibilities for crop protection from fungal pathogens.

* Tel./fax: ++381(11)3076-133 E-mail: ivanapotocnik@yahoo.com

Introduction

Cultivated Mushrooms

World production of mushrooms (*Agaricus bisporus* (Lange) Imbach) in the begining of the 21st century was over 2 million tonnes, earning over 15 billion dollars. It indicated an increase rate of world annual production growth of 5-6% (Rai and Ahlawat, 2005). China dominates in edible mushroom production with a share of 64%. However, European countries have 50% or 1 million tonnes of the total world production of *A. bisporus*, of which 875,000 tonnes belong to EU countries and 175 000 tonnes to countries outside the European Union (Chang, 1999a, 1999b). Button mushroom (*A. bisporus*), oyster mushroom (*Pleurotus* sp.) and shii-take (*Lentinula edodes*) are the most commonly cultivated basidiomycetes worldwide (Hatvani *et al.*, 2008). With the popularization of mushroom farming, mushroom production continues to increase. The button mushroom *A. bisporus* is one of the most extensive cultivated mushrooms in the world. Its production exeeds $1.8x10^6$ tonnes per year, $8\text{-}9x10^5$ of these are cultivated in Europe. The Netherlands and France are countries with highest production of mushrooms (2.4 and $1.6x10^5$ tonnes, respectively), followed by the UK with an annual production of 10^5 tonnes (Fan *et al.*, 2006).

Commercial strains of cultivated mushrooms produce substantial yield and exhibit attractive morphology and texture, but they are susceptible to a variety of viral, bacterial and fungal diseases. Mushroom pathogens have a significant effect on yield and quality (Fletcher *et al.*, 1989). The mushroom industry in Europe has undergone much change in the past 10 to 15 years. In Britain and France production has decreased by approximately 50%, in the Netherlands and Ireland it has decrased by about 20% while in Poland it has incrased by 100%. In 2006, Poland become the second largest producer of mushrooms in Europe (216,000 tonnes) after The Netherlands (225,000 tonnes). This changing pattern of production across Europe has resulted in large reductions in growers number in many countries (Grogan, 2008).

Diseases of Cultivated Mushrooms

Fungal pathogens such as *Lecanicillium (Verticillium) fungicola*, *Mycogone perniciosa, Cladobotryum* (*Dactylium*) spp. and *Trichoderma* spp. are known since the early days of commercial production (Kligman, 1950; Atkins & Atkins, 1971). They couse the most serious fungal diseases, respectively, dry and wet bubble, cobweb disease and green mould (Harvey *et al.*, 1982). Several diseases of cultivated mushrooms are coused by pseudomonads bacteria and viruses (Grogan *et al.*, 2003; Geels *et al.*, 2008).

Fourty years ago, pathogenic fungus *L. fungicola*, *M. perniciosa* and *Cladobotryum* spp., were damaged production of *A. bisporus* by 9.1 million dollars a year in Pennsylvania, US (Forero *et al.*, 1974), while losses from the three pathogenic fungi in the UK amounted to 5% of the total yield (Gaze & Fletcher, 1975). With the appearance of cobweb epidemic proportions in the mid-1990's, the costs are significantly increased in yield losses of up to 40%. *Cladobotryum* spp. becomes a pathogen that causes the greatest damage in the Republic of Ireland, the UK and the US (Gaze 1995a, 1995b, 1995c, 1996, McKay *et al.*, 1998). Brown blotch disease was responsible for losses of 8-10%, worldwide (Olivier *et al.*, 1997). Recently, major *A. bisporus* diseases are caused by mushroom virus X complex (Grogan *et al.*, 2003) and vairous *Trichoderma* species, fungi that are causal organisms of green mould. They both count for approximately 25% of the total production value, being the major pathogens of all three cultivated mushrooms (Soković & Van Griensven, 2006).

The spread of pathogens from one country to another through imported goods and machinery increases the risk of unwanted introduction of pathogens. In the mid-1990's, dry and wet bubbles were largely reduced quality and yield of *A. bisporus* in Serbia (Potočnik, 2006; 2008b; 2010b). The last ten years, cobweb disease causes significant damage in mushroom cultivation in this region. A few years ago, disease with simptoms resembling green mould becomes serious problem in Serbian mushroom industry (Potočnik, 2006, 2009).

Dry Bubble Disease

Causal agent of dry bubble disease of button and oyster mushrooms, *Verticillium fungicola*, was recently renamed as *Lecanicillium fungicola* (Zare & Gams, 2008). There are two varieties, *L. fungicola var. fungicola* in Europe and *L. fungicola* var. *aleophilum* in North America. Reports on the disease originate mainly from these both localities (Berendsen *et al.*, 2010). A *L. fungicola* isoalate identified in Mexico in 2002 appears to be var. *fungicola*, like Eruropean isolates, and the question was raisen about the posissibility of it having been introduced into Mexico on machinery originating in Europe (Largeteau *et al.*, 2004; Largeteau & Savoie, 2008). Recently, a highly aggressive *L. islae* has been isolated in Australia that does not appear to be controlled by fungicide prochloraz (Grogan, 2008).

The disease presents three types of symptoms designed spotty cap, stipe blowout and fruiting body malformation, called dry bubble (Largeteau & Savoie, 2010). The disease severity and symptoms depend on the stage of mushroom development at the time of infection (North & Wuest, 1993, Grogan *et al.*, 2000).

L. fungicola is devastating pathogen in the mushroom industry and causes significant losses in commercial production of *A. bisporus*. Annual costs for mushroom growers are estimated at 2-4% of total revenue. The primary source of dry bubble infection is casing (Ware, 1933, Wong & Preece, 1987). Ware (1933) had observed that dry bubble disease was associated with the presence of insects. Dispersal by splashing water was found to be very effective (Cross & Jacobs, 1968), and spreading by employees and equipment has also been reported to be important (Fekete, 1967; Wong & Preece, 1987). Infected mushrooms, either in the wild or growing on mushroom farms, could be an important source of inoculation (Zare & Gams, 2008). The anaerobic conditions and low pH are unfavourable for *L. fungicola*. In the UK in 1988, surface peat was used mainly (Visscher, 1988) and *L. fungicola* may survive on basidiomycete species that grow in. Peat, which is taken from lower peat layers seem to be a less likely habitat for *L. fungicola*. Replacing casing mixtures of clay, loam and humus by mixtures of sphagnum peat, sand and carbonate, resulted in a considerable reduction in dry bubble incidence in the mushroom industry in Denmark (Bech & Riber-Rasmussen, 1967).

Wet Bubble Disease

Mycogone perniciosa causes wet bubble, a disease of *A. bisporus,* which is very contagious and results in severe crop losses. The mycelium, phialoconidia and aleuriospores of *M. perniciosa*, all proven to be infectious, are easily spread by vectors. The mycopathogen grows superficially but causes dramatic cytological changes inside the host. Exudation of accumulated extracellular fluid results in a tear-drop phenomenon, hence wet bubble. When young primordia, up to 6 mm are infected, large, very irregular, nodular and tumorous fungal masses are formed. Despite the large size of these fungal masses, no signs of cell differentiation and organogenesis can be found (Umar *et al.*, 1999, 2000). Beside *A. bisporus*, only *Pleurotus eryngii* and *P. nebrodensis* were susceptible to *M. perniciosa*. They showed carpophore malformations resembling some morphological alterations observed in the past and thought to be of viral origin (Sisto *et al.*, 1997).

Cobweb Disease

Cladobotryum spp., soil-inhabiting cosmopolitan fungi, is causal agents of cobweb disease of button mushroom. The disease symptoms are: cottony fluffy white or yellowish to pink colonies on mushroom casing, rapid colonization of casing surface, covering of host basidiomata by mycelia, and their decay. Infection intensity depended on the development stage of fruiting bodies. Oposite from the spores of *L. fungicola*, *M. perniciosa* and *Trichoderma* sp. that are sticky and spread by insects, *Cladobotryum* conidia are dry, large and aerodynamic and spread by ventilation system to all parts of the farms (Bourgin, 1946; Adie & Grogan, 2000; Adie, 2006; Adie *et al.*, 2006).

Grogan and Gaze (2000) have classified all *Cladobotryum* spp. isolates from the UK into four groups: *C. mycophilum* Group 1 (Type A), slow-growing, sensitive to fungicide thiabendazole, smell of camphor; *C. mycophilum* Group 2 (Type B1), a fast-growing, resistant to thiabendazole, does not smell like camphor, (Type B2), fast growing; sensitive to tiabiendazol, smell of camphor; *C. mycophilum* Group 3 (Type C), a fast-growing; sensitive to tiabiendazol, smell of camphor; and C. dendroides, slow-growing, sensitive to thiabendazole, does not smell like camphor. Until the 1990's C. dendroides was ocasional pathogen of crops, usualy late in the cycle, but after that, serous cobweb epidemics started to occur, first in Europe then later, further afield in Australia and U.S.A. Major cause was fungicide-resistant strain of *C. mycophilum* (McKay et al., 1999). It has a faster growth rate, more profuse and earlier sprorulation, no distinct odour and spores which were regularly two, three and four-celled rather than the two celled spores characteritic of *C. mycophilum* (Adie, 2000; Gorgan & Gaze, 2000). Cobweb disease caused by *C. dendroides* is currently one of the most serious diseases in Serbian mushroom farms (Potočnik, 2009; Potočnik *et al.*, 2009a).

Green Mould Disease

Trichoderma species such as *T. koningii* and *T. viride* were long associated with musrhoom cultivation, their impact on productin was relatively minor. In the mid-1980's and

1990's, *T. aggressivum* emerged as an aggressive compost mould that decimated mushroom production if it got into freshly spawned Phase II compost. It had a higher temperature optimum than other *Trichoderma* spp. and grew rapidly on the carbohydrates in the grain supporting the mushroom spawn. Once *T. aggressivum* encountered mushroom cultivation facilities, it found a niche ideally suited to its characteristics: it is a successful fungus producing enzymes capable of growing on carbohydrat-rich substrates with a fast growth rate and temperature optimum close to that of *Agaricus*. Poor hygiene at Phase 2 emptying and spawning either through ineffective disinfection of equipment or through the ingress of contaminated air into spawning halls is the likely rout of entry into a crop or Phase 3 tunnel. Infected crops show a dark green colour indicative of the *T. aggressivum* sporulation on the compost surface. Infected areas produce no mushrooms and are surrounded by mushrooms infested by red-pepper mites (*Pygmephorus* spp.) (Krupke *et al.*, 2003).

In the 1980's, a mixture of strains at first identified as *T. harzianum* was found to cause deleterious infections in mushrooms farms of *A. bisporus*, *P. ostreatus* and *L. edodes* (Seaby, 1987) with losses between 30% and 100% (Seaby 1998). Since then, green mold disease has spread all over the world in *A. bisporus* (Komon-Zelazowska *et al.*, 2007b) and was shown to be mainly caused by *Trichoderma aggressivum* fsp. *europaeum* (Ta2) in Europe, and *T. aggressivum* fsp. *aggressivum* (Ta4) in US and Canada (Ospina-Giraldo *et al.*, 1998; Castle *et al.*, 1998; Savoie *et al.*, 2001; Samuels *et al.* 2002b). In tha past few years severe green mould infections coused by *T. pleurotum*, *T. pleuroticola* of cultivated *P. ostreatus* have been reported in South Korea (Park *et al.*, 2004; 2006), Italy (Woo *et al.*, 2004), Hungary (Hatvani *et al.*, 2006) and Romania (Kredics *et al.*, 2006), which indicate a worlwide threat. Different species of *Trichoderma* inculding *T. viride*, *T. harzianum*, *T. polysporum*, commonly cause injury to mycelial growth and fruit body formation of *L. edodes* (Tokimoto & Komatsu, 1979).

Also, *T. harzianum*, *T. longibrachiatum*, *T. ghanense*, *T. asperellum*, and *T. atroviride* have been detected in *Agaricus* compost and *Pleurotus* substrates (Hatvani *et al.*, 2007), but aggressive colonization of the substrate has not been proven for these species. This treat to commercial mushroom production has recently also led to the development of methods for rapid and specific detection and identification of these fungi in cultivation substrates (Muthumeenakshi *et al.*, 1998; Samuels *et al.*, 2002b; Kredics *et al.*, 2009).

Bacterial Diseases

Brown blotch is a bacterial disease caused by *Pseudomonas tolaasii* and *P. reactans* and is responsible for severe outbreaks throughout the world (Paine, 1919). When bacterial brown blotch is detected in a farm, it is very difficult to erradicate. Although bacterial spoilage does not cause any health problems, it can result in a loss in consumer appeal in the marcet place. The disease has been detected and described all over the world since many years and it affects not only the button mushroom market in the USA, Australia, Israel and Turkey, but also *Pleurotus, Lentinula* etc. in Japan, Korea, China and India (Tolaas, 1915; Nair, 1975; Guleria, 1976; Bashan & Okon, 1981; Cutri *et al.*, 1984; Suyama & Fujii, 1993; Ozaktan & Bora, 1994; Kim *et al.*, 1995). Colonisation of mushroom caps by the bacterium results in development of unappealing brown or cream lesions on pileus and stipe. These lesions are slightly concave spots, round, or spreading in many directions (Olivier *et al.*, 1978). The spots may be small, 1-4 mm diameter, and pale brown. When the damage is more intense, the spots are darker and sunken. Lesions may coaleasce to cover the entire mushroom surface (Rainey *et al.*, 1992).

P. agarici is responsible for drippy gills symptoms but also for brown discouloration. *P. gingeri* is the causal agent of ginger blotch, *P. aeruginosa* and *P. fluorescens* of mummy disease (Geels *et al.*, 2008).

Mushroom Viruses

La France virus disease was probably the first major pathogen that affected mushroom cultivation, impacting severely on the mushroom industry in the USA and UK in the 1960's. La France virus charactrized by slow-growing, appressed vegetative mycelia and abnormal fruit bodies. It was also detected in mycelial cultures of *L. edodes* (Rytter *et al.*, 1991). Epidemiological research indicated that it was predominantely spread by spores from infected mushrooms and also by infected muycelium carried over from one crop to the next (Schisler *et al.*, 1967). Control was achived by preventing the spread of spores by filtration and not growing flat mushrooms and by cooking out crops at the end cycle. The disease is now rare in the UK (Grogan *et al.*, 2003; Grogan, 2008).

Mushroom Virus X (MVX) is a disease which emerged in the late 1990's and the symptoms were consistently associated with a variable number of viral dsRNAs (Gaze *et al.*, 2000). MVX of *A. bisporus* encompasses a range of disease symptoms. These include bare cropping areas (primordial disruption), crop delay, premature veil opening, off- or brown-colored mushrooms and sporophore malformations. The various symptoms can occur either singularly or in combination but are all associated with loss of crop yield and product quality. MVX first appeared on UK farms but rapidly become prevalent (Gaze *et al.*, 2000). Estimates of direct and indirect losses to the UK industry in 2000 were 50 million punds and several British farms were alleged to have closed as a result of MVX. La France virus is most problematical when virus infected spores or mycelium infect the crop at spawning and is less problematic if infection takes place later. In contrast, epidemiological research indicates, that MVX infected mycelium can cause severe effects when minute quatitites of infected mycelium are introduced at either spawning or casing. Early infection of a crop often produces no brown mushroom symptoms wheares infection of a crop at casing time produces the brown symptomatic mushroom more consistently. MVX was brought under control in Britain through improvements in farm hygiene aimed at preventing the contamination of crops with potentially infected spores and mycelial debris at spawning or casing (Grogan, 2008).

Disease Control of Cultivated Mushrooms by Fungicides

In recent decades, a common method of pathogen control on farms worldwide is application of various fungicides. Ongoing pesticide reviews in Europe have resulted in many chemicals being no longer approved for use. Thus, a major challenge for mushroom growers in the 21st Century is disease control with few or no chemicals. Other chellenges include fungicide resistance among pathogen populations, managing existing available pesticides and the emergence of new pathogens (Garod, 2004; Anonymous, 2005, 2006; Grogan, 2008).

Calcium chloride and chlorinated compounds are the most commonly utilised chemicals for brown blotch disease control (Solomon *et al.*, 1991). Several chemicals, disinfectants and antibiotics have been studied for their ability to control bacterial blotch, but none has been found fully effective and non-toxic to humans (Soler-Rivas *et al.*, 1999; Abosriwil & Clancy, 2002).

For improvement of crop protection and reduction of production costs, the effects of some new fungicides are being tested. However, fungicide efficiency depends on frequency of usage (Bonnen & Hopkins, 1997), as well as on the persistence of fungicides in high concentrations in the casing during cultivation (Grogan & Jukes, 2003). Since studies of fungicides efficacy on cultivated mushrooms by agrochemical companies are very rare, only few fungicides are officially recommended in mushroom industry: prochloraz in EU countries, and chlorothalonil, thiabendazol and tiophanate-methyl in US and Canada (Beyer & Kremser, 2001, 2004; Grogan, 2008). The Australian mushroom industry has access to three registrated fungicides to use against fungal diseases: prochloraz manganese, carbendazim and thiabendazole (Allan *et al.*, 2008).

Decreased sensitivity of *L. fungicola* to prochloraz was noted (Gea *et al.*, 2005). Also, inefficiency of this fungicide was recorded in experimental growing room, at a level of spotting symptoms of cobweb disease (Grogan, 2006). This combination of circumstances makes controlling disease outbreaks more challenging (Grogan, 2008).

At the beginning of usage of fungicides from the group of methyl benzimidazole carbamates (MBC) in the late 1960's, excellent control of many fungal diseases was noted (Gaze & Fletcher, 1975; Delp, 1987; Gaze 1995a, 1995b, 1995c, 1995c). MBC fungicides affect the division of cell nuclei by disrupting the assembly of tubulin into microtubules (Smith, 1994). After several years of their intensive usage, resistant strains of *L. fungicola* occurred (Wuest *et al.*, 1974; Bollen & Van Zaayen, 1975; Fletcher & Yarham, 1976; Bonnen & Hopkins, 1997; Desrumeaux *et al.*, 1998; Potočnik *et al.*, 2008b).

The resistance of cobweb disease agents to MBC fungicides, especially to benomyl and carbendazim, was reported after their extensive usage in the Republic of Ireland in 1992 (McKay *et al.*, 1998). In 1994/95, when cobweb disease took epidemic proportions in British mushroom farms, the majority of *C. mycophillum* isolates were strongly resistant to thiabendazole and weakly resistant to carbendazim (Grogan & Gaze, 2000; Potočnik *et al.*, 2008a) Prochloraz manganese was still very toxic for the Serbian *C. dendroides* isolates (Potočnik *et al.*, 2009a). Recently, *T. aggressivum* have also developed resistance to benzimidazole fungicides (Romaine *et al.*, 2005). Only *M. perniciosa* has remained sensitive. In Europe, the bnezimidazole fungicide carbendazim is now withdrawn for use on mushrooms leaving no benzimidazoles approved for controlling sensitive pathogens such as *Mycogone*. Fungicide imazalil applied to spawn or at casing, was proposed by Shamshad *et al.* (2008) as an alternative to carbendazim, currently under review for use in Australian mushroom farms. Chlorothalonil was widely applied in Europe and North America (Van Zaayen & Rutjens, 1978; Van Zaayen, 1979).

Imidazole demethylation inhibitors (DMI's), one of which is prochloraz manganese, provide effective prevention of *L. fungicola*, *M. pernicosa* and *Cladobotryum* spp. by inhibition of the demethylation step in ergosterol biosythesis, an essential compound responsible for stability and functioning of lipoprotein membranes in many fungi (Van

Zaayen & Van Adrichem, 1982; Whitehead, 2002; Chrysayi-Tokousbalides *et al.*, 2006). Prochloraz manganese is the most effective fungicide in prevention of *Cladobotryum* spp., but it has been found that this fungicide is no longer able to control the spotting symptoms of cobweb disease (Grogan, 2006). Sensitivity tests of Serbian *C. dendroides* isolates have shown that although they were weakly resistant to thiophanate-methyl, cross resistance to neither benomyl nor carbendazim was observed (Potočnik *et al.*, 2009a). Cross resistance to MBC fungicides occured in the mushroom pathogen *L. fungicola*, which was highly resistant to thiabendazole and benomyl (Grogan & Gaze, 2000).

Trifloxystrobin had satisfactory fungitoxicity to *L. fungicola*, and it is recomended as safe to operators, consumers and the environment (Chrysayi-Tokousbsaliedes *et al.*, 2006; Diamantopoulou *et al.*, 2006). However, Serbian *C. dendroides* isolates were significantly resistant to this fungicide. Kresoxim-methyl, another fungicide from the strobilurin group, was even less toxic to the tested *C. dendroides* isolates than trifloxystrobin (Potočnik *et al.*, 2009b). Captan and metalaxyl had satisfactory effect in the control of mushroom disease in India, UK and Serbia but less than prochloraz manganese (Fletcher *et al.*, 1983; Bhatt & Singh, 1992; Potočnik *et al.*, 2009a, 2009b). Spanish isolates of *L. fungicola* were resistant to iprodione, and Serbian ones were moderately sensitive to this fungicide (Gea *et al.*, 1996; Potočnik *et al.*, 2009b).

The fungicide prochloraz is the only effective chemical to control *L. fungicola* in European countries which tend to become more tolerant. Majority of *L. fungicola* isolates, 70%, from Great Britain and Spain were moderately sensitive to prochloraz manganese (Grogan *et al.*, 2000; Gea *et al.*, 2005), and some farms reported unsatisfactory levels of control by that fungicide, which was explained by the fact that resistance was developed (Gea *et al.* 2005). This tolerance has not been associated with any major loss of control probably becasue of the complex control mechanism of resistance and the fact that the sexual state of *Lecanicillium* has not been encountered so far, reducing the risk of increase resistance due to sexual recombinatioan (Grogan, 2008).

The pH optimum for the growth of *Pleurotus* was alkaline (pH 8-9) whereas *Trichoderma* preferred acidic-neutral conditions (pH 5-7). This finding suggests that adjusting the pH of the substrate to 8-9 might slow the growth of *Trichoderma*, resulting in a reduction in the spread of the infection (Woo *et al.*, 2004). The inhibitory effect of several fungicides commonly used in agriculture (prochloraz, thabendazole, dichloran, benomyl, propiconazole, thiofanatomethyl) was also tested in this study, and both prochloraz and thiabendazole, the pesticides allowed in edible mushroom production, were found to inhibit the growth of the aggresive *Trichoderma* isolates without having a negative effect on *Pleurotus*. In order to prevent the contamination from spreading it was suggested the application of calcium hydroxide onto the affected area, and the use of fungicides benomyl, thiabendazole and prochloraz was also reported to be effective (Hatvani *et al.*, 2008).

The effect of various fungicides was examined on spore germination and the mycelial growth of *Trichoderma* isolates aggressive to *Pleurotus*. Prochloraz was shown to be the most effective fungicide for inhibition of the mycelial growth of the green moulds. Captan was toxic to *P. ostreatus*. Prochloraz, benomyl and probineb were found to inhibit spore germination of benomyl-susceptible isolates, while chlorothalonil was effective for that of benomyl-resistant strains. Prochloraz, thiabendazole and benomyl residues in the fruit bodies of oyster mushroom harvested from chemical-treated substrates and mushroom beds were below permitted values for fungicide residues in the mushroom (Hatvani *et al.*, 2008).

Rinker and Alm (2008) evaluated *in vitro* and *in vivo* toxicity of fungicides against *Trichoderma* in *A. bisporus*. Thiabendazole was the most effective, followed by benomyl and thiophanate methyl. Chlorothalonil was ineffective. Adjustement of casing pH is another approach to *Trichoderma* management. In agar, high pH reduced spore germination and retarded linear growth. Romaine *et al.* (2008) reported of benomyl and thiophanate-methyl resistance within the Ta2 pathogen in US. Imazalil sulfat, an imidazole, was found to be effective as a spawn and supplement treatment for benzimidazole-resistant Ta2. Initially, benzimidazole fungicides applied to the spawn gave good control of the problem but in the US resistance has now emerged leaving imporved hygiene as the only options to cotrol this pathogen (Romaine *et al.*, 2005).

It was found that thiabendazole concentration in the casing soil remained high throughout the cropping period, wheares the concentration of carbendazim and prochloraz manganese had dropped considerably by the end of the second flush (Grogan & Jukes, 2003). Despite the very high organic matter content of mushroom casing soil, these trends are similar to data from mineral soils (Helweg, 1977; Yarden *et al.*, 1990; Höllrigl-Rosta *et al.*, 1999). From the high levels of thiabendazole that were recoreded, good control of mushroom diseases, involving thiabendazole-sensitive isolates, would be expected with this fungicide. However, the thiabendazole form that was approved for use on mushroom crops in the UK is no longer available for commercial reasons, although thiabendazole itself continues to be available for other uses. This represent a serous loss to the UK mushroom industry, given its better persistance in mushroom casing soil compared with carbendazim and prochloraz manganese (Grogan & Jukes, 2003). Benomyl and carbendazim are both known to be susceptible to microbial degradation in soils previosly exposed to them (Flethcer *et al.*, 1980; Yarden *et al.*, 1990) and this can lead to reduced control of pathogens which are still sensitive to these fungicides. Biodegradation of pesticides is an environmentlaly desirable trait so that toxic chemicals do not accumulate in the environment such as was the case with DDT. But there has to be a balance between the time frame within which the chemical is effective against the target pathogen and the ultimate degradation of the chemical to non-toxic components (Grogan, 2008). Recent research has shown that prochloraz too can to be degraded by microbes present in prochloraz-treated casing (Papadopoulos, 2006). Under normal growing conditions on a mushroom unit, there was a rapid decline in the concentration of the active ingredient with less than 15% remaining by day 45. Further research by Papadopoulos (2006) indicated that ^{14}C-labelled prochloraz was degraded much more rapidley in the presence of residual liquid from a fungicide spray tank compared with fresh casing indicating that the source of prochloraz-degrading microbes in casing may be the tank used to apply the chemical.

Bernardo *et al.* (2002, 2004) noted that prochloraz manganese alters structure of the cell wall, as well as the ratio of cell wall components and their structure in both *L. fungicola* and *A. bisporus*, which can be attributed to the fungicide's inhibitory effect on sterol biosynthesis. Owing to everything mentioned above and despite the existence of some efficient fungicides for *L. fungicola* control, special attention is now being paid to genetic reduction of pathogen virulence by generation of mutants with diminished ability to utilize chitin as a carbon source, as well as to the possibility of using natural products (such as essential oils of different plants) to inhibit pathogen activity (Soković *et al.*, 2006).

Disease Tolerance of Cultivated Mushrooms

Mushroom diseases have been recognized as a threat to mushroom crops for more than a century yet there are no spawn strains of *A. bisporus* sold on the basis of disease resistance (Wuest, 1991).

Generally, the strains that were commonly cultivated across Europe in the 1990's and more recently, seem to be more tolerant to fungicides *in vitro* than the earlier commercial strains. *A. bisporus* sensitivity to fungicides depended on the strain used in a test (Chalen *et al.*, 1985; Chalaux *et al.*, 1993). Flusilasole and cyproconazole were the most toxic fungicides for *A. bisporus* F56 according to Potočnik *et al.* (2009a, 2009c) but Chalaux *et al.*, (1993) reported that flusilazole did not significantly limit the growth of *A. bisporus* mycelium. Strain F56 and X22 of *A. bisporus* had low sensitivity to carbendazim and intermediate to prochloraz manganese (Chrysayi-Tokousbaliedes *et al.*, 2007; Potočnik *et al.*, 2009a). The selectivity indexes indicated that thiophanate-methyl, cyproconazole and flusilazole had much less selective fungitoxicity to both *C. dendroides* and *A. bisporus* than benomyl, carbendazim and prochloraz manganese (Potočnik *et al.*, 2009a, 2009c). Benomyl showed satisfactory selective fungitoxicity to both *C. dendroides* and *A. bisporus*, but this fungicide has been withdrawn from the market. Metalaxyl, captan and chlorothalonil were toxic to strain F56 of *A. bisporus*. Low toxicity to *A. bisporus* F56 was displayed by mancozeb and iprodione (Potočnik *et al.*, 2009b). Metalaxyl and captan had little inhibitory effect on the *A. bisporus* growth in India (Bhatt & Singh, 1992). Chlorothalonil induce a toxicity problem in *A. bisporus* mycelial growth *in vitro* to tested *A. bisporus* strains B62, B98, and U3 (Chalaux *et al.*, 1993).

Standardisation of compost composition and composting processes has considerably reduced the occurrence of these moulds in mushroom crops. Especially, computer-assisted management of temperature kinetic and moisture during substrate pasteurisation has considerably improved destruction of competitor and pathogen spores. Selection of productive spawn, fungicide application to the spawn and temperature control during spawn run was found important in disease prevention (Seaby 1996; Rinker & Alm, 2000).

Due to the role of mushroom components and observations of differences in susceptibility between mushroom strains, the molecular and genetic origins of *A. bisporus* susceptibility to *P. tolaasii* have been investigated. Considering the healthy mushroom and the susceptibility, a linear model depending on the type of mushroom strain was proposed to describe mushroom susceptibility to bacterial blotch in relation with the chemical composition, brown mushroom were always found more resistant than white strains (Olivier *et al.*, 1997; Moquet *et al.*, 1999). Also was found that brown cultivars of *A. bisporus* are partial resistant to *L. fungicola* (Dragt *et al.*, 1995). Brown mushroom strains showed sifnificantly smaller yield decreases in comparison with the wihite strains after infections with *Trichoderma aggressivum* fsp. *europaeum* (Sobieralski *et al.*, 2009). Anderson *et al.* (2001) evaluated that hybrid white strains were extremily susceptible, hybrid off-white strains exhibited intermediate susceptibility, brown strain were highly resistant to disease. Gea *et al.* (2005) noted that 26-47% of fruiting bodies on *A. bisporus* farms and only 4-12% on *A. bitorquis* farms were infected by *L. fungicola*. Olivier *et al.* (1997) found that wild strans of *A. bisporus* were significantly more resistant than B62, commercial strain. Therefore, natural resistance could be used in a mushroom breeding program.

Savoie and Largeteau (2004) tested different strains of *A. bisporus* and found a negative correlation between the susceptibility of *A. bisporus* strains to *L. fungicola* and hydrogen peroxide levels in dry bubbles of these strains, whereas there was no such correlation in healthy sporocarps. This indicates that hydrogen peroxide is involved in the defence against *L. fungicola* (Largeteau *et al.*, 2006; Savoie *et al.*, 2004). In addition to peroxide, antibiotics that are produce by *A. bisporus* may play a role in the defence against *L. fungicola* (Mamoun *et al.*, 1995; Largeteau *et al.*, 2006). Efficiency of self-defense depends on the level of resistance of *Agaricus* species to *L. fungicola*, as well as on sensitivity of the pathogen to host metabolites (Savoie & Largeteau, 2004; Dragt *et al.*, 1995). The role of antibiotic production has been demonstrated in the interaction of *A. bisporus* with the green mould *T. harzianum*. An unidentified metabolite extracted from *A. bisporus* vegetative mycelium and fruiting bodies was able to inhibit the growth of two biotypes of *T. harzianum*, but stimulated the growth of *T. aggressivum* fsp. *europaeum* (Mumpuni *et al.*, 1998). Antibacterial effects of extracts of *A. bisporus* fruiting bodies have also been demonstrated (Tambekar *et al.*, 2006). However, the inhibitory effect of *A. bisporus* strains in this test did not correlate with their dry bubble resistance (Mamoun *et al.*, 1995). A significant correlation between the time needed by *A. bisporus* strains to form their first fruiting bodies and the susceptibility to *L. fungicola* was reported; earlier fruiting strains were significantly less diseased in a casing inoculation experiment (Largeteau *et al.*, 2004b). In addition, there was no correlation of the susceptibility of different *A. bisporus* strains to *L. fungicola*, *P. tolaasii* and *T. aggressivum* (Largeteau *et al.*, 2004b). This suggests that different pathogens are differentially recognized or affected by different defence mechanisms.

BIOLOGICAL DISEASE CONTROL OF CULTIVATED MUSHROOMS

However, with regard to resistance development, harm to the environment and human health, as well as an increase in production costs, special attention should be focused on inventing and developing alternative biological methods of disease control. The introduction of new fungicides of biological origin creates new possibilities for crop protection from fungal pathogens. Environmental awareness has grown, with the result that the enforcement of pollution control laws has become more and more effective. This situation forces the mushroom industry to develop technologies that ensure production with the least possible harmful effects on the environment (Levanon *et al.*, 1988).

Krupke *et al.* (2003) suggested that lactonase-producing bacteria should be examined for green mould prevention. Despite *T. aggressivum* resisting numerous bacteria isolated from the compost, it was affected by some strains (Savoie *et al.*, 2001), belonging predominantly to *Bacillus* species. Based on these observations, Serenade® biofungicide (active ingredients: *Bacillus subtilis* and its lipopeptides), is approved against *T. agressivum* in French mushroom farms (Vedie & Rousseau, 2008). Rekanovic *et al.* (2007) noted that significant amount of research work is currently focused on the use of the mycopathogenic fungus *Pythium oligandrum* in the control of *Verticillium* spp.

Many compounds have been tested as control agents against edible mushroom diseases. Plant extracts, essential oils and their components have demonstrated strong fungistatic effects (Potočnik *et al.*, 2005; 2010b; Džamić *et al.*, 2008; Glamočlija *et al.*, 2008; Soković *et al.*, 2009; Tanović *et al.*, 2009). Although fungitoxic activity of oils is not strong, they could be used as a supplement to commercial preparations for disease control, which will minimize the quantity of fungicides used.

Essential oils *Origanum vulgare* and *Thymus vulgaris* and their respective major components, carvacrol and thymol, had very strong activity against T. *aggressivum* fsp. *europaeum*, *T. harzianum* and *T. atroviride*. *Mentha piperata* which produces a major essential oil, menthol, also shows strong inhibitory activity against *Trichoderma* (Soković & Van Griensven 2006). Addition of *Melaleuca alternifolia* (tea tree) essential oil to *Pleurotus* substrate resulted in strong to total inhibition of *T. harzianum* (Angelini *et al.*, 2008). Tea tree oil was considerably less toxic than prochloraz manganese *in vitro* to *C. dendroides* isolates and *A. bisporus* F56 strain. Tea tree oil, active ingredient of biofungicide Timorex 66 EC, applied at the standard product application rate caused a significant reduction in cobweb disease levels in the *A. bisporus* experimental growing room (Potočnik *et al.*, 2010a).

Conclusion

Agaricus, *Pleurotus* and *Lentinula* are the most commonly cultivated basidiomycetes worldwide. With the popularization of mushroom farming, mushroom production continues to increase. Commercial strains of cultivated mushrooms produce substantial yield and exhibit attractive morphology and texture, but they are susceptible to a variety of viral, bacterial and fungal diseases. The most serous diseases of cultivated mushroom are caused by mushroom virus X, pseudomonads bacteria and fungi *Lecanicillium*, *Mycogone*, *Cladobotryum* and *Trichoderma*. Important part of disease control can be achived through improvements in farm hygiene, aimed at preventing the contamination of crops with potentially infected spores and mycelial debris at spawning or casing, by preventing spread of spores by filtration and by cooking out crops at the end cycle.

A common method of pathogen control on farms worldwide is application of various fungicides. Ongoing pesticide reviews in Europe have resulted in many chemicals being no longer approved for use. A major challenge for mushroom growers in the 21st Century is disease control with few or no chemicals. Other chellenges include fungicide resistance among pathogen populations, managing existing available pesticides and the emergence of new pathogens. Fungicide efficiency depends on frequency of usage, as well as on the persistence of fungicides in high concentrations in the casing during cultivation. Few fungicides are officially recommended in mushroom industry: prochloraz manganese in EU countries, chlorothalonil, thiabendazol and tiophanate-methyl in US and Canada and prochloraz manganese, carbendazim and thiabendazole in Australia. The most effective fungicide in mushroom disease control is prochloraz. Decreased sensitivity of pathogenic fungi *Lecanicillium* and *Cladobotryum* to prochloraz has been already recorded. Special attention is now being paid to genetic reduction of pathogen virulence by generation of mutants with diminished ability to utilize chitin as a carbon source, as well as to the possibility of using natural products, such as essential oils of different plants, to inhibit pathogen activity. Tea tree oil, active ingredient of biofungicide Timorex, caused a significant reduction in cobweb disease in *Agaricus* and *Pleurotus*. It was found that brown cultivars of *Agaricus* are partial resistant to bacterial blotch, dry bubble disease and green mould.

Antibiotics that are produce by *A. bisporus* may play a role in the defence against bacteria, *Lecanicillium* and *Trichoderma*.

With regard to resistance development, harm to the environment and human health, as well as an increase in production costs, special attention should be focused on inventing and developing alternative biological methods of disease control. The introduction of new fungicides of biological origin creates new possibilities for crop protection from pathogens. Based on these observations, Serenade biofungicide, based on *Bacillus subtilis*, is approved against *T. agressivum* in French mushroom farms.

REFERENCES

Abosriwil, S.O. & Clancy, C.J. (2002). A protocol for evaluation of the role of disinfectants in limiting pathogens and weed moulds in commercial mushroom production. *Pest Management Science*, *58*, 282-289.

Adie, B.A.T. (2000). The biology and epidemiology of the cobweb disease pathogen (*Cladobotryum* spp.) infecting the cultivated mushroom (*Agaricus bisporus*). Ph.D. Thesis. Imperial College, University of London, UK.

Adie, B.A.T. & Grogan, H. (2000). The liberation of cobweb (*Cladobotryum mycophilum*) conidia within a mushroom crop. In: Van Griensven (Ed.), Science and Cultivation of Edible Fungi (pp. 595-600). Rotterdam, Netherland: Balkema.

Adie, B.; Grogan, H.; Archer, S.; Mills, P. (2006). Temporal and Spatial Dispersal of *Cladobotryum* Conidia in the Controlled Environment of a Mushroom Growing Room. *Applied and Environmental Microbiology*, *72*, 7212-7217.

Allan, J.; Shah, F.A.; Khan, I. (2008). Establishing a Baseline for Fungicide Sensitivity of Three Major Mushroom Pathogens in Australia. In: M. Van Greuning (Ed.), Science and cultivation of edible and medicinal fungi: Mushroom Science XVII, Proceeding of the 17th Congress of the International Society for Mushroom Science (pp. 565-569). Cape Town, South Africa: ISMS.

Andersen, M.G.; Beyer, D.M.; Wuest, P.J. (2001). Yield comparision of hybrid *Agaricus* Mushroom Strains as a Measure of Resistance to *Trichoderma* Green Mold. *Plant Disease*, *85(7)*, 731-734.

Angelini, P.; Pagiotti, R.; Granetti, B. (2008). Effect of Antimicrobial Activity of *Melaleuca alternifolia* Oil on Antagonistic Potential of *Pleurotus* Species Against *Trichoderma harzianum* in Dual Culture. *World Journal of Microbiology and Biotechnology*, *24*, 197-202.

Anonymous, (2005). Index Phytosanitaire Acta, 41st (Ed.), Association De Coordination Technique Agricole. Paris, France: ACTA.

Anonymous, (2006). Crop Specific Protocol – Mushrooms (Crop ID: 22), Assured Produce, Food Standards, Control Document No: 00035/06, 1-38.

Bashan, Y. & Okon, Y. (1981). Integrated control of bacterial blotch in Israel. *Mushroom Journal*, *97*, 29-33

Atkins, P. & Atkins, F.C. (1971). Major Diseases of the Cultivated White Mushroom. In: W.A. Hayes (Ed.), The Mushroom Growers Association Bulletin (pp. 1-20). London, England, UK: MGA.

Bech, K. & Riber-Rasmussen, C. (1967). Experiments with soil disinfectants for casing material and their effect on yield. *Mushroom Science, 6*, 515-521.

Berendsen, R.L.; Baars, J.J.P.; Kalkhove, S.I.C.; Lugones, L.G.; Wösten, H.A.B.; Bakker, P.A.H.M. (2010). *Lecanicillium fungicola*: Causal Agent of Dry Bubble Disease in White-button Mushroom. *Molecular Plant Pathology, 11(5)*, 585-595.

Bernardo, D.; Novaes-Ledieu, M.; Perez Cabo, A.; Gea Alegria, F. J.; Garcia Mendoza, C. (2002). Effect of the Fungicide Prochloraz-Mn on the Cell Wall Structure of *Verticillium fungicola*. *International Micorbiollogy, 5*, 121-125.

Bernardo, D.; Perez Cabo, A.; Novaes-Ledieu, M.; Garcia Mendoza, C. (2004). *Verticillium* disease or »dry bubble« of cultivated mushrooms: the *Agaricus bisporus* lectin recognizes and bind the *Verticillium fungicola* cell wall glucogalactomannan. *Canadian Journal of Microbiology, 50*, 729-735.

Beyer, D.M. & Kremser, J.J. (2001). Possible development of fungicide resistance by cobweb diseases on mushrooms. *Phytopathology, 91*, 58.

Beyer, D.M. & Kremser, J.J. (2004). Evaluation of fungicide tolerance and control for three fungal diseases of mushrooms. In: Romaine, C.P., Keil, C.B., Rinker, D.L., Royse, D.J. (Eds), Science and cultivation of edible and medicinal fungi. Mushroom Science XVI. (pp- 521-529). Pennsylvania, PA, USA: Penn State University, University Park.

Bhatt, N. & Sing, R.P. (1992). Cobweb disease of *Agariucus bisporus*: Incidence, losses and effective management. *Indian Journal of Mycology and Plant Pathology, 22*, 178-181.

Bollen, G.J. & Van Zaayen, A. (1975). Resistance to benzimidazole fungicides in pathogenic strains of *Verticillium fungicola*. *Netherlands. Journal of Plant Pathology, 81*, 157-167.

Bonnen, A.M. & Hopkins, C. (1997). Fungicide resistance and population variation in *Verticillium fungicola*, a pathogen of the button mushroom, *Agaricus bisporus*. *Mycological Research, 101*, 89-96.

Bourgin, V.G. (1946). La culture du Champignon de Couche. (Suit) VIII. *Dactylium dendroides* parasite du Champignon de Cuche. *Reviu Mycologie N.S., 11*, 4-6.

Castle, A.; Speranzini, D.; Rghel, N.; Alm, G.; Rinker, D.L.; Bissett, J. (1998). Morphological and molecular identification of *Trichoderma* isolates on North American farms. *Applied Environmental Microbiology, 61*, 133-137.

Chalaux, N.; Savoie, J.M.; Olivier, J.M. (1993). Growth inhibition of *Agaricus bisporus* and associated thermophilic species by fungicides used in wheat cultivation. *Agronomie, 13*, 407-412.

Challen, M.P. & Elliott, T.J. (1985). The *in vitro* reponses to a range of fungicides of two strains of the mushroom *Agaricus bisporus* and the pathogen *Verticillium fungicola*. *Mycopathologia, 90*, 161-164.

Chang, S.-T. (1999a): Global impact of edible and medicinal mushrooms on human welfare in the 21st Century: Non-green revolution. *International Journal of Medicinal Mushrooms, 1*, 1-7.

Chang, S.-T. (1999b): World production of cultivated edible and medicinal mushrooms in 1997 with emphasis on *Lentinus edodes* (Berk.) Sing. in China. *International Journal of Medicinal Mushrooms, 1*, 291-300.

Chrysay-Tokousbalides, M.; Kastanias, M.A.; Coward, S.; Philippoussis, A.; Diamantopoulou, P. (2006). Residue Evaluation of Famoxadone and Trifloxystrobin in Cultivated Mushrooms. *Journal of Environmental Science and Health, Part B: Pesticides, Food Contaminants, and Agricultural Wastes, 41(5)*, 571-583.

Chrysay-Tokousbalides, M.; Kastanias, M.A.; Philippoussis, A.; Diamantopoulou, P. (2007). Selective Fungitoxicity of Famaxadone, Tebuconazole and Trifloxystrobin between *Verticillium fungicola* and *Agaricus bisporus*. *Crop Protection, 26*, 469-475.

Cross, M.J. & Jacobs, L. (1967). Some observations on the biology of spores of *Verticillium malthousei*. *Mushroom Science, 7*, 239-244.

Cutri, S.S.; Macauley, B.J.; Roberts, W.P. (1984). Characteristic of pathogenic non-fluorescent (smooth) and non-pathogenic fluorescent (rough) forms of *Pseudomonas tolaasii* and *Pseudomonas gingeri*. *Journal of Applied Bacteriology, 57*, 291-298.

Delp, C.J. (1987). Benzimidazole and releted fungicides. In: Prof. Dr.H. Lyr, (Ed.), Modern Selective Fungicides: Properties, Applications, Mechanisms of Action (pp. 233-244). New York, USA: Longman Scientific and Technical: John Wiley and Sons.

Desrumeaux, B.; Sedeyn, P.; Webrouck, A.; Lannoy, P. (1998). Resistance de la moele seche au sporgon (*Verticillium fungicola* var. *fungicola*). *Le bulletin de la Federation Nationale des Syndicats Agricoles de Cultivateur de chamipignons, Nouvelle serie, 77*, 677-681.

Diamantopoulou, P.; Philippoussis, A.; Kastanias, M.; Flouri, F.; Chrysayi-Tokousbalides, M.A. (2006). Effect of Famoxadone, Tebuconazole and Trifoxystrobin on *Agaricus bisporus*, Productivitiy and Quality. *Scientia Horticulturae, 109*, 190-195.

Dragt, J.W.; Geels, F.P.; Rutjens, A.J.; Van Grirensen, L.J.L.D. (1995). Resistance in wild types of *Agaricus bisporus* to the mycoparasite *Verticillium fungicola* var. *fungicola*. *Mushroom Science, 14*, 679-683.

Džamić, A.; Soković, M.; Ristić, M.; Grujić-Jovanović, S.; Vukojević, J.; Marin, P.D.; (2008). Chemical composition and antifungal activity of *Origanum heracleoticum* essential oil. *Chemistry of Natural Compounds, 44*, 659-661.

Fan, L.; Pan, H.; Soccol, A.T.; Pandey, A.; Soccol, C.R. (2006). Advaces in Mushroom Research in the Last Decade. *Food Technology and Biotechnology, 44(3)*, 303-311.

Fekete, K. (1967). Über Morphologie, Biologie und Bekämpfung von *Verticillium malthousei,* Einem Parasiten des Kulturchampignons. *Phytopathologische Zeitschrift, 59*, 1-32.

Fletcher, J.T.; Connolly, G.; Moungield, E.I.; Jacobs, L. (1980). The disappearance of benomyl from mushroom casing. *Annals of Applied Biology, 95*, 73-82.

Fletcher, J. T.; Hims, M. J.; Hall, R. J. (1983). The control of bubble diseases and cobweb disease of mushrooms with prochloraz. *Plant Pathology, 32,* 123-131.

Fletcher, J.T.; White, P.F.; Gaze, R.H. (1989). Mushrooms - Pest and Disease Control (2nd edition). Newcastle upon Tyne, UK: Intercept.

Fletcher, J.T. & Yarham, D.J. (1976). The incidence of benomyl tolerance in *Verticillium fungicola*, *Mycogone perniciosa* and *Hypomyces rosellus* in mushroom crops. *Annual Applied Biollogy, 84*, 343-353.

Forer, L.B.; Wuest, P.J.; Wagner, U.R. (1974). Occurence and economic impact of fungal diseases of mushrooms in Pennsylvania. *Plant Diesease Reporter, 54*, 987-991.

Garrod, E. (2004). Pesticides – The latest position. Mushroom Journal, 649: 20-23.

Gaze, R.H. (1995a). Dactylium or Cobweb. *Mushroom Journal, 546*, 23-27.

Gaze, R.H. (1995b). Dactylium or Cobweb II. *Mushroom Journal, 548*, 13.

Gaze, R.H. (1995c). Dactylium or Cobweb in conclusion. *Mushroom Journal, 549*, 26.

Gaze, R.H. (1996). The past year. Dactylium or Cobweb, *Mushroom Journal, 552*, 24-25.

Gaze, R.H. & Fletcher, J.T. (1975). ADAS survey of mushroom diseases and fungicide usage 1974/75. *Mushroom Journal, 35*, 370-376.

Gaze, R.H.; Calvo-Bado, L.; Challen, M.P.; Adie, B.; Romaine, C.P. (2000). A new virus disease of *Agaricus bisporus*? *Mushroom Science, 15*, 701-705.

Geels, F.P.; Hessen, L.P.W.; Van Griensven, L.J.L.D. (2008). Brown discoloration of mushrooms caused by *Pseudomonas agarici. Journal of Phytopathology, 140*, 249-259.

Gea, F.J.; Navarro, M.J.; Tello, J.C. (2005). Reduced Sensitivity of the Mushroom Pathogen *Verticillium fungicola* to Prochloraz-manganese *in Vitro. Mycological Research, 109*, 741-745.

Gea, F. J.; Tello J. C.; Honrubia M. (1996). *In vitro* sensitivity of *Verticillium fungicola* to selected fungicides. *Mycopathologia, 136*, 133-137.

Glamočlija, J.; Soković, M.; Vukojević, J.; Milenković, I.; Van Griensven, L.J.L.D. (2006). Chemical composition and antifungal activities of essential oils of *Satureja thymbra* L. and *Salvia pomifera* ssp. *calcyna* (Sm.) Hayek. *Journal of Essential Oil Research, 18*, 115-117.

Grogan, H.M. (2006): Fungicide Control of Mushroom Cobweb Cisease Caused by *Cladobotryum* Strains with Different Benzimidazole Resistance Profiles. *Pest Management Science, 62(2)*, 153-161.

Grogan, H.M. (2008): Challenges Facing Mushroom Disease Control in the 21st Century. In: Lelley, J.I., Buswell, J.A. (Eds.), Proceeding of the Sixth International Conference on Mushroom Biology and Mushroom Products, 29th September - 3rd October 2008 (pp. 120-127). Bonn, Germany: WSMBMP.

Grogan, H.M.; Adie, B.A.T.; Gaze, R.H.; Chalen, M.P.; Mills, P.R. (2003). Double-stranded RNA elements associated with the MVX disease of *Agaricus bisporus. Mycological Research, 107(2)*, 147-154.

Grogan, H.M. & Gaze, R.H. (2000): Fungicide resistance among *Cladobotryum* spp. – causal agents of cobweb disease of the edible mushroom *Agaricus bisporus. Mycological Research, 104(3)*, 357-364.

Grogan, H. M. & Jukes, A. A. (2003): Persistence of the fungicides thiabendazole, carbendazim and prochloraz-Mn in mushroom casing soil. *Pest Management Science, 59*, 1225-1231.

Grogan, H. M.; Keeling, C.; Jukes, A.A. (2000): *In vivo* response of the mushroom pathogen *Verticillium fungicola* (dry bubble) to prochloraz-manganese. In: Proceeding of Brighton Crop Protection Conference: Pests & Diseases (1, 273-278). Farnham, Surrey, UK: BCPC.

Guleria, D.S. (1976). A note on the occurrence of brown blotch of cultivated mushrooms in India. *Indian Journal of Mushrooms, 2*, 1-25.

Harvey, C.L.; Wuest, P.J.; Schisler, L.C. (1982). Diseases, weed molds, indicator molds and abnormalities of the commercial mushroom. In: P.J. Wuest & G.D. Bengston (Eds.). Penn State handbook for commercial mushroom growers (pp. 19-33). Pennsylvania, PA, USA: The Pennsylvania State University, State College.

Hatvani, L.; Antal, Z.; Manczinger, L.; Szekeres, A.; Druzhina, I.S.; Kubicek, C.P.; Nagy, A.; Nagy, E.; Válvögyi, E.; Kredics, L. (2006). Green mold diseases of *Agaricus* and *Pleruotus* are caused by related but phylogenetically different *Trichoderma* species. *Phytopathology, 97*, 532-537.

Hatvani, L.; Antal., Z.; Nagy, E.; Válvögyi, C.; Kredics, L. (2007). Green mold diseases of *Agaricus* and *Pleurotus* spp. are caused by releted but phylogeniticially different *Trichoderma* species. *Phytopathology, 97*, 532-537.

Hatvani, L.; Kocsubé, S.; Mnczinger, L.; Antal, Z.; Szekeres, A.; Druzhina, I.S.; Komoń-Zelazowska, M.; Kubicek, C.P.; Nady, A.; Csaba, V.; Kredics, L. (2008). The green mould disease global threat to the cultivation of oyster mushroom (*Pleurotus ostreatus*): a review. In: M. Van Greuning (Ed.), Science and cultivation of edible and medicinal fungi: Mushroom Science XVII, Proceeding of the 17th Congress of the International Society for Mushroom Science (pp. 485-495). Cape Town, South Africa: ISMS.

Helveg, A. (1977). Degradation and adsorption of carbednazim and 2-amnobenzimidazole in soil. *Pesticide Sciences, 8*, 71-78.

Höllrigl-rosta, A.; Kreuyig., R.; Bahadir, M. (1999). Investigations on the metabloic fate of prochloraz in soil under field and laboratory conditions. *Pesticide Sciences, 55*, 531-538.

Kim, J.; Kwon, S.; Kang, H.; Kim, J.W.; Kwon, S.I.; Kang, H.J. (1995). Studies on the pathogenic Psedomonas cauing bacterial diseases of cultivated mushrooms in Korea. 2. Bacteriological characteristics of *Pseudomonas tolaasii* causing mushroom brown blotch and white line reacting organisms. *Korean Journal of Plant Pathology, 11*, 353-360.

Kligman, A.M. (1950). Handbook of Mushroom Culture. (2nd edition.). Pennsylvania, PA, USA: J.B. Swayne, Kennet Square.

Komon-Zelazowska, M.; Bissett, J.; Zafari, D.; Hatvani, L.; Manczinger, L.; Woo, S.; Lorito, M.; Kredics, L.; Kubicek, C.P.; Druzhinina, I.S. (2007b). Genetically colosely related but phenotypically divergent *Trichoderma* species cause green mold disease in oyster mushoom farms woldwide. *Applied Environmental Microbiology, 73*, 7415-7426.

Kredics, L.; Hatvani, L.; Antal, Z.; Manczinger, L.; Druzhina, I.S.; Kubicek, C.P.; Szekeres, A.; Nagy, A.; Válvögyi, C.; Nagy, E. (2006). Green mould disease of oyster mushroom in Hungary and Transylvania. *Acta Microbiologica et Immunologica Hungarica, 53*, 306-307.

Kredics, L.; Kocsube, S.; Nagy, L; Komon-Zelazowska, M.; Manczinger, L.; Sajben, E.; Nagy, A.; Vagvolgyi, C.; Kubicek, C.P.; Druzhinina, I.S.; Hatvani, L. (2009). Molecular identification of *Trichoderma* species associated with *Pleurotus ostreatus* and natural substrates of the oyster mushroom. *FEMS Microbiology Letters, 300*, 58-67.

Krupke, O.A.; Castle, A.J.; Rinker, D.L. (2003). The North American mushroom competitor, *Trichoderma aggressivum* f. *aggressivum*, produces antifunal compounds in mushroom compost that inhibit mycelial growth of the commercial mushroom *Agaricus bisporus*. *Mycological Research, 107*, 1467-1475.

Largeteau, M.L.; Baars, J.P.P.; Regnault, R.C.; Savoie, J.M. (2006). Molecular and physiological diversity among *Verticillium fungicola* var. *fungicola*. *Mycological Research, 110*, 431-440.

Largeteau, M.; Mata, G.; Savoie, J.M. (2004a). *Verticillium fungicola* v. *fungicicola* affects *Agaricus bisporus* cultivation in Mexico. *FEMS Microbiology Letters, 236*, 191-196.

Largeteau, M.L.; Rodier, A.; Rrousseau, T.; Juarez del Carmen, S.; Védie, R.; Savoie, J.M. (2004b). Agaricus susceptibility to *Verticillium fungicola*. *Mushroom Science, 16*, 504-522.

Largeteau, M. & Savoie, J.-.M. (2008). Effect of the fungal pathogen *Verticillum fungicola* on fruiting initiation of its host, *Agaricus bisporus*. *Mycological Research, 112*, 825-828.

Largeteau, M.L. & Savoie, J.-M. (2010). Microbially induced diseases of *Agaricus bisporus*: biochemical mechanisms and impact on commercial mushroom production. *Applied Microbiology and Biotechnology, 86*, 63-73.

Levanon, D.; Danai, O.; Masaphy, S. (1988). Chemical and physical parameters in recycling organic wastes for mushroom production. *Biological Wastes, 26*, 341-348.

Mamoun, M.; Olivier, J.M.; Vedie, R. (1995). Discussion on assessment of artificial infections with *Verticillium fungicola* for breeding programmes. *Mushroom Science, 14*, 669-677.

McKay, G. L.; Egan, D.; Morris, E.; Brown, A.E. (1998). Identification of benzimidazole resistance in *Cladobotryum dendroides* using a PCR-based method. *Mycological Research 102(6)*, 671-676, 1998.

McKay, G.J.; Egan, D.; Morris, E.; Scott, C.; Brown, A.E. (1999). Genetic and morphological characterization of *Cladobotryum* species causing cobweb disease of mushrooms. *Applied and Environmental Microbiology, 65(2),* 606-610.

Moquet, F.; Desmerger, C.; Mamoun, M.; Ramos-Guedes-Lafargue, M.; Olivier, J.-M. (1999). A quantitative trait locus of *Agaricus bisporus* resistance to *Pseudomonas tolaasii* is closely linked to natural cap color. *Fungal genetics and biology, 28(1),* 34-42.

Mumpuni, A.; Sharma, H.S.S.; Brown, A.E. (1998). Effect of metabolites produced by *Trichoderma harzianum* biotypes and *Agaricus bisporus* on their respective growth radii in culture. *Applied Environmental Microbiology, 64*, 5053-5056.

Muthumeenakshi, S.; Brown, A.E.; Mills, P.R. (1998). Genetic comparison of the aggressive weed mould strains of *Trichoderma harzianum* from mushroom compost in North America and the British Isles. *Mycological Research, 102*, 385-390.

Nair, N.G. (1975). Observations on three important diseases of the cultivated mushroom in New South Wales, *Plant Disease Survey*, 30-32.

North, L.H. & Wuest, P. (1993). The infection process and symptom expression of *Verticillium* disease of *Agaricus bisporus*. *Canadian Journal of Plant Pathology, 15,* 74-80.

Olivier, J.M.; Guillaumes, J.; Martin, D. (1978). Study of a bacterial disease of mushroom caps. In: Proceedings of 4th International Conference on Plant Pathogenic Bacteria (pp. 903-916). Angers, FRA: INRA.

Olivier, J.M.; Mamoun, M.; Munsh, P. (1997). Standardization of a method to assess mushroom blotch resistance in cultivated and wild *Agaricus bisporus* strains. *Canadian Journal of Plant Pathology, 19*, 36-42.

Ospina-Giraldo, M.D.; Royse, D.J.; Thon, M.R.; Chen, X.; Romaine, C.P. (1998). Phylogenetic relationships of *Trichoderma harzianum* casing mushroom green mold in Europe and North America to other species of *Trichoderma* from world-wide sources. *Mycologia, 90*, 76-81.

Ozaktan, H. & Bora, T. (1994). Studies on identification of bacterial microflora of mushroom in Turkey. *Journal of Turkish Phytopathology, 23*, 73-78.

Paine, S.G. (1919). Studies in bacteriosis II: a brown blotch disease of cultivated mushrooms. Ann. Appl. Biol., 5, 206-219.

Papadopoulos, G. (2006). The fate of prochloraz in mushroom casing. Ph.D. Thesis, University of Reading, UK.

Park, M.S.; Bae, K.S.; Yu, S.H. (2004). Molecular and morphological analyisis of *Trichoderma* isolates associated with green mold epidemic of oyster mushroom in Korea. *Journal of Huazhong Agricultural University, 23*, 157-164.

Park, M.S.; Bae, K.S.; Yu, S.H. (2006). Two new species of *Trichoderma* associated with green mold of oyster mushroom cultivation in Korea. *Mycobiology, 34*, 11-113.

Potočnik, I. (2006): Causal agents of bubble diseases of cultivatad mushroom (*Agaricus bisporus* (Lange) Imbach) and its sensitivity to fungicides. University of Belgrade, Faculty of Biology, 1-63.

Potočnik, I. (2009). *Cladoobtryum dendroides* (Bull.) W. Gams & Hooz., causal agent of cultivated mushroom (*Agaricus bisporus* (Lange) Imbach) and its sensitivity to fungicides and biofungicides. University of Belgrade, Faculty of Biology, 1-131.

Potočnik I.; Milijašević S.; Rekanović E.; Todorović B.; Stepanović, M. (2008a). Sensitivity of *Verticillium fungicola* var. *fungicola, Mycogone perniciosa* and *Cladobotryum* spp. to fungicides in Serbia. In: M. Van Greuning (Ed.), Science and cultivation of edible and medicinal fungi: Mushroom Science XVII, Proceeding of the 17th Congress of the International Society for Mushroom Science (pp. 615-627). Cape Town, South Africa: ISMS.

Potočnik I.; Vukojević, J.; Stajić, M.; Rekanović, E.; Milijašević, S.; Todorović, B.; Stepanović, M. (2009a). *In vitro* toxicity of selected fungicides from the groups of benzimidazoles and demethylation inhibitors to *Cladobotryum dendroides* and *Agaricus bisporus. Journal of Environmental Science and Health, Part B, 44(4),* 365-370.

Potočnik I.; Vukojević, J;, Stajić, M.; Rekanović, E.; Milijašević, S.; Stepanović, M.; Todorović, B. (2009b). Toxicity of Fungicides with Different Modes of Action to *Cladobotryum dendroides* and *Agaricus bisporus. Journal of Environmental Science and Health, Part B, 44(8)*, 823-827.

Potočnik, I.,; Rekanović, E.; Milijašević, S.; Todorović, B.; Stepanović, M. (2009c). *In vitro* toxicity of fungicides of different mode of action to *Agaricus bisporus* (Lange) Imbach. *Pesticides and Phytomedicine, 24(1),* 29-34.

Potočnik I.; Vukojević, J.; Stajić, M.; Rekanović, E.; Stepanović, M.; Milijašević, S.; Todorović, B. (2010a). Toxicity of biofungicide Timorex 66 EC to *Cladobotryum dendroides* and *Agaricus bisporus. Crop Protection, 29*, 290-294.

Potočnik, I.; Vukojević, J.; Stajić, M.; Tanović, B.; Rekanović, E. (2010b). Fungicide and essential oils sensitivity of *Mycogone perniciosa* originated from *Agaricus bisporus* farms. *International Journal of Medicinal Mushrooms, 12(1),* 91-98.

Potočnik, I.; Vukojević, J.; Stajić, M.; Tanović, B.; Todorović, B. (2008b). Fungicide sensitivity of selected *Verticillium fungicola* isolates from *Agaricus bisporus* farms. *Archives of Biological Sciences, Belgrade, 60(1),* 151-158.

Potočnik I.; Tanović B.; Milijašević S.; Rekanović E.; Todorović B. (2005). Response of the mushroom pathogen *Verticillium fungicola* (Preuss) Hasebrauk (dry bubble) to some essential oils. ESNA Meeting, Amiens, France – *Revue de Cytologie et de Biologie Végétales - Le Botaniste, 28 (s.i.)*, 388-392.

Rai, R.D.I. & Ahlawat, O.P. (2005). Edible fungi: Biotechnological approaches. In: G. G. Khachatourians, D.K. Arora, R. M. Berka (Eds.), Applied Mycology and Biotehnology (pp. 87-123). Maryland Heights, USA: Elsevier.

Rainey, P.B.; Brodey, C.L.; Johnston, K. (1992). Biology of *Pseudomonas tolaasii*, causal of brown blotch desease of the cultivated mushrooms. In: Andrews, J.H., Tommerum, I. (Eds.), Advanced Planth Pathology (pp. 95-117). New York, USA: Accademic Press.

Rekanović, E.; Milijašević, S.; Todorović, B.; Potočnik, I. (2007). Posibilities of Biological and Chemical Control of Verticillium Wilt in Pepper. *Phytoparasitica* 35(5): 436-441.

Rinker, D.L. & Alm, G. (2000). Management of green mould disease in Canada. *Mushroom Science, 15,* 617-623.

Rinker, D.L. & Alm, G. (2008). Management of casing *Trichoderma* using fungicides. In: M. Van Greuning (Ed.), Science and cultivation of edible and medicinal fungi: Mushroom Science XVII, Proceeding of the 17th Congress of the International Society for Mushroom Science (pp. 496-509). Cape Town, South Africa: ISMS.

Romaine, C.P.; Royse, D.J.; Schlagnhaufer, C. (2005). Superpathogenic *Trichoderma* resistant to TopsinM found in Pennsylvania and Delaware. *Mushroom News, 53*, 6-9.

Romaine, C.P.; Royse, D.J.; Schlagnhaufer, C. (2008). Emergence of benzimidazole-resistant green mould *Trichoderma aggressivum*, on cultivated *Agaricus bisporus* in North America. In: M. Van Greuning (Ed.), Science and cultivation of edible and medicinal fungi: Mushroom Science XVII, Proceeding of the 17th Congress of the International Society for Mushroom Science (pp. 510-523). Cape Town, South Africa: ISMS.

Rytter, J.L.; Royse, D.J.; Romaine, C.P. (1991). Incidence and diversity of double-stranded RNA in *Lentinula edodes*. *Mycologia, 83(4),* 506-510.

Samuels, G.J.; Dodd, S.L.; Gams, W.; Castlebury, L.A.; Petrini, O. (2002a). *Trichoderma species* associated with the green mold epidemic of commercially grown *Agaricus bisporus*. *Mycologia, 94*, 146-170.

Savoie, J.-M.; Iapicco, R.; Largeteau-Mamoun, M. (2001). Factors influencing the competitive saprophytic ability of *Trichoderma harzianum* Th2 in mushroom (*Agaricus bisporus*) compost. *Mycological Research, 105*, 1348-1356.

Savoie, J.-M. & Largeteau, M.L. (2004). Hydrogen peroxide concentrations detected in *Agaricus bisporus* sporocarps and relation with their susceptibility to the pathogen *Verticillium fungicola*. *FEMS Microbiology Letters, 237*, 311-315.

Seaby, D.A. (1987). Infection of mushroom compost by *Trichoderma* species. *Mushroom Journal, 179*, 355-361.

Seaby, D.A. (1996). Investigation of the epidemiology of green mould of mushroom (*Agaricus bisporus*) compost caused by *Trichoderma harzianum*. *Plant Pathology, 45*, 913-923.

Seaby, D.A. (1998). *Trichoderma* as a weed mould or pathogen in mushroom cultivation. In: Harman G.E., Kubicek C.P. (Eds.) *Trichoderma* & *Gliocladium* (pp. 267-288). London, UK: Taylor & Francis.

Shamsad, A.; Clift, A.; Butler, R. (2008). Studies on dry bubble disease caused by *Verticillium fungicola*. In: M. Van Greuning (Ed.), Science and cultivation of edible and medicinal fungi: Mushroom Science XVII, Proceeding of the 17th Congress of the International Society for Mushroom Science (570-579). Cape Town, South Africa: ISMS.

Schisler, L.C.; Sinden, J.W.; Sigel, E.M. (1967). Etiology, symptomology and epidemiology of a virus disease of cultivated mushrooms. *Phytopathology, 57*, 519-526.

Sisto, D.; Faggiano, S.; Rana, G.L. (1997). *Mycogone perniciosa*, a potential threat for cultivated mushrooms in southern Italy. *Petria, 7(3),* 159-164.

Smith, C. M. (1994). History of benzimidazole use and resistance. In: C.J. Delp (Ed.), Fungicide Resistance in North America (pp. 23-24). St. Paul, Minnesota, USA: The American Phytopathological Society.

Sobieralski, K.; Siwulski, M.; Frużyńska-Jóźwiak, D.; Górski, R. (2009). Impact of *Trichoderma aggressivum* fsp. *europaeum* Th2 on the yielding of *Agaricus bisporus*. *Phytopathologia, 53*, 5-10.

Soković, M.; Van Griensven, LJ.L.D. (2006). Antimicrobial activity of essential oils and their components against the three major pathogens of the cultivated button mushroom, *Agaricus bisporus*. *European Journal of Plant Pathology, 116*, 211-224.

Soković, M.; Vukojević, J.; Marin, P.; Brkić, D.; Vajs, V.; Griensven, L.J.L.D. (2009). Chemical composition of essential oils of *Thymus* and *Mentha* species and their antifungal activities. *Molecules, 14*, 238-249.

Soler-Rivas, C.; Jolivet, S.; Arpin, N.; Olivier, J.M.; Wichers, H.J. (1999). Biochemical and physiological aspects of brown blotch disease of *Agaricus bisporus*. *FEMS Microbiology Reviews, 23*, 591-614.

Solomon, J.M.; Beelman, R.B.; Bartley, C.E. (1991). Addition of calcium chloride and stabilized chlorine dioxide to irrigation water to improve quality and shelf life of *Agaricus bisporus*. *Mushroom Science, 13*, 695-701.

Suyama, K. & Fujii, H. (1993). Bacterial disease occurred on cultivated mushroom in *Japanese Journal of Agricicultural Science, 38*, 35-50.

Tambekar, D.H.; Sonar, T.P.; Khodke, M.V.; Khante, R.S. (2006). The novel antibacterials from the two edible mushrooms: *Agaricus bisporus* and *Pleurotus sajor caju*. *International Journal of Pharmaceutics, 2*, 584-585.

Tanović, B.; Potočnik, I.; Delibašić, G.; Ristić, M.; Kostić, M.; Marković, M. (2009). *In vitro* effect of essential oils from aromatic and medicinal plants on mushroom pathogens: *Verticillium fungicola* var. *fungicola*, *Mycogone perniciosa*, and *Cladobotryum* sp. *Archives of biological sciences, Belgrade, 61(2),* 231-238.

Tokimoto, K. & Komatsu, M. (1979). Effect of carbon and nitrogen sources in media on the hyphal interference between *Lenditunus edodes* and some species of *Trichoderma*. *Annals of the Phytopathological Society of Japan, 45*, 261-264.

Tolaas, A.G. (1919). A bacterial disease of cultivated mushrooms. *Phytopathology, 5,* 51-54.

Umar, M. H.; Geels, F. P.; Van Griensven L.J.L.D. (2000). Pathology and pathogenesis of *Mycogone perniciosa* infection of *Agaricus bisporus*. In: Proceeding of the 15th International Congress on the Science and Cultivation of Edible Fungi 2, 15-19 May 2000, (pp. 561-567). Maastricht, Netherlands: ISMS.

Umar, M. H. & Van Griensven L.J.L.D. (1999): Studies on the morphogenisis of *Agaricus bisporus*: dilema of normal versus abnormal fruit body development. *Mycological Research, 103*, 1235-1244.

Van Zaayen, A. (1979). Daconil and intermediate mushroom cultivars. *Champignoncultuur, 23*, 217.

Van Zaayen, A. & Rutjens, A.A. (1978). The application of Daconil for the control of bubble. *Champignoncultuur, 22*, 121-123.

Van Zaayen, A. & Van Adrichem, J.C.J. (1982). Prochloraz for control of fungal pathogens of cultivated mushrooms. *Netherlands Journal of Plant Pathology, 88(5),* 203-213.

Védie, R. & Rousseau, T. (2008). Serenade biofungicide: une innovation mjeure dans les champignonnières françaises pour lutter contre *Trichoderma aggressivum*, agent de la moisissure verte du compost. *La Lettre du CTC, 21*, 1-2.

Visscher, H.R. (1988). Casing soil. In: Van Griensven, L.J.L.D. (Ed.), The Cultivation of Mushrooms (pp. 73-90). Waalwijk, Holland: Grafidrukkerij Waalwijk.

Ware, W.M. (1933). Annual Report, Department of Mycology, *Journal of South Eastern Agricultural College, Wye, Kent, UK, 31*, 13-21.

Whitehead, R. (2002). The e-UK Pesticide Guide. British Crop Protection Council, Wallingford, UK: CABI Publishing.

Wong, W.C. & Preece, T.F. (1987). Sources of *Verticillium fungicola* on a commercial mushroom farm in Englend. *Plant Pathologogy, 36*, 577-582.

Woo, S.L.; Di Benedetto, P.; Senatore, M.; Abadi, K.; Gigante, S.; Soriente, I.; Ferraioli, S.; Scala, F.; Lorito, M. (2004). Identification and charactrization of *Trichoderma* species aggressive to *Pleurotus* in Italy. *Journal of Zhejiang University (Agriculture and Life Sciences), 30*, 469-470.

Wuest, P.J. (1991). Nature of disease resistance to compost-borne and airborne pathogens of *Agaricus bisporus*. In: L.J.L.D. van Griensven (Ed.), Genetics and breeding of *Agaricus bisporus*. First International Seminar on Mushroom Science (pp. 99-102). Wageningen, The Netherlands: ISMS.

Wuest, P.J.; Cole, H.; Sanders, P.L. (1974). Tolerance of *Verticillium malthousei* to Benomyl. *Phytopathology, 64*, 331-334.

Yarden, O.; Salomon, R.; Katan., J.; Aharonson, N. (1990). Involvent of fungi and bacteria in enhanced and nonenhanced biodegradation of carbendazim and other benzimidazole compounds in soil. *Canadian Journal of Microbiology, 36*, 15-23.

Zare, R. & Gams, W. (2008). A revision of th *Verticillium fungicola* species complex and its affinity with the genus *Lecanicillium*. *Mycological Research, 112*, 811-824.

In: Mushrooms: Types, Properties and Nutrition
Editors: S. Andres and N. Baumann
ISBN 978-1-61470-110-1

Chapter 4

PRODUCTION OF *PLEUROTUS OSTREATUS* (OYSTER MUSHROOM) GROWN ON SUGAR CANE BIOMASS (TRASH, BAGASSE AND PITH)

Noé Aguilar-Rivera**, *Adolfo Castillo Moran, Daniel Arturo Rodríguez Lagunes and Joaquín Murguia Gonzalez
Universidad Veracruzana, Facultad de Ciencias Biológicas y Agropecuarias, Km. 1 Carretera Peñuela Amatlan de los Reyes S/N. C.P. 94945, Córdoba, Veracruz México

ABSTRACT

Mushroom culture is a biotechnological process that recycles ligninocellulosic wastes and can be produced on natural materials from agriculture, woodland, animal husbandry, and manufacturing industries and the spent substrates can be used in different ways (soil conditioner, compost, cattle feed). Edible mushrooms are highly appreciated and have a commercial potential in many countries. They have a chemical composition, which is attractive from the nutritional point of view for protein. Mushroom yields and biological efficiency vary according to biological factors, environmental conditions and growing substrates.

Mushroom cultivation is a well-established and profitable agribusiness carried out worldwide on a large or small scale. Cultivation on sugarcane substrates is an option to be considered due to low cost as substrate, physic-chemical structure, and their positive economical, social and environmental factors for the sugarcane cultivation areas in a sugarcane diversification concept in comparison with other agricultural by-products. However, successful technology transfer programs to increase yields and reduce production costs of edible mushrooms are limited. The sustainable model for production of edible mushrooms represents a strategy that allows large-scale, small-scale, and domestic cultivation to promote regional development.

This article discusses the development of an agro-industry of fresh and dried edible mushrooms with sugarcane production, major fundamental technologies involved: Spawn technology, Mushroom production technology, and processing technology and how their

* Tel./Fax: (01) 271 71 6 73 92
Email: naguilar@uv.mx

economic and nutritional potentials will increase the productivity of mushrooms on sugarcane substrates with better flavor, appearance, texture, nutritional qualities at low cost.

INTRODUCTION

The diet in developed countries should be changed radically and urgently. This is because of the high consumption of animal products, which are too rich in saturated fats and cholesterol. Dietary changes in developed countries should be initiated by health programs that could promote the partial substitution of animal protein with plant protein, such as mushrooms as a source of non-starchy carbohydrate and protein. The specialized literature of recent years is characterized by looking for new sources of proteins and essential amino acids (Jwanny, 1995).

Mushrooms have been widely cultivated, eaten and appreciated for their flavor, economic and ecological values, and medicinal properties for many years. Since the 1700's and presently more than 30 unknown species are cultivated as foods (Ibekwe, 2008). By 2009, world mushroom production had reached 3.6 million tons of edible mushrooms (FAOSTAT, 2010). Now, mushrooms are being cultivated in more than 100 countries of the world by mushroom farming and/or industrialization for producing protein-rich food by the concept of health food, functional food and nutraceuticals; thus mushroom cultivation can usually be viewed as an effective means to extract resources left behind in agricultural solid wastes (Siu-Wai Chiu, 2000).

In recent years, there has been an increasing trend towards more efficient utilization of agro-industrial residues. Wastes from agriculture and agro-industries are the excesses and residues from growing and processing raw agricultural products such as grain, fruits, poultry, fish and trees (Detroy, 1978). Several processes have been developed that utilize these as raw materials for the production of bulk chemicals and value-added fine products such as ethanol, single cell protein (SCP), mushrooms, enzymes, organic acids, amino acids, biologically active secondary metabolites, etc. (Pandey, 2000).

In developing countries, the importance of edible mushrooms within consumer preferences and perceptions has not been studied. During the past decades, major efforts in the cultivation of edible mushrooms have been focused on technological developments, cultivation, fruiting, genetic breeding systems and yield on a large and small (rural) scale; which makes the use of agricultural and industrial residues possible as carbon sources by recycling them as substrates for biorefinery or mushroom production, because more than 70% of agricultural and forest materials are nonproductive and have been wasted in the processing, consequently resulting in low-cost production and a continuous market. However, the knowledge on the composition of lignocellulosic materials is important for their effective utilization, thus the mushroom science is a relatively new applied science and the mushroom industry is still small compared to many plant crops, so the investment is limited. As a consequence, scientific research on mushrooms (Cytogenetic study, Breeding strategies, Encoded genes of the principal mushrooms, Cross breeding, Transgenic breeding, Active Compounds of Mushrooms) generally lags behind that of plant and animal (Melo de Carvalho, 2010,Apati et al, 2010, Rashad, 2009, Gregori, 2007, Fan et al, 2008, 2006).

Identification of suitable substrate material, both biologically and economically, is critical for successful cultivation. Climate is another factor in successful mushroom cultivation. Indoor cultivation methods, which allow for precise climate control, are generally very expensive to establish and operate (Tisdale et al. 2006).

However, in a global economy, the future of mushroom cultivation agribusiness will also domestically depend on a thorough understanding of consumption trends worldwide. Despite the social, economic, and ecological importance of commercial mushroom production in developing countries, several aspects are virtually unattended: (1) organization of growers (large, small); (2) improvement of the channels of distribution; (3) integral marketing strategies; (4) analysis of the consumer needs and (5) each species requires specialized treatment and growing is labor intensive to produce consistent yields of high-quality, marketable mushrooms (Mayett et al. 2006, 2004).

This suggests that the potential of mushroom cultivation as an alternative source of protein, with a high potential employment and income generation within a short period of time (they require shorter growth time when compared to other edible mushrooms), few environmental controls, their fruiting bodies are not very often attacked by diseases and pests and they can be cultivated in a simple and cheap way; they could be a possible offer to alleviate poverty and develop the life style of the vulnerable people in developing countries where population is directly or indirectly engaged in a wide range of agricultural activities (Apati et al, 2010, Imtiaj, 2008).

Mushrooms could become an important addition to farmers looking for a value-added product and a way to supplement farm income while making use of byproducts or co-products from other crops. Animal producers also could benefit from the addition of spent substrate into their herds' diets. This could benefit mushroom producers by giving them a market for the substrate after the production cycle.

PLEUROTUS OSTREATUS MUSHROOMS

Mushrooms are fleshy, spore-bearing reproductive structures of fungi. Mushrooms have been eaten and appreciated for their flavor, economic and ecological values and medicinal properties for many years. Commercial production of fresh edible mushrooms is a rapidly growing industrial activity (Contreras, 2004, Fan, 2006, 2000).

On the other hand, mushroom cultivation represents one of the most economically and cost effective organic recycling processes which could also be regarded as a functional food and have long been consumed for their high protein content (crude protein represents roughly 20–30% of the dry matter) and rich B vitamins (thiamin, riboflavin and niacin), but they also have low fat and are virtually cholesterol free (Imtiaj, 2008, Wai Chiu, 2000, Velazquez-Cedeño, 2000, Stamets, 1984).

Chang (2007) and Lindequist et al. (2005) demonstrated that many mushrooms produce a range of metabolites of intense interest to the nutriceutical and pharmaceutical (e.g. anti–tumor, immunomodulation agents, and hypocholesterolaemic agents) and food (e.g. flavor compound) industries. Most mushroom species, if not all, contain biologically active polysaccharides. Fruiting bodies and submerged broth are sources of the bioactive compounds.

The *Pleurotusostreatus* is also known as the *Oyster mushroom*, edible mushroom or giant mushroom, or"hiratake", "shimeji", "houbitake" by Asians. In these countries, mushroom cultivation is still regarded as an art more than it is a science because of the flexibility in human adjustment and very extensive experience (Siu-Wai Chiu, 2000). The Latin *Pleurotus* means "beside the ear" and *ostreatus* means "oystershaped" (Cohen 2002). Their flavor is typically described as oysterlike, hence the name oyster mushrooms (Croan, 2000). It is one of the most known and examined mushrooms worldwide and are currently the second most produced edible mushrooms, representing approximately 25% of the total world production of cultivated mushrooms (Rizki, 2010).Commonly consumed in Asian communities and havinga commercial potential in many countries, *Pleurotus* species have penetrated markets worldwide(Rühl, 2008).

It generally is eaten as a substitute of meat or eggs, like a food of high nutritional value (protein tends to be present in an easily digested form and on a dry weight basis a mushroom normally ranges between 20 and 40% protein which is better than many legume sources like soybeans and peanuts, and protein-yielding vegetable foods), source of dietary fiber and other valuable nutrientsthat has allowed it to have a high level of commercialization (Dundar, 2009). The Saprophyte mushroom is considered as a highly nutritious food for human consumption; besides its pleasant flavor and texture, among the volatile compounds that constitute edible mushroom flavor, 1-octen-3-ol is considered to be the major contributor and has been described as having a mushroom-like flavor (Cohen, 2002).

But, low yield level, inconsistency in flush appearance, texture, color and taste affects its adoption as the favorite mushroom for cultivation (Gupta, 2010). However, it has been cultivated in different substrates with higher profitability than any other mushroom.It can make use of the largest variety of waste substrates with its fast mycelial growth and its multilateral enzyme system that biodegrades nearly all types of available wastes. Itscultivation technique is spreading nowadays through out the world (Kumari, 2008).

The genus *Pleurotus*(oyster mushroom) comprises some the most popular edible mushrooms (fruiting bodies) due to their favourable organoleptic and medicinal properties, vigorous growth and undemanding cultivation conditions. It is grown under natural conditions, on living trees as a parasite or dead woody branches of trees as a saprophyte and primary decomposer (Bhatti et al., 2007). For many reasons the fungi of the *Pleurotus* genus have been intensively studied in many different parts of the world. The first attempts to grow *Pleurotus ostreatus* for human consumption were made during the First World War in Germany (Eger, 1976).

With the exception of traditional edible mushroom cultivation, large-scale practical implementation of the biotechnological potential, based on the unique properties and enzymatic systems of *Pleurotus* spp, has not yet been commercially developed. The areas covered here could be of enormous benefit to the medicinal, environmental and food industries (Cohen, 2002).

Various *Pleurotus* species have been shown to possess a number of medicinal properties, such as antitumour, immunomodulatory, antigenotoxic, antioxidant, anti-inflammatory, hypocholesterolaemic, antihypertensive, antiplatelet aggregating, antihyperglycaemic, antimicrobial and antiviral activities. These therapeutic activities are exhibited by extracts or isolated compounds from *Pleurotus* spp. fermentation broth, mycelia and fruiting bodies. *Pleurotus*spp can also be cultivated in submerged liquid cultures for the production of fungal protein, or as a source of spawn or flavoring agents (Gregori, 2008, Cohen, 2002, Guzmán, 1993, Stamets, 1984).

Pleurotus species are white-rot fungi known to degrade the lignin polymer present in lignocellulosic biomass, while utilizing free sugars and cohydrolyzed oligosaccharides for growth of their fruiting bodies. Substrate processed by fungal conditioning have several advantages that include: forgoing significant particle size reduction, improving susceptibility of cellulosic and hemicellulosic polysaccharides to subsequent hydrolysisand potential to reduce severity of the thermochemical pretreatment step that could save substantial amounts of energy and chemicals (Balan et al, 2008).

P. ostreatus or oyster mushroom is interesting for other biotechnological and environmental applications such as the upgrade of agricultural wastes for animal feed, paper pulp bleaching, treatment of a pulp and paper mill and agroindustrial farm effluent, biodegradation of organopollutants, xenobiotics and industrial contaminants or bioremediation, degradation of industrial dyes, phenols and PAHs, H_2O_2-generating enzymes, cosmetics, and the pharmaceutical industry. These different applications have fueled research on specific aspects of *Pleurotus* biochemistry and molecular biology (Rodrıguez Perez S. 2008, Kumari, 2008, Ragunathan 2004, Cohen 2002, Larraya, 2000).

SUBSTRATES FOR PLEUROTUS OSTREATUS CULTIVATION

Mushroom cultivation is a direct utilization of their ecological role in the bioconversion of solid wastes generated from industry and agriculture into edible biomass as an alternative to other methods of disposal or managing organic wastes due to a large fraction of the carbon fixed by photosynthesis of plants being deposited as lignocelluloses.

To achieve this, mushroom hyphae (*i.e.*, mycelium) produce a wide range of extracellular enzymes capable of degrading complex organic material. In general, there are three major fundamental technologies involved: 1) Spawn technology, 2) Mushroom production technology, and 3) Processing technology. A variety of methods and techniques have been developed and described in detail for each technology:

1) Spawn technology includes the isolation of strains from wild mushrooms growing in nature. When a selected strain is available, spawn preparation is carried out using cereal grains (*e.g.*, wheat, rye, millet, rice, sorghum) or other organic substrates (*e.g.*, coffee pulp, straw, cotton waste, sawdust) sterilized in glass jars or polypropylene plastic bags. These organic materials are inoculated, and incubated to be completely colonized by the mushroom mycelium and then used as spawn.
2) Mushroom production technology starts with construction of the farm, according to local environmental conditions and species requirements. Suitable substrates are then prepared using organic raw materials, easily available at low costs in each region. These substrates, selective for mushroom growing, are spawned and incubated in the farm. Production of fruit bodies varies according to each species, spawn quality, substrate quality and environmental conditions.
3) Processing technology is fundamental for commercial production and requires specific facilities. After harvesting, mushrooms are normally cooled down to retard

> fruit-body metabolism, and then sent to the fresh market. Alternative methods are available, such as cooling, vacuum cooling, cooling with positive ventilation, and ice-bank cooling with positive ventilation. Further processing, *e.g.* canning, drying, or irradiation, can also be carried out depending on marketing strategies (Martínez-Carrera, 2002, 2000).

Previous research has shown great potential for using some lignocellulosic materials as raw material for the production of *Pleurotus ostreatus*. Lignocellulosic materials, agro-waste or biomass are the world's most widely available low-cost renewable resources to be considered for production of organic fuels, food, chemicals, and new-generation materials of economic and social interest (Dundar, 2009, Kumari, 2008).

Mane et al. (2007), Mandeel et al. (2005), Ragunathan 2004, and Poppe (2004, 1995) listed the results of a worldwide survey on agro-forestry waste that can be used as a substrate in the cultivation of oyster mushrooms. These wastes are produced in big volumes during production of agricultural products every year causing lots of environmental problems in many countries, and only a very small part of these agro-wastes has been properly converted into useful or high-value products. Typically 80–90% of the total biomass of agricultural production is discarded as waste (Chang, 1999).

The substrate used depends on its availability in a specific region. To facilitate growth on such lignocellulosic substrates, white rot-fungi secrete different types of oxidative enzymes for lignin degradation they produce lignocellulase enzymes, mainly laccase (LAC as Broad role in lignin biodegradation) and Mn-peroxidase (MnP), which convert these lignocellulosic residues into food (Melo de Carvalho, 2010, Hassan etal. 2010, Rashad, 2009, Fan et al. 2008, 2006, 2000, Šnajdr, 2008, Norouzi, 2008, Mane, 2007, Mora, 2007, Mandeel, 2005, Royse et al., 2004, Moncaio Moda, 2005, Contreras, 2004; Sánchez, 2004, Job, 2004, Howard, 2003,Escalona *et al.*, 2001, Velazquez-Cedeño, 2000, Guzmán, 1993). After harvest, mushrooms offer the possibility ofharvestingspent mushroom substrate (SMS) enzymes for biotechnological applications as an extra benefit from mushroom cultivation (Rühl, 2008).

This type of mushroom, when in a normal cultivation system, gets its nourishment degrading organic material that comes from lignocellulosic byproducts. Several research approaches are being carried out in order to assess the possibility of increasing value from alternative and available feedstocks (consisting of different sizes, shapes, forms, and densities like straw, stalks, sticks, leaves, haulms, fibrous materials, roots, branches, and twigs) in fields and factories, like agroforestry by-products (generated during wood and industrial processing), bamboo leaves, weeds, various grasses and reed stems, wheat straw, cassava stalks, pineapple leaves, rice straw, pangola grass, groundnut haulms, sorghum stem, alfalfa, horse chestnut, peanut shells , soy straw, hull/meal, agave wastes, corn straw, corncob, paddy straw, barley straw, bean straw, maize leaves and stover, coffee husk, coffee dregs, coffee pulp, coffee spent-ground, cottonseed hulls, coconut shell, grape stalks, crushed sugar-cane trash and bagasse, paddy, banana tree straw, oilseed rape straw, lemon pulp, fruit waste, orange residue,wood chip and sawdust on different formulations. Some of these residues represent an abundant, cheap and readily available source of raw-material to be converted into edible mushrooms under favorable environments (temperature, relative humidity, luminosity and supplements such as wheat bran, rice and soy).

Rice, barley and wheat straw (cereal straws) appeared to be the best substrate for *P. ostreatus* mushroom cultivation (maximum protein, fat, ash, P, K and Na content) when compared to banana leaves, maize stover, corn husks, rice husks and elephant grass, etc. (Patil et al., 2010, Philippoussis, 2009,2001, Bhatti, 2007, Bonatti, 2004,Obodai, 2003, Ragunathan 1996); but recently, farmers have been searching for an alternative substrate that may be more readily available and more cost effective, or that may provide higher yields and better mushroom quality (Rizki, 2010; Aguilar, 2010, Pant et al. 2006, Mora 2007, Moncaio 2005, Sánchez, 2004, Pandey, 2000), Ortega (1992) concluded that traditional substrates could be replaced by sugarcane bagasse and trash as substrates due to their abundant availability, chemical composition, and they are the major by-product of the sugar cane industry.

Philippoussis, (2009), Taurachand, (2004), Ruegger et al. (2001), Fan et al. (2008) Klibansky et al. (1993), Ragunathan (1996) and Ortega et al. (1993) have identified sugar cane agrowastes as a feasible substrate for *Pleurotus* spp. production with yields and biological efficiencies comparable, and to some extent better, than others reported with conventional lignocellulosic residues such as cereal straw. However, there are not official numbers about production in sugarcane producing countries (Mora, 2007, Moncaio, 2005, Hernandez, *et al*; 2003 and Grodzinskaya et al; 2002).

SUGARCANE BY-PRODUCTS

Worldwide a great number of investigations have confirmed sugarcane's wide adaptability and its suitability for developing countries that have vast areas under tropical and sub-tropical zones and produce crop residues in abundance (Kumari, 2008).

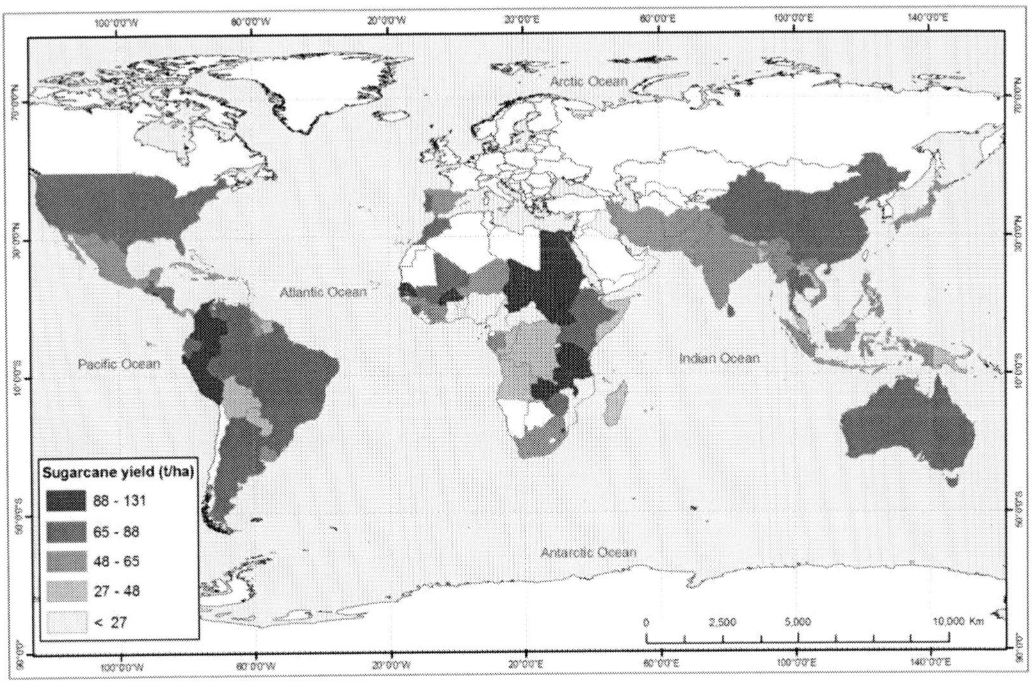

FAOSTAT, 2010.

Figure 1. Sugarcane producers and productivity.

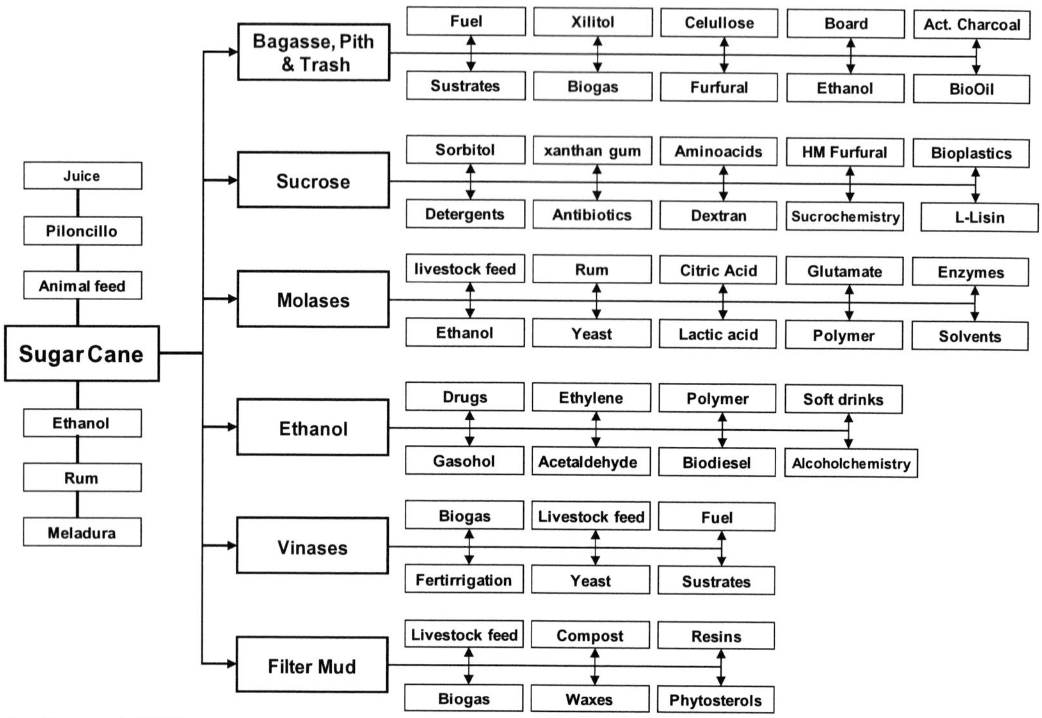

Aguilar et al. 2010.

Figure 2. Sugar, co-products and by-products on sugarcane agribusiness diversification.

Sugarcane (*Saccharum officinarum*) as a renewable feedstock is a C4 plant, is worldwide cultivated in 130 countries in 23,727,751 ha with a production of 1,682,577.8 MT. The top ten sugarcane producers are: Brazil, India, China, Thailand, Pakistan, Mexico, Colombia, Australia, Argentina and United States. Approximately 79% of total sugar production is made from cane grown primarily in the tropical and sub-tropical zones of the southern hemisphere with an average productivity or yield of 70 TM/ha with individual productivity by country (Figure 1).

The sugarcane cycle is semi-perennial and begins with the planting of a stem cutting that grows for about 12 months (year sugarcane) or 18 months (year-and-half sugarcane). After the first harvest, the ratoons are harvested annually for a period of about 5 to 7 years or more. Successive harvests lead to a gradual yield loss until the crop is no longer economically profitable. At this point, the cycle is interrupted, and the area is renovated with the planting of new stem cuttings. Should the ripe sugarcane not be harvested, it will keep on growing to be harvested in the following year of crop season (zafra).

In sugarcane production the whole parts of the plants are used. The average composition of sugarcane is 65-75% water, 11-18% sugars, 8-14%fibers and 12-23% soluble solids. The cane basically consists of juice and fiber. After extracting the juice, the material – called bagasse – is obtained, consisting of fiber (46%) (Cellulose, hemicelluloses), water (50%), and dissolved solids (4%) (ashes). The polyphenolic fraction, called lignin, is a complex structure formed by the polymerization of aromatic alcohols that, when combined with the hemicellulose, covers the cellulose matrix, conferring resistance to enzymatic and chemical degradation (Vargas Betancur, 2010). The amount obtained ranges from 240 kg to 280 kg of bagasse (fiber and pith) per ton of cane and the sugar (sucrose) inside them represents one of the losses of the process. Currently, the available surplus bagasse for hydrolysis or other purposes varies from 7 to 10% of the total bagasse. Sugarcane trash and bagasse have great potential for being used to improve and to diversify the food and energy matrix of the sugarcane regions worldwide (VazRossell, 2006, Sun, 2004). However, these values differ with crop, plant part, season, soil moisture as well as other factors that affect plant growth. Additionally, with the increase in cane production in response to the growth of global ethanol demand more bagasse and trash will be available (Vargas Betancur, 2010).

On this concept, the sugarcane diversification has been questioned in several major reports since the 1970′s as a complement and profitable future viability of the sugar industry. Agroindustrial diversification of the sugar industry provides substantial increases in regional sustainable development and income while at the same time reduces the vulnerability of the industry to change, bringing flexibility to future business strategies increasing the resilience of a sugar industry facing decreasing sugar prices. The structural and chemical composition of sugarcane, coproducts and byproducts (Figure 2) makes it particularly appealing for transformation into valuable products through chemical or biotechnical processes in industrial processing with a high aggregate value of interest to the market.

Sugarcane field residues or sugarcane trash are materials that remain after cutting, harvesting and processing sugarcane (140 Kg of trash, tops and green and dry leaves per ton of cane), to obtain sugar and ethanol. Cane tops residues are generally divided into two types, namely: green tops, representing 80-90% of the waste; and trash, representing 10-20% of residues. According to Rípoli *et al.*, (2000), green leaves, dry leaves and tops left unburned in the field have an average moisture content of 50%. In this way, the volume of trash remaining in fields after green harvesting would be considered as a function of the amount of available trash prior to cane harvesting and the harvesting process itself.

The dry leaves and stems (trash) that are left over are spread over the cane roots to protect them during the winter season. It is also customary practice for a significant portion of the dried trash to be used in rural stoves or for household heating in rural communities, while the remainder is burnt in the field. While there is little economic cost to the farmer from the burning of these residues in the field, the impacts of open burning of sugarcane stalks and biomass is an important contributor to black smog and air pollution (Ashmawy, 2007).

Variability in trash yield is very high due to several factors: varieties, production levels, crop age, management variability, differences in harvesting system, and operational efficiency. In this way, the quantity of trash remaining in fields after green harvesting would be considered as a function of the amount of available trash prior to cane harvesting and the harvesting process itself.

Green cane harvesting is a widespread practice in many countries and the sugar industry facing changes and progressively moving from the burning of sugarcane plantations (prior to harvest and after harvest) towards green cane management and trash can play an important role in improving soil conditions (such as organic matter and moisture content), energy production and other uses, such as animal feed and mushroom production. Eliminating cane burning in management systems entails harvesting green cane mechanically, developing and implementing technological alternatives for vegetal residues utilization (Romero et al. 2009);that with the increment of cane yield/ha worldwide, the potential trash and bagasse available per ha increased, thus the potential amount of dry trash and bagasse increased linearly and significantly with cane yield and diversification process.

However, regardless of the great production of sugarcanebyproducts(bagasse, tops, dry and green leaves) they are being burned mainly during harvest as partial fuel in sugar mills in developing countriesand can be converted into value-added products such as edible mushrooms (Solid-state Fermentation cultivation, SSF) for a human diet (Mandeel, 2005), (Moncaio, 2005) and (Souza Dias, 2003).

An estimation of potential sugarcane residues available to be used as substrates are very important in order to define agroindustrial management strategies that ensure sustainable development of both agriculture and food. In addition, for a substrate to be used for commercial purposes, the nutrient composition of the substrate is one of the factors limiting the saprobiotic colonization of cultivated mushrooms, and particularly the fruiting of Pleurotus spp (Naraian et al. 2009; Oei 2005).

SUGARCANE BIOMASS COMPOSITION

The sugarcane bagasse has been extensively studied, but only focused on its use in the energy industry, pulp and paper, and other processes of energy generation. Moreover, to our knowledge, the literature does not provide enough information about physical, chemical and other properties for sugarcane trash (Rossell, 2006). Likewise, there are some reports in theliterature about its composition (Tables 1 to 6 and Figures 3 and 4).

Table 1. Chemical composition (dry weight) of lignocellulosic sugarcane by-products

Components (%)	**Trash (straw)**	**Depithed bagasse**	**Bagasse from mills**	**Sugarcane Biomass**
Extractives ETOH-Toluene	3.4	3.8	4.6	4.2
Extractives ETOH	2.7	2.5	7.5	5.1
Extractives H_2O	3.7	4.2	1.3	2.7
Lignin	24.5	20.7	20.3	20.5
Holocellulose	78.7	76	74.7	75.4
α cellulose	47	46.5	45.7	46.1
Pentoses	25.6	25.2	22.4	23.9

Aguilar et al. 2010.

Table 2. Chemical composition (dry weight) of sugarcane biomass

Components (bagasse)	(%)	Components (%)	bagasse	Trash
Glucose (hexose)	42.9	C	45.2	46.34
Xylose (pentose)	23.1	H	5.8	4.67
Arabinose (hexose)	3.3	O	43.3	32.61
Galactose (hexose)	1.2	S	0.03	0.1
Reducing sugars	1.85	N	0.28	0.35

Area, 2008; VazRossell, 2006; Woytiuk, 2006; Sanjuan, 2001.

Table 3. Gross heat and Net (useful) heat combustion from sugarcane biomass

Material	Gross heat (kJ kg-1)	Net heat (kJ kg-1)
sugar cane trash (straw)	18,870	13,551
sugar cane green leaves	18,439	4,930
sugar cane bagasse (49% moisture)	18,322	8,638
sugar cane tops	18,138	3,327

Rípoli et al. 2000, De Oliveira 1999.

Table 4. Bromatologic Analysis of Lignocellulosic Sugarcane byproducts

Components (%)	Moisture	Ashes	Fat	Crude fiber	Protein	Carbohydrates
DepithedBagasse	11.49	13.33	0.98	34.11	6.69	33.40
Sugarcane Trash (straw)	11.58	2.40	0.28	45.00	2.69	38.05
Bagasse from Mills	25.10	7.58	1.24	38.51	1.75	25.82
Sugarcane Biomas	19.46	10.46	1.11	36.31	4.22	29.63

Aguilar et al. 2010.

Table 5. Ashes composition from sugarcane biomass

Componet (%)	$S\acute{I}O_2$	AI_2O_3	Fe_2O_3	TiO_2	CaO	MgO	SO_3	K_2O	Na_2O	P_2O_5
Bagasse	72.74	5.26	3.92	0.32	7.99	2.78	0.13	3.47	0.84	1.59
Trash	59.06	4.75	3.18	0.34	19.59	2.25	1.37	4.75	0.73	1.67

Romero, 2009, Martirena etal., 2000.

Table 6. Morphological composition of sugarcane bagasse

Componet	Dimensional characteristics
Fiber Length	1.1088 mm.
Fiber Width	18.873 μm
Lumen Width	10.85 μm.
Wall Thickness	8.023 μm
Length Vessel	1.23718 mm
Cell diameter	0.10793 mm

Aguilar, 2006.

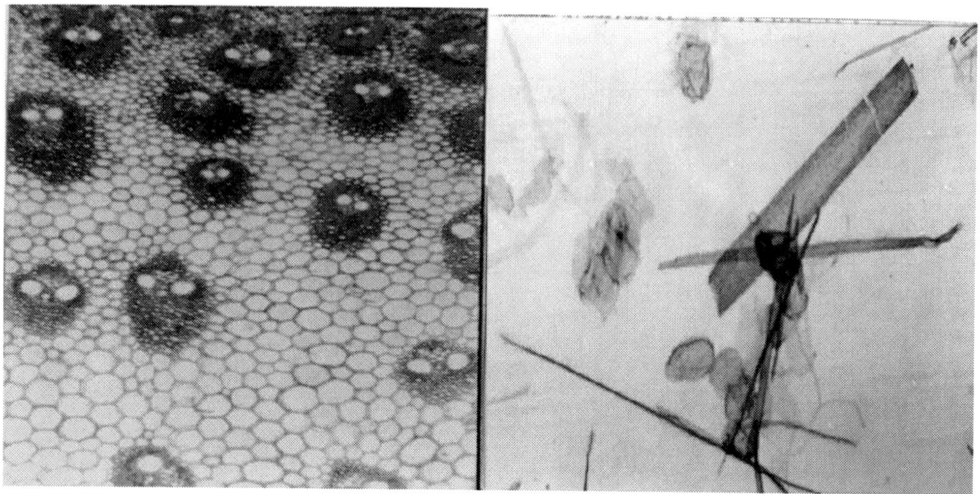

Aguilar, 2006.

Figure 3. Sugarcane and Sugarcane bagasse.

Saska, 2006.

Figure 4. Prepared sugarcane biomass (with a centimeter scale). Left: trash (leaf matter). Right: bagasse fiber and pith.

The composition of sugarcane straw (tops, dry and green leaves) is somewhat similar to that of bagasse in terms of cellulose and hemicellulose. However, the lignin content in the straw is approximately 30% lower than that of bagasse or its component fractions.Although no data concerning the shape and size of straw are available, visual inspection shows that it is highly heterogeneous, since it comes from different fractions of the sugarcane plant and was not prepared beforehand (VazRossell, 2006).

These materials, according to their chemical and bromatological composition, are ideal for *Pleurotus ostreatus* production and development. This is because these mushrooms are common primary degraders in wood or forest byproducts, and are able to metabolize lignin and cellulose to obtain nutrients and energy.

PLEUROTUS OSTREATUS CULTIVATION IN SUGAR CANE BIOMASS

The whole process of mushroom production (mycelial growth, spawn running, primordial initiation, fruit body yield and its biological efficiency) requires continued and careful attendance or good management practices. Mushroom cultivation involves several different operations each of which must be performed properly if the enterprise is to be successful.

For some substrates, such as sugarcane biomass, the materials are generally not composted prior to mushroom inoculation and they are pasteurized to diminish contamination and increase selectivity for the oyster mushroom.

Todd et al. (2003) concluded that the organic materials must be blended, pasteurized and inoculated with oyster mushroom spawn (*Pleurotus ostreatus*). The substrate is placed in plastic or polypropylene bags punched with holes and placed in a mushroom incubator room. When the bags are fully colonized by the mushroom mycelium they are transferred to a grow room for fruiting and harvest. After harvest the remaining substrate has the potential to be used as a high quality animal feed for livestock. In the process of mushroom production the vegetative forms of fungi colonize the straw and spent grains and produce essential amino acids. The spent mushroom substrate is converted to enriched compost.

Traditionally, oyster mushrooms are cultivated using sterile substrates packed in a polythene bag with spawns distributed at regular intervals followed by storing the bag in a humid environment (>80%) at 25°C for a period of 35–40 days (Balan et al, 2008).

Based on information published in the literature, mushroom culture on sugarcane substrates involves several different operations divided into 5 stages, each of which must be carefully performed:

1) Preliminary substrates chemical and bromatological analysis
2) *Pleurotus ostreatus* spawn production
3) Mushroom cultivation
4) Mushroom dehydration
5) Final substrates and fructification bodies bromatological analysis

The yield of the *P. ostreatus* on the different substrates is determined by the number and size of the fruit bodies produced. The fruit bodies are harvested at the end of the crop cycle and number of fruit bodies, diameter, fresh and dry weights of fruit bodies are measured using conventional techniques. A maximum yield value in substrates confirms that it can be easily used in the production of mushrooms (Kumari, 2008).

All these stages are crucial but the last important factor for mushroom growing is providing an appropriate physicochemical environment (temperature, humidity, light and ventilation) both for vegetative and reproductive growth for the production of mushrooms with better flavour, appearance, texture, and nutritional qualities at low cost (Imtiaj, 2008, Tisdale et al. 2006).The main difficulty for industrial *Pleurotus* sp. cultivation on sugarcane biomass (derivedfromresidualsucrose intheir cellular structure)is the substrate disinfestation stage, performed by pasteurization or sterilization (MoncaioModa, 2005) because the rate of production of oyster mushroom from mycelia stage to production is significantly influenced by the carbohydrate source (type of substrate), nutrient composition of the substrate, nitrogen sources, availability of light, initial pH of substrate, incubation period and the control of contamination (Ibekwe, 2008).

Figures 5. Pleurotus ostreatus production in sugar cane biomass in Cordoba Veracruz Mexico.

On sugarcane biomass, each substrate is ground in order to reduce its particles' size, soaked in water for 24 hours to moisten and then are pasteurized at 85° C for 45 minutes by immersion in hot water. Each pasteurized sugarcane substrate must be cooled, andthe excessive water removed to get 65-75% moisture level and homogeneously packaged in 40 x 60 cm transparent plastic (polyethylene) bags. These bags must have 10 cm substrate layers to be inoculated by hand with the spawn (*Pleurotus ostreatus*) in a 5% w/w proportion of fresh substrate weight.

The treatments, derived from different combinations of materials (bagasse from mills, fiber or bagasse depithed, trash and mixtures) thus prepared are taken to the growing-room under complete darkness, where they will remain during the incubation and harvest periods. The average temperature mustbe 25 °C and relative humidity > 60% during 60-90 days after which the colonized substrates will be harvested. The humidity of bags must be accomplished by spraying of water on them twice a day (Figure 5) (Aguilar et al, 2010).

After harvest, remnants of fruit bodies on substrate should be removed from the bags to prevent contamination (MoncaioModa 2005, Taurachand, 2004).

YIELD OF OYSTER MUSHROOM

The main parameters used to evaluate mushroom yield is called biological efficiency (BE) and yield. It mainly depends on the characteristics of the material and the enviroment in which the growth process occurs. It is defined as the ratio of weight (kg) of fresh mushrooms harvested to dry weight (kg) of the substrate (Norouzi, 2008):

Yield = kg fresh mushrooms harvested at maturity/ kg moist substrate weight.

(BE) [weight fresh mushrooms harvested/dry weight substrate at spawning] 100 per each bag or experimental unit and yield/ha/year.

On sugarcane biomass, the biggest production reported in the literature is in trash as a substrate and mixed with bagasse (50-50) (Table 7).

Table 7. Biological efficiency (BE)

Substrates	**Biological efficiency (%)(kg 100kg^{-1} substrate)**
Depithed Bagasse	66.64
Sugar Cane Trash	106.64
Bagasse from mills	70.09
Mixture	103.50

Aguilar, et al, 2010.

Apati et al., (2010) said that the mushrooms of the *Pleurotus* genus are more delicate and sensitive than others and they start deteriorating immediately within one day after the harvest. Once deteriorated, these fruiting bodies can cause severe gastrointestinal discomfort. Under ideal climatic conditions, shelf life of these mushrooms is about 10 days, their quality being affected predominantly by storage temperature. The shelf life can be reduced from 9 days at 2 °C to 3 days at 18 °C. Therefore, cooling the fresh mushrooms can be an alternative regarding their distribution and sale, thus increasing their shelf life. Dehydration is a classic method of food conservation, based on the principle that the reduction of the water activity of the product must be conducted until defined levels that guarantee the microbiological and physicochemical stability. The drying temperature is a very important process variable in mushroom drying. The best temperature for *P. ostreatus* fruiting bodies' drying process is 40 ºC, taking into consideration the best rehydration of the fruiting bodies dried at this temperature. The rehydration could be done in water at room temperature during 30 minutes, or 3 hours at 35 °C average using direct solar radiation (Sun drying (SD) mushrooms) (Aguilar et al, 2010).This processing technology also permits increasing the value added to mushrooms, to standardize mushroom quality, to highlight certain culinary properties of mushrooms, and to develop marketing strategies at the local or national level (Vetter, 2003).

When comparing the obtained results from sugarcane biomass and cereal straw with the protein content for every sample they arevery similar and equivalent with other types of food (Table 8 and Figure 6); the protein content of edible mushrooms can be considered as their main nutritional attribute (Martínez-Carrera, 2000). Besides, there was a variation of the carbohydrate content and ash due to the fact that these were consumed by the mushroom during the fermentation process and different substrates affected the nutritional composition of the mushroom; also the nature of protein in the substrate influences the protein content of the fruiting bodies. These results are in accordance with the values reported earlier for several substrates (Patil et al. 2010, Philippoussis, 2009).

Table 8. BromatologicAnalysis of Edible mushroom *Pleurotus ostreatus* (%dry weight)

	Components (%)	Ashes	Fat	Crude fiber	Protein	Carbohydrates
Substrates	Cereal straw	8.3	2.2	12.08	20.56	40.96
	Depithed Bagasse	6.28	1.48	12.83	17.18	44.78
	Sugarcane Trash	7.82	1.79	12.39	19.68	41.2
	Bagasse from Mills	5.81	1.94	12.89	15.81	45.14
	Mixture (RAC-Bagasse)	6.64	1.85	12.65	21.25	40.41

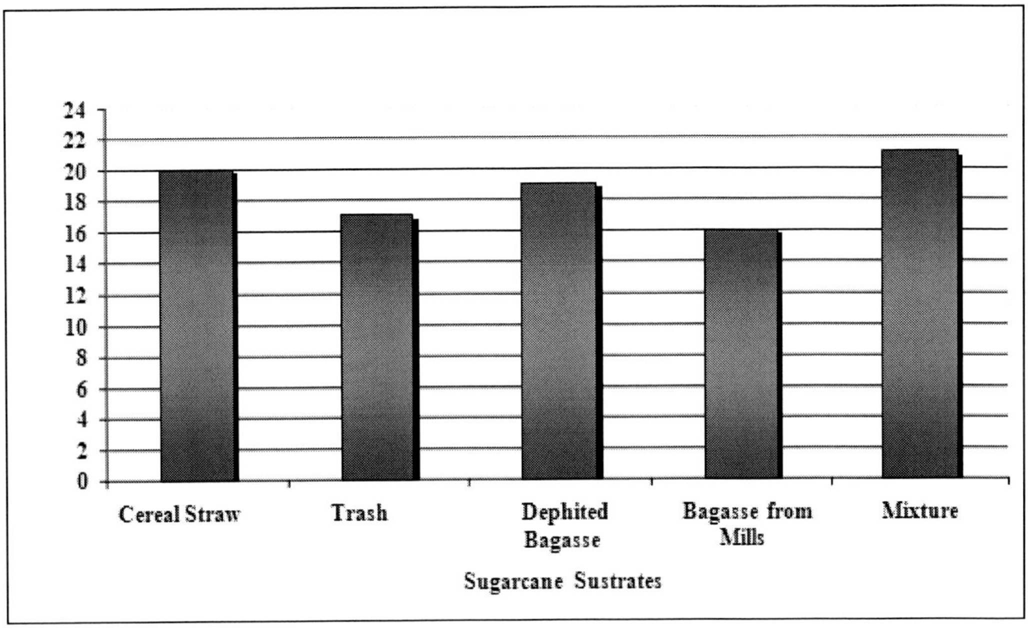

Figure 6. Protein from mushroom *Pleurotus ostreatus*(%dry weight).

The treatment of sugarcane lignocellulosic biomass with offers a promising means to convert low quality biomass into a high protein food. They are viable substrates for mushroom cultivation alone or in combinations and increase mushroom yield for *Pleurotus ostreatus* for massive production in sugar cane farms, providing better levels of inclusion in human diet and a better usage of its remnants, due to their easy digestibility in cattle diet.

Spent Mushroom Substrate

Mushroom culture is a biotechnological process that recycles ligninocellulosic wastes, since mushrooms are food for human consumption. However, mushroom production generates an enormous amount of used "spent substrate", which might also be a source of environmental contamination. Currently, several uses for spent substrate are being evaluated and some of them have already been established (Sánchez, 2004).

For each metric tonne of mushrooms produced, at least an equivalent amount of spent mushroom substrate (also called spent mushroom compost) is generated, and this needs to be disposed. Because of its high mineral contents, phosphate content, and high porosity, it is a good soil conditioner and soil fertilizer for stimulating seed germination. This spent substrate can be a good source of lignocellulolytic enzymes for biodegrading various xenobiotics and persistent pollutants; bioremediation process including pentachlorophenol, polyaromatic hydrocarbons, and azo dyes also has a good pH buffering capacity (Siu-Wai Chiu, 2000).

Some authors (Fan et al., 2000) suggest that straw spent mushroom compost (SMC) degraded by enzymes during mushroom production can be more easily digested by ruminants. It contains more free sugars, more protein, less cellulose and lignin, with an increased content of ash compared with beginning material, and is a potentially useful source of S, K, Ca and

Mg for plants. It contains i.e. 1.3–1.4% sulphur (S), 0.7–2.1% potassium (K), 5.9–16.0% calcium (Ca) and 0.2–0.5% magnesium (Mg) (Stewart, 1998). Advantages of such a process include recycling of the pulp, the rapid and easy production of a food rich in protein and low in fat and an end-product that can be used as an organic fertilizer or as cattle feed (Rodriguez Perez, et al. 2008).

Biochemical changes take place in the substrates because of the mushroom growth (Table 9). A decrease in cellulose, hemicellulose, crude fibre, carbohydrate lignin and tannin content is observed and also improves the straw quality;while an increase in protein, ash and mineral content in spent straw is recorded making it ideal for animal feed or composted, either aerobic composting or vermicomposting, to produce an organic fertilizer or soil conditioner for crop soils (Patil et al. 2010, Ortega, 1992).

Table 9. Spent Lignocellulosic Substrates Sugarcane (% dry weight)

Components (%)	**Moisture**	**Ashes**	**Fat**	**Crude fiber**	**Protein**	**Carbohydrates**
Depithed Bagasse	10.03	22.46	0.89	25.92	3.38	37.32
Sugarcane Trash	10.59	3.28	0.49	40.70	4.19	40.75
Bagasse from Mills	10.15	14.63	0.73	27.07	4.19	43.22
Mixture (RAC-Bagasse)	9.97	22.06	0.99	29.98	4.75	32.25

Aguilar et al. 2010.

Sugarcane by-products can be considered practical and economically feasible due to their availability throughout the year at little or no cost in large quantities.To make the mushroom cultivation an environmentally friendly industry, the basic biology of mushrooms and the cultivation technology in sugar cane lignocellulosic by-products must be researched and developed. The mushroom inoculation on substrates increases biodegradation of the components in their organic matter (fiber, protein and carbohydrates) and it takes place at the very beginning of the inoculation itself, with the mushroom development and its incorporation of carbon main sources to its metabolism (Mandeel, 2005, Moncaio, 2005, Royse 2004, Murrieta, 2002 and Escalona *et al.*, 2001) where the change in the composition increased at the end of the colonization and continued with the development of the fructification bodies. However, these can be influenced as well by the substrate, type of stock that is used and the conditions of the crop in the different substrates when increasing or decreasing their values. The hemicelluloses have to present an increased degradation by the mushroom, due to the easiness of hydrolysis of the xiloses, followed by the cellulose and the lignin.

SOCIOECONOMIC ASPECTS AND SUSTAINABILITY FROM MUSHROOM PRODUCTION ON SUGARCANEBIOMASS

The technology of artificial cultivation of mushrooms is a somewhat recent innovation; incorporation of non-conventional crops in existing agricultural systems can help in improving the social as well as economic status of small farmers.

Table 10. Indicators for sustainability of the model for rural production of edible mushrooms using rustic technologies

Category	Indicator	Critical value	Factor(s) to evaluate
Environmental	Biological efficiency	> 32%	Spawn, yields, substrates
	Degradation rate	> 31%	Spent substrates, potential organic fertilizer
	Contamination rate	< 20%	Raw materials, growing systems and technology, spawn, environmental conditions, hygiene, labour skills
	Energy-use efficiency	< 12%	Energy consumption
	Water-use efficiency	> 28 L/kg	Water consumption to produce mushrooms
	Temperature	15 °C 33 °C	Minimum temperature Maximum temperature
	Relative humidity	> 70%	Environmental moisture
Economic	Cost-benefit ratio	> 1.0	Gross incomes, production costs, profits
Social	Mushroom consumption	0.914 kg	*Per capita* (Minimum mushroom consumption per year required to maintain the mushroom farm
		> 4.100 kg	*Per household* (Minimum mushroom consumption per year required to maintain the mushroom farm)
	Labour efficiency	< 3	Number of workers in the farm
	Market	As market trends depend on social, economic, and political circumstances the market can be local, national or international;	

Martinez-Cabrera *et al.* 2002, 2000.

Adopting the technology is a simpler way to invest in a small scale; the mushroom cultivation may reduce poverty and improve the life style of many sugar cane farmers having the following elements: 1) Characteristics and life circumstances from family members; 2) sugar cane and others crop plants; 3) Livestock; 4) Gathering; 5) The making of produces; 6) Trading; and 7) Labouring 8) Distance to the market;considering the complex social, economic, and ecological factors involved, it is fundamental to develop sustainable strategies for mushroom production (Imtiaj, 2008, Pellicer-González, 2002). Mushroom cultivation is highly labour intensive, a short duration crop, land saving and can be welcomed by the poor farmers (Shah et al., 2004).

Four aspects for profibality of crop yield must be evaluated: (1) economic yield (fresh weight of mushrooms in first three flushes), (2) biological efficiency (BE), (3) flush number, and (4) crop period (sum of incubation and fruiting periods).

Main production and operation costs from commercial and rural production using rustic technologies in the sugar cane regions in developing countries are: salaries, raw materials (Spawn, Plastic bags and Miscellaneous) and energy, travelling expenses, maintenance, and regional marketing (Table 10).

Mushroom farming is both a science and an art. The science, in its broad form, will come through training workshops/courses. The art will come through practical involvement and experience, and will have to be modified in accordance with the prevailing conditions on site. A comprehensive training programme for both researchers and mushroom growers will need to be formulated.

According to Oteng et al. (2008), the objectives of growing mushrooms should ideally cover: (a) improvement of living conditions of farmers by increasing production and productivity, (b) promoting self-reliance among members in respect toplanning, production and marketing, (c) finding solutions collectively to constraints which smallholders cannot personally achieve, (d) assisting members in obtaining modern technologies and products which are capable of increasing agricultural production and materials necessary for their livelihoods easily at fair prices and (e) improvement and production of value addition on mushroom chains.

The sustainable model for rural production of edible mushrooms represents a strategy that allows large-scale, small-scale, and domestic cultivation to promote regional development (Martínez-Carrera *et al.*, 2002, 1998). However, mushroom exploitation is an economic concept to provide full-time occupation and primary source of income, as a part-time occupation and secondary income, to utilise existing facilities such as buildings or cold stores not currently in use, to maximise labour efficiency in a complementary business, to utilise more fully existing marketing and distribution systems. In developing cultivation methods in a sugar mill, this means modernization of the traditional methods,integration of mushroom cultivation and a profitable marketing scheme due to the existing marketing scheme, in rural production, which consists of four main agents thatare the mushroom producers in a farm, the village's middlepersons, the outsider middlepersons and the retail business sector including local markets and supermarkets (Yoshiro K., 1999).

In sugar mill mushroom production, the economic questions are: What finance is required? What knowledge and training are necessary? What problems may be encountered in growing, management wastes and marketing? What isthe overall production cost?

Joined to the previous questions, enough capital investment is necessary to develop an efficient technology for large-scale processing, in which surplus bagasse or trash must be stored in appropriate containers to be raw material in the mushroom industry (storage cost) (Szmidt, 1997).

The results obtained in several investigations contribute to expand the knowledge on this fungus, with respect to BE, mushroom quality and crop cycle time. Therefore, sugarcane biomass might be an interesting alternative for growth of *Pleurotus* spp in a sugar mill or farm. However, to make the mushroom cultivation an environmentally friendly industry, whichcould be a Profitable, Sustainable, and Environmentally Friendly Solid-state Fermentationthe basic biology of mushrooms and the cultivation technology and Mathematical Model for Growth in Solid-State Fermentation Systems must be researched and developed (Siu-Wai Chiu, 2000, Sarikaya, 1997).

The commercial applications of *Pleurotus ostreatus* require quantitative models for prediction of growth and substrate utilization. Some researchers have investigated the possibility of growing edible mushrooms in submerged fermentation in a bioreactor. On the other hand, the morphology of the organism itself can be modified by mechanical or chemical influences dependingon the culture conditions (Marquez-Rocha et al. 1999). A mushroom plant coupled with a biorefinery could help generate both food and fuel ultimately benefiting the rural economy (Balan et al., 2008). Advances in cultivation technology guarantee the control of product quality and consistency and permit modification as required meeting market demands (Siu-Wai Chiu, 2000).

Additionally, the sugar mill will have to consider whether to convert radically existing buildings from other uses or develop a specific mushroom production site (Growing rooms adapted cost), laboratory for spawn production (cost of mushroom spawn), and the cost of labor as the most significant investment. Other considerations include the method and scale of cultivation, management of growing environments (composting cost) and how the mushrooms will be marketed orientingtowards a significant increase in the regional or national consumption of processed mushrooms (fresh or dried) while the surplus bagasse prices or sugar cane trash are readily available (price had a negative impact on the financial viability of the project) for very high profitability.

REFERENCES

Aguilar-Rivera N., J. De Jesús- Merales (2010). Edible mushroom *Pleurotus ostreatus* production on cellulosic biomass of sugar cane, N., Sugar Tech 12(2):25-28.

Aguilar N.R., A. Castillo M., A. Herrera S., D. A. Rodriguez. (2010). Productive diversification from sugar cane lignocellulosic byproducts *Proceedings* of the *XXVII ISSCT*International Society of Sugar Cane Technologists ISSCT CONGRESS March 2010.

Aguilar R.N. 2006. Producción de celulosa y papel a partir de bagazo de caña de azúcar Revista Residuos Bilbao España XVI (93):70-80.

Ashmawy K.H. (2007).Socioeconomic and environmental aspects of women labor in the Egyptian agricultural sector: Case study of sugar crops", *American-Eurasian Journal of Agriculture and Environment Science*, 2(3): 255-260.

Apati G.P., S.A. Furlan and J.B. Laurindo. (2010). Drying and Rehydration of Oyster Mushroom.*Braz. Arch. Biol. Technol.* 53(4): 945-952.

Balan V, L. da Costa Sousa, P. S. Chundawat, R. Vismeh. (2008). Mushroom spent straw: a potential substrate for an ethanol-based Biorefinery. *J Ind Microbiol Biotechnol.* 35:293–301

Bhatti M.I., M.M. Jiskani, K. H. Wagan, M.A. Pathan and M.R. Magsi. (2007). growth, development and yield of oyster mushroom, *pleurotus ostreatus*(jacq. ex. fr.) kummer as affected by different spawn rates *Pak. J. Bot.*, 39(7): 2685-2692.

Bonatti, M., Karnopp, P., Soares, H.M., Furlan, S.A., (2004). Evaluation of *Pleurotus ostreatus*and *Pleurotus sajor-caju* nutritional characteristics when cultivated in different lignocellulosic wastes. *Food Chem.*, 88(3): 425-428.

Cohen R., L. Persky, Y. Hadar. (2002). Biotechnological applications and potential of wood-degrading mushrooms of the genus Pleurotus .*Appl Microbiol Biotechnol.* 58:582–594.

Contreras E. P., Sokolov M., Mejia G.and Sanchez J. E. (2004). Soaking of substrate in alkaline water as a pretreatment for the cultivation of Pleurotus ostreatus. *Journal of Horticultural Science & Biotechnology*, 79 (2) 234-240.

Chang, S.T., (2007). Mushroom cultivation using the "ZERI" principle: potential for application in Brazil, *Micologiaaplicada international*, 19(2): 33-34.

Chang, S. T. (1999) Global impact of edible and medicinal mushrooms on human welfare in the 21st century: Nongreen revolution. *Int. J. Med. Mushrooms*, 1, 1–7.

Croan S.C. 2000. Conversion of Wood Waste into Value-Added Products by Edible and Medicinal Pleurotus (Fr.) P. Karst. Species (Agaricaless. l., Basidiomycetes. *International Journal of Medicinal Mushrooms*, 2:73-80

De Oliveira M. W. (1999). decomposição e liberação de nutrientes da palhada de cana-de-açúcar em campo. Pesq. agropec. bras., Brasília, 34(12):2359-2362

Detroy, R.W. Hesseltine, C.W.(1978). Availability and utilization of agricultural and agro industrial wastes *Process biochemistry*. 13 (9):2-8

Dundar A. (2009). Effect of using different lignocellulosic wastes for cultivation of *Pleurotus ostreatus*(Jacq.) P. Kumm. On mushroom yield, chemical composition and nutritional value. *African Journal of Biotechnology*, 8 (4):662-666

Eger G. (1976).Pleurotusostreatus-Breeding Potential of a New Cultivated Mushroom Theoretical and Applied Genetics 47, 155-163 (1976)

Escalona l.; Ponce, A.I. Estrada, G. Solano, A. Mojena O. Cutido M. (2001). Inoculación de una mezcla de residuales azucareros con una cepa de *Pleurotus ostreatus*, Universidad y Ciencia, Tabasco México, 17 (33).

Fan L., C. R. Soccol and A. Pandey (2008). Mushroom Production. Current Developments in Solid-state Fermentation Part 2, 253-274.

Fan L. H. Pan, A. ThomazSoccol, A. Pandey and C. R. Soccol. (2006). Advances in Mushroom Research in the Last Decade.*Food Technol. Biotechnol.* 44 (3) 303–311

Fan L, Pandey A &Soccol CR, (2000), Solid-state cultivation - an efficient method to use toxic agro-industrial residues, *Journal of Basic Microbiology*, 40, 177-187.

FAOASTAT. 2009. Estadísticas de producción. http://faostat.fao.org/site/567/DesktopDefault.aspx?PageID=567#ancor

Guzmán, G., Mata, G., Salmones, D. and Soto- Velazco, C (1993). El cultivo de los hongos comestibles. Instituto Politécnico Nacional. México. 42-44, 1era edición.

Gregori, A. (2007).CultivationTechniques and Medicinal Properties of Pleurotus spp *Food Technol. Biotechnol.* 45 (3) 236–247.

Grodzínskaya A. A., Infante H.y Nickolai M. P. (2002). Cultivo de hongoscomestibles utilizando de sechosagrícolas e industriales. *Agronomía Tropical.* 52(4):427-447 pp.

Gupta B., B. P. Niranjan Reddy. (2010). Molecular characterization and mating type analysis of oyster mushroom (Pleurotus spp.) using single basidiospores for strain improvement. *World J MicrobiolBiotechnol.*

Hassan, F.R.H. GhadaM .Medany, S.D. Abou Hussein (2010).Cultivation of the King Oyster Mushroom (*Plerrotus eryngii*) in Egypt.*Australian Journal of Basic and Applied Sciences*, 4(1): 99-105

Hernandez D,. Sanchez J. E., Yamasaki K. (2003). A simple procedure for preparing substrate for Pleurotusostreatus cultivation *Bioresource Technology*, 90 145–150 pp. Elsevier Ltd.

Howard RL, Abotsi E, Jansen van Rensburg EL (2003) Lignocellulose biotechnology: issues of bioconversion and enzyme production. *African J Biotechnol.*2(12): 602–619.

Ibekwe, V.I., P.I. Azubuike, E.U. Ezeji and E.C. Chinakwe. 2008. Effects of Nutrient Sources and Environmental Factors on the Cultivation and Yield of Oyster Mushroom (*Pleurotus ostreatus*) *Pakistan Journal of Nutrition*, 7 (2): 349-351.

Imtiaj A, (2008). Economic viability of mushrooms cultivation to poverty reduction in Bangladesh. *Tropical and Subtropical Agroecosystems*, 8: 93 – 99.

Job, D. (2004). La utilización de la borra del café como sustrato de base para el cultivo de *Pleurotusostreatus* (Jacq.:Fr.) Kummer. Revista Iberoamericana de Micologia Bilbao España.195-197 pp.

Jwanny E. W., 1995. Solid-State Fermentation of Agricultural Wastes into Food Through Pieurotus Cultivation.*Applied Biochemistry and Biotechnology*, 50:71-78.

Klibansky, M. M., Mansur, M., Gutierrez, I. and González, L. (1993), Production of *Pleurotusostreatus* mushrooms on sugar cane agrowastes. *ActaBiotechnologica*, 13: 71–78.

Kumari D. V. Achal (2008). Effect of different substrates on the production and non-enzymatic antioxidant activity of Pleurotus ostreatus (Oyster mushroom). Life Science Journal, 5(3): 73-76.

Larraya,L.M. 2000. Genetic Linkage Map of the Edible Basidiomycete *Pleurotus ostreatus. applied and environmental microbiology*, 66(12): 5290–5300.

Lindequist U., T. H. J. Niedermeyer and W. Julich. 2005. The Pharmacological Potential of Mushrooms. *Evidence-Based Complementary and Alternative Medicine*, eCAM 2005;2(3)285–299.

Mandeel Q.A., A.A. Al-Laith and S.A. Mohamed (2005). Cultivation of oyster mushrooms (Pleurotus spp.) on various lignocellulosic wastes *World Journal of Microbiology & Biotechnology*, 21:601–607.

Mane V. P. P. S. Sopanrao, S. A. Ahmed, B. M. Mushtaq Vaseem. (2007). Bioconversion of low quality lignocellulosic agricultural waste into edible protein by *Pleurotussajor-caju* (Fr.) Singer.*Journal of Zhejiang University SCIENCE B.*8(10):745-751.

Martirena-Hernández J. F. S. Betancourt-Rodríguez, B. Middendorf, A. Rubio, L. Martinez, I.Machado-López, R. Gonzalez-López. (2000). *Pozzolanic properties of residues of sugar industries (first part)* Materiales de construcción, 50(260): 71-78.

Martínez-Carrera D. (2002). Current development of mushroom biotechnology in Latin America. *Micologia Aplicada internacional.* 14 (002): 61.74 pp.

Martínez-Carrera, D., A. Aguilar, W. Martínez, M. Bonilla, P. Morales & M. Sobal, (2000). Commercial production and marketing of edible mushrooms cultivated on coffee pulp in

Mexico. Chapter 45, pp. 471-488. *In*: Sera, T., C. Soccol, A. Pandey& S. Roussos (Eds.). *Coffee biotechnology and quality*.Kluwer Academic Publishers, Dordrecht, The Neterhlands.

Martínez-Carrera, D., A. Aguilar, W. Martínez, P. Morales, M. Sobal, M. Bonilla, A. Larqué-Saavedra. (1998). A sustainable model for rural production of edible mushrooms in Mexico.*Micol. Neotrop. Apl. 11*: 77-96

Márquez-Rocha F. J., G. K. Guillen, J. E. Sánchez and R. Vazquez-Duhalt. (1999). Growth characteristics of *Pleurotusostreatus*in bioreactors *Biotechnology Techniques* 13: 29–32,

Mayett Y., D. Martínez-Carrera, M. Sánchez, A. Macías, S. Mora, A. Estrada-Torres 2006. Consumption Trends of Edible Mushrooms in Developing Countries:The Case of Mexico. *Journal of International Food & Agribusiness Marketing*, 18(1/2): 151-76.

Mayett, Y., Martínez-Carrera, D., Sánchez, M., Macías, A., Mora, S. & Estrada, A., 2004. Consumption of edible mushrooms in developing countries: the case of Mexico. *In*: C. P. Romaine, C. B. Keil, D. L. Rinker & D. J. Royse (Eds.), *Science and cultivation of edible and medicinal fungi* (pp. 687-696). Penn State University Press, U.S.A.

Melo de Carvalho C.S. 2010. mushrooms of the *pleurotus*genus: a review of cultivation techniques. I*nterciencia*. 35 (3):177-182.

Moncaio M. E.; J. Horii; M. H. Fillet S. (2005). Edible mushroom *Pleurotussajor-caju*production on washed and supplemented sugarcane bagasse *Sci. Agric.*(Piracicaba, Braz.).62(2): 127-132.

Mora, V. M. & D. Martínez-Carrera. (2007). Investigaciones básicas, aplicadas y socioeconómicas sobre el cultivo de setas (*Pleurotus*) en México. Capítulo 1.1, 17 pp. *In*: *El Cultivo de Setas* Pleurotus spp. *en México*. J. E. Sánchez, D. Martínez-Carrera, G. Mata & H. Leal (Eds.). ECOSUR-CONACYT, México, D.F.

Murrieta H.D.M. (2002). Cambios en la producción de Lacasa por el hongo Pleurotusostreatus cultivado en pulpa de café. Foresta Veracruzana. 4 (1):47-52

Naraian R, Sahu K, Kumar S, Garg SK, Singh CS, Kanaujia RS (2009) Influence of different nitrogen rich supplements during cultivation of Pleurotusflorida on corn cob substrate. Environmentalist 29:1–7.

Norouzi A. (2008). Oilseed rape straw for cultivation of oyster mushroom *Mj. Int. J. Sci. Tech.* 2008, *2*(03), 502-507.

Oei P (2005) Small-scale mushroom cultivation: oyster, shiitake and wood ear mushrooms. First edition of Digigrafi, Wageningen, The Netherlands. ISBN Agromisa: 90-8573-038-4.

Oteng' B.B., A. J. Sigot, P. M. Kisiangani, L. Mudavadi (2008).mushroom production techniques and utilization for enhanced food security in lake Victoria basin *International Journal for Disaster Management & Risk Reduction* 1 (2).

Pandey A. (2000). Biotechnological potential of agro-industrial residues. I: sugarcane bagasse *Bioresource Technology*, 74(1):69-80

Patil S.S., S. A. Ahmed, S. M. Telang, M. V. Baig (2010). The nutritional value of pleurotusostreatus (jacq.:fr.) kumm cultivated on different lignocellulosicagrowastes. *Innovative Romanian Food Biotechnology*,.7: 66-76.

Obodai, M., J. Cleland-Okine, K.A. Vowotor, Comparative study of the growth and yield of *Pleurotusostreatus*on different lignocellulosic by-products, *J. Ind. Microbiol.Biotechnol.30* (2003) 146–149.

Ortega, G.M.; Martinez, E.O.; Betancourt, D.; Gonzalèz, A.E.; Otero, M.A. 1992. Bioconversion of sugar cane crop residues with white-rot fungi *Pleurotus*sp. *World Journal of Microbiology and Biotechnology*, 8:402-405

Ortega G. M., E. O. Martinez, P. C. González, D. Betancourt and M. A. Otero 1993.Enzyme activities and substrate degradation during white rot fungi growth on sugar-cane straw in a solid state fermentation. *World Journal of Microbiology and Biotechnology*, 9(2): 210-212

Pandey A. 2000. Biotechnological potential of agro-industrial residues. II: cassavaBagasse.*Bioresource Technology*, 74:81-87.

Pant D., U. Gangi Reddy and A. Adholeya. (2006). Cultivation of oyster mushrooms on wheat straw and bagasse substrate amended with distillery effluent *World Journal of Microbiology & Biotechnology*, 22:267–275.

Pellicer-González, E., D. Martínez-Carrera, M. Sánchez, M. Aliphat & A. Estrada-Torres (2002). Rural management and marketing of wild edible mushrooms in Mexico. Proceed. IV International Conference on Mushroom Biology and Mushroom Products, Cuernavaca, Mexico. 433-443 pp.

Philippoussis A.N. (2009). Production of Mushrooms Using Agro-Industrial Residues as Substrates Chapter 9 In: Biotechnology for Agro-Industrial Residues

Utilisation. Poonam Singh nee' Nigam · Ashok Pandey Editors Springer Science+Business Media B.V.163-197 p.

Philippoussis, A., Zervakis, G. &Diamantopoulou, P. (2001). Bioconversion of agricultural lignocellulosic wastes through the cultivation of the edible mushrooms Agrocybeaegerita, Volvariellavolvaceaand Pleurotusspp. World Journal of Microbiology and Biotechnology 17:191–200.

Poppe J. (2004). Oyster Mushroom Cultivation,*Chapter* 5, Substrate. Mushroom Growers′ Handbook 1: Oyster Mushroom Cultivation, Published by MushWorld – Heineart Inc., Seoul, Korea. 75-86 pp.

Poppe, J.A. &Hofte, M. (1995) Twenty wastes for twenty cultivated mushrooms. In Proceedings of the 14th International Congress on the Science and Cultivation of Edible fungi, Oxford, pp. 171–179.

Ragunathan R. and K. Swaminathan. 2004. Biological treatment of a pulp and paper industry effluent by Pleurotus spp. *World Journal of Microbiology & Biotechnology*, 20: 389–393.

Ragunathan R, Gurusamy R, PalaniswamyM(1996) Cultivation of *Pleurotus*spp. on various agroresidues.FoodChem 55(2): 139–144.

Rashad M. M. 2009. Nutritional Analysis and Enzyme Activities of *PleurotusOstreatus*Cultivated on *Citrus Limonium*and *Carica Papaya* Wastes Australian Journal of Basic and Applied Sciences, 3(4): 3352-3360.

Rípoli T. C. C.; W. F. Molina Jr.; M. L. C. Rípoli.(2000). Energy potential of sugar cane biomass in brazil. *Scientia Agricola*, 57(4):677-681.

Rizki M., Y. Tamai 2010. Effects of different nitrogen rich substrates and their combination to the yield performance of oyster mushroom (Pleurotusostreatus) *World J MicrobiolBiotechnol*. 1-8.

Rodrıguez Perez S. (2008). Decolourisation of mushroom farm wastewater by Pleurotusostreatus.*Biodegradation*, 19:519–526.

Romero E. R., J. Scandaliaris, P. A. Digonzelli. (2009). Effect of variety and cane yield on sugarcane potential trash.Rev. Ind. y Agríc.de Tucumán 86 (1): 9-13.

Rühl M., 2008. Ligninolytic enzyme activities alternate with mushroom production during industrial cultivation of *Pleurotusostreatus*on wheatstraw-based substrate. *Current Trends in Biotechnology and Pharmacy*, 2 (4): 478 -492.

Sánchez, C. (2004). Modern aspects of mushroom culture technology. *ApplMicrobiolBiotechnol*. (2004) 64: 756–762.

Sanjuan, R.; Anzaldo, J.; Vargas, J.; Turrado, J.; Patt, R. 2001. Morphological and chemical composition of pith and fibers from Mexican sugarcane bagasse HolzalsRoh- und Werkstoff v. 59(6) p. 447-450.

Sarikaya ., M. R. Ladisch 1997. An Unstructured Mathematical Model for Growth of *Pleurotusostreatus*on Lignocellulosic Material in Solid-State Fermentation Systems. *Applied Biochemistry and Biotechnology*, 62:71-85.

Saska, M. and M. Gray (2006).Pre-treatment of sugarcane leaves and bagasse pith with lime-impregnation and steam explosion for enzymatic conversion to fermentable sugars.*28th Symposium on Biotechnology for Fuels and Chemicals*. Nashville, TN.

Saska M., C. Martin. 2006. Production of fuel ethanol from sugarcane bagasse and sugarcane trash. *IX. Congress on Sugar and Sugar Cane Derivatives, Havana, Cuba, June 19 - 22, 2006.*

Shah, Z.A. M. Ashraf and M. Ishtiaq Ch.(2004). Comparative Study on Cultivation and Yield Performance of Oyster Mushroom (*Pleurotusostreatus*) on Different Substrates (Wheat Straw, Leaves, Saw Dust) *Pakistan Journal of Nutrition*, 3 (3): 158-160

Šnajdr J., (2008). Temperature Affects the Production, Activity and Stability of Ligninolytic Enzymes in *Pleurotusostreatus*and *Trametesversicolor. Folia Microbiol.* 52 (5), 498–502.

Siu-Wai Chiu, Shui-Chee Law, Mei-LunChing, Ka-Wan Cheung, and Ming-Jie Chen (2000).Themes for mushroom exploitation in the 21st century: Sustainability, waste management, and conservation *J. Gen. Appl. Microbiol*., 46, 269–282

Szmidt R., (1997). Advice for prospective mushroom growers. SAC receives financial support from the Scottish Office Agriculture, *Environment and Fisheries Department*, pp. T453-T457.

Souza Dias E, Koshikumo E, Da Silva R. (2003). Cultivation of the mushroom *Pleurotussajor-caju*in different agricultural residues.CiêncAgrotec, Lavras; 27:1363-1369 pp.

Sun, J.X., Sun, X.F., Sun, R.C. and Su, Y.Q. Fractional extraction and structural characterization of sugarcane bagasse hemicelluloses. *Carbohydrate Polymers*. 56, 195-204.

Stamets, P. and Chilton, J.S.(1984). Mushroom Cultivator: A Practical Guide to Growing Mushrooms at Home. Agarikon Press, Olympia, First Edition. Washington. 415 p.

Stewart D.P.C. (1998). Release of sulphate, potassium, calcium and magnesium from spent mushroom compost under laboratory conditions. *BiolFertil Soils*, 26:146–151

Royse D.J,. Rhodes T.W, Ohga, S. Sanchez J.E. (2004) Yield, mushroom size and time to production of *Pleurotuscornucopiae* (oyster mushroom) grown on switch grass substrate spawned and supplemented at various rates, *Bioresource Technology*, 91 85–91 Elsevier Ltd.

Ruegger J.S. S. M. TaukTornisielo; V. L. Ramos Bononi; M. Capelari (2001).Cultivation of the edible mushroom *oudemansiellacanarii*(jungh.) höhn.in lignocellulosic substrates Brazilian *Journal of Microbiology*, 32:211-214

Vargas Betancur G. J., N. Pereira. (2010). Sugar cane bagasse as feedstock for second generation ethanol production. Part I: Diluted acid pretreatment optimization. *Electronic Journal of Biotechnology*, 13 (3).

VazRossell C.E., 2006 Conversion of lignocellulose biomass (bagasse and straw) from the sugar-alcohol industry into bioethanol in: industrial perspectives for bioethanol Edited by T. F. Teixeira UNIEMP São Paulo – SP pp. 123-142

Velazquez-Cedeño M.A. (2000).Waste-reducing cultivation of Pleurotusostreatus and Pleurotuspulmonariuson coffee pulp: changes in the production of some lignocellulolytic enzymes. *World Journal of Microbiology & Biotechnology*, 18: 201–207.

Vetter J. (2003).Chemical composition of fresh and conserved Agaricusbisporus mushroom.*Eur Food Res Technol*. 217:10–12.

Taurachand D. 2004. Chapter 5 sugarcane bagasse. Mushroom Growers′ Handbook 1: Oyster Mushroom Cultivation, Edited by MushWorld (non-profit organization), Published by MushWorld – Heineart Inc., Seoul, Korea. pp. 112-115.

Woytiuk K. 2006. Sugar cane trash processing for heat and power production http://epubl.ltu.se/1653-0187/2006/57/LTU-PB-EX-0657-SE.pdf

Yoshiro K. (1999). Mushroom Production Project. In: http://www.gsid.nagoya-u.ac.jp/project/fieldwork/ofw/OFW_Report/1999/OFW%20Report%20WG4.pdf

Tisdale T.E. S. C. Miyasaka, and D. E. Hemmes (2006).Cultivation of the oyster mushroom (Pleurotusostreatus) on wood substrates in Hawaii.*World Journal of Microbiology & Biotechnology*, 22:201–206.

Todd J., J.G. E. Brown, E. Wells (2003). Ecological design applied. *Ecological Engineering*, 20:421–440.

In: Mushrooms: Types, Properties and Nutrition
Editors: S. Andres and N. Baumann
ISBN 978-1-61470-110-1

Chapter 5

SUSTAINABLE USE OF MICROBIAL ENDOPHYTES

Devendra Kumar Choudhary*
Department of Science, Faculty of Arts, Science & Commerce (FASC), Mody Institute of Science & technology (MITS), Lakshmangarh-332311, Sikar (Rajasthan), INDIA

ABSTRACT

In the past two decades, a great deal of information on the potential role of endophytes in nature has been collected. Endophytes, microorganisms that colonize internal host tissues of living plants are a relatively unstudied and dependable source of bioactive and chemically novel compounds with potential for exploitation in a wide variety of medical and pharmacological areas. The mechanism through which endophytes exist and respond to their surroundings must be better understood. In most cases their relationship with the host plant is symbiotic and probably mutualistic. Many are capable of synthesizing bio-active compounds that can be used by the plant for defense against pathogenic fungi and bacteria. Some of these compounds have proven useful for novel drug discovery. The opportunity to find new and interesting endophytes among myriad plants is great. Endophytes represent a huge diversity of microbial adaptations that have developed in special and sequestered environments, and their diversity and specialized habituation make them an exciting field of study in the search for new medicines. The hunt for new drugs is particularly important in view of the fact that so many diseases are developing immunity to some of the current treatments. This review will concentrate on what has been discovered, and what is still unknown about endophytes that synthesize chemicals with bioactive properties.

Keywords: Biodiversity; Endophytic fungi; Endophytic bacteria; Sustainable production; Biological control; Bioremediation.

* For Correspondence: Department of Science, Faculty of Arts, Science & Commerce (FASC), Mody Institute of Science & technology (MITS), Lakshmangarh-332311, Sikar (Rajasthan), INDIA E-mail: devmicro@rediffmail.com

An old adage in science refers to over-specialization as a process of knowing more and more about less and less, until eventually a person knows everything about nothing! The tacit assumption underlying this statement is that as the area of interest becomes narrower or smaller, the complexity of the system becomes less. This certainly does not appear to hold true when applied to the chemistry of plants, because as one delves into the details of plant structure and metabolism, the subject becomes much more complex than expected, and as a result it often seems that a researcher appears to know less and less about more and more!

Virtually all plants are hosts to endophytes that include bacteria and fungi. Bacon and White [1] define endophytes as "microbes that colonize living internal tissues of plants without causing any immediate, overt negative effects". This definition implies a mutualistic relationship between the host plant and the endophytic microbe, whereas Strobel [2] suggests that the relationship can range from mutualistic to bordering on pathogenic. The most frequently encountered endophytes are fungi and there may be over a million different fungal species on this Earth, of which only a small fraction (ca. 5%) have been identified. Hawksworth and Rossman [3] have estimated that there may be as many as 1 million different fungal species, although so far only about 100,000 have been described. As more evidence accumulates, estimates keep rising as to the actual number of fungal species. For instance, Dreyfuss and Chapela [4] estimate there may be at least 1 million species of endophytic fungi alone. It seems obvious that endophytes are a rich and reliable source of genetic diversity and novel, undescribed species.

The existence of endophytes has now been known for over one hundred years. Botanists have carried out much research into the plant/endophyte relationship, especially for grasses viz., tall fescue, where it has been shown that endophytes produce toxins that prevent the attack of phytophagous insects and other grazing animals [5]. It wasn't until the past decade or so, however, that endophytes have been studied for their potential as novel sources of effective new drugs (Table 1). There is a general call for new antibiotics, chemotherapeutic agents, and agrochemicals that are highly effective, possess low toxicity, and have a minor environmental impact. This search is driven by the development of resistance in infectious microorganisms (e.g., species of *Staphylococcus*, *Mycobacterium*, and *Streptococcus*) to existing compounds and by the menacing presence of naturally resistant organisms.

The new diseases such as AIDS and severe acute respiratory syndrome require the discovery and development of new drugs to combat them. Furthermore, others who are immunocompromised (e.g., cancer and organ transplant patients) are at risk of opportunistic pathogens, such as *Aspergillus* spp., *Cryptococcus* spp., and *Candida* spp. In addition, more drugs are needed to efficiently treat parasitic protozoan and nematodal infections, such as malaria, leishmaniasis, trypanosomiasis, and filariasis. Malaria alone is more effective in claiming lives each year than any other single infectious agent with the exception of the AIDS virus and *Mycobacterium tuberculosis*. Finally, because of safety and environmental problems, many synthetic agricultural agents have been and currently are being targeted for removal from the market, which creates a need to find alternative ways to control farm pests and pathogens [6].

Table 1. Endophytes and their bioactive compounds

S. No.	Endophyte	Plant host	Metabolite	Functional role	Reference
1.	*Cryptosporiopsis quercina*	*Tripterigeum wilfordii*	Cryptocin	Antimycotic	41
2.	*Pseudomonas viridifava*	Grass species	Ecomycins	Antimycotic	42
3.	*Pseudomonas syringae* MSU 16H	Banana	Pseudomycins	Antifungal	43
4.	*Pestalotiopsis microspora*	*Torreya taxifolia*	Pestaloside	Antifungal	44
5.	*P. microspora*	*T. brevifolia*	Pestalotiopsins A and B	Antifungal	45
6.	*P. jesteri*	*Maytenus ilicifolia* Mart. ex. Reiss	Jesterone and hydroxyjesterone	Antifungal	47
7.	*Muscodor albus*	*Cinnamomum zeylanicum*	Volatiles	Antimicrobial	49, 50
8.	*Taxomyces andrenae*	*Taxus brevifolia*	Paclitaxel (Taxol^{TM})	Anticancer	52
9.	*Seimatoantlerium tepuiense, Periconia* sp., *S. nepalense*	*Maguireothamnus speciosus*	paclitaxel	Anticancer	54
10.	*Pestalotiopsis guepinii*	*Wollemia nobilis* (Wollemi pine)	Taxol	Anticancer	55
11.	*Tubercularia* sp. strain TF5	*Taxus mairei*	Taxol	Anticancer	56
12.	*Paecilomyces* sp. and *Aspergillus clavatus*	*T. mairei, Cephalataxus fortunei*, and *Torreya grandis*	Brefeldin A	Anticancer	56
13.	*Entrophospora infrequens*	*Nothapodytes foetida*	Camptothecin	Anticancer	59
14.	*Pestalotiopsis* spp., *Acremonium* sp., *Mucor* sp., and *Verticillium* sp.	*Tripterygium wilfordii* H, f., (TWHf)	Subglutinol A and B	Immunosuppressive	60
15.	*Fusarium subglutinans*	*Tripterygium wilfordii*	Subglutinol A and B	Immunosuppressive	61

Novel natural products and the organisms that make them offer opportunities for innovation in drug and agrochemical discovery. Natural products are naturally derived metabolites and/or by-products from microorganisms, plants, or animals [7]. These products have been exploited for human use for thousands of years wherein plants have been the chief source of compounds used for medicine. Even today the largest users of traditional medicines are the Chinese, with more than 5,000 plants and plant products in their pharmacopoeia [8].

In fact, the world's best known and most universally used medicinal is aspirin (salicylic acid), which has its natural origins from the glycoside salicin, found in many species of the plant genera *Salix* and *Populus*. In fact, nearly 3,000 years ago, the Mayans used fungi grown on roasted green corn to treat intestinal ailments. More recently, the Benedictine monks (800 AD) began to apply *Papaver somniferum* as an anesthetic and pain reliever as the Greeks had done for years before [9]. Exciting possibilities thus exist for those who are willing to venture into the wild and unexplored territories of the world to experience the excitement and thrill of engaging in the discovery of endophytes, their biology, and their potential usefulness [10].

Even with untold centuries of human experience behind us and a movement in to a modern era of chemistry and automation, natural-product-based compounds have had an immense impact on modern medicine since about 40% of prescription drugs are based on them. Furthermore, 49% of the new chemical products registered by the U.S. Food and Drug Administration are natural products or derivatives thereof [9]. Excluding biologics, between 1989 and 1995, 60% of approved drugs and pre new drug application candidates were of natural origin. From 1983 to 1994, over 60% of all approved cancer drugs and those at the pre-new drug application stage were of natural origin, as were 78% of all newly approved antibacterial agents [11]. In fact, the world's first billion-dollar anticancer drug, paclitaxel (Taxol), is a natural product derived from the yew tree [12]. Many other examples abound that illustrate the value and importance of natural products in modern civilizations.

The goal of this essay has been to introduce the reader to the biochemical and medicinal implications of endophytes; to outline the potential of this new area of research; and to enumerate some of the important issues associated with bio-active endophytes that are still unresolved. This is an exciting new area of research with many new discoveries to be made, and with considerable potential for discovering novel and commercially exploitable compounds.

Endophyte Diversity

One of the early publications describing an endophytic fungus was by Freeman [13] where he recovered first endophytic fungus from Persian darnel- an annual grass that considered a troublesome weed by many wheat farmers. Grasses are probably the plants that have been most extensively studied as far as endophytes are concerned, wherein high endophytic contents provide resistant to attack by certain phytophagous insects [14]. In addition, however, considerable work [15] has also been carried out on a variety of different plants, including many trees and plants of horticultural importance viz., apple and kiwifruit.

Analysis of any plant material may reflect discovery of a range of different endophytic fungi and bacteria, and several of the fungi would likely to be specific to the particular host. There is no evidence that endophytic bacteria exhibit the same host specificity other than bacterial pathogens. This fungal-host specificity was found to be true for the fescue and rye grasses where the endophytes complete their life cycle within the above-ground portions of the grass host forming nonpathogenic, symbiotic associations [16].

In addition, environmental conditions viz., soil temperature and humidity, would also be expected to affect the nature and the population of endophytes [17]. Those plants in unique environments that struggle to compete with other living organisms, or that need as much assistance as possible to survive, are likely to host endophytes which can generate secondary metabolites that will assist the plant [18]. Schutz [19] notes that several microbial metabolites seem to be characteristic of certain biotopes, on both an environmental as well as organismal level. Accordingly, it appears that the search for novel secondary metabolites should center on organisms that inhabit unique biotopes. Thus, the investigator should carefully study and select the biological source before proceeding, rather than to have a totally random approach in the use of biological source material. It is also anticipated that organisms and their biotopes that are subjected to constant metabolic and environmental interactions should produce even more secondary metabolites [19].

Maciá-Vicente *et al.* [20] reported fungal endophytes in roots of 24 plant species growing at 12 sites (coastal and inland soils, both sandy soils and salt marshes) under either water/salt stress in the Alicante province, Southeast Spain. All plant species examined were colonized by endophytic fungi wherein a total of 1,830 fungal isolates were recovered and characterized by phonetic and molecular [internal transcribed spacer (ITS) and translation elongation factor-1α gene region (TEF-1α) sequencing] techniques. *Fusarium* and *Phoma* species were the most frequent genera, followed by *Aspergillus*, *Alternaria*, and *Acremonium*. Fungal root endophytic communities were influenced by the soil type where their respective host plants grew, but not by location.

Davis *et al.* [21] studied the distribution of Xylariaceous endophytes of liverworts and angiosperms from Jamaica and North Carolina, and found a broad host range for several species of *Xylaria*. In addition, members of the family Xylariaceae have been reported from nearly very continent together with conifers, dicots, monocots, ferns, and lycopsids, and they are often very closely related to each other [22].

Several temperate and tropical plant species from different habitats have been examined for their endophyte assemblages; there is no comprehensive study that assesses endophyte diversity and host range in different tropical forest communities, except for a mangrove community [23]. Suryanarayanan *et al.* [24] reported the occurrence and distribution of foliar endophytes in trees of four types of tropical forests, dry thorn forest, dry deciduous forest, moist deciduous forest, and semi-evergreen forest, in the Western Ghats of Southern India. Several endophytes viz., *Cladosporium cladosporioides*, *Corynespora cassiicola*, *Lasiodiplodia theobromae*, *Pestalotiopsis* spp., *Phomopsis* spp., *Phyllosticta* sp., *Sporormiella intermedia*, *Xylaria* spp., and the two sterile forms were recovered from trees of all the tropical forests, indicating that these endophytes had a wide host range.

Arnold *et al.* [25] reported a remarkable diversity among endophytes in the tropics in a lowland Panamanian forest, wherein the endophyte communities of trees growing in a semi-evergreen forest exhibited the highest species diversity and species richness. This forest receives the maximum rainfall in this region and has the highest host species diversity [26]. Interestingly, several earlier studies showed that endophyte colonization is positively correlated with rainfall and it is a major environmental factor that determines the density of endophyte infection, whereas it reduced in leaves of temperate trees where they are shielded from rain [27].

Actinomycetes are bacteria known to constitute a large part of the rhizosphere microbiota. The actinomycetes usually grow and colonise the soil and consequently it is easy to isolate them from internal cortical tissues of roots in which they grow [28]. Their isolation is an important step for screening of new bioactive compounds. Mundt and Hinckle [29] recovered the genera *Nocardia* and *Streptomyces* from seed and ovules of 27 plant species. Sardi *et al.* [28] isolated and observed endophytic actinomycetes from the roots of 28 plant species from the Northwestern Italy wherein most of the isolates were classified as *Streptomyces* together with *Streptoverticillium*, *Nocardia*, *Micromonospora*, and *Streptosporangium*.

In addition, endophytic bacteria have been found in virtually every plant studied and most of them appear to originate from the rhizosphere/phyllosphere, however, several may be transmitted through the seed. Endophytic bacteria have been recovered from both monocotyledonous and dicotyledonous plants, ranging from woody tree species, such as oak and pear, to herbaceous crop plants such as sugarcane, sugar beet and maize. Velázquez *et al.* [30] reported genetic diversity of 29 endophytic bacterial strains isolated from apoplastic sap of the medullary parenchyma of the stem of healthy sugarcane plants grown in Cuba. Analysis of 16S rRNA sequences showed that the isolates belong to a wide variety of phylogenetic groups being closely related to species of genera *Bacillus*, *Staphylococcus*, *Microbacterium*, *Micrococcus*, *Kokuria*, *Rhizobium*, *Gluconacetobacter*, *Comamonas*, *Xanthomonas*, *Acinetobacter*, and *Pantoea*.

Citrus variegated chlorosis (CVC) is caused by *Xylella fastidiosa*, a phytopathogenic bacterium that infect all *Citrus sinensis* cultivars. The endophytic bacterial communities of healthy, resistant, and CVC-affected citrus plants were studied [31] employing denaturing gradient gel electrophoresis (DGGE) and fatty-acid methyl ester (FAME) analysis, wherein dominant species included, *Bacillus pumilus*, *Curtobacterium flaccumfaciens*, *Enterobacter cloacae*, *Methylobacterium* spp., *Nocardia* sp., *Pantoea agglomerans*, and *Xanthomonas campestris*. *Methylobacterium* species, which have the capacity to fix nitrogen [32], have recently been isolated from citrus [33], Scotch pine [34], and crotalaria [32], showing the capacity of member of this bacterial genus to colonize the plant habitat. *Methylobacterium* was the most frequently recovered genus in citrus plants and there was a positive association with the occurrence and intensity of symptoms of CVC. This suggests that CVC could play a role in the establishment of endophytic bacteria in the host plant or that endophytic *Methylobacterium* spp. could trigger CVC by a synergistic interaction with *X. fastidiosa* [31].

In addition, the composition of endophytic bacteria colonizing the aerial parts of poplar was studied [35] using a multiphasic approach. Detailed analysis of endophytic bacteria using cultivation methods in combination with cloning of 16S rRNA genes amplified from plant tissue revealed a high phylogenetic diversity with a total of 53 taxa at the genus level that included Proteobacteria, Actinobacteria, Firmicutes, and Bacteroidetes. The most abundant genera among the isolates were *Pseudomonas* and *Curtobacterium*.

ENDOPHYTES: BIOPROSPECTING

Most of us are aware that bacteria are ubiquitous viz., the human mouth, for example, is home to a large number of different strains. However, only some of us are aware that bacteria and fungi can be found not only on the surface of plants (as epiphytes), but also as endophytes, colonizing the tissues between plant cells and possibly within plant cells [36].

Endophytes are relatively poorly understood but are potential sources of novel natural products for exploitation in medicine, agriculture, and industry. Of the nearly 300,000 plant species that exist on the earth, each individual plant is likely to be host to one or more endophytes. However, only a few of the existing plant species have so far been completely studied with respect to their endophytic biology. Consequently, the opportunity to find new and interesting endophytic microorganisms among myriads of plants in different settings and ecosystems is indeed large [10].

A large number of compounds (secondary metabolites) have been extracted, isolated and characterized from endophytic microbes, and these are detailed with extensive references in the reviews of Tan and Zou [37] and Strobel and colleagues [2, 38]. Many of these compounds are bioactive, and the range includes alkaloids, steroids, terpenoids, peptides, polyketones, flavonoids, quinols and phenols as well as some chlorinated compounds. The compound Taxol (currently used as an anti-cancer agent) was first found in the bark of the Pacific Yew tree. Later, it was discovered that a fungus associated with the Yew tree also produces taxol [39].

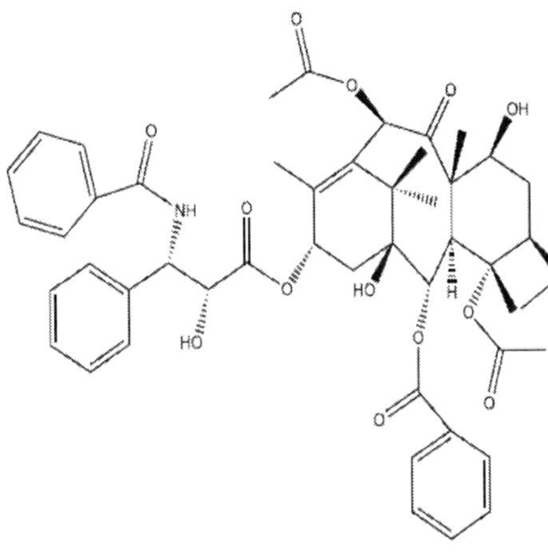

Figure 1. Taxol - an anti-cancer compound synthesized by the Pacific Yew tree and by a fungus associated with the tree.

Further analysis found that the genes for taxol production are found in both the tree and the fungus, which leads to several intriguing questions that have yet to be conclusively answered. For example: 1. Was the compound first synthesized by the tree or the fungus? 2. Could there be a transfer of genetic information between the two? 3. Are microbes capable of communicating with each other and communicating genetic information with other organisms?

A previously unknown compound showing significant bio-activity has been extracted and isolated from an endophytic fungus found in the leaves of a plant from the highlands of Papua New Guinea. The plant (*Desmodium uncinatum*) is used by the indigenous people for healing wounds and body sores. The compound shows anti-fungal and anti-bacterial effects and has been shown to destroy HeLa cervical cancer cells with an IC_{50} value of 0.9 μg/μL [36].

Traditional medicines are extensively used in China, and there are over 5,000 plants and plant products in their pharmacopeia, and work has already been initiated by Tan and others to investigate the endophytes in these plants [25]. For example, one Chinese plant that has become important in the treatment of malaria (*Artemisia annua*) also hosts endophytic fungi that produce several indole-based metabolites, some of which show activity against Gram-positive and Gram-negative bacteria. Related fungi from another Artemisia species (*A. mongolica*) also produce anti-microbial metabolites [40].

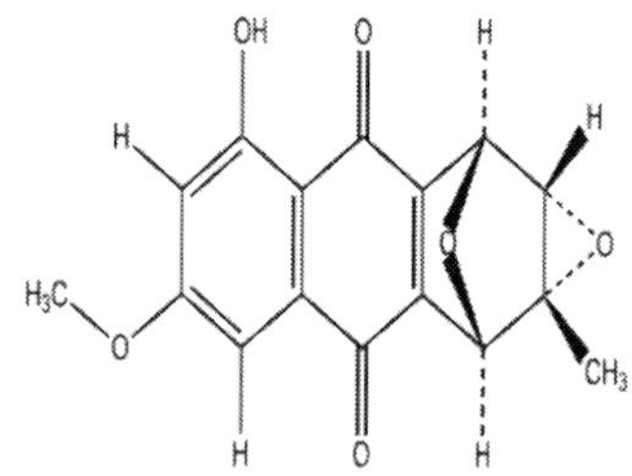

Figure 2. A new bio-active compound from a fungal endophyte of *Desmodium uncinatum*, a medicinal plant growing in Papua New Guinea.

Antimicrobial Compounds

Natural products from endophytic microbes have been observed to inhibit or kill a wide variety of harmful disease-causing agents, as well as bacteria, fungi, viruses, and protozoans that affect humans and animals. *Cryptosporiopsis quercina,* the imperfect stage of *Pezicula cinnamomea*, a fungus commonly associated with hardwood species in Europe, was isolated as an endophyte from *Tripterigeum wilfordii*, a medicinal plant native to Eurasia. A unique antimycotic peptide, termed cryptocin, was isolated and characterized from *C. quercina* and is currently being considered for use against dermatophytes [41]. The ecomycins are produced by *Pseudomonas viridifava*; a plant-associated fluorescent bacterium generally associated with the leaves of grass species and is located on and within the tissues [42]. The ecomycins are active against such human-pathogenic fungi as *Cryptococcus neoformans* and *Candida albicans*.

Another group of antifungal compounds is the pseudomycins, produced by a plant-associated pseudomonad [43]. The pseudomycins represent a family of lipopeptides that are active against a variety of plant- and human-pathogenic fungi that include *C. albicans*, *C. neoformans*, and a variety of plant-pathogenic fungi, including *Ceratocystis ulmi* (the Dutch elm disease pathogen) and *Mycosphaerella fijiensis* (the causal agent of Black Sigatoka disease of banana).

A strain of *Pestalotiopsis microspora*, an endophytic fungus was also isolated from the endangered tree *Torreya taxifolia* and produces several compounds that have antifungal activity, including pestaloside, an aromatic β-glucoside, and two pyrones: pestalopyrone and hydroxypestalopyrone [44]. These products also possess phytotoxic properties. Other newly isolated secondary products obtained [45] from *P. microspora* (endophytic on *T. brevifolia*) include two new caryophyllene sesquiterpenes-pestalotiopsins A and B. A newly described species of *Pestalotiopsis*, namely, *Pestalotiopsis jesteri*, from the Sepik River area of Papua New Guinea, produces jesterone and hydroxyjesterone, which exhibit antifungal activity against a variety of plant-pathogenic fungi [46].

Recently, Gomes-Figueiredo *et al.* [47] reported antimicrobial property of thirteen endophytic fungal strains of the genus *Pestalotiopsis* that recovered from the medicinal plant *Maytenus ilicifolia* Mart. ex. Reiss (known as "espinheira santa"). Two isolates were successful in inhibiting the growth of the pathogenic microorganisms, *Escherichia coli*, *Klebsiella pneumoniae*, *Micrococcus luteus*, *Staphylococcus aureus*, and methicillin-resistant *Staphylococcus aureus* (MRSA).

Another fascinating use of antibiotic products from endophytic fungi is the inhibition of viruses [48]. Two novel human cytomegalovirus protease inhibitors, cytonic acids A and B, have been isolated from the solid-state fermentation of the endophytic fungus *Cytonaema* sp. *Muscodor albus* is a newly described endophytic fungus obtained [49] from small limbs of *Cinnamomum zeylanicum* (cinnamon tree). This xylariaceaous (non-spore-producing) fungus effectively inhibits and kills certain other fungi and bacteria by producing a mixture of volatile compounds [50].

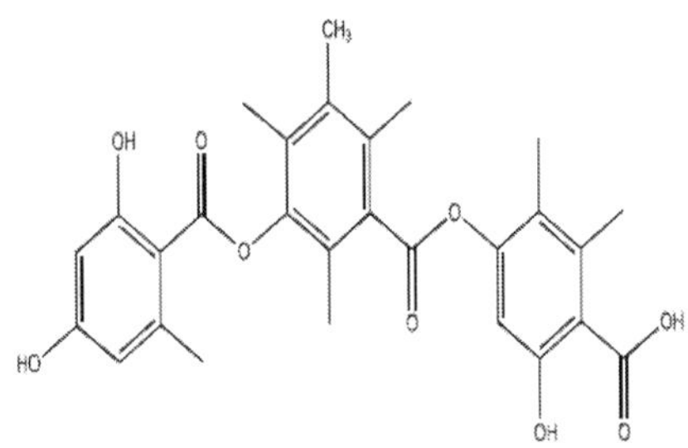

Figure 3. Colletotric acid - an anti-microbial compound from an endophytic fungus in *Artemisia mongolica*, a Chinese medicinal plant.

Anticancer Agents

The research activities in the field of secondary metabolism of endophytic microorganisms were increased after the finding of Paclitaxel (Taxol^{TM}) production by *Taxomyces andrenae*, a fungus found living in association with *Taxus brevifolia* [51]. Paclitaxel, a highly functionalized diterpene, is found in each of the world's yew (*Taxus*) species, with strong anticancer activity mainly against breast and ovarian cancer [52]. The mode of action of paclitaxel is to preclude tubulin molecules from depolymerizing during the processes of cell division [53]. This compound is the world's first billion-dollar anticancer drug. It is used to treat a number of other human tissue-proliferating diseases as well. Many other fungi species associated with *Taxus* have shown paclitaxel production [54].

Surprisingly, reports of paclitaxel production by fungi isolated from plants which are not taxol producers have been published recently such as, a rubiaceous plant, *Maguireothamnus speciosus* has yielded a novel fungus, *Seimatoantlerium tepuiense*, which produces paclitaxel.

Furthermore, fungal paclitaxel production has also been observed in a *Periconia* sp. and in *Seimatoantlerium nepalense*, another novel endophytic fungal species [54]. Beside, the taxol-producing fungus *Pestalotiopsis guepinii* was isolated as an endophyte from *Wollemia nobilis* (Wollemi pine), an araucareaceous plant occurring in the Wollemi National Park in Australia [55]. Wang *et al.* [56] screened 45 endophytic fungi isolated from the inner bark of *Taxus mairei* and found that *Tubercularia* sp. strain TF5 produced taxol. The cytotoxin brefeldin A was found to be produced by *Paecilomyces* sp. and *Aspergillus clavatus*, from plants *T. mairei*, *Cephalataxus fortunei*, and *Torreya grandis* [56]. Caruso *et al.* [57] screened 150 fungal strains isolated from *Taxus baccata* and *T. brevifolia* wherein 15 strains produced taxanes. Likewise, Huang *et al.* [58] screened 172 endophyte isolates from *T. mairei*, *Cephalataxus fortunei*, and *Torreya grandis* to analyze the antitumor activities. Amna *et al.* [59] screened 52 strains of endophytic fungi for their ability to produce the anticancer alkaloid camptothecin, from *Nothapodytes foetida* growing in the Jammu and Mahabaleshwar regions in India. Out of 52, only one isolate *Entrophospora infrequens*, recovered from inner bark tissue of the *N. foetida*, was found to produce detectable quantity of camptothecin [59].

Endophyte and Immunosuppressive Drugs

Immunosuppressive drugs are used today to prevent allograft rejection in transplant patients, and in the future they could be used to treat autoimmune diseases such a rheumatoid arthritis and insulin-dependent diabetes. An array of endophytic fungi viz., *Pestalotiopsis* spp., *Acremonium* sp., *Mucor* sp., and *Verticillium* sp., which habitat in a Chinese medicinal plant, *Tripterygium wilfordii* H, f., (TWHf) were recovered and tested for their suppressive activity on phytohemaglutinin (PHA) stimulated proliferation of human peripheral blood mononuclear cells (PBMC). Likewise, endophytic fungus *Fusarium subglutinans*, isolated from *Tripterygium wilfordii*, produces the immunosuppressive but noncytotoxic diterpene pyrones subglutinol A and B [60, 61].

Endophytes: Sustainable Use and Application

Exploitation of endophyte-plant interactions can result in the promotion of plant health and can play a significant role in low-input sustainable agriculture applications for both food and non-food crops. An understanding of the mechanisms enabling these endophytes to interact with plants will be worthwhile to fully achieve the biotechnological potential of efficient plant-endophyte partnerships for a range of applications. The most important and promising area of research for future studies is developing endophytes to promote the sustainable production of biomass and bioenergy crops in conjunction with phytoremediation of soil contamination. In addition, the ability of endophytes to confer stress tolerance to plants may provide a novel strategy for mitigating the impacts of global climate change on agricultural and native plant communities [62]. In this section stress has been given on the role of endophytes in securing crop protection and phytoremediation because of their persistence upon the maintenance of biological integrity and diversity together with habitat-adapted symbiosis.

Stress Tolerance in Plants through Endophyte Symbiosis

There are several outcomes of symbiotic interactions which reflect by the fitness benefits realized by each partner [63]. It is fairly straightforward to determine the impact of symbiosis on host fitness, while it is more challenging to determine the benefits for endophytes. Mutualistic benefits for endophytes involve acquiring nutrients from hosts, abiotic and biotic stress avoidance and dissemination by seed transmission [64].

Rodriguez *et al.* [62] reported that native grass species from coastal and geothermal habitats require symbiotic fungal endophytes for salt and heat tolerance, respectively. Symbiotically conferred stress tolerance is a habitat-specific phenomenon with geothermal endophytes conferring heat but not salt tolerance, and coastal endophytes conferring salt but not heat tolerance. The same fungal species isolated from plants in habitats devoid of salt/heat stress did not confer these stress tolerances. Moreover, fungal endophytes from agricultural crops conferred disease resistance and not salt/heat tolerance.

Studies indicate that plant-fungal symbiosis might have been in existence for ≥ 400 MYA (million year ago), and might have played an important role in the movement of plants from an aquatic arena to comparatively dry terrestrial habitats. When this happened, plants must have been confronted with intermittent water stress. It is tempting to speculate that the ability to confer drought tolerance may simply be a legacy shared by all endophytes [65].

Endophyte as Repository Material

Plant exploration as a primary mechanism for finding and adding new plant germplasm to *ex situ* repositories is also a means for collecting microbial germplasm in the form of *Neotyphodium* fungal endophytes of temperate grasses, tall fescue (*Festuca arundinacea*) for storage in repositories [66]. There is a widespread interest in *Neotyphodium* endophytes because of their presence in temperate grasses is linked to enhanced plant fitness, such as drought tolerance and resistance to insect and mammalian herbivores [67]. Toxicity to insects and mammals is the result of specific metabolites (alkaloids) produced by the endophytes in association with their hosts [68]. Preserving this new *Neotyphodium* germplasm is important because it may provide an expanded pool of endophyte and alkaloid diversity needed for developing new endophyte-tall fescue combinations for specific needs such as pest resistance and livestock production.

Endophyte-Enhanced Tissue Culture

Compared to conventional planting material, tissue culture plants give higher yield, and earlier and more vigorous sucker production. Rapid and easy mass production by tissue culture technology allows for facilitated distribution of improved cultivars, and compensate for planting material shortages. Besides, tissue culture plantlets reduce damage by pests and diseases. Dubois *et al.* [69] made an effort to develop endophyte-enhanced tissue culture technology at the International Institute of Tropical Agriculture (IITA), Uganda, using banana as a source.

Cooking banana is a key staple food throughout East and Central Africa, while dessert banana is an important source of income for many farmers. However, accelerated yield declines of banana have been associated with increasing incidence of soil borne pests and diseases. The banana weevil, *Cosmopolites sordidus*, is the most important insect pest of highland bananas and plantains along with burrowing nematode, *Radopholus similis*, in East Africa. Hence, the use of clean planting material is essential in overcoming such constraints, especially when establishing new field. Pest infestation/reinfestation, however, remains a vital concern. Fungal endophyte, *Fusarium oxysporium*, when inoculated into banana tissue culture plants, extend the benefits of clean planting material. Endophyte-enhanced tissue culture technology is being developed at the International Institute of Tropical Agriculture (IITA), Uganda, and it has so far yielded a wealth of information [69].

Endophytes and Plant Growth-Promotion Activity

Endophytes have the ability to promote plant growth by a number of several mechanisms. These include phosphate solubilization activity, indole acetic acid production, and the production of siderophore [70, 71, 72]. In addition, endophytes can supply essential vitamins to plants [73] together with osmotic adjustment, stomatal regulation, modification of root morphology, enhanced uptake of minerals and alteration of nitrogen accumulation, and metabolism [74].

Endophytes as Biological Control Agents (Bcas)

The capability of colonizing internal host tissues has made endophytes valuable for agriculture as a tool to improve crop performance [75]. Suppression of plant diseases due to the action of endophytic microorganisms has been demonstrated in several pathosystems [76]. Several mechanisms may control this suppression, either directly on the pathogen inside the plant by antibiosis, and competition for nutrients, or indirectly by induction of plant resistance response [77]. Diseases of fungal, bacterial, viral origin, insect, and nematodes can be reduced following prior inoculation with endophytes [78].

The most important effects of endophytic fungi have been observed against wild life, particularly insects feeding on green tissues of the plant. Bio-insecticides are only a small part of the insecticide field, but their market is increasing [79]. Several endophytes are known to possess anti-insect properties. Webber [80] was the first researcher to report plant (elm tree) protection against the beetle *Physocnemum brevilineum* employing an endophytic fungus, *Phomopsis oblonga*, followed by Funk *et al.* [81], who showed protection of the perennial ryegrass *Lolium perenne* L. against the sod webworm. Kindler *et al.* [82] reported that *Acremonium* (*Neotyphodium*) reduced the attacks of the aphid *Diuraphis noxia* in *Lolium* as well as in *Festuca*.

Nodulisporic acids, novel indole diterpenes that exhibit potent insecticidal properties against the larvae of the blowfly, have been isolated from an endophyte, a *Nodulisporium* sp., from the plant *Bontia daphnoides*. Patterson *et al.* [83] reported in *Lolium* and *Festuca* that the production of alkaloids by Acremonium reduced attacks of the Japanese beetle *Popilla japonica*. This is the case of the peramine toxin, produced in *L. perenne* by *A. lolii*. Two new active toxins against the Spruce budworm *Choristoneura fumiferana* were reported [84] in an unidentified endophytic fungus infecting the wintergreen *Gaultheria procumben*. Miles *et al.* [85] exhibited that endophytic isolates of *Neotyphodium* sp. produce N-formilonine and a paxiline analogous in the host *Echinopogum ovatus*, show insecticidal activity against *Listronotus bonariensis* and other insects. Azevedo *et al.* [75] elaborately described the role of endophytic microorganisms in insect control wherein the mechanisms by which endophytic fungi control insect attacks are listed and include toxin production as well as the influence of these compounds on plant and livestock and how their production may be affected by genetic and environmental conditions.

Beside endophyte influence on insects, there are several repots of endophyte activity against fungal diseases, viral pathogens, and nematode herbivores viz., a biocontrol strain *Paenibacillus polymyxa*, isolated from wheat, lodge pine, green beans, and *Arabidopsis thaliana*, produced [86] antifungal agent fusaricidin A-D against phytopathogen *Fusarium* sp. The basidiomycete fungus *Crinipellis perniciosa* (Stahel) Singer is the causal agent of Witches' Broom Disease of cacao (*Theobroma cacao* L.), which is one of the main limiting factor for the cacao production in South America and Caribbean islands, and has been considered as one of the most important pathogen for the crop [87]. In order to evaluate the potential of endophytes as a biological control agent of this phytopathogen, the endophytic fungal community was identified [87] by morphological traits and rDNA sequencing, which include genera *Acremonium*, *Blastomyces*, *Botryosphaeria*, *Cladosporium*, *Colletotrichum*, *Cordyceps*, *Diaporthe*, *Fusarium*, *Geotrichum*, *Gibberella*, *Gliocladium*, *Lasiodiplodia*, *Monilochoetes*, *Nectria*, *Pestalotiopsis*, *Phomopsis*, *Pleurotus*, *Pseudofusarium*, *Rhizopycnis*, *Syncephalastum*, *Trichoderma*, *Verticillium*, and *Xylaria*. Among these, some were identified as potential antagonists, but only one fungus (*Gliocladium catenulatum*) reduced the incidence of Witches' Broom Disease in cacao seedling to 70%.

Effect of endophytic fungi on endoparasitic nematodes has been taken the most attention after insects. *Meloidogyne marylandi*, a root nematode can be deterred when endophyte exists in above ground tissues of perennial ryegrass [88]. A similar result was obtained [89] for *Pratylenchus scribneri*. Pine wood nematode, *Bursaphelenchus xylophilus*, causes pine wilt disease in a number of *Pinus* species, which is one of the most serious plant diseases in forest, Korea. Kwon *et al.* [90] reported nematicidal substances (bikaverin and fusaric acid) from endophytic fungi, *Fusarium oxysporium* EF119, which showed the strongest activity to *B. xylophilus*.

In addition, bacterial endophytes promote plant growth and improve the host's capacity to withstand pathogen attack by causing organism competition, antibiosis, and by inducing systemic resistance (ISR) [91]. Bacterial blight of cotton caused by *Xanthomonas axonopodis* pv. *malvacearum* (*Xam*) is a major yield constraint in cotton crops. Rajendran *et al.* [91] reported the effect of inducing resistance against *Xam* in cotton with a talc-based bioformulation of the endophytic *Bacillus* strains EPCO 102 and EPCO 16 and *Pseudomonas fluorescens* strain Pf1 with or without the addition of chitin. The addition of chitin to the formulation reduced disease incidence. The bacterial strains also caused higher level of chitinase, peroxidase, polyphenol oxidase, phenylalanine ammonia-lyase and phenol in cotton, besides reduction of blight incidence.

Sabaratnam and Beattie [92] have been isolated many plant-growth promoting endophytic (PGPE) strains from the internal tissues of various crops and tested against diseases caused by phytopathogens. As a result of a greater ISR in plants, growth was enhanced and disease reduced in many crops [93]. Benhamou *et al.* [94] reported that the endophytic bacterium *Serratia plymuthica* enhanced ISR by increase in levels of phenolics in cucumber roots, affording protection against *Pythium ultimum*.

Kim *et al.* [95] reported that the bacterium *Bacillus lentimorbus* produces the antifungal substances alpha- and beta-glucosidase, with an inhibitory action against the development of *Botrytis cinerea*, while Sadfi *et al.* [96] reported that this bacterium is capable of releasing volatile substances that contribute to the inhibition of *Fusarium sambucinum* Fuckel in potato tubers. In addition, *B. cereus* can produce various chitinases, against several phytopathogens, such as *F. sambucinum*, *Rhizoctonia solani*, *Helminthosporium solani*, *Sclerotium rolfsii*, *F. oxysporum*, and *Pythium aphanidermatum* [97]. *B. cereus* has also been reported [98] as endophyte in cotton (*Gossypium hirsutum* L.), sweet corn (*Zea mays* L.), and citrus plants (*Citrus* spp.). Lodewyckx *et al.* [99] reported the endophytic association of bacteria of the genera *Clavibacter* and *Klebsiella* with some agronomic crops, such as corn, grapevine, rice, cotton, and some crucifers.

Endophytes and Phytoremediation

Endophytes have the capacity to occupied plants that grown in soil contaminated with xenobiotics with the necessary contaminant-degrading genes. Genes encoding for nitro-aromatic compound degradation were more prevalent in endophytic strains in the field sites contaminated with nitro-aromatics [100]. A phyto-symbiotic strain of *Methylobacterium*, isolated from hybrid Poplar tree (*Populus deltoids* x *nigra*), exhibited biodegradation of nitro-aromatic compounds such as 2,4,6-trinitrotoluene [101].

Table 2. Xenobiotic compounds that have been associated with endophyte phyto-remediation strategies

Xenobiotic compound	Plant	Endophyte	Reference
Volatile organic compounds (VOCs)	Yellow lupine (*Lupinus luteus* L.)	*Burkholderia cepacia* G4	109
TCP and PCB	Wheat (*Triticum aestivum*)	*Herbaspirillum* sp. K1	110
Toluene	Poplar (*Populus*)	*Burkholderia cepacia* Bu61 (pTOM-Bu61)	111
NDAB, BTEX,	*Populus* cv. *Hazendans* and cv. *Hoogworst*	*Pseudomonas* sp.	112
TNT, RDX, HMX	*Populus deltoidesnigra* DN34	*Methylobacterium Populi* BJ001	101
Methane	*Populus deltoidesnigra* DN34	*Methylobacterium Populi* BJ001	101
2,4-D	*Populus* and Willow (*Salix*)	*P. putida* VM1450	113
Mono- & dichlorinated benzoic acids	Wild rye (*Elymus dauricus*)	*Pseudomonas aeruginosa* strain R75	114

Genetically engineered strains also have the capability to degrade xenobiotic compounds viz., endophytes of yellow lupine that were genetically constructed for nickel resistance, were able to increase the nickel accumulation and tolerance of inoculated plants [102]. Table 2 describes the potential role of endophytes in phytoremediation.

TCP (2,3,4,6-tetrachlorophenol); PCB (polychlorinated biphenyl); NDAB (aliphatic nitramine 4-nitro-2,4-diazabutanal); BTEX (benzene, toluene, ethylbenzene, and xylene); TNT (2,4,6-trinitrotuluene); RDX (hexahydro-1,3,5-trinitro-1,3,5-triazene); HMX (octahydro-1,3,5,7-tetranitro-1,3,5-tetrazocine); 2,4-D (2,4-dichlorophenoxyacetic acid).

In addition, quantitative gene expression of pollutant catabolic genes within the endophytic populations could be a useful monitoring tool for assessing the efficiency of the remediation process. A major advantage being where genetic engineering of a xenobiotic degradation pathway is required, bacteria are easier to manipulate than plants [103]. Taghavi *et al.* [104] demonstrated that endophyte niche is considered a hot spot for horizontal gene transfer (HGT). In this study, the degradative plasmid, pTOM-Bu61, was found to have transferred naturally to a number of different endophytes *in planta*. This approach maay reflect practical applications in equipping the natural endophyte populations with the capacity to degrade a pollutant and does not require long-term establishment of the inoculant strain.

Endophytes and Biomaterials

The search for and exploitation of natural products and their properties has been the mainstay of biotechnology research. Bioplastics are biomaterials that are receiving attention for commercial interest together with eco-friendly environment. Earlier, Lemoigne [105], first described a bioplastic, poly-3-hydroxybutyrate (PHB) recovered from *Bacillus megaterium*. Among bioplastics, the most widely produced microbial bioplastics are poly-3-hydroxyalkanoate (PHA) and PHB. Based on genomic analysis Kalia *et al.* [106] reported that many species of bacteria have the potential to produce bioplastics. A diazotrophic endophyte, *Herbaspirillum seropedicae*, has the capability to colonize a diverse variety of higher plants and accumulates significant level of PHB, when grown on a range of individual carbon sources [107]. The design and development of bacteria and higher plants may help to streamline cost-effective production and to produce novel biomaterials (heteropolymers) for a range of applications [108].

FUTURE PERSPECTIVES

Although many interesting and potentially valuable bio-active products synthesized by endophytes have been identified and characterized, there is much more to be discovered. The search for novel compounds should also be directed towards plants that commonly serve indigenous populations for medicinal purposes. A thorough comprehensive study of the endophytes of each of the medicinally important plants in the world should be made. Microbes associated with plants, and other substrates merit a higher profile in future research programs related to the understanding, management, and sustainable use of biodiversity at a level appropriate both to their numbers and to their economic and environmental importance.

An understanding of the mechanisms enabling endophytes to interact with plants will be essential to fully achieve the biotechnological potential of efficient plant-endophyte partnerships for a range of applications. Historically, the study of endophytes has been the purview of botanists, but the recent discovery of interesting chemical compounds generated by certain endophytes has opened the field to structural chemists and biochemists. Real success in this area will probably only be achieved through an interdisciplinary approach, whereby the knowledge and insight of the botanists is combined with the technology and structural tools of the chemists.

REFERENCES

[1] Bacon CW & White JF (2000) Microbial endophytes. Marcel Dekker Inc, New York.

[2] Strobel GA (2002) Rainforest endophytes and bioactive products *Crit Rev Biotechnol.* 22:315-333.

[3] Hawksworth DC & Rossman AY (1987) Where are the undescribed fungi? *Phytopathology*, 87:888–891.

[4] Dreyfuss MM & Chapela IH (1994) Potential of fungi in the discovery of novel, low-molecular weight pharmaceuticals. In: The discovery of natural products with therapeutic potential, (Gullo VP ed). Butter-worth-Heinemann, London, UK, pp 49–80.

[5] Bultman TL & Murphy JC (2000) Do fungal endophytes mediate wound-induced resistance? In: Microbial endophytes (Bacon CW & White JF eds), Marcel Dekker, New York, pp 421-452.

[6] Demain AL (2000) Microbial natural products: a past with a future. In: Biodiversity: new leads for pharmaceutical and agrochemical industries (Wrigley SK, Hayes MA, Thomas R, Chrystal EJT & Nicholson N eds). The Royal Society of Chemistry, Cambridge, UK, pp 3–16.

[7] Baker D, Mocek U & Garr C (2000) Natural products vs. combinatorials: a case study. In: Biodiversity: new leads for pharmaceutical and agrochemical industries (Wrigley SK, Hayes MA, Thomas R, Chrystal EJT & Nicholson N eds). The Royal Society of Chemistry, Cambridge, UK, pp. 66–72.

[8] Bensky D & Gamble A (1993) Chinese herbal medicine, Materia medica, new ed, Eastland Press Inc., Seattle, Wash.

[9] Grabley S & Thiericke R (1999) Drug discovery from nature, Springer-Verlag, Berlin, Germany.

[10] Strobel G & Daisy B (2003) Bioprospecting for microbial endophytes and their natural products. *Microbiol Mol Biol Rev*. 67:491-502.

[11] Concepcion GP, Lazaro JE & Hyde KD (2001) Screening for bioactive novel compounds. In: Bio-exploitation of filamentous fungi (Pointing SB & Hyde KD eds). Fungal Diversity Press, Hong Kong, Hong Kong, pp 93–130.

[12] Wani MC, Taylor HL, Wall ME, Goggon P & McPhail AT (1971) Plant antitumor agents, VI. The isolation and structure of taxol, a novel antileukemic and antitumor agent from *Taxus brevifolia*. *J American Chem Soc*. 93:2325–2327.

[13] Freeman EM (1904) The seed fungus of *Lolium temulentum* L., the darnel. *Phil Trans*. 196:1-27.

[14] Redlin SC & Carris LM (eds) (1996) Endophytic fungi in grasses and woody plants: systemics, ecology and evolution, APS Press, St Paul, USA.

[15] Pennycook SR & Samuels GJ (1985) *Botryosphaeria* and *Fusicoccum* species associated with ripe fruit rot of *Actinidia deliciosa* (Kiwifruit) in New Zealand Mycotaxon 24:445-458.

[16] Bacon CW & De Battista J (1991) Endophytic fungi of grasses. In Handbook of applied mycology (Arora DK, Rai B, Mukerji KG & Knudsen G eds). Marcel Dekker, New York, pp 231-356.

[17] Hata K, Futai K & Tsuda M (1998) Seasonal and needle age-dependent changes of the endophytic microbiota in *Pinus thunbergii* and *Pinus densiflora* needles Canadian J Bot 76:245-250.

[18] Bills G, Dombrowski A, Pelaez F, Polishook J & An Z (2002) Recent and future discoveries of pharmacologically active metabolites from tropical fungi. In Tropical mycology: Micromyceyes (Watling R, Frankland JC, Ainsworth AM, Isaae S, & Robinson CH eds). CABI Publishing, New York, pp 165-194.

[19] Schutz B (2001) Endophytic fungi: a source of novel biologically active secondary metabolites. In: British Mycological Society, international symposium proceedings: bioactive fungal metabolites-impact and exploitation. University of Wales, Swansea, Wales, pp 20.

[20] Maciá-Vicente JG, Jansoon HB, Abdullah SK, Descals E, Salinas J & Lopez-Llorca LV (2008) Fungal root endophytes from natural vegetation in Mediterranean environments with special references to *Fusarium* spp. *FEMS Microbiol Ecol*. 64:90-105.

[21] Davis EC, Franklin JB, Shaw AJ & Vilgalys R (2003) Endophytic *Xylaria* (Xylariaceae) among liverworts and angiosperms: phylogenetics, distribution, and symbiosis. *American J Bot*. 90:1661-1667.

[22] Brunner F & Petrini O (1992) Taxonomic studies of *Xylaria* species and xylariaceous endophytes by isozyme-electrophoresis. *Mycol Res*. 96:723-733.

[23] Kumaresan V & Suryanarayanan TS (2001) Occurrence and distribution of endophytic fungi in a mangrove community. *Mycol Res*. 105:1388-1391.

[24] Suryanarayanan TS, Murali TS & Venkatesan G (2002) Occurrence and distribution of fungal endophytes in tropical forests across a rainfall gradient. *Canadian J Bot*. 80:818-826.

[25] Arnold AE, Maynard Z, Gilbert GS, Coley PD & Kursar TA (2000) Are tropical fungal endophytes hyperdiverse? *Ecol Lett*. 3:267-274.

[26] Suresh HS, Bhat HR, Dattaraja HS & Sukumar R (1999) Flora of Mudumalai wildlife sanctuary, Nilgiris, Tamil Nadu, Tech. Rep. 64, Centre for Ecological Sciences, Indian Institute of Science, Bangalore, India.

[27] Wilson D (2000) Ecology of woody plant endophytes. In: Microbial endophytes (Bacon CW, White JF eds). Marcel Dekker Inc, New York, pp. 389-420.

[28] Sardi P, Saracchi M, Quaroni B, Borgonovi GE & Merli S (1992) Isolation of endophytic *Streptomyces* strains from surface-sterilized roots. *Appl Environ Microbiol*. 58:2691-2693.

[29] Mundt JO & Hinckle NF (1976) Bacteria within ovules and seeds. *Appl Environ Microbiol*. 32:694-698.

[30] Velázquez E, Rojas M, Lorite MJ, Rivas R, Zurdo-Piňeiro JL, Heydrich M & Bedmar EJ (2008) Genetic diversity of endophytic bacteria which could be find in the apoplastic

sap of the medullary parenchyma of the stem of healthy sugarcane plants J Basic Microbiol 48:118-124.

[31] Araújo WL, Marcon J, Maccheroni W Jr, van Elsas JD, van Vuurde JWL & Azevedo JL (2002) Diversity of endophytic bacterial populations and their interaction with *Xylella fastidiosa* in citrus plants. *Appl Environ Microbiol*. 68:4906-4914.

[32] Abdoulaye SY, Giraud E, Jourand P, Garcia N, Willems A, de Lajudie P, Prin Y, Neyra M, Gillis M, Bivin-Masson C & Dreyfus B (2001) Methylotrophic *Methylobacterium* bacteria nodulate and fix nitrogen in symbiosis with legumes J Bacteriol 183:214-220.

[33] Araújo WL, Saridakis HO, Barroso PAV, Aguilar-Vildoso CI & Azevedo JL (2001) Variability and interactions between endophytic bacteria and fungi isolated from leaf tissues of citrus rootstocks. *Canadian J Microbiol*. 47:229-236.

[34] Pirttilä AM, Laukkane H, Pospiech H, Myllylä R & Hohtola A (2000) Detection of intracellular bacteria in buds of Scotch pine (*Pinus sylvestris* L.) by *in situ* hybridization *Appl Environ Microbiol*. 66:3037-3077.

[35] Ulrich K, Ulrich A & Ewald D (2008) Diversity of endophytic bacterial communities in poplar grown under field conditions FEMS Microbiol Ecol 63:169-180.

[36] Owen NL & Hundley N (2004) Endophytes-the chemical synthesizers inside plants. *Sci Prog*. 87:79-99.

[37] Tan RX & Zou WX (2004) Endophytes: a rich source of functional metabolites. *Nat Prod Rep*. 18:448-459.

[38] Strobel GA, Daisy B, Castillo U & Harper J (2004) Natural products from endophytic microorganisms. *J Nat Prod*. 67:257-268.

[39] Li JY, Sidhu RS, Ford EJ, Long DM, Hess WM & Strobel GA (1998) The induction of taxol production in the endophytic fungus: *Periconia* sp from *Torreya grandifolia. J Indus Microbiol Biotechnol*. 20:259-264.

[40] Zou WX, Meng JC, Lu H, Chen GX, Shi GX, Zhang TY & Tan RX. *J Nat Prod*. 63:1529-1530.

[41] Li JY, Strobel GA, Harper JK, Lobkovsky E & Clardy J (2000) Cryptocin, a potent tetramic acid antimycotic from the endophytic fungus *Cryptosporiopsis cf. quercina* Org Lett 2:767–770.

[42] Miller RV, Miller CM, Garton-Kinney D, Redgrave B, Sears J, Condron M, Teplow D & Strobel GA (1998) Ecomycins, unique antimycotics from *Pseudomonas viridiflava. J Appl Microbiol*. 84:937–944.

[43] Ballio A, Bossa F, Di Giogio P, Ferranti P, Paci M, Pucci P, Scaloni A, Segre A & Strobel GA (1994) Structure of the pseudomycins, new lipodepsipeptides produced by *Pseudomonas syringae* MSU 16H FEBS Lett 355:96–100.

[44] Lee JC, Yang X, Schwartz M, Strobel G & Clardy J (1995) The relationship between an endangered North American tree and an endophytic fungus Chem Biol 2:721–727.

[45] Pulici M, Sugawara F, Koshino H, Uzawa J, Yoshida S, Lobkovsky E & Clardy J (1996) Pestalotiopsin-A and pestalotiopsin-B: new caryophyllenes from an endophytic fungus of *Taxus brevifolia. J Org Chem*. 61:2122–2124.

[46] Li JY & Strobel GA (2001) Jesterone and hydroxy-jesterone antioomycet cyclohexenenone epoxides from the endophytic fungus *Pestalotiopsis jesteri* J Phytochem 57:261–265.

[47] Gomes-Figueiredo J, Pimentel IC, Vicente VA, Pie MR, Kava-Cordeiro V, Galli-Terasawa L, Pereira JO, de Souza AQL & Glienke C (2007) Bioprospecting highly

diverse endophytic *Pestalotiopsis* spp. with antibacterial properties from *Maytenus ilicifolia*, a medicinal plant from Brazil. *Canadian J Microbiol.* 53:1123-1132.

[48] Guo B, Dai J, Ng S, Huang Y, Leong C, Ong W & Carte BK (2000) Cytonic acids A and B: novel tridepside inhibitors of hCMV protease from the endophytic fungus *Cytonaema* species. *J Nat Prod.* 63:602–604.

[49] Worapong J, Strobel GA, Ford EJ, Li JY, Baird G & Hess WM (2001) *Muscodor albus* gen. et sp. nov., an endophyte from *Cinnamomum zeylanicum* Mycotaxon 79:67–79.

[50] Strobel GA, Dirksie E, Sears J & Markworth C (2001) Volatile antimicrobials from a novel endophytic fungus. *Microbiol.* 147:2943–2950.

[51] Stierle A & Strobel G (1995) The search of taxol-producing microorganism among the endophytic fungi of the pacific yew, *Taxus brevifolia* J Nat Prod 58:1315-1324.

[52] Holmes FA, Kudelka AP, Kavanagh JJ, Huber MH, Ajani JA & Valero V (1995) Taxane anticancer agents: basic science and current status. In: ACS Symposium Series 583 (Georg GI, Chen TT, Ojima I & Vyas DM eds). American Chemical Society. Washington, DC, pp 31.

[53] Schiff PB & Horowitz SB (1980) Taxol stabilizes microtubules in mouse fibroblast cells. *Proc Natl Acad Sci.* 77:1561–1565.

[54] Bashyal B, Li JY, Strobel GA & Hess WM (1999) *Seimatoantlerium nepalense*, an endophytic taxol producing coelomycete from Himalayan yew (*Taxus wallichiana*) Mycotaxon 72:33–42.

[55] Strobel GA, Hess WM, Li JY, Ford E, Sears RS & Summerell B (1997) *Pestalotiopsis guepinii*, a taxol producing endophyte of the Wollemi pine, *Wollemia nobilis* Australian J Bot 45:1073-1082.

[56] Wang J, Li G, Lu H, Zheng Z, Huang Y & Su W (2000) Taxol from *Tubercularia* sp. strain TF5, an endophytic fungus of *Taxus mairei*. *FEMS Microbiol Lett.* 193:249-253.

[57] Caruso M, Colombo AL, Fedeli L, Pavesi A, Quaroni S, Saracchi M & Bentrella G (2000) Isolation of endophytic fungi and actinomycetes taxane producers. *Annl Microbiol.* 50:3-13.

[58] Huang Y, Wang J, Li G, Zheng Z & Su W (2001) Antitumor and antifungal activities in endophytic fungi isolated from pharmaceutical plants *Taxus mairei, Cephalotaxus fortunei,* and *Torreya grandis. FEMS Immunol Med Microbiol.* 31:163-167.

[59] Amna TS, Puri C, Verma V, Sharma JP, Khajuria RK, Musarrat J, Spiteller M & Quazi GN (2006) Bioreactor studies on the endophytic fungus *Entrophospora infrequens* for the production of an anticancer alkaloid camptothecin Canadian J Microbiol 52:189-196.

[60] Borel JF & Kis ZL (1991) The discovery and development of cyclosporine. *Transplant Proc.* 23:1867–1874.

[61] Kumar DSS, Cheung HY, Lau CS, Chen F & Hyde KD (2004) *In vitro* studies of endophytic fungi from *Tripterygium wilfordii* with anti-proliferative activity on human peripheral blood mononuclear cells. *J Ethnopharmacol.* 94:295-300.

[62] Rodriguez RJ, Henson J, Volkenburgh EV, Hoy M, Wright L, Beckwith F, Kim Y-O & Redman RS (2008) Stress tolerance in plants via habitat-adapted symbiosis The ISME J 2:404-416.

[63] Lewis DH (1985) Symbiosis and mutualism: crisp concept and soggy semantics. In: The biology of mutualism (Boucher DH ed). Croom Helm Ltd., London, pp 29-39.

[64] Schardl SL, Leuchtmann A & Spiering MJ (2004) Symbioses of grasses with seed-borne fungal endophytes. *Annu Rev Plant Biol.* 55:315-340.

[65] Krings M, Taylor TN, Hass H, Kerp H, Dotzler N & Hermsen EJ (2007) Fungal endophytes in a 400-million-yr-old land plant: infection pathways, spatial distribution, and host responses. *New Phytol.* 174:648-657.

[66] Clement SL, Elberson LR, Youssef NN, Davitt CM & Doss RP (2001) Incidence and diversity of *Neotyphodium* fungal endophytes in tall fescue from Morocco, Tunisia, and Sardinia. *Crop Sci.* 41:570-576.

[67] Bacon CW, Richardson MD & White JF Jr (1997) Modification and uses of endophyte-enhanced turfgrasses: A role for molecular technology. *Crop Sci.* 37:1415-1425.

[68] Porter JK (1994) Chemical constituents of grass endophytes. In: Biotechnology of endophytic fungi of grasses (Bacon CW & White JF Jr). CRC Press, Boca Raton, F.L., pp 103-123.

[69] Dubois T, Gold CS, Paparu P, Athman S & Kapindu S (2006) Tissue culture and the *in vitro* environment. Enhancing plants with endophytes: potential for ornamentals? In: Floriculture, ornamental and plant biotechnology: advances and topical issues (Da Silva JT ed). Global Science Books, London.

[70] Wakelin S, Warren R, Harvey P & Ryder M (2004) Phosphate solubilization by *Penicillium* spp. closely associated with wheat roots. *Biol Fertl Soils*, 40:36-43.

[71] Lee S, Flores-Encarnacion M, Contreras-Zentella M, Garcia-Flores L, Escamilla JE & Kennedy C (2004) Indole-3-acetic acid biosynthesis is deficient in *Gluconoacetobacter diazotrophicus* strains with mutations in cytochrome C biogenesis genes. *J Bacteriol.* 186:5384-5391.

[72] Costa JM & Loper JE (1994) Characterization of siderophore production by the biological control agent *Enterobacter cloacae. Mol Plant Microbe Inter.* 7:440-448.

[73] Pirttilä A, Joensuu P, Pospiech H, Jalonen J & Hohtola A (2004) Bud endophytes of Scots pin produce adenine derivatives and other compounds that affect morphology and mitigate browning of callus cultures. *Plant Physiol.* 121:305-312.

[74] Compant S, Reiter B, Sessitsch A, Nowak J, Clément C & Barka EA (2005) Endophytic colonization of *Vitis vinifera* L., by a plant growth-promoting bacterium, *Burkholderia* sp. Strain. *PsJN Appl Environ Microbiol.* 71:1685-1693.

[75] Azevedo JL, Maccheroni W Jr, Pereira JO, de Araújo WL (2000) Endophytic microorganisms: a review on insect control and recent advances on tropical plants. *Electronic J Biotechnol.* 3:40-65.

[76] Narisawa K, Tokumasu S & Hashiba T (1998) Suppression of clubroot formation in Chinese cabbage by the root endophytic fungus, *Heteroconium chaetospira* Plant Pathol 47:206-210.

[77] Sturz AV, Christie BR & Matheson BG (1998) Association of bacterial endophyte populations from red clover and potato crops with potential for beneficial allelopathy. *Canadian J Microbiol.* 44:162-167.

[78] Sturz AV, Christie BR & Nowak J (2000) Bacterial endophytes: potential role in developing sustainable systems of crop production. *Crit Rev Plant Sci.* 19:1-30.

[79] Sabbour MM & Shadia E-Abd-EI-Aziz (2007) Efficiency of some bioinsecticides against broad bean beetle, *Bruchus rufimanus* (Coleoptera: Bruchidae) Res J Agric Biol Sci 3:67-72

[80] Webber J (1981) A natural control of Dutch elm disease Nature 292:449-451.

[81] Funk CR, Halisky PM, Johnson MC, Siegel MR, Stewart AV, Ahmad S, Hurley RH & Harvey IC An endophytic fungus and resistance to sod webworms: association in *Lolium perenne*. *Biores Technol*. 1:189-191.

[82] Kindler D, Breen JP & Springer T (1990) Russian wheat aphid resistance in cool-season grasses. *J Econ Entomol*. 84:685-692.

[83] Patterson CG, Potter DA & Fannin FF (1992) Feeding deterrency of alkaloids from endophyte-infected grasses to Japanese beetle grubs. *Entomol Expl Appl*. 61:285.

[84] Findlay JA, Buthelezi S, Li GQ, Seveck M & Miller JD (1997) Insect toxins from an endophytic fungus from wintergreen. *J Nat Prod*. 60:1214-1215.

[85] Miles CO, di Mena ME, Jacobs SWL, Garthwaite I, Lane GA, Prestidge RA, Marshal SL, Wilkinson HH, Schardl CL, Ball OJP & Latch CM (1998) Endophytic fungi in indigenous Australasian grasses associated with toxicity to livestock. *Appl Environ Microbiol*. 64:601-606.

[86] Li J, Beatty PK, Shah S & Jensen EE (2007) Use of PCR-targeted mutagenesis to disrupt production of Fusaricidin-type antifungal antibiotics in *Paenibacillus polymyxa*. *Appl Environ Microbiol*. 73:3480-3489.

[87] Rubini MR, Silva-Ribeiro RT, Pomella AWV, Maki CS, Araújo WL, dos Santos DR & Azevedo JL (2005) Diversity of endophytic fungal community of cacao (*Theobroma cacao* L.) and biological control of *Crinipellis perniciosa*, causal agent of Witches Broom Disease Internl J Biol Sci 1:24-33.

[88] Ball OJP, Bernard EC & Gwinn KD (1997) Effect of selected *Neotyphodium lolii* isolates on root knot nematode (*Meloidogyne marylandi*) numbers in perennial ryegrass. In: Proceeding of 50th New Zealand Plant Protection Society Conference. Palmerston North, NZ, pp 65-68.

[89] Kimmons CA, Gwinn KD & Bernard EC (1990) Nematode reproduction on endophyte-infected and endophyte-free tall fescue. *Plant Dis*. 74:757-761.

[90] Kwon RK, Son SW, Han HR, Choi GJ, Jang KS, Choi YH, Lee S, Sung ND & Kim J-C (2007) Nematicidal activity of Bikaverin and Fusaric acid isolated from *Fusarium oxysporum* pine wood nematode, *Bursaphelenchus xylophilus*. *Plant Pathol*. 23:318-321.

[91] Rajendran L, Saravanakumar D, Raghchander T & Samiyappan R (2006) Endophytic bacterial induction of defense enzyme against bacterial blight of cotton. *Phytopathol Medit*. 45:203-214.

[92] Sabaratnam S & Beattie GA (2003) Differences between *Pseudomonas syringae* B728a and *Pantoea agglomerans* BRT98 in epiphytic and endophytic colonization of leaves. *Appl Environ Microbiol*. 69:1220-1228.

[93] Bacon CW & Hinton DM (2002) Endophytic and biological control potential of *Bacillus mojavensis* and related species. *Biol Control*. 23:274-284.

[94] Benhamou N, Gagne S, Ie Quere D & Dehbi L (2000) Bacteria-mediated induced resistance in cucumber: beneficial effect of the endophytic bacterium *Serratia plymuthica* on the protection against infection by *Pythium ultimum* Phytopathol 90:45-56.

[95] Kim KJA, Yang YJ & Kim J (2002) Production of alpha-glucosidase inhibitor by beta-glucosidase inhibitor producing *Bacillus lentimorbus* B-6 J Microbiol Biotechnol 12:895-900.

[96] Sadfi N, Cherif M, Fliss I, Boudabbous A & Antoun H (2001) Evaluation of bacterial isolates from salty soils and *Bacillus thuringiensis* BRT98 strains for the biocontrol of Fusarium dry rot of potato tubers. *Plant Pathol.* 83: 101-118.

[97] Ryder MH, Yan Z, Terrace TE, Rovira AD, Tang WH & Correll RL (1999) Use of strains of *Bacillus* isolated in China to suppress take-all and *Rhizoctonia* root rot, and promotes seedling growth of glasshouse-grown wheat in Australian soils. *Soil Biol Biochem.* 31:19-29.

[98] Di Fiore S & Del Gallo M (1995) Endophytic bacteria: their possible role in the host plants. In: *Azospirillum* VI and related microorganisms (Fendrik I, Del Gallo M, Vanderleyden J & De Zamarocyz M). Springer-Verlag, Berlin, pp 169-187.

[99] Lodewyckx C, Vangrosveld J, Porteous F, Moore ERB, Taghavi S, Mezgeay M & van der Lelie D (2002) Endophytic bacteria and their potential applications. *Crit Rev Plant Sci.* 21:583-606.

[100] 100. Siciliano SD, Fortin N, Mihoc A, Wisse G, Labelle S, Beanmier D, Ouellette D, Roy R, Whyte CG, Banks MK, Schwab P, Ken L & Greer CW (2001) Selection of specific bacterial genotypes by plants in responses to soil contamination. *Appl Environ Microbiol.* 67:2469-2475.

[101] Van Aken B, Peres C, Doty S, Yoon J & Schnoor J (2004) *Methylobaacterium populi* sp. nov., a novel aerobic, pink-pigmented, facultatively methylotrophic, methane-utilising bacterium isolated from poplar trees (*Populus deltoides* x *nigra* DN34). *Evol Microbiol.* 54:1191-1196.

[102] Lodewyckx C, Taghavi S, Mergeay M, Vangronsveld J, Clijsters H & van der Lelie D (2001) The effect of recombinant heavy metal resistant endophytic bacteria in heavy metal uptake by their host plant. *Internatl J Phytoremed.* 3:173-187.

[103] Newman L & Reynolds C (2005) Bacteria and phyto-remediation: new uses for endophytic bacteria in plants Trends Biotechnol 23:6-8.

[104] Taghavi S, Barac T, Greenberg B, Borremans B, Vangronsveld J & van der Lelie D (2005) Horizontal gene transfer to endogenous endophytic bacteria from poplar improved phyto-remediation of toluene. *Appl Environ Microbiol.* 71:8500-8505.

[105] Lemoigne M (1926) Produit de déshydratation et de polymérisation de l'acide β-oxybutyrique. *Chem Biol.* 8 :770-782.

[106] Kalia VC, Chauhan A, Bhattacharyya G (2003) Genomic databases yield novel bioplastic producers. *Nature Biotechnol.* 21:845-846.

[107] Catalán AI, Ferreira F, Gill PR & Batista S (2007) Production of polyhydroxyalkanoate by *Herbaspirillum seropedicae* grown with different sole carbon sources and on lactose when engineered to express the *lac Z, lac Y* genes. *Enzyme Microbial Technol.* 40:1352-1367.

[108] Aldor LS & Keasling JD (2003) Process design for microbial plastic factories: metabolic engineering of polyhydroxyalkanoates. *Curr Opin Biotechnol.* 14:475-483.

[109] Barac T, Taghavi S, Borremans B, Provoost A, Oeyen L, Colpaert JV, Vangronsveld J & van der Lelie D (2004) Engineered endophytic bacteria improve phyto-remediation of water-soluble, volatile, organic pollutants. *Nature Biotechnol.* 22:583-588.

[110] Mannisto MK, Tiirola MA & Puhakka JA (2001) Degradation of 2,3,4,6-tetrachlorophenol at low temperature and low dioxygen concentrations by phylogenetically different groundwater and bioreactor bacteria *Biodegradation*, 12:291-301.

[111] Taghavi S, Barac T, Greenberg B, Borremans B, Vangronsveld J & van der Lelie D (2005) Horizontal gene transfer to endogenous endophytic bacteria from poplar improved phyto-remediation of toluene *Appl Environ Microbiol.* 71:8500-8505.

[112] Porteous- Moore F, Barac T, Borremans B, Oeyen L, Vangronsveld J, van der Lelie D, Campbell D & Moore ERB (2006) Endophytic bacterial diversity in poplar trees growing on a BTEX-contaminated site: the characterization of isolates with potential to enhance phytoremediation. *Syst Appl Microbiol.* 29:539-556.

[113] Germaine K, Liu X, Cabellos G, Hogan J, Ryan D & Dowling DN (2006) Bacterial endophyte-enhanced phyto-remediation of the organochlorine herbicide 2,4-dichlorophenoxyacetic acid. *FEMS Microbiol Ecol.* 57:302-10.

[114] Siciliano SD, Goldie H & Germida JJ (1998) Enzymatic activity in root exudates of dahurian wild rye (*Elymus dauricus*) that degrades 2-chlorobenzoic acid. *J Agric Food Chem.* 46:5-7.

In: Mushrooms: Types, Properties and Nutrition
Editors: S. Andres and N. Baumann

ISBN 978-1-61470-110-1

Chapter 6

Chemical Composition and Nutritional Value of European Species of Wild Growing Mushrooms

Pavel Kalač*
Department of Applied Chemistry, Faculty of Agriculture, University of South Bohemia, CZ-370 05 České Budějovice, Czech Republic

Abstract

Tens of wild growing mushroom species are widely consumed as a delicacy in a part of Europe. Knowledge of their nutritional value has so far been fragmentary mainly due to the very limited information on the bioavailability of their constituents. Dry matter content varies usually between 80 and 140 g kg^{-1}. Usual medians of crude protein, lipid and ash content are about 25, 3 and 8 g per 100 g of dry matter, respectively. Various carbohydrates form the rest. However, great variations occur. Energy is low, about 150 kJ per 100 g of fresh mushrooms. The proportion of essential amino acids seems to be nutritionally favorable, while the content of n-3 fatty acids is negligible. Chitin, glycogen, mannitol and trehalose are typical carbohydrate constituents. Potassium is the highly prevailing element within minerals. Relatively high proportion of fiber, health-promoting β-glucans, compounds with antioxidation activity and flavor constituents are the topics provoking an increasing interest of both researchers and consumers. Nevertheless, several popular species accumulate high levels of cadmium, mercury and lead if growing on heavily polluted soils.

Keywords: edible mushrooms; nutritional value; proximate composition; proteins; lipids; carbohydrates; minerals; vitamins; flavor constituents; detrimental compounds.

* E-mail: kalac@zf.jcu.cz; Phone: +420 387 772 657; Fax: +420 385 310405.

1. Introduction

The collection of wild mushrooms can be seen also as a relic drive of bygone hunters and gatherers. In some parts of the world, e.g. in Slavonic countries of central and eastern Europe, it has been a lasting part of cultural heritage. Mushroom picking in forests and grasslands has been a highly evaluated recreational activity* and mushrooms in their natural habitat are regarded for their aesthetic value.

Mushrooms have been mostly collected as a delicacy for pickers own consumption however the collection has been an economic activity for a part of rural population. For instance, the picking is a "national hobby" in the Czech Republic interesting about 70% of the population with a statistical mean of 5.6 kg of fresh mushrooms per household yearly (Šišák, 2007). Nevertheless, some individuals consume over 10 kg yearly.

Specific aroma and texture are appreciated in mushrooms for culinary use. Tens of edible species can be variously preserved and culinary-processed. The knowledge of composition and nutritional value of wild growing mushrooms has been limited as compared with vegetables. This is understandable, because the mushrooms are perceived only as a delicacy. Moreover, their consumption in many developed countries has been marginal and thus off an interest of researchers.

Data on chemical composition and nutritional value of European edible mushroom species were reviewed (Bernaś et al., 2006; Kalač, 2009). Thus, overall knowledge from the reviews supplemented with current information will be given in following sections. Only the weightiest articles until 2007 will be cited, further papers from that period are referred in the reviews.

Medicinal mushrooms, mostly originating from East Asia, will not be referred here. Interesting information on many compounds participating in chemical defense strategies of mushrooms is available from a review of Spitteler (2008).

Various weight units will be used in the following sections according to common convention, e.g. mg per kg of dry matter for trace elements, but mg per 100g of fresh matter for vitamins.

2. Mycological Terms

Some mycological terms used in this chapter are given in Figure 1. The term mushroom will be used for a fruit body, mostly aboveground, of higher fungi. A fruit body is formed from spacious underground mycelia by the process of fructification. The lifetime of the bulk of fruit bodies is only 10-14 days.

According to their nutritional strategy, mushrooms can be divided into three groups. Mycorrhizal (symbiotic) species form a close, mutually favorable relationship with their host plant, mostly a tree. Saprotrophic species (or saprophytes) live on and metabolically consume organic matter. Parasitic species live on other species in a non-symbiotic relationship.

* Forests are freely accessible with some limitations (nature reserves, military areas, etc.)

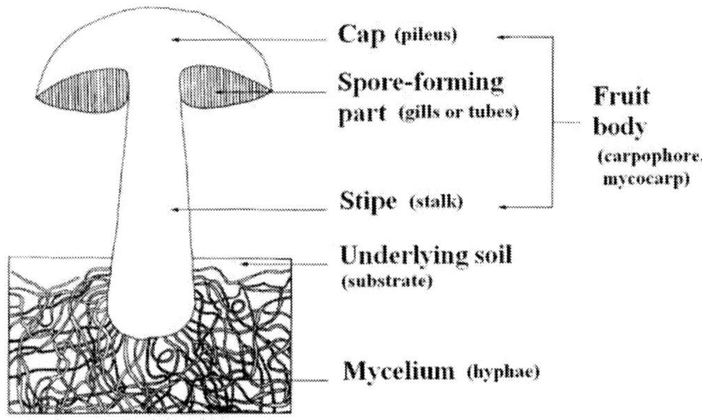

Figure 1. A sketch of a typical mushroom.

3. Dry Matter, Proximate Composition, Energy Value and Bioavailability

Dry matter (DM) of mushrooms is very low, usually in the range of 80-140 g kg^{-1}. In case that factual DM content is unknown, value 100 g kg^{-1} (10%) is commonly used for the calculations. High water content and activity participate in the short shelf life of fruit bodies. Proportion of DM increases during mushroom cooking due to water shrinkage. Dried mushrooms are known for their hygroscopicity.

Proximate composition of dry matter of selected mushroom species is given in Table 1. Carbohydrates content is calculated as [100 – (moisture + crude protein + lipids + ash)]. Considerable differences in the composition are evident not only among species but also within the species among data of different laboratories. Such differences can be partially explained by varied stage of fruit bodies maturity and also by different analytical methods. The data of Table 1 should be thus interpreted as a general information.

As results from Table 1, carbohydrates and crude protein (median 27.5 g $100g^{-1}$ DM) are the two main components of mushroom dry matter. Low dry matter and lipid contents (median only 2.79 g $100g^{-1}$ DM) result in low energy values (median 1,552 kJ $100g^{-1}$ DM). If the median is expressed per "a standard mushroom" containing 100 g DM per kg of fresh matter, 100 g of such fresh mushroom has energy 155.2 kJ or 37.2 kcal. Thus, mushrooms are a delicacy of low energy value.

Information on digestibility and bioavailability of mushroom nutritional constituents has been scarce. A high proportion of indigestible chitin (see section on carbohydrates) apparently limits availability of other components. Nevertheless, credible data for wild growing mushrooms have been lacking.

Moreover, most data deal with fresh mushrooms. Information has been scarce on the changes of the individual constituents during various preservation methods, under storage and during different cooking processes.

Table 1. Dry matter content (g $100g^{-1}$), proximate composition (g $100g^{-1}$ dry matter) and energy (kJ $100g^{-1}$ dry matter) of mushroom fruit bodies. Content of carbohydrates is calculated: [100 – (moisture + crude protein + lipids + ash)]

Species	n	Dry matter	Crude protein	Lipids	Ash	Energy	Reference
Agaricus campestris	3	14.9	38.9	2.70	3.50	1,570	Beluhan and Ranogajec, 2011
Armillariella mellea	15	12.8	17.15	2.10	7.95	-	Ouzuni et al., 2009
	3	11.7	16.4	5.56	6.78	1,670	Vaz et al., 2011b
	1	16.2	14.8	6.08	3.16	1,741	Colak et al., 2009
Boletus aereus	15	12.4	19.05	4.47	6.25	-	Ouzuni et al., 2009
Boletus edulis	3	12.2	36.9	2.92	5.30	1,488	Beluhan and Ranogajec, 2011
Calocybe gambosa	3	13.9	36.65	1.34	7.98	1,398	Beluhan and Ranogajec, 2011
	3	9.1	15.5	0.83	13.9	1,453	Vaz et al., 2011b
Cantharellus cibarius	3	7.6	53.7	2.89	11.5	1,541	Barros et al., 2008c
	15	17.4	15.1	2.88	9.44	-	Ouzuni et al., 2009
	3	14.2	30.9	1.90	8.80	1,488	Beluhan and Ranogajec, 2011
	1	12.1	23.9	1.40	7.78	1,533	Colak et al., 2009
Craterellus cornucopioides	3	10.1	47.2	4.87	10.1	1,730	Beluhan and Ranogajec, 2011
	1	10.4	35.1	5.89	10.3	1,618	Colak et al., 2009
Hydnum repandum	1	6.7	23.9	8.80	11.4	1,810	Colak et al., 2009
Laccaria laccata	3	11.8	62.8	3.76	20.7	1,401	Heleno et al., 2009
Lepista nuda	3	6.2	59.4	1.77	18.5	1,395	Barros et al., 2008c
	15	8.7	24.1	3.23	6.03	-	Ouzuni et al., 2009
Lycoperdon perlatum	3	11.4	17.1	4.41	31.9	1,145	Barros et al., 2008c
	1	30.0	31.5	10.6	2.00	1,854	Colak et al., 2009
Macrolepiota procera	3	13.2	24.2	2.23	5.37	1,630	Beluhan and Ranogajec, 2011
Pleurotus ostreatus	2	-	30.3	1.10	13.2	-	Akyüz and Kirbağ, 2010
	3	11.7	24.9	2.08	7.62	1,552	Beluhan and Ranogajec, 2011
Ramaria botrytis	3	10.2	39.9	1.37	8.80	1,549	Barros et al., 2008c

n ... number of samples. Contents of crude protein were recalculated for data from papers where factor of 6.25 was used. Factor of 4.38 is thought to be credible for mushrooms.

4. Proteins and Amino Acids

The informative data on crude protein content in fourteen mushroom species are given in Table 1. The values are comparable with those reported by Bauer-Petrovska (2001) in 47 species of Macedonian edible mushrooms. She determined the mean crude protein content of 22.8% with wide variations of 9.8-36.5% in DM. A similar value of 24.9 ±1.75 % in DM reported Uzun et al. (2009) for 30 species of Turkish mushrooms. Crude protein variability within a species can be seen from Table 1 (e.g. for *C. cibarius*).

For the calculation of crude protein in mushrooms, if determined by the Kjeldahl method, a specific converting factor has to be used due to the high proportion of non-protein nitrogen, particularly in chitin. The factor of 4.16 was recommended by Bauer-Petrovska (2001), who observed the mean proportion of 33.4% of non-protein nitrogen (from total nitrogen) in numerous samples. The factor of 4.38 has been mostly used in recent publications. This value was used for the correction of data from papers using factor of 6.25 collated in Table 1 and also of both the papers in the previous paragraph.

A unique paper on protein fractions in 24 mushroom species reported mean levels of 24.8, 11.5, 7.4, 11.5, 5.7, 5.3 and 33.8% from total proteins of albumins, globulins, glutelin-like material, glutelins, prolamins, prolamine-like material and residues, respectively (Bauer-Petrovska, 2001).

The information on changes of protein content and digestibility during fruit bodies development and on the distribution within a fruit body remains unclear. Moreover, limited data have been available also on amino acid composition of proteins. It seems from limited data (Kalač, 2009) that the composition of mushroom proteins is of higher nutritional value than that of most plant proteins.

Protein content, expressed as a proportion in DM, remained nearly stable during air-drying of mushrooms at 40 °C or on freezing to –20 °C, while boiling of fresh mushrooms caused a significant decrease (Barros et al., 2007). In cultivated *Pleurotus ostreatus* (oyster mushroom, hiratake), freezing processes followed by 12-month storage resulted in significant decrease of alanine, glutamine, cysteine and tyrosine contents, while arginine, glycine, serine, methionine and threonine contents decreased in the canned mushroom (Jaworska et al., 2011).

The reported contents of total free amino acids vary. While Ribeiro et al. (2008a) determined range between 0.15 and 2.27 % in DM for *Fistulina hepatica* and *Boletus edulis,* respectively, Beluhan and Ranogajec (2011) reported between 4.39 and 7.20 % for *Entoloma clypeatum* and *B. edulis*, respectively. Dembitsky et al. (2010) found arginine, alanine, glutamine and glutamic acid to be the major free amino acids in 15 species of genus *Boletus* and related genera.

The contents of free indispensable (essential) amino acids in various species are given in Table 2. Their nutritional contribution is limited, but some of free amino acids participate in the taste of mushrooms.

Wild mushrooms were formerly called "meat of poverty" in central Europe. However, usual protein content of about 2.5 % in fresh matter, limited knowledge on amino acid composition and lacking information on digestibility challenge the often used characterization of mushrooms as a source of valuable proteins.

5. Lipids

The contents of total lipids (crude fat) ranges mostly from 2% to 6% of dry matter (Tables 1 and 3). Higher values have been reported rarely. Neutral lipids are usually extracted with hexane or another non-polar solvent.

Information on neutral and polar lipid contents in ten species of mushrooms growing in eastern Canada is available (Pedneault et al., 2008). Polar lipids accounted for more than 50% in most species. For instance, the proportions of neutral and polar lipids were 3.3% and 3.1% in DM, respectively, in *Boletus edulis,* while 1.4% and 4.9% in DM, respectively, in *Agaricus arvensis.*

Several tens of fatty acids were identified in mushroom lipids. Nevertheless, only three acids markedly prevail: nutritionally desirable polyunsaturated linoleic acid (C18:2c,n-6), monounsaturated oleic acid (C18:1c,n-9) and undesirable saturated palmitic acid (C16:0) (Table 3). However, low proportion of oleic acid seems to be characteristic for the species of genus *Agaricus* and in a limited extent also for *Cantharellus cibarius* (Table 3). The proportions of nutritionally neutral saturated stearic acid (C18:0) and especially of essential α-linolenic acid (C18:3,n-3) are low. Only selected species are collated in Table 3. More information is available in the cited references.

Other fatty acids are only minor components of mushroom lipids. The contents of odd- and branched-chain fatty acids and hydroxy fatty acids are negligible. Nutritionally undesirable elaidic acid, being *trans* isomer of oleic acid, was firstly reported by Pedneault et al. (2008), however, at low level of <0.1% to 0.3% of total fatty acids.

Barros et al. (2007) observed some changes in fatty acid profiles of dried mushrooms, particularly a decrease of linoleic acid proportion and an increase of monounsaturated acids.

Overall, the nutritional value of lipids of wild growing mushrooms is limited due to low total lipid content and a low proportion of desirable n-3 fatty acids. Linoleic acid is nonetheless an important precursor for the development of attractive smell of dried mushrooms (see section 9).

6. Carbohydrates and Fiber

Carbohydrates constitute the prevailing component of mushroom dry matter, usually about 50-60%. This group comprises various compounds – monosaccharides, their derivatives and oligosaccharides (commonly called sugars) and both reserve and construction polysaccharides (glycans).

Mannitol and α,α-trehalose (Table 4) are the main representatives of the polyols and oligosaccharides, respectively. The contents of both the sugars vary widely both among species and probably also within the individual species, as may be supposed from data for *Cantharellus cibarius* determined in the same laboratory. Drying and freezing decreased the contents of mannitol and trehalose in a lower extent than cooking of mushrooms (Table 5). Unfortunately, the contents in fresh mushrooms prior to the preservation or cooking were not reported.

Table 2. Content of free indispensable amino acids (mg per 100g dry matter) in selected mushrooms

Species	Val	Leu	Ile	Thr	Met	Lys	Phe	Trp	Reference
Agaricus campestris	119	24	41	79	71	53	28	8	Beluhan and Ranogajec, 2011
Amanita rubescens	16	50	25	48	-	16	29	18	Ribeiro et al., 2008a
Boletus edulis	45	43	29	95	-	52	57	72	Ribeiro et al., 2008a
	ND	58	48	19	41	217	104	ND	Tsai et al., 2008
	141	47	12	91	74	55	19	3	Beluhan and Ranogajec, 2011
Calocybe gambosa	127	45	ND	57	101	419	8	9	Beluhan and Ranogajec, 2011
Cantharellus cibarius	ND	14	10	18	-	36	10	15	Ribeiro et al., 2008a
	134	21	ND	90	41	57	6	2	Beluhan and Ranogajec, 2011
Craterellus cornucopioides	172	105	ND	456	16	469	85	12	Beluhan and Ranogajec, 2011
Macrolepiota procera	139	38	19	58	69	411	45	9	Beluhan and Ranogajec, 2011
Pleurotus ostreatus	121	29	ND	70	12	465	9	1	Beluhan and Ranogajec, 2011
Russula cyanoxantha	119	36	45	49	-	20	40	17	Ribeiro et al., 2008a
Suillus granulatus	65	13	50	30	-	12	19	22	Ribeiro et al., 2008a
Suillus luteus	54	7	11	14	-	15	18	13	Ribeiro et al., 2008a
Tricholoma equestre	42	102	71	68	-	252	76	20	Ribeiro et al., 2008a

Val...valine; Leu...leucine; Ile...isoleucine; Thr...threonine; Met...methionine; Lys...lysine; Phe...phenylalanine; Trp...tryptophan; ND ... not detected

Table 3. Lipids (crude fat) content (% of dry matter) and proportion of major fatty acids (% of total fatty acids) in selected mushroom species

Species	Lipids	Palmitic acid	Stearic acid	Oleic acid	Linoleic acid	α-Linolenic acid	Reference
Agaricus arvensis	6.2	20.2	5.5	4.5	58.7	0.1	Pedneault et al., 2008 *
Agaricus bisporus	0.92	10.0	4.1	1.3	75.7	0.1	Barros et al., 2008b
Agaricus campestris	13.6	16.5	3.0	4.0	68.8	0.2	Pedneault et al., 2008 *
Agaricus silvaticus	2.05	11.7	1.4	6.7	74.8	0.1	Barros et al., 2008b
Agaricus silvicola	2.43	10.0	2.6	3.5	76.5	0.1	Barros et al., 2008b
Armillariella mellea	5.56	11.0	3.5	47.7	27.7	Traces	Vaz et al., 2011b
Amanita rubescens	10.6	14.5	4.5	58.0	19.0	0.1	Pedneault et al., 2008 *
	-	22.7	16.1	40.4	21.1	Traces	Ribeiro et al., 2009
Boletus edulis	4.60	10.0	2.8	39.7	44.3	0.1	Barros et al., 2008b
	6.4	10.5	3.2	36.2	42.7	0.2	Pedneault et al., 2008 *
	-	17.3	10.1	37.9	24.7	Traces	Ribeiro et al., 2009
	-	8.9	3.0	29.4	51.7	1.2	Dembitsky et al., 2010
Calocybe gambosa	1.05	15.2	2.1	18.1	57.8	0.5	Barros et al., 2008b
	0.83	13.6	3.2	32.5	43.9	0.9	Vaz et al., 2011b
Cantharellus cibarius	4.49	13.1	6.5	10.8	53.6	0.1	Barros et al., 2008b
	2.89	7.2	3.3	8.1	50.0	0.1	Barros et al., 2008c
	-	1.9	0.8	17.8	78.8	Traces	Ribeiro et al., 2009
Craterellus cornucopioides	4.88	6.7	7.8	51.9	23.7	0.1	Barros et al., 2008b
Laccaria laccata	3.76	11.6	2.0	60.7	20.5	0.4	Heleno et al., 2009
Leccinum scabrum	-	9.7	3.4	31.7	45.8	3.1	Dembitsky et al., 2010
Lepista nuda	1.77	11.8	2.4	29.5	51.5	0.2	Barros et al., 2008c
Lycoperdon perlatum	4.41	12.9	3.0	4.6	70.7	0.2	Barros et al., 2008c
Marasmius oreades	2.99	13.8	1.7	28.5	50.7	0.1	Barros et al., 2008b
Ramaria botrytis	1.37	9.9	2.4	43.9	38.3	Traces	Barros et al., 2008c

Table 3. (Continued)

Suillus granulatus	9.9	13.0	3.0	33.7	44.6	0.3	Pedneault et al., 2008 *
	-	3.7	3.6	62.8	35.4	Traces	Ribeiro et al., 2009
Suillus luteus	-	6.3	1.8	61.9	36.7	Traces	Ribeiro et al., 2009
	-	7.6	4.1	37.8	44.1	0.7	Dembitsky et al., 2010
Tricholoma imbricatum	1.88	7.4	4.1	51.5	33.0	0.2	Heleno et al., 2009
Xerocomus badius	-	15.1	2.1	36.5	38.2	1.6	Dembitsky et al., 2010

*... Data on fatty acid proportion deal with neutral lipids fraction.

Table 4. Content of soluble sugars (g $100g^{-1}$ dry matter) in selected mushroom species

Species	Mannitol	Maltose	Trehalose	Total sugars	Reference
Agaricus bisporus	19.6	ND	0.8	20.9	Barros et al., 2008b
Agaricus silvaticus	2.7	0.4	0.3	3.7	Barros et al., 2008b
Agaricus silvicola	6.1	0.6	0.7	7.8	Barros et al., 2008b
Boletus edulis	3.5	ND	9.7	13.5	Barros et al., 2008b
Calocybe gambosa	0.3	ND	8.0	9.1	Barros et al., 2008b
Cantharellus cibarius	8.3	ND	6.1	14.5	Barros et al., 2008b
	1.1	ND	0.8	1.9	Barros et al., 2008c
Craterellus cornucopioides	10.7	ND	0.1	10.8	Barros et al., 2008b
Laccaria laccata	0.6	-	5.8	6.5	Heleno et al., 2009
Lepista nuda	Traces	0.1	0.8	0.9	Barros et al., 2008c
Lycoperdon perlatum	Traces	ND	0.3	0.3	Barros et al., 2008c
Marasmius oreades	2.4	ND	10.5	13.5	Barros et al., 2008b
Ramaria botrytis	1.2	ND	0.2	1.4	Barros et al., 2008c
Tricholoma imbricatum	10.5	-	6.6	17.1	Heleno et al., 2009

ND...not detected

Table 5. Soluble sugars content (g $100g^{-1}$ dry matter) in preserved and cooked mushrooms

Species	Mannitol			Trehalose			Total sugars		
	D	F	C	D	F	C	D	F	C
Lactarius deliciosus	15.4	13.9	10.2	0.9	3.5	2.2	16.3	17.4	12.4
Macrolepiota mastoidea	7.8	4.6	7.5	4.1	5.1	ND	12.0	9.7	7.5
Macrolepiota procera	4.7	6.5	2.3	2.9	7.6	1.2	7.7	14.1	3.5
Sarcodon imbricatus	19.6	25.3	11.8	6.0	5.0	3.4	25.6	30.3	15.2

D...dried at 40 °C; F...frozen at –20 °C; C...cooked with olive oil, salt and onions; ND...not detected
Aadapted from Barros et al., 2007.

Mannitol participates in volume growth and firmness of fruit bodies. Other water-soluble sugars, namely glucose, arabinose, maltose and melezitose, are reported in some papers as minor components.

The reserve polysaccharide of mushrooms is glycogen. The usual content is 5-10% of dry matter. Glycogen is a common component of widely consumed meat and liver and its low intake from mushrooms thus seems to be of low nutritional importance.

Chitin is a water-insoluble structural N-containing polysaccharide, accounting for up to 80-90% of dry matter in mushroom cell walls. Data on its content in both wild growing and cultivated mushrooms have been scarce. The contents about 5-10% of dry matter seem to be taken into consideration. Chitin contents increased with the maturation of several cultivated species and decreased during cooking of fruit bodies (Dikeman et al., 2005). Chitin is indigestible for humans, and apparently decreases the digestibility of other mushroom components.

It is interesting that mushrooms contain as the main polysaccharides glycogen and chitin, occurring in animals, not starch and cellulose as plants.

Great attention has recently been focused on mushroom β-glucans, due to their health-positive effects. β-Glucan research and application have been successful, mainly in the east-Asian countries and deal mostly with cultivated mushroom species. Moreover, numerous mushroom polysaccharides are regarded as a potential source of prebiotics. Recent knowledge is auspicious particularly for cultivated *Pleurotus ostreatus* and *P. eryngii* (Aida et al., 2009).

Information on dietary fiber content has also been very limited. Data resulting from several reports (see Kalač, 2009) are about 4-9% and 22-30% for soluble and insoluble fiber, respectively. It is apparent that mushrooms contain other structural polysaccharides in addition to chitin. A single information reported hemicelluloses and pectic substances. Thus, relatively high content of particularly insoluble fiber seems to be a nutritional advantage of mushrooms. Further research into mushroom fiber has been however needed. Moreover, information on changes of polysaccharides during various preservation treatments has been lacking.

7. Mineral Composition

Ash content of mushrooms is usually 5-12% of dry matter (Table 1) and its variability seems to be lower than that of crude protein and carbohydrates. Numerous data on the contents of major elements, and particularly on trace elements, have been available.

Information on inorganic anions has been very rare. Isildak (2009) reported contents of eight anions in eight wild growing mushroom species. The decreasing order of medians was sulfate (2.26 % of DM), nitrate (2.04 % of DM) and chlorate (0.36 % of DM). Contents of phosphate, chloride and nitrite were considerably lower and fluorides and bromides were undetectable. The initial data on inorganic iodine were recently published by Vetter (2010). Wild growing mushrooms contained only low level of 284 ± 211 μg kg^{-1} DM. The highest iodine mean contents of 601, 559, 451 and 449 μg kg^{-1} DM were determined in *Calvatia excipuliformis*, *Macrolepiota procera*, *Amanita rubescens* and *Lepista nuda*, respectively. Commonly, mycorrhizal species had higher iodine content than the wood-decaying ones.

7.1. Major Elements

Usual contents of major elements are given in Table 6. The data are collated from original papers cited in a review (Kalač, 2009). Potassium is the prevailing element followed by phosphorus, while calcium and sodium are at the opposite end. However, exceptions do exist. For instance, calcium content was 0.07-0.52 % of DM in seven species of genus *Morchella* (Gursoy et al., 2009).

Table 6. Usual contents (% of dry matter) of major elements in wild growing mushrooms

Element	Content
Sodium	0.01 – 0.04
Potassium	2.0 – 4.0
Calcium	0.01 – 0.05
Magnesium	0.08 – 0.18
Phosphorus	0.5 – 1.0
Sulfur	0.1 – 0.3

Adapted from Kalač, 2009.

Potassium is not distributed evenly within fruit bodies. Its content decreases in the order cap > stipe > spore-forming part > spores. Potassium is highly accumulated. Its levels are between 20- and 40-fold higher in fruit bodies than in the underlying soils. Similar accumulating abilities were observed for phosphorus. On the contrary, bioaccumulation was not reported for sodium and calcium. Magnesium contents in fruit bodies are even lower than those in surface, mostly organic horizons of soils, from which mycelium uptakes the nutrients.

Overall, ash content of mushrooms is somewhat higher or comparable with most of vegetables. Mushrooms contain elevated levels of phosphorus and potassium. However, the availability of these elements remains unknown.

7.2. Trace Elements

Several hundreds of original papers have been published on many trace element contents in wild growing mushrooms since the 1970s. Overall information is available in a review with numerous references therein (Kalač, 2010). The basic information for 15 trace elements is given in Table 7. The usual contents refer to values reported for non-accumulating species from unpolluted rural areas.

The contents are considerably higher, often by one and sometimes by even two orders of magnitude than those in Table 7, in some accumulating species or in mushrooms growing in heavily contaminated sites, such as in the vicinity of metal smelters, in mining areas or within cities. Such values are markedly higher than those observed in contaminated plant products. Extremely high bioaccumulation from underlying soils is characteristic for cadmium and mercury. However, the relationship between soil contamination with a trace element and its content in fruit body is not tight enough to enable usage of a mushroom species as a reliable bioindicator of local pollution.

Relatively high selenium content in the attractive group of *Boletus* mushrooms, particularly in king bolete (*B. edulis*) and closely related boletes (Costa-Silva et al., 2011), seemed to be promising as a source of this element deficient throughout Europe. However, low availability was reported in the 1980′s.

Table 7. Usual content of 15 trace elements (mg kg^{-1} dry matter) in fruit bodies of mushrooms from unpolluted areas, and accumulating genera and species

Element	Content	Accumulators
Aluminum	20 - 150	*Amanita rubescens, Leccinum scabrum, Xerocomus chrysenteron*
Antimony	< 0.1	*Suillus* spp.
Arsenic	0.5 - 5	*Laccaria amethystea,* saprotrophic genera
Barium	2 - 4	
Cadmium	1 - 5	*Agaricus* spp. (group *flavescentes*)
Cobalt	< 0.5	*Ramaria largentii, Hygroporus eburneus, Cantharellus cibarius*
Copper	20 – 100	*Agaricus macrosporus, A. silvicola, Macrolepiota procera, M. rhacodes*
Chromium	0.5 - 5	*Morchella elata, Armillariella mellea, Macrolepiota procera, Marasmius oreades*
Iron	50 - 300	*Suillus variegatus, S. luteus, Armillariella mellea*
Lead	< 1 - 5	some saprotrophic species
Manganese	10 - 60	*Boletus edulis, Macrolepiota procera*
Mercury	< 0.5 - 5	*Boletus edulis, B. pinophilus, Calocybe gambosa*
Nickel	< 1 - 15	*Laccaria amethystina, Coprinus comatus*
Selenium	< 2 - 20	*Albatrellus pes-caprae, Boletus edulis, B. pinophilus, B. reticulatus (syn. B. aestivalis)*
Zinc	25 - 200	*Calvatia utriformis, Lycoperdon perlatum*

Adapted from Kalač, 2010.

Most of the elements are distributed unevenly within a fruit body. Usually, the highest contents are observed in the spore-forming part of cap (but not in spores), lower in the rest of cap and the lowest in stipe.

A part of detrimental elements can be leached away. Short-term boiling is more efficient than soaking at ambient temperature. The elements are more easily leached from destroyed tissues of frozen mushroom slices than from fresh or dried slices or particularly from intact fruit bodies. Such culinary treatments naturally decrease content of valued taste components.

There exists a consensus that most of edible mushroom species from unpolluted areas do not pose a heath risk. Nevertheless, picking mushrooms from polluted sites should be limited. The most risky elements are cadmium, mercury and lead. Unfortunately, very limited information is available on the chemical forms (species) of the elements and on their bioavailability in humans. Only insufficient data on methylmercury and arsenic and chromium species were published so far.

Contents of risk elements in cultivated mushrooms have been low.

8. Vitamins and Provitamins

An initial information on vitamin contents in European wild growing mushrooms is available only since 2008. Therefore, the partial data will be combined with information available for the cultivated species.

Ascorbic acid (vitamin C) content of mostly 20-30 mg $100g^{-1}$ fresh matter (FM) was reported (Barros et al., 2008c; Jaworska and Bernaś, 2009a). Lower values of 3-4 mg $100g^{-1}$ FM were determined in three *Agaricus* spp., while 40 mg $100g^{-1}$ FM in *Cantharellus cibarius* (Barros et al., 2008b).

Low contents of 0.29 and 0.33 mg $100g^{-1}$ FM of thiamin (vitamin B1) and riboflavin (vitamin B2), respectively, were quantified in *B. edulis* (Jaworska and Bernaś, 2009a). These values are considerably higher than data reported by Çağlarirmak (2009) for cultivated *A. bisporus* (brown; portobello).

The contents of all three water-soluble vitamins decreased during *B. edulis* soaking or blanching followed by frozen storage (Jaworska and Bernaś, 2009a). Similar trends were observed also in *Pleurotus ostreatus* (Jaworska and Bernaś, 2009b).

Data for tocopherols (vitamin E) originate from a Portuguese laboratory (Barros et al., 2008a,b,c; Heleno et al., 2010). Usual contents of total tocopherols were 0.05-0.25 mg $100g^{-1}$ FM. The highest value of 0.80 mg $100g^{-1}$ FM was reported for *Laccaria laccata.* The prevailing isomer was β-tocopherol, while δ-tocopherol, tocotrienols and in some species also γ-tocopherol were not detected.

Earlier data on ergosterol, the provitamin ot ergocalciferol (vitamin D2), were reviewed by Kalač (2009). The contents of tens mg per 100g FM were usual. As was observed in both white and brown cultivated *A. bisporus*, ergosterol contents were higher in early growth stages. It was distributed evenly in caps and stipes during early developmental stages but accumulated more in the caps after maturation (Shao et al., 2010). The irradiation with UV is necessary for the conversion of ergosterol to vitamin D2.

The contents of β-carotene, a provitamin A, were mostly 0.2-1.3 mg $100g^{-1}$ FM (Barros et al., 2008b,c).

The reported vitamin contents are comparable or somewhat lower than those in the bulk of common vegetables. However, much more data are necessary for the credible comparison.

9. Flavor and Taste Components

The characteristic flavor of many mushroom species, mainly dried, is highly valued by many consumers. Until now, hundreds of odorous compounds were identified. According to their chemical structure, they can be classified as derivatives of octane and octenes, lower terpenes, aldehydes, sulfur and heterocyclic compounds and various others.

The very characteristic group of mushroom aroma is formed by the derivatives of octane, 1-octene and 2-octene, particularly alcohols, their esters with volatile fatty acids, and ketones (e.g. 3-octanone). The extraordinary role is ascribed to "mushroom alcohol" 1-octen-3-ol with its "raw mushroom odor". It occurs in fresh mushrooms at level of up to several mg per 100g with elevated content in *B. edulis* and related boletes. Its content increases during mushroom drying because free linoleic acid is oxidized under catalysis of lipoxygenase and hydroperoxide lyase. Such a process is characteristic just for mushroom drying. Moreover, the increase of 1-octen-3-ol was observed also in *B. edulis* preserved in other ways. The contents of 1.41, 1.94 and 6.43 mg per 100g were determined in the fresh, boiled and canned *B. edulis* (Misharina et al., 2009, 2010). As there was proved in cultivated *A. bisporus*, tissue disruption had a major impact on the profile of volatiles, both qualitatively and quantitatively.

Linoleic acid oxidation and the emissions of eight-carbon volatiles decreased with fruit body development and storage (Combet et al., 2009).

The recent state of knowledge of eight-carbon volatiles formation and properties in mushrooms was thoroughly reviewed by Combet et al. (2006).

Among non-volatile taste components, initial data on free 5′-nucleotides of adenosine, guanosine, inosine, uridine and xanthosine are now available. Tsai et al. (2008) determined total content of 2.76 mg g^{-1} DM in air-dried *B. edulis*. A wide range between 2.0 and 35.4 mg g^{-1} DM was reported for 10 freeze-dried wild mushroom species with *Cantharellus cibarius* and *Craterellus cornucopioides* having the border values (Beluhan and Ranogajec, 2011). The equivalent umami concentrations (EUC), expressed as g of monosodium glutamate (MSG) per 100g of mushroom DM, were calculated from the 5′-nucleotide contents. Values of 10.5, 2.0 and 35.4 g MSG per 100g DM for *B. edulis, C. cibarius* and *C. cornucopioides*, respectively, were given.

10. Pigments

Mushroom pigments belong to numerous chemical groups. The topic was recently covered with comprehensive reviews (Zhou and Liu, 2010; Velíšek and Cejpek, 2011), particularly from the points of view of biochemists and food chemists, respectively.

As compared with plants, chlorophylls and anthocyanins are not present and carotenoids are not widespread in mushrooms (they are characteristic e.g. for genus *Cantharellus*). Color changes following a mechanical damage of tissues of some mushroom species (e.g. becoming dark or blue) are usually caused by enzymically catalyzed oxidation of various polyphenols to quinones.

11. Health-Promoting Constituents

Traditions of folk medicine and the extensive recent research in the east-Asian countries prove various preventive and therapeutic properties of many mushroom species (so-called medicinal mushrooms). Over one hundred of medicinal effects are thought to be produced by such mushrooms (for an overview see Wasser, 2010). Knowledge of European species has so far fallen behind. The overviews of the very wide pharmacological potential of mushrooms were published by Lindequist et al. (2005) and Zhong and Xiao (2009). Recently, reviews dealing with a partial topic are available, namely on mushroom anticarcinogens (Ferreira et al., 2010), β-glucans (Rop et al., 2009) and lectins (Singh et al., 2010).

11.1. Antioxidants

Under the situation of increasing excess of free radicals, mainly of reactive oxygen species (ROS) and reactive nitrogen species (RNS), which may disrupt normal functions of a human organism due to the damage of cellular lipids, proteins and DNA, the interest in antioxidant abilities of various foods has increased for years. This trend includes recently also wild growing mushrooms. The topic has been the most dynamic within mushroom research during the last years.

A thorough review on the topic with numerous references until mid-2008 is available (Ferreira et al., 2009). Thus, only selected current papers will be given here.

It is not easy to draw a conclusion from the available data on mushroom antioxidants. A number of compounds, particularly various phenolics, tocopherols, ascorbic acid and carotenoids differing in their antioxidant potential (scavenging effect and reducing power), participate in total antioxidation effects. Numerous other compounds can also act as antioxidants. A Quantitative Composition-Activity Relationships (QCAR) model was therefore constructed for the prediction of the reducing power of mushrooms (Froufe et al., 2009). Moreover, several assays have been used for the determination of antioxidation capacity and the results have been expressed in various units, which also complicates the comparability of the published data.

Among widely consumed species, the high antioxidant activity was reported for *Cantharellus cibarius*, *Lactarius deliciosus* and *Agaricus arvensis* (Barros et al., 2009).

Cooked mushrooms had lower antioxidant activities than their dried or frozen counterparts (Barros et al., 2007). Water-soluble antioxidants of three mushroom species were more resistant to boiling, frying, grilling and microwave heating than methanol-soluble antioxidants (Soler-Rivas et al., 2009).

Wild growing mushrooms have been commonly consumed as mixtures of various species. As reported Queirós et al. (2009) from a study *in vitro*, synergism of species in the inhibition of lipid peroxidation was the effect prevailing over additive and negative synergistic effects. *Marasmius oreades* was present in the mixtures with higher antioxidant properties and synergistic effects, while *Cantharellus cibarius* in the mixtures with the lowest antioxidant properties and negative synergistic effects.

11.1.1. Phenolic Compounds

There exists consensus of various researchers, that phenolics, mainly phenolic acids, are the main antioxidants of mushrooms. Phenolic compounds include various subgroups, particularly phenolic acids, flavonoids, stilbenes, lignans and tannins, distinguishing by a large structural diversity.

Phenolic acids can be divided into two major groups, hydroxybenzoic acids and hydroxycinnamic acids. Among the phenolic acids detected in many mushroom species, *p*-hydroxybenzoic, protocatechuic, gallic, gentisic, vanillic and syringic acids belong to the former group, while *p*-coumaric, caffeic and ferulic acids to the latter one. The acids are mostly bound in various complex structures. Very wide contents were reported in mushrooms, from non-detectable level to several hundreds mg kg^{-1} DM (Barros et al., 2009; Vaz et al., 2011a, b). The highest levels of 357 and 284 mg kg^{-1} DM were determined in *Ramaria botrytis* and *Agaricus arvensis*, respectively (Barros et al., 2009).

Occurrence of flavonoids in mushrooms was reported only sporadically (Barros et al., 2008b, c; Gursoy et al., 2009). The opinion that this group of nutritionally favorable compounds is lacking has been more frequent.

11.1.2. Other Compounds with Antioxidant Properties

Tocopherols, ascorbic acid and carotenoids, all being mentioned in the section 8, are mushroom components participating in the antioxidant activities. Their contribution has been assessed as limited in comparison with phenolic compounds.

Numerous mushroom species, preferably cultivated ones including mycelium, seem to be a promising source for the isolation of various antioxidants and their use as nutraceuticals. The extensive research has therefore proceeded to find the optimal solvents and conditions for their extraction.

12. Detrimental Components of Edible Mushrooms

The dangerous constituents of toxic mushrooms have been extensively studied for many decades. Nevertheless, some detrimental compounds occur also in some edible species.

12.1. Natural Components

Hydrazines with a potential pro-carcinogenic activity occur in *Agaricus* spp. (agaritine) and in *Gyromitra esculenta* (gyromitrin). Abundant literature on agaritine was collated by Andersson and Gry (2004). According to Roupas et al. (2010), the available current evidence suggests that agaritine from consumption of cultivated *A. bisporus* poses no known toxicological risk to healthy humans.

As reported Schulzová et al. (2009), whereas the cultivated *A. bisporus* commonly contains 200-500 mg agaritine per kg FM, 24 of the 53 analyzed wild growing *Agaricus* species contained above 1,000 mg kg^{-1} FM. The highest level of up to 10,000 mg kg^{-1} FM was found in *A. elvensis.* No correlation was observed between agaritine content and size of the mushroom, season, year or site of the sampling. Agaritine was also detected in some species of genera *Leucoagaricus* and *Macrolepiota.*

Tricholoma equestre (syn. *T. flavovirens*) was identified as a cause of several outbreaks of rhabdomyolysis and also as the species with cardio- and hepatotoxic effects (Nieminen et al., 2008).

Consumption of several *Coprinus* spp. has been known to induce ethanol intolerance. The mushrooms contain coprine, a free amino acid, which is converted to toxic cyclopropanone hydrate, blocking oxidation of acetaldehyde, produced from ethanol, to acetic acid.

A significant content of nicotine was detected in some mushrooms, particularly *B. edulis*, imported to the European Union several years ago. As a consequence, temporary maximum residue levels of 0.036, 0.17 and 2.3 mg kg^{-1} for fresh, dried mushrooms and dried ceps (boletes) were proposed. In a survey of Italian both wild and cultivated mushrooms, contents of 0.01-0.04 and 0.1-4.5 mg kg^{-1} were determined in fresh/frozen and dried samples (Cavalieri et al., 2010). The reasons for surprisingly high level of nicotine in dried mushrooms have not yet been explained.

A short shelf life is characteristic for mushrooms. Biogenic amines thus occur among the products of protein degradation. In 17 species of fresh wild mushrooms, putrescine was the amine with the highest level, sometimes exceeding 150 mg kg^{-1} FM, followed by phenylethylamine. Toxicologically the most significant amines histamine and tyramine were not detected or were present at very low contents (Dadáková et al., 2009).

12.2. Contaminants

Relatively high formaldehyde contents of 119-494 and 110-240 mg kg^{-1} FW in cultivated *Lentinus edodes* (shiitake) were reported from China and the UK, respectively (Liu et al., 2004 and Mason et al., 2004). Initially, formaldehyde used for the disinfection and disinfestation of substrate was indicated as a source of the contamination. Nevertheless, in samples of the UK and Chinese shiitake, verified as produced without any formaldehyde treatments, similar formaldehyde levels of 100-320 mg kg^{-1} FW were determined. Storage for zip to 10 days had no effect on formaldehyde content, whereas frying significantly reduced the level (Mason et al., 2004). Due to lower formaldehyde content in mushrooms than in many widely consumed foods and the carcinogenity of formaldehyde only through inhalation and not by ingestion, the dietary exposure from mushrooms is not a cause for concern (Claeys et al., 2009).

Negligible contents of dioxins (PCDDs/PCDFs) and polychlorinated biphenyls (PCBs) were determined in cultivated *Flammulina velutipes* (enokitake), *Grifola frondosa* (maitake) and *Lentinus edodes* (Amakura et al., 2003). However, up to 13.1 μg of PCBs per kg of lipids was found in 10 wild species in Poland (Kotlarska et al., 2010). Unfortunately, further data are lacking. A risk of the translocation of PCB congeners from substrate seems to be low. Moeder et al. (2005) found in their study with *Pleurotus ostreatus*, cultivated on straw contaminated with PCBs, only below 0.1% of the applied amount in the fruit bodies.

Radioactivity of mushrooms is described in chapter 10.

13. Other Compounds

Ribeiro et al. (2008b) determined mean content of carboxylic acids of 10.4, 48.8, 97.3 and 119.5 g per kg DM in *B. edulis*, *Suillus granulatus*, *Russula cyanoxantha* and *Amanita rubescens*, respectively. The acids were present preferably in caps. The contents of prevailing acids decreased in order malic > fumaric ≈ citric >oxalic acid.

Polyamines spermidine and spermine are ubiquitous compounds of living organisms. Dietary polyamines participate in numerous physiological processes, e.g. cell growth. Thus, they can participate in tumor growth, and to the contrary, in wound healing or in intestinal mucosa development. Information on their level in foods is therefore needed. The contents of spermidine in 17 wild mushroom species were usually tens mg per kg of FM, sporadically above 100 mg kg^{-1}. Mushrooms should be thus included among food items with high spermidine level. The contents of spermine were considerable lower than those of spermine (Dadáková et al., 2009).

Within non-hallucinogenic indole derivatives, 6-7 compounds were detected in cultivated *A. bisporus* and three wild species. Serotonin contents of 52.1, 184, 296 and 317 mg kg^{-1} DM were determined in *A. bisporus*, *Lactarius deliciosus*, *Cantharellus cibarius* and *Leccinum rufum*, respectively. The contents of other compounds were considerable lower except for 392 mg kg^{-1} DM of kynureine sulfate in *L. deliciosus* (Muszyńska et al., 2011).

The intake of purine compounds needs to be restricted in the individuals afflicted by gout. Total contents of purine bases in seven Japanese cultivated mushroom species varied from 9.5 to 142 mg per 100 g FM with the highest contents in *Pleurotus ostreatus* and *Grifola frondosa*. The values are comparable with vegetables (Kaneko et al., 2008). Unfortunately, information for wild growing species has been lacking.

A steroidal derivative with cytotoxic activity, ergosterol peroxide, seems to be a common compound in mushrooms. After its detection in several cultivated and medicinal species it was found also in dried wild edible mushrooms (Krzyczkowski et al., 2009). The contents of 126, 134, 173 and 293 mg kg^{-1} DM were reported in *Xerocomus badius*, *Morchella esculenta*, *Suillus bovinus* and *B. edulis*.

14. Concluding Remarks

A part of the European population widely consumes tens of wild growing mushroom species as a delicacy. The credible evaluation of their nutritional value has so far been limited, however, the information increases dynamically during the last years. Unfortunately, very poor knowledge on the bioavailability of the individual constituents remains. Moreover, changes of the constituents during various preservation methods, storage and cooking treatments have been so far poorly understood.

Wild mushrooms have been valued preferably for their characteristic flavor and also for their texture. From the nutritional point of view, low energy, a relatively high fiber and ergosterol contents and antioxidation activity seem to be contributing. Some constituents with propitious effects (e.g. β-glucans) wait for researchers.

In reverse, several widely consumed species (e.g. among wild *Agaricus* spp.) accumulate high levels of several trace elements, mainly cadmium and mercury, if growing on heavily polluted soils. Their consumption should be thus very limited.

Acknowledgement

Financial support of the project MSM 6007665806 of the Ministry of Education, Youth and Sports of the Czech Republic was highly acknowledged.

References

Aida, F.M.N.A., Shuhaimi, M., Yazid, M., & Maaruf, A.G. (2009). Mushroom as a potential source of prebiotics: a review. *Trends in Food Science and Technology, 20*, 567-575.

Akyüz, M., & Kirbağ, S. (2010). Nutritive value of wild edible and cultured mushrooms. *Turkish Journal of Biology, 34*, 97-102.

Amakura, Y., Tsutsumi, T., Sasaki, K., & Maitani, T. (2003). Levels and congener distributions of PCDDs, PCDFs and Co-PCBs in Japanese retail fresh and frozen vegetables. *Journal of the Food Hygienic Society of Japan, 44*, 294-302.

Andersson, H.C., & Gry, J. (2004). Phenylhydrazines in the cultivated mushroom (*Agaricus bisporus*) – Occurrence, biological properties, risk assessment and recommendations. Copenhagen: Nordic Council of Ministers, 123 pp.

Barros, L., Baptista, P., Correia, D.M., Morais, J.S., & Ferreira, I.C.F.R. (2007). Effects of conservation treatment and cooking on the chemical composition and antioxidant activity of Portuguese wild edible mushrooms. *Journal of Agricultural and Food Chemistry, 55*, 4781-4788.

Barros, L., Correia, D.M., Ferreira, I.C.F.R., Baptista, P., & Santos-Buelga, C. (2008a). Optimization of the determination of tocopherols in *Agaricus* sp. edible mushrooms by a normal phase liquid chromatographic method. *Food Chemistry, 110*, 1046-1050.

Barros, L., Cruz, T., Baptista, P., Estevinho, L.M., & Ferreira, I.C.F.R. (2008b). Wild and commercial mushrooms as source of nutrients and nutraceuticals. *Food and Chemical Toxicology, 46*, 2742-2747.

Barros, L., Ventuizini, B.A., Baptista, P., Estevinho, L.M., & Ferreira, I.C.F.R. (2008c). Chemical composition and biological properties of Portuguese wild mushrooms: A comprehensive study. *Journal of Agricultural and Food Chemistry, 56*, 3856-3862.

Barros, L., Dueñas, M., Ferreira, I.C.F.R., Baptista, P., & Santos-Buelga, C. (2009). Phenolic acids determination by HPLC-DAD-ESI/MS in sixteen different Portuguese wild mushrooms species. *Food and Chemical Toxicology, 47*, 1076-1079.

Bauer Petrovska, B. (2001). Protein fraction in edible Macedonian mushrooms. *European Food Research and Technology, 212*, 469-472.

Beluhan, S., & Ranogajec, A. (2011). Chemical composition and non-volatile components of Croatian wild edible mushrooms. *Food Chemistry, 124*, 1076-1082.

Bernaś, E., Jaworska, G., & Lisiewska, Z. (2006). Edible mushrooms as a source of valuable nutritive constituents. *Acta Scientiarum Polonorum, Technologia Alimentaria, 5*, 5-20.

Çağlarirmak, N. (2009). Determination of nutrients and volatile constituents of *Agaricus bisporus* (brown) at different stages. *Journal of the Science of Food and Agriculture, 89*, 634-638.

Cavalieri, C., Bolzoni L., & Bandini, M. (2010). Nicotine determination in mushrooms by LC-MS/MS with preliminary studies on the impact of drying on nicotine formation. *Food Additives and Contaminants A, 27*, 473-477.

Claeys, W., Vleminckx, D.A., Huyghebaert, A., Hofte, M., Daenens, P., & Schiffers, B. (2009). Formaldehyde in cultivated mushrooms: a negligible risk for the consumer. *Food Additives and Contaminants A, 26*, 1265-1272.

Colak, A., Faiz, Ö., & Sesli, E. (2009). Nutritional composition of some wild edible mushrooms. *Turkish Journal of Biochemistry, 34*, 25-31.

Combet, E., Henderson, J., Eastwood, D.C., & Burton, K.S. (2006). Eight-carbon volatiles in mushrooms and fungi: Properties, analysis, and biosynthesis. *Mycoscience, 47*, 317-326.

Combet, E., Henderson, J., Eastwood, D.C., & Burton, K.S. (2009). Influence of sporophore development, damage, storage, and tissue specificity on the enzymic formation of volatiles in mushrooms (*Agaricus bisporus*). *Journal of Agricultural and Food Chemistry, 57*, 3709-3717.

Costa-Silva, F., Marques, G., Matos, C.C., Barros, A.I.R.N.A., & Nunes, F.M. (2011). Selenium contents in Portuguese commercial and wild edible mushrooms. *Food Chemistry, 126*, 91-96.

Dadáková, E., Pelikánová, T., & Kalač, P. (2009). Content of biogenic amines and polyamines in some species of European wild-growing edible mushrooms. *European Food Research and Technology, 230*, 163-171.

Dembitsky, V.M., Terent'ev, A.O., & Levitsky, D.O. (2010). Amino and fatty acids of wild edible mushrooms of the genus *Boletus*. *Records of Natural Products, 4*, 218-223.

Dikeman, C.L., Bauer, L.L., Flickinger, E.A., & Fahey, G.C. (2005). Effects of stage of maturity and cooking on the chemical composition of select mushroom varieties. *Journal of Agricultural and Food Chemistry, 53*, 1130-1138.

Ferreira, I.C.F.R., Barros, L., & Abreu, R.M.V. (2009). Antioxidants in wild mushrooms. *Current Medicinal Chemistry, 16*, 1543-1560.

Ferreira, I.C.F.R., Vaz, J.A., Vasconcelos, M.H., & Martins, A. (2010). Compounds from wild mushrooms with antitumor potential. *Anti-Cancer Agents in Medicinal Chemistry, 10*, 424-436.

Froufe, H.J.C., Abreu, R.M.V., & Ferreira, I.C.F.R. (2009). A QCAR model for predicting antioxidant activity of wild mushrooms. *SAR and QSAR in Environmental Research, 20*, 579-590.

Gursoy, N., Sarikurkcu, C., Cengiz, M., & Solak, M.H. (2009). Antioxidant activities, metal contents, total phenolics and flavonoids of seven *Morchella* species. *Food and Chemical Toxicology, 47*, 2381-2388.

Heleno, S.A., Barros, L., Sousa, M.J., Martins, A., & Ferreira, I.C.F.R. (2009). Study and characterization of selected nutrients in wild mushrooms from Portugal by gas chromatography and high performance liquid chromatography. *Microchemical Journal, 93*, 195-199.

Heleno, S.A., Barros, L., Sousa, M.J., Martins, A., & Ferreira, I.C.F.R. (2010). Tocopherols composition of Portuguese wild mushrooms with antioxidant capacity. *Food Chemistry, 119*, 1443-1450.

Isildak, O. (2009). Determination of inorganic anions in mushrooms by ion chromatography with potentiometric detection. *Journal of Analytical Chemistry, 64*, 1242-1246.

Jaworska, G., & Bernaś, E. (2009a). The effects of preliminary processing and period of storage on the quality of frozen *Boletus edulis* (Bull:Fr.) mushrooms. *Food Chemistry, 113*, 936-943.

Jaworska, G., & Bernaś, E. (2009b). Qualitative changes in *Pleurotus ostreatus* (Jacq.:Fr.) Kumm. mushrooms resulting from different methods of preliminary processing and periods of frozen storage. *Journal of the Science of Food and Agriculture, 89*, 1066-1075.

Jaworska, G., Bernaś, E., & Mickowska, B. (2011). Effect of production process on the amino acid content of frozen and canned *Pleurotus ostreatus* mushroom. *Food Chemistry, 125*, 936-943.

Kalač, P. (2009). Chemical composition and nutritional value of European species of wild growing mushrooms: A review. *Food Chemistry, 113*, 9-16.

Kalač, P. (2010). Trace element contents in European species of wild growing edible mushrooms: A review for the period 2000-2009. *Food Chemistry, 122*, 2-15.

Kaneko, K., Kudo, Y., Yamanobe, T., Mawatari, K., Yasuda, M., Nakagomi, K., & Fujimori, S. (2008). Purine contents of soybean-derived fopods and selected Japanese vegetables and mushrooms. *Nucleosides, Nucleotides, and Nucleic Acids, 27*, 628-630.

Kotlarska, M.M., Pietrzak-Fiecko, R., Smoczyński, S., & Borejszo, Z. (2010). [The level of polychlorinated biphenyls in edible mushrooms available at the market in the region of Warmia and Masuria]. *Żywność – Nauka, Technologia, Jakość, 17,* 49-57 (in Polish).

Krzyczkowski, W., Malinowska, E., Suchocki, P., Kleps, J., Olejnik, M., & Herold, F. (2009). Isolation and quantitative determination of ergosterol peroxide in various mushroom species. *Food Chemistry, 113*, 351-355.

Lindequist, U., Niedermeyer, T.H.J., & Jülich, W.-D. (2005). The pharmacological potential of mushrooms. *Evidence-based Complementary and Alternative Medicine, 2*, 285-299.

Liu, J., Peng J., Chi, Y., & Jiang, G. (2005). Determination of formaldehyde in shiitake mushroom by ionic liquid-based liquid-phase microextraction coupled with liquid chromatography. *Talanta, 65*, 705-709.

Mason, D.J., Sykes, M.D., Panton, S.W., & Rippon, E.H. (2004). Determination of naturally occurring formaldehyde in raw and cooked shiitake mushrooms by spectrophotometry and liquid chromatography-mass spectrometry. *Food Additives and Contaminants, 21*, 1071-1082.

Misharina, T.A., Muhutdinova, S.M., Zharikova, G.G., Terenina, M.B., & Krikunova, N.I. (2009). The composition of volatile components of cepe (*Boletus edulis*) and oyster mushrooms (*Pleurotus ostreatus*). *Applied Biochemistry and Microbiology, 45* (2), 187-193.

Misharina, T.A., Muhutdinova, S.M., Zharikova, G.G., Terenina, M.B., Krikunova, N.I., & Medvedeva, I.B. (2010). Formation of flavor of dry champignons (*Agaricus bisporus* L.). *Applied Biochemistry and Microbiology, 46*, 108-113.

Moeder, M., Cajthaml, T., Koeller, G., Erbanová, P., & Šašek, V. (2005). Structure selectivity in degradation and translocation of polychlorinated biphenyls (Delor 103) with a *Pleurotus ostreatus* (oyster mushroom) culture. *Chemosphere, 61*, 1370-1378.

Muszyńska, B., Sułkowska-Ziaja, K., & Ekiert, H. (2011). Indole compounds in fruiting bodies of some edible *Basidiomycota* species. *Food Chemistry, 125*, 1306-1308.

Nieminen, P., Kärjä, V., & Mustonen, A.-M. (2008). Indications of hepatic and cardiac toxicity caused by subchronic *Tricholoma flavovirens* consumption. *Food and Chemical Toxicology, 46*, 781-786.

Ouzuni, P.K., Petridis, D., Koller, W.-D., & Riganakos, K.A. (2009). Nutritional value and metal content of wild edible mushrooms collected from West Macedonia and Epirus, Greece. *Food Chemistry, 115*, 1575-1580.

Pedneault, K., Angers, P., Gosselin, A., & Tweddell, R.J. (2008). Fatty acid profiles of polar and neutral lipids of ten species of higher basidiomycetes indigenous to eastern Canada. *Mycological Research, 112*, 1428-1434.

Queirós, B., Barreira, J.C.M., Sarmento, A.C., & Ferreira, I.C.F.R. (2009). In search of synergistic effects in antioxidant capacity of combined edible mushrooms. *International Journal of Food Sciences and Nutrition, 60*, 160-172.

Ribeiro, B., Andrade, P.B., Silva, B.M., Baptista, P., Seabra, R.M., Valentão, P. (2008a). Comparative study on free amino acid composition of wild edible mushroom species. *Journal of Agricultural and Food Chemistry, 56*, 10973-10979.

Ribeiro, B., Lopes, R., Andrade, P.B., Seabra, R.M., Gonçalves, R.F., Baptista, P., Quelhas, I., & Valentão, P. (2008b). Comparative study of phytochemicals and antioxidant potential of wild edible mushroom caps and stipes. *Food Chemistry, 110*, 47-56.

Ribeiro, B., de Pinho, P.G., Andrade, P.B., Baptista, P., & Valentão, P. (2009). Fatty acid composition of wild edible mushroom species: A comparative study. *Microchemical Journal, 93*, 29-35.

Rop, O., Mlcek, J., & Jurikova, T. (2009). Beta-glucans in higher fungi and their health effects. *Nutrition Reviews, 67*, 624-631.

Roupas, P., Keogh, J., Noakes, M., Margetts, C., & Taylor, P. (2010). Mushrooms and agaritine: A mini-review. *Journal of Functional Foods, 2*, 91-98.

Schulzová, V., Hajslova, J., Peroutka, R., Hlavacek, J., Gry, J., & Andersson, H.C. (2009). Agaritine content of 53 *Agaricus* species collected from nature. *Food Additives and Contaminants, 26*, 82-93.

Shao, S., Hernandez, M., Kramer, J.K.G., Rinker, D.L., & Tsao, R. (2010). Ergosterol profiles, fatty acid composition, and antioxidant activities of button mushrooms as affected by tissue part and developmental stage. *Journal of Agricultural and Food Chemistry, 58*, 11616-11625.

Singh, R.S., Bhari, R., & Kaur, H.P. (2010). Mushroom lectins: Current status and future perspectives. *Critical Reviews in Biotechnology, 30*, 99-126.

Šišák, L. (2007). [The importance of mushroom picking as compared to forest berries in the Czech Republic.] *Mykologický Sborník, 84*(3), 78-83 (in Czech).

Soler-Rivas, C., Ramírez-Anguiano, A.C., Reglero, G., & Santoyo, S. (2009). Effect of cooking, *in vitro* digestion and Caco-2 cells absorption on the radical scavenging activities of edible mushrooms. *International Journal of Food Science and Technology, 44*, 2189-2197.

Spitteler, P. (2008). Chemical defence strategies of higher fungi. *Chemistry – A European Journal, 14*, 9100-9110.

Tsai, S.Y., Tsai, H.L., & Mau, J.L. (2008). Non-volatile taste components of *Agaricus blazei, Agrocybe cylindracea* and *Boletus edulis*. *Food Chemistry, 107*, 977-983.

Uzun, Y., Gençcelep, H., Tunçtürk, Y., & Demirel, K. (2009). Determination of protein and nitrogen fractions of wild edible mushrooms. *Asian Journal of Chemistry, 21*, 2769-2776.

Vaz, J.A., Barros, L., Martins, A., Morais, J.S., Vasconselos, M.H., & Ferreira, I.C.F.R. (2011a). Phenolic profile of seventeen Portuguese wild mushrooms. *LWT – Food Science and Technology, 44*, 343-346.

Vaz, J.A., Barros, L., Martins, A., Santos-Buelga, C., Vasconselos, M.H., & Ferreira, I.C.F.R. (2011b). Chemical composition of wild edible mushrooms and antioxidant properties of their water soluble polysaccharidic and ethanolic fractions. *Food Chemistry, 126*, 610-616.

Velíšek, J., & Cejpek, K. (2011). Pigments of higher fungi: A review. *Czech Journal of Food Sciences, 29*, 87-102.

Vetter, J. (2010). Inorganic iodine content of common edible mushrooms. *Acta Alimentaria, 39*, 424-430.

Wasser, S.P. (2010). Medicinal mushroom science: History, current status, future trends, and unsolved problems. *International Journal of Medicinal Mushrooms, 12*, 1-16.

Zhong, J.-J., & Xiao, J.-H. (2009). Secondary metabolites from higher fungi: Discovery, bioactivity, and bioproduction. In: Advances in Biochemical Engineering/Biotechnology, vol. 113. Berlin, Springer, 79-150.

Zhou, Z.Y., & Liu, J.K. (2010). Pigments of fungi (macromycetes). *Natural Products Report, 27*, 1531-1570.

In: Mushrooms: Types, Properties and Nutrition
Editors: S. Andres and N. Baumann
ISBN 978-1-61470-110-1

Chapter 7

PLEURAN: IMMUNOMODULOTOR POLYSACCHARIDE FROM *PLEUROTUS OSTREATUS*, STRUCTURE, PRODUCTION AND APPLICATION

***Hesham El Enshasy*[1,2*], *Parisa Maftoun*[1] and *Roslinda Abd Malek*[1]**
[1]Institute of Bioproducts Development (IBD),
Universiti Teknologi Malaysia (UTM), Johor, Malaysia
[2]Mubarak City for Scientific Research and Technology Applications (MuCSAT),
New Burg Al Arab, Alexandria, Egypt

ABSTRACT

Traditionally mushrooms are considered in daily diet, especially in the orient as they are rich in nutrition and have abundant therapeutic properties. Nowadays large scale mushroom cultivation takes attention of many pharmaceutical companies around the world due to their rich and variety of medicinal compounds. Extensive research in isolation, characterization and production of these biological components is conducted in last decades. Among them numerous studies are dedicated to polysaccharide production as they exhibit immunomodulator and anti-tumor activities. Glucan with different active unit linkage such as (1→3), (1→6)-β-glucan and (1→3)-α- glucans constitute mushroom polysaccharides which perform immunomodultor activity as they are biological response modifiers (BRMs). Beta-glucan obtained from the *Pleurotus ostreatus (*Oyster mushroom*)* is known in the scientific literature as pleuran. It is structural cell wall compound. It stimulates the body´s defense system against infections (bacterial, viral, yeast, and parasitic) and modulates the blood producing activity of bone marrow. In this chapter we are going to focus on the structure of pleuran, followed by its production strategies. Furthermore the therapeutic application of pleuran will be demonstrate and are for further researched will be suggested.

1. Introduction

Mushroom cultivation stems from historical tradition in the Far East due to its nutritious compounds. In recent years, several types of bioactive compounds have been isolated from mushroom species exhibiting antitumor, anticancer and immunomodulatory effects (Sarangi et al., 2006; Lindequist et al., 2005; Wasser, 2002; Karacsonyi and Kuniak, 1994). Among different types of bioactive molecules, polysaccharides were the most abundant. These polysaccharides are considered as an innovative source of biomedicines compounds by many pharmaceutical companies. Many polysaccharide bound proteins produced by basidiomycete fungi have been classified as anti-tumour chemicals by the US National Cancer Institute (Smith et al., 2002). During recent decades, antitumor and immunostimulating exopolysaccharide (EPS) compounds are known for 651 mushroom species and several of them is highly purified until now (Smith et al., 2002).

In general, polysaccharide is polymeric carbohydrate which is formed by glycosidic linkage between monosaccharide residues. Despite, diverse range of combinations is theoretically feasible; the majority of polysaccharides isolated from mushrooms are (1→3), (1→6)-β-glucans and (1→3)-α-glucans. These polysaccharides are known as Biological Response Modifiers (BRM) due to their active physiological properties (Novak and Vetvicha, 2008; Sarangi et al., 2006). BRM classified to two groups according to their effect on cytokine and immunodulators. Immunomodulators can manipulate the immune response of the host organism either positively such as imunopotentions or negatively such as immune suppression (Novak and Vetvicha, 2009; Novak and Vetvicha, 2008; Sarangi et al., 2006).

In recent years, different soluble and insoluble polysaccharides were extracted from basidomycota groups and mainly from fungal cell wall. Extensive research was conducted to have better understanding on the chemical and biological characterization of this unique polysaccharides. Pleuran is an alkali-insoluble exo-polysaccharide isolated from *Pleurotus ostreatus*. It stimulates the body′s immune system and help in defense mechanism against various types of bacterial, viral, fungal, and parasitic infections. It also modulates the blood producing activity of bone marrow. This chapter will be focused on the structure of pluran followed by different types of systems used for production of this important polysaccharides. Furthermore, the therapeutic application of pleuran will be demonstrate and area for further researched will be suggested.

2. Chemical Structure of the Pleuran

Among a variety of polysaccharides identified and isolated during last decades, β-glucans attract the attention of innumerable research. They can be isolated from different natural sources such as fungal cell wall, yeast and seaweeds (Novak and Vetvicka, 2008). β-glucans extracted from group of *basidiomycetes* and *accomycetes*, performed notable immunomodulatury effects (Karácsonyi and Kuniak, 1994). Fungi from genus *Pleurotus* synthesize β-glucans with almost similar primary structure. One of these β-glucans is pleuran which was isolated from *Pleurotus ostreatus*. However, pleuran form *P. ostreatus* along with lentinan from *Lentinus edodes* are of the most commercially used polysaccharides (Rop et al., 2009).

The widespread research was done on the isolation and structural characterization of pleuran from *P. ostreatus*. From 1985, it is accounted for significant drug properties, especially to against antitumor that is exerted by these types D-glucan polysaccharide. The famous type of extraction, is a hot water extracts from the edible mushroom was used for several authors as traditional medicine for cancer therapy. The research on the chemical structure of this polysaccharide investigated numerous parameters including the molecular weight, degree of substitution, degree of branching, chain conformation in solution, sugar component and the structures of main and branches till now (Surenjav et al., 2006).

Pleuran is a carbohydrate which contains a large number of β-glucose. Pleuran is beta-(1,3/1,6)-D-glucans with a typical structure of beta glucans which enclosed in cell wall (Novak and Vetvicha, 2008). It is an alkali insoluble exopolysaccharide (EPS) with molecular formula of $(C_6H_{12}O_6)_n$ and molecular weight of 762 kD (Frederick et al.,1995); Karácsonyi and Kuniak, 1994). The molecular weight of 2.4 x 10^4 Da was stated by Sun and Liu, 2009 for this water-soluble polysaccharide. This water-soluble polysaccharide compositions included total carbohydrate was determined to be 95% (Gutierrez et al., 1996; Sun and Liu, 2009) and 95.6% (Tong et al., 2009).

The biological activity of polysaccharides strongly associated to their structure and conformation (Novak and Vetvicha, 2008; Rop et al., 2009;). Thereby, various studies attempted to demonstrate pleuran structure (Karácsonyi and Kuniak, 1994). As illustrated in Figure 1, a main chain of pleuran consisting of backbone (1→3)–linked β-D glucose polymers units. A β (1,6) or β (1,4) bonds link a β-D-glucosyl side at the -0-6 position of every fourth anhydroglucose unit (Karacsonyi and Kuniak, 1994). The main chain of pleuran consist of triple helix coiled which joined with single or double filaments of gluco-pyranoses. A triple helix conformation will be occurred when C2- position bonded with 3-H. Side chains assist the stabilazation of conformation. The triple helix can be found in beta-glucans with molcular weight more than 90 kDa. Polysacharides with almost similar chemical structure is reported in *P. eryngii*, *P. ostreatoroseus* and *P. florida* (Carbonero et al; 2006; Rout et al., 2006). This high value polysaccharide structure and conformation pleuran was observed by NMR and IR spectra (Gutierrez et al., 1996; El-Enshasy et al., 2010).

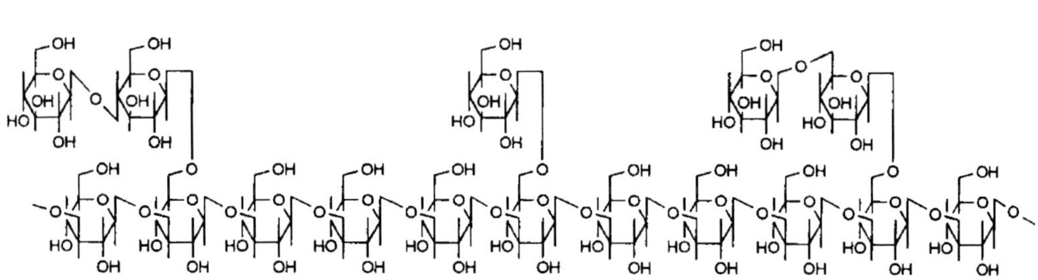

Frederick et al., 1995.

Figure 1. Chemical structure of Pleuran.

The solubility of β-glucans depends on various factors such as the complexity of structure, degree of branching (DB) and origin (Rop et al., 2009). DB of pleuran is 0.25 (Yoshioka et al., 1985; Kogan, 2000; Novak and Vetricha, 2008; Novak and Vetricha, 2009). With respect to degree of branching pleuran appears to be closely related to the water soluble (1-3)-β-D-glucan isolated from the same fungus (Yoshioka et al., 1985). Although DB of pleuran is near to that of soluble polysaccharides, pleuran is alkali insoluble due to high order of structure of triple helix conformation. The alkalin pH and high temperature can interrupt the H bond, following by conformation disruption, lead to alkalin inslobility. In addition it was stated by Bohn and BeMiller, 1995 and supported also by Rop et al., 2009; the most active polymers have degrees of branching (DB) between 0.20 and 0.33 which represent high anti-tumor and immunodulator activity. Pleuran showed to have 55% and 10 to 15% of water insoluble and water soluble pleuran fraction (Bobek et al., 1997). β-glucan showed to have 62.2% and 37.8% of water insoluble and water soluble in *P. ostreatus*, respectively in (mg.100/g DM) (Manzi and Pizzoferrato, 2000; Rop et al., 2009).

Pleuran from the cell wall of *P. ostreatus* was subjected to investigation by the procedure of sugar analysis. Several methods for sugar determination was applied to structural analysis of pleuran such as partial hydrolysis, methylation analysis, periodate oxidation and Smith degradation (periodate oxidation, reduction by $NaBH_4$, and subsequent partial hydrolysis). The completed partial pleuran acid hydrolysis was determined glucose and gluco-oligosaccharides, which were resolved on Gel filtration on Bio Gel column. As it is previously mentioned, a total carbohydrate composition was 95% (Sun and Liu, 2009; Gutierrez et al., 1996) and 95.6% (Tong et al., 2009). The compositions of carbohydrate are of D-galactose and D-glucose as identified by GC in the ratio of 2:1.

Periodate oxidation and methylations are two different methods commonly applied to determine the bond-type of polysaccharide followed by hydrolysis. The monosaccharide of pleuran was revealed the presence of 2,3,4,6-tetra, 2,4,6-tri, and 2,4-di-O-methyl-glucose with 2,3,4- and 2,3,6-tri-O-methyl-glucose (Karacsonyi and Kuniak, 1994). Methylation analysis was used to determine the mode of linkages of the polysaccharide. These results indicate by Karacsonyi and Kuniak, (1994) revealed that the polysaccharide contain a (1-3)-β-D-glucan highly branched at 0-6 at d branching ¼, was successfully characterized using methylation analysis, periodate peroxidation, NMR spectrometry and bioassay.

Smith degradation procedure was also carried out with the polysaccharide to determine the sequences of sugar residues present in the pleuran. It is believed recently the main component of pleuran from the fruiting body cell wall is a slightly branched, high molecular (1-3)-β-D-glucan with about 3% of β-(1-6) branching.

3. Production

For centuries, mushrooms have been used as nourishment worldwide due to their rich nutrients such as proteins, carbohydrates, minerals, vitamins and lipids (Smith et al., 2002). Therefore, many mushroom metabolites are of major concern in many industries as food additives, fermented beverages, antibiotics, pigments, pharmaceuticals, industrial enzymes, vitamins, organic and fatty acids and sterols (Yildiz et al., 1998). In addition to food value, mushrooms health benefit was recognized in the orient from several thousand years ago and medicinal properties of edible and non-edible mushroom are used in traditional Chinese medicine (Smith et al., 2002). Therefore, many pharmaceutical companies in the Far East are viewing the medicinal mushrooms as a rich source of innovative biomedical molecules (Smith et al., 2002). Currently, *P. ostreatus* is the third largest commercially produced mushroom in the world, based on its wide applications in food, nutraceutical and medicinal fields (Yildiz et al., 2006). Recently, extensive research on this fungus is conducted with the aim of extracting biological active substances. As previously mentioned, pleuran is one of active ingredient of this fungus which attracts attention of many researches due to its potential medicinal application as immunomodulator.

The conventional method for *P. ostreatus* cultivation is solids state fermentation (SSF). Solid state fermentation was defined as the interaction of homogenous microorganism in controlled condition including temperature and humidity, but in the absence of water (Tellez-Tellez et al., 2008). Cultivation of fruiting bodies of mushroom is time consuming in SSF system. Moreover, controlling product quality from batch to batch is challenging. Considering high demands for natural medicinal active ingredient in the pharmaceutical and nutraceutical market, it is critical to supply large amount of products with high quality. Thereby, submerged liquid fermentation became appropriate alternative method for pleuran production.

It has been proven that yield of enzymes production is better in SSF system than SMF, but for extracting polysaccharides, SMF showed better result (Tellez-Tellez et al., 2008). The concentration of active metabolite ingredients increase when SMF is used. Besides, the possibility of contamination in SSF along with low yield of polysaccharide makes it not suitable for large scale production of pleuran. Therefore, submerged culture of mushrooms for efficient production of pleuran is studied by several authors (Rosado et al., 2003; Wu et al., 2004; Zhong and Tang, 2004; Wang et al., 2005; Lin et al., 2007; Gern et al., 2008; Tellez-Tellez et al., 2008; El-Enshasy et al., 2010).

3.1. Solid State Fermentation

For centuries, solid state fermentation (SSF) was used to cultivate saprophytic edible mushroom to enhance the mycelia and fruiting body on substrate, as a source of food. SSF was define as contribution non soluble material as a source of nutrients in absence of water during fermentation. In SSF, the microorganism can inhibit along fermentation due the low moisture content. Although SSF offers few advantages over SMF according to simpler and cheaper techniques, precise control of parameter such as nutrition, pH, aeration, oxygen transfer and moisture is almost infeasible. The other drawback of SSF is the low yield of pleuran which is also fluctuated. Furthermore long cultivation period of several months, cause contaminations occur owing to open cultivation system (Gregori et al., 2007). Thereby for producing metabolite products such as pleuran is not a preferred method compare to SMF. However, SSF give notable yield when applied for enzymes production such as laccase as it a lingocellulose degraders. Therefore, it can use wood as substrate for mycelia growth fruiting body production (Silveira et al., 2008).

For large scale SSF cultivation, several operations steps should be performed precisely to have accepted yield. These operations consist of obtaining pure mycelia culture and its maintenance, preparing spawn (inoculums) followed by substrate preparation and inoculation of spawn at the end. *P. ostreatus*, were mainly cultivated on variability low cost of solid substrate under controlled temperature and humidity in green house, such as apple pomace, rice straw, rice brain, saw dust and others.

The pure *P. ostreatus* mycelia culture is mainly provided form reputable cell bank centers by growers and proceed for spawn preparation. For large scale production of mushrooms, large amount of inoculum (5-10% of final product medium) is required. These inoculums are known as spawn in mushroom cultivation. The mycelia grow through sterile solid organic matrixes such as grains under control environment condition for spawn production. The grains will be covered with mycelia during spawn production and can be act as mycelia capsules. The sterile technique should be applied for all steps involve in spawn production to prevent possibility of contaminations (Smith et al., 2002). The spawn used to inoculate the growing bags.

The prepared spawn can be applied in two different types of culture, log culture or saw dust culture. The culture should be in active stage and substrate act as carrier of the spawn in an appropriate medium that can be used to inoculate the beds. Too low spawn level would result in lower mushroom yield and the longer growth was obtained, which lead to poor process economy. Log culture or grains spawn was successfully done using cooked sorghum, rye and wheat, mixed with calcium carbonate. Selection of spawn cultivation method either in form of wood log or saw dust depends on bioconversion efficiency, biological efficiency and substrate dry matter (Zhang et al., 2002).

P. ostreatus cultivation in rice straw and wheat straw was achieved by using two different strategies. First, the rice straw and wheat straw was cut into pieces and secondly was grinded to get the powder. Higher growth rate were observed in *P. ostreatus* cultivated in cut straw and the yield achieved 10% than wheat straw was observed in *P. ostreatus* cultivated in cut straw (Zhang et al., 2002). A similar study was also carried out by Kalmis et al. (2008) using *P. ostreatus*, cultivated on wheat straw as substrate supplemented with wheat bran and olive oil. Another research showed also that wheat straw used as vehicle is an excellent alternative for *P. ostreatus* cultivation in SSF (Sainos et al. (2006).

3.2. Submerged Fermentation

As it is previously mentioned, the preferred method for pleuran and other EPS productions is submerged culture. Predominantly, SMF can be applied to produce liquid spawn, biomass or exomethabolites (Lomberh, 2002). The pharmaceutical, food additive and dietary industries can make use of the biometabolites from SMF cultivation (Lomberh, 2002). SMF is one of the promising technologies for production of pleuran, as the environmental condition can be controlled and optimized to have high cell biomass or metabolites (Hui Xiao et al., 2010). Besides, harvesting the mycelia and extracting of bio metabolites is easier than SSF (Wang et al., 2005). The most serious problem in large scale submerged culture of mushroom might be oxygen supply, shear stress and scale up as mushrooms form pellets which are shear sensitive and lead to increase in culture viscosity during the process (Zhong and Tang, 2004; El Enshasy, 2007). Difficulty of oxygen transfer as a result of EPS production, which leads to high viscosity, is also reported by Gern et al. (2008) in submerged culture of *P. ostreatus.*

In recent study of El- Enshasy et al. (2010), *P. ostreatus* submerged cultivation is improved regarding EPS production using pilot level 15-L bioreactor. At first, cells were cultivated in batch culture and glucose limitation is observed, then fed batch culture was used with intermittent glucose addition and higher cell activity was observed than batch culture. However, it show decrease in oxygen consumption and CO_2 production which represent reduction in metabolic activity under glucose limitation. In another approach they use fed batch culture with on-line glucose feeding strategy using out-gas analysis data and the results indicate that glucose is not limiting factor for growth but for EPS production (El-Enshasy et al., 2010).

Carbon as macronutrient is one of the main source of energy for *P. ostreatus* growth and EPS production. Beside this, it is needed as structural element of fungal cells in combination with hydrogen, oxygen and nitrogen. It is vital for synthesis of polysaccharides. Carbone sources on earth have abundant diversity such as glucose, sucrose and lactose (Kavanagh, 2005). However mushrooms consume glucose as their primary carbone source for growth and EPS production. It should be noted that applying glucose in high concentration inhibits mycelia growth. Several authors investigate effect of carbon sources on *P. ostreatus* with the aims of evaluating its effect on the growth and its bioactive compounds production (El-Enshasy et al., 2010; Tellez-Tellez et al., 2008; Stajic et al., 2006; Rosado et al., 2003).

Rosado et al. (2003) evaluate effect of glucose as carbon source for EPS production by *P. ostreatus*. The results show decrease in the glucose medium concentration and increase in total carbohydrate content during the fungal growth which indicate the use of available carbon source for growth and EPS production (Rosado et al., 2003). In addition, fed batch cultivation of *P. ostreatus* by El-Enshasy et al. (2010) reveal that glucose is not limiting factor for growth but for EPS production.

Beside carbon source, nitrogen is another essential component for the growth and EPS production by *P. ostreatus*. This based on the fact that, fungi are non-diastrophic (cannot fix nitrogen) and they should be supplied by nitrogen containing compounds. It could be applied to the medium in inorganic form such as ammonium salts or in organic from including yeast extract, peptone, tryptone, casein peptone, beef extract, soybean steep liquor and wheat bran steep (Kavanagh, 2005). It have been also observed that after 20 days cultivation of *P. ostreatus* var florida, the exopolysaccharides and biomass production were limited, even with the presence of glucose in the medium, probably due to formation of a polymer layer surrounding the cell (Burns et al., 1994; Tellez-Tellez et al., 2008).

Another research conducted by Gern et al. (2008) showed also that the organic nitrogen source (yeast extract or corn steep liquor) has a significant effect on the maximum concentration of biomass in *P. ostreatus*. Also the substitution of yeast extracts by corn steep liquor was investigated by Gern et al. (2008) and provides better results. However the initial pH was about 25% lower than those containing yeast extract which should be consider in evaluation of the substitution. The maximum specific growth rate, the polysaccharides concentration and the productivity in polysaccharides were negatively affected by corn steep liquor. An increase in the organic nitrogen concentration had a significant positive effect on biomass production. The organic nitrogen concentration at the higher level also favored the increase in the concentration and the productivity of polysaccharides (Tellez-Tellez et al., 2008).

In other study, addition of inorganic nitrogen (ammonium sulfate) led to a decrease of all evaluated parameters (Tawiah and Martin., 1986). It was also suggested that when ammonium sulfate was employed in the absence of an organic nitrogen source, *P. ostreatus* does not use this nitrogen source. Moreover the polysaccharide productivity also decreases when ammonium sulfate is used (Tellez-Tellez et al., 2008).

The interaction between the organic nitrogen source and the ammonium sulfate concentration is significant factor for the polysaccharides production, with better results being achieved when yeast extract is used in the presence of ammonium sulfate. It was also found that when the nitrogen source is corn steep liquor, the ammonium sulfate always has a negative effect on the biomass and polysaccharides productivities (Tellez-Tellez et al., 2008).

In a similar study Burns et al. (1994) studied the effect of organic and inorganic nitrogen on the biomass and exopolysaccharides production by *Pleurotus sp. Florida* and media containing a low nitrogen concentration found to be the most appropriate for the production of polysaccharides (Gern et al., 2008). Similar result was found by Rosado et al. (2003) with *P. ostreatus*" Florida".

The effect of different nitrogen sources on EPS production of other fungus such as *Cordyceps taii* reveal that the organic nitrogen source preferred than inorganic one in term of cell growth, while applying inorganic sources (ammonium sulfate) improve the EPS production. However, poor cell growth which is observed in the present of ammonium sulfate. For this reason, composition of organic and inorganic nitrogen sources is suggested for further optimization (Lin and Chen, 2007).

Carbon to nitrogen (C/N) ratio play crucial role in trend of growth curve and this factor is studied for optimization of growth and productivity of various fungi. Several studies evaluate the different applied C/N ratios to various fungi growth and their productions, but the accurate function of C/N ratio level is still unclear regarding polysaccharide production (Karacsonyi and Kuniak, 1994). Besides, to our knowledge there is no report looking through effect of different C/N ratios on growth and EPS production in *P. ostreatus*. However in an assay done by Wu *et al* (2004) this is investigated on *P. tuber-regium* and it is found that increase in C/N ratio is favored for glucan content (Synytsya et al., 2009). Also it is proven that C/N ratio and pH have a significant influence on water-soluble polysaccharide production in *P.citrinopileatus* (Wang et al. 2005). In addition, Lavi et al. (2006) demonstrated that maximum polysaccharide production was achieved in synthetic medium of carbon/nitrogen 32:1 after7 days cultivation.

In other studies, endopolysacharide appeared to be more affected by variable C/N ratio than EPS in *Ganoderma applanatum* (Wu et al., 2004). Furthermore, it is reported that EPS production in *Pediococcus damnosus* was mainly enhanced by increase in glucose concentration but not by increase in nitrogen concentration (Wu et al., 2004). In other study, high C/N ratio was favorable for both the mycelial growth (5.41 g/l) and EPS production (0.55 g/l) in *Antrodia cinnamomea* culture.

Gibriel et al. (1996) evaluate growth rate of *P. ostreatus* in five natural and synthetic media (PDA, Czapeck's agar, glucose, yeast extract agar, nutrient agar and malt extract agar) and result illustrated that malt extract agar and Potato dextrose agar (PDA) were the best medium respectively. In addition PDA performed the highest dry weight of the mycelium when liquid form of aforementioned media is used. However in this case, malt extract agar did not perform significant result. Moreover, PDA stimulated the growth of *P. ostreatus* significantly (Wang et al., 2005).

In another approach, mycelia yield of *P. ostreatus* is studied using thinned apple, pear and peach on submerged culture by Jung et al. (2003) and thinned peach give the highest mycelia dry weight. Moreover, *P. ostreatus* produce 2-3 times more dry weight in 20-liter pilot scale fermentor with 10 liter of liquid medium containing 3% of thinned fruit compared to control medium without thinned fruit addition (Stajic et al., 2006).

Monitoring the pH of the culture medium is essential for improving the growth and productivity of mushrooms as different pH result in different product yield and microbial growth. Regarding *P. ostreatus* submerge cultivation; considerable reduction is reported by

EL- Enshasy et al. (2010) in shake flask level mainly due to acid formation. Furthermore insufficient oxidation and formation of large pellets can also lead to drop in the pH. It is notworthy also to mention that, acidification of the culture medium by *Pleurotus* strain in submerged culture was also reported by (Rosado et al., 2003; Gern et al., 2008).

However, as mentioned previously, one of the main barriers for large scale production of mushrooms is oxygen supply as they have tendency to form pellets which decrease mass transfer inside the biopellet especially when *P. ostreatus* cultivated in shake flask (El Enshasy, 2007; El Enshasy *et al.,* 2010). Another studies showed also that, limitation of oxygen supply within *Gnoderame Lucidum* mycelia when it is larger than critical size is also reported by Tang and Zhong (2003) and Rosado et al. (2003).

3.3. Extraction and Purification

The extraction of water-soluble compounds from biological materials with hot water is an appropriate procedure and widely used approach for the isolation of valuable natural products. The decoctions and essences of mushroom hot water extract was used in traditional Chinese medicine in the treatment various diseases and it is depends on the cell wall structure. The extraction methods used are varied based on the structure and water-solubility of polysaccharides. This method is significant to break the cell wall from outer layer to the inner layer with mild to strong extraction conditions by controlling the pH and temperature (Zhang et al., 2007). Recently, increased interest has been observed on using the *P. ostreatus* mushroom extract from fruiting bodies for cancer treatment, especially pleuran as new safe and high therapeutic compounds. The hot water extraction method is widely used approach, however, with a comparable proficiency. The chosen conditions of the hot water treatment lead to a significant less toxicity of product.

Among the several techniques used for extraction of mushrooms, two type of extractions were successfully reported. Extractions with hot water to get water-soluble polysaccharide and the extraction with alkali solution to get the water-insoluble fractions are widely used (Zhang et al., 2007; Karacsonyi and Kuniak, 1994). The water-soluble polysaccharide isolated from fruiting bodies of *Pleurotus* species, was reported by several authors (Sun and Liu, 2009; Maity et al., 2010 ; Pramanik et al., 2005; Rout et al., 2006).

Pleuran was obtained after the fruiting bodies of *P. ostreatus* were washed with distilled water at the first stage to minimum contamination from fruiting body during cultivation at controlled environment, followed extraction with boiled distilled water for 3 hour at 75°C for cell wall swelling. This process was effective for cell wall breakdown and validated as proper process for pleuran production using electron micrographs.

During this boiling technique, the primary medicinal extracts were collected. The extraction then cooled overnight at 4°C and filtered using linen cloth. The filtrate then centrifuged at 10000 rpm at low temperature, and the supernatant could be easily precipitated using cold ethanol. The precipitated crude polysaccharide can be collected through centrifugation and dried by either vacuum drying or lyophilization. The obtained polysaccharide is pleuran in sermi-purified form. The dried pleuran it was again dissolved in water. The mixture of soluble polysaccharide then dialyzed to remove low molecular weight carbohydrate molecules, through DEAE cellulose bag against distilled water for 4h. After the extraction process, successfully extracted polysaccharide could be further purified using several techniques, including ethanol precipitation for excludes the impurities from the polysaccharides and fractional precipitation, acidic precipitation using acetic acid, ion-exchange chromatography (for separation of acidic and neutral polysaccharides), gel filtration and affinity chromatography.

Extraction of polysaccharide from mycelial culture cultivation from SMF is different from the process for extraction from fruiting body. The fermentation broth should first centrifuged and the resulting culture filtrate mixed with equal volume of cold ethanol, stirred vigorously and cools overnight at 4°C. The presence of precipitate exo-biopolymer is centrifuged at 10000g for 20 minute. The precipitate crude polysaccharide of pure EPS is then washed separately with ethanol, acetone and ethyl ether followed by lyophilization step (Enshasy et al., 2010; Bae et al., 2000). The final precipitate only harvesting after the lyophilisation cycle process was completely after 4 days and weighed.

Preparation clean fruiting bodies of *Pleurotus ostreatus*

↓

Extraction with boiled distilled water for 3 hours at 75°C

↓

Cooling overnight at 4°C

↓

Filtration using linen cloth

↓

Centrifugation at 10000 rpm at 4 °C

↓

Precipitation using cold ethanol

↓

Separation of the precipitated polysaccharide

↓

Drying using vacuum drying or lyphilization

↓

Pleuran powder

Figure 2. Simple scheme for extraction and purification of the pleuran from the *P. ostreatus* fruit bodies.

Submerged liquid broth

↓

Separation of biomass by centrifugation

↓

Mixing of culture filtrate mixed with an equal volume with ethanol

↓

Vigerous stirring and overnight cooling at 4°C

↓

Separation of EPS by centrifugation (10000g/20 minute)

↓

Precipitate crude polysaccharide of pure EPS
and washed separately with ethanol/ acetone/ ethyl ether

↓

Lyophilization

↓

Pleuran powder

Bae et al., 2000; El-Enshasy et al., 2010

Figure 3. Simple scheme of extraction of polysaccharide from mycelial culture from the *P. ostreatus*.

4. APPLICATIONS

During recent years, increased interest on therapeutic compounds derived from natural resources has increased. Mushrooms are considered as a vast prospects source of nutritional and medicinal value, regarding production of high therapeutic biometabolite (Aida et al., 2009; Lindequist et al., 2005; Smith et al., 2002). In recent decades the medicinal value from mushroom extract, in particular, *G.luscidome*, *P. ostreatus and L. endodes* is revealed by different authors (Morris et al., 2003; Gao et al., 2004; Jayakumar et al., 2008; Hearst et al., 2009). Bioactive components from these mushrooms can act as immunodulers, immonusuppressant, anti-bacterial, antiviral, anti-tumor and anticancer (Chihara, 1993; Borchers et al 2004; Gao et al., 2004; Gregori et al., 2007; Daba, 2006). Several polysaccharides of this type such as pleuran and lentinan with therapeutic property successfully isolated and characterized (Chihara, 1993; Karacsony and Kuniak, 1994). They performed different characteristic such as high and low molecular weights, water-soluble or insoluble which make each one appropriate for specific applications (Shamtsyan et al., 2004;

Akramiene et al., 2007; Mantovani et al., 2008; Rop et al., 2009;). These compounds could be extracted from fruting bodies, scerotium or mycelia (Akramiene et al., 2007 and Zhang et al., 2007). Pleuran obtained from *P. ostreatus* is BRM which performed notable immunomodulatay effect *in vivo* studies. Moreover, anti-tumor and anti-cancer properties of pleuran investigated in some trials by different authors (Yoshioka et al., 1985; Bobek and Galbavy, 2001; Bobek et al., 2001). They applied pleuran in several types such as powder, hydrogel or adjuvant compounds (Bobek et al., 2001; Bobek and Galbavy, 2001; Bobovcak et al., 2010). However, recent research indicated that *P. ostreatus* extract showed positive effect on treatment of different types of cancer (Bobovcak et al., 2010). Table 1 summarizes different studies which used pleuran as therapeutic polysaccharides.

Table 1. A brief description of trials which apply Pleuran as therapeutic compounds

Application	**Cell type/ disease**	**In vivo / In vitro**	**Result**	**References**
Anti tumor	Sarcoma 180 cancer cell	Female mice	Inhibition ratio of 95 ±3% at a dose of 0.2 mg/kg and 74 ± 10% at a dose of 0.1 mg/kg.	Yoshioka et al., (1985)
Anti lipoperoxidative effect and antioxidant enzymes activity	Serum and liver cells	Male rats	No effect	Bobek et al., (1997)
Immunomodulatory	Colitis	Male rats	Decline in colonic damage score and colonic wet weight	Nosalova et al., (2001)
Anti oxidant defence	Colitis	Rats	Decline in colonic damage score	Bobek et al., (2001)
Anti oxidant	Precancerous aberrant crypt foci (ACF) lesions in the colon	Male rats	>50% Reduction of ACF lesions	Bobek and Galbavy, (2001)
Prebiotic	*Lactobacillus ssp. Bifidobacterium sp, Enterococcus faecium*	*in vitro*	Support probiotic growth of *Lactobacillus ssp.*	Synytsya et al., (2009)
Immunomodulatory after intensive exercise	Blood cells	Human	Modulate the suppression of immune cells	Bobovcak et al., (2010)
keratinocyte stimulator	Human keratinocyte primary cultures	*in vitro*	Increase in pro-matrix metalloproteinase-9 over the range of 2 to 200 μg/ml,	Bobek et al., (2001)

β-glucans from basidomycoat group, especially *Pleurotus* species seems to have most potent, anti-tumor and immunomodulatory effect (Zhang et al., 2007; Rop et al., 2009). In general, β-glucans interact with the lectin-like membrane receptors on the immune components such as macrophages, neutrophils and natural killer cells (Sarangi et al., 2006; Akramiene et al., 2007; Rop et al., 2009). The biological activity of these glucans is affected by ability of binding with these receptors. Various factors determine such ability including conformation, degree of branching, water solubility, growing condition, method of extraction and the degree of fruiting body maturity (Wasser, 2002; Daba and Ezeronye, 2003; Akramiene et al., 2007; Zhang et al., 2007; Mantovani et al., 2008; Rop et al., 2009). It is reported that water-soluble polysaccharide, with high molecular weight perform higher immunomodulatory and anti-tumor activity (Sarangi et al., 2006; Mantovani et al., 2008; Rop et al., 2009).

Mechanism of immunomodulatory, anti-tumor and anti-viral activity of β-Glucans such as pleuran is one of the major areas of research during the last years. However the exact mechanism is still unclear and need more investigations. Mainly β-Glucans increase the immune defense by activation of macrophages. They play vital role in the immune system and attack to any invading cells or pathogens which could pass through epithelial barrier (Akramiene et al., 2007). Phagocytosis of macrophages leads to digestion of pathogens through lysosomal enzymes. In addition they release cytokines and inflammatory mediators which assist the defense against pathogens. The produced cytokines move around from place they bind with glucans. Thus, glucan has an overall immune response as well as local activation (Shamtsyan et al., 2004; Mantovani et al., 2008).

Despite, cancer cells is one of invading cells and macrophages are activated against them, natural killer cells are the immune cells which are responsible for killing the tumor cells. Therefore, activation of this types of cells indicates the anti-tumor activity of compounds in particular (Sarangi et al., 2006). However, β-(1→3) linkage in the back bone along with additional branch of β-(1→6) branch which is believed to be needed for anti-tumor activity, is present in pleuran (Wasser, 2002; Sarangi et al., 2006). Ordinary, β-glucans do not target cancer cells directly, but they enhance the immune defense system and assist adaptation of host to biological stress (Daba and Ezeronye, 2003; Zhang et al., 2007).

It is believed that antiviral and anti-bacterial property of pleuran as a β- glucans is due to activation of monocytes and neutorphils. Pleuran can be applied to a patient under major surgeries for decreasing their infection reduction (Akramiene et al., 2007). They secrete reactive oxygen and nitrogen intermediates which are bactericidal compounds (Sarangi et al., 2006; Akramiene et al., 2007).

The immunomodulating activity of pleuran, make it suitable for people suffering from stress as the close relationship between the immune and the nervous systems has been well documented and immune disorders can have a negative effect on mental health, and vice versa. Furthermore the positive effects of taking pleuran have been demonstrated in top-level sportsmen, whose immune system is permanently weakened as a result of exposure to frequent and long-term physical stress. Two group of randomized athletes consumed 100mg of pleuran (IMUNOGLUKAN®) or placebo supplements, once a day for 2 months. This clearly showed that glucan accelerates the body's ability to adapt to different time zones and promotes recovery after injuries. The cell count of blood samples which was collected before and after 2 month supplementation indicates that the reduction of natural killer cells' activity was not significant in pleuran groups, while 28% reduction was observed when placebo supplementation is consumed. Other immune cells compounds such as, erythrocyte, and lymphocytes, subpopulations of lymphocytes, granulocytes and monocytes activities do not performs considerable differents (Bobek et al., 2001).

One of the first investigations on antitumor property of pleuran was done by Yoshioka et al. (1985). Female mice were transplanted to sarcoma 180 (~5 × 10^6 cells) subcutaneously. After 24 hour, β-glucan samples from *P. ostreatus* were applied every day for 10 days. They observed that, notable anti-tumor activity of 95 +3% at dose of 0.2 mg/kg is performed by β-glucan. The efficient dose of 0.1 mg/kg is indicated with antitumor activity of 74± 10% (Yoshioka et al., 1985)

In another approach, the effect of pleuran on inflammation of colon (colitis) which caused by intraluminar installation of 4% acetic acid was studied using experimental rats. The experimental rats went on a diet of 2% food components or 0.44% of hydrogel drins for 4 weeks. Both diet regimes prevent the increase of colon tumore wet weight and notable reduction in macroscopic damage score of colons observed in hydrogel drinking and food component diets, respectively. Both diet and control groups showed decrease in the activity of lysosomal enzymes, acid phosphatase and cathepsin in the injured site, however both pleuran regimes moderate this status. Moreover, diminution of conjugated dienes in the erythrocytes, liver and colonic tissue is observed as a result of both pleuran regimes, whereas they were not influence the antioxidant enzymes activity. In addition, the pleuran hydrogel drink enhances the activity of superoxide dismutase in colon and glutathione transferase in erythrocytes. It could be conclude that dietary pleuran can influence the simulation of antioxidant defence (Bobek et al., 2001).

It should be noted that pluran modulate the skin's wound-healing processes. In an *in-vitro* study, human keratinocyte primary cultures stimulated by pleuran and after elapse of 24 h, secreted matrix of metalloproteinase 9 (MMP-9) and metalloproteinase 2 (MMP-2) measured through gelatine zymography. As a result, pro-MMP-9 which treated by pleuran raise with range of 2 to 200 μg/ml. In contrast pro-MMP-2 stayed at constant level. The potent of pleuran as keratinocyte stimulator of pro- MMP-9 release was demonstrated in this experiment.

5. Further Research Areas

Medicinal properties of different mushrooms are detected by several *in-vitro* and *in-vivo* experiment. Furthermore, mushroom bioactive compounds were isolated, characterized, purified formulated, human use and passed through hard clinical trials. Medicinal property of *P. ostreatus* extract is studied by different authors (Jedinak and Sliva, 2008). However, there is less studies focused on the potent therapeutic activities of pleuran and other bioactive compounds from it. Most of these studies conducted with the aim of investigating the effect of pleuran on colitis, despite many experiment can be designed for detecting the antitumor and immunomodulatory effect of pleuran on different cancer cell types and tissues.

It is clear that the activity of polysaccharides is highly dependent on the produced molecule structure in term of molecular weight and degree of branching. Moreover, the method of extraction, purification and storage condition is also critical. Although some methods is suggested in this review but as far as we know the efficient methods for purification is not suggested and the existing methods is not optimized for pleuran purification. In addition efforts for biomedical modification of pleuran can be done to increase its medicinal activity and assisting its commercialization. These include sulfations which increase solubility. Sulfated β-glucan from *Pleurotus tuber regium* present showed increase in antitumor and antiviral activity (Mantovani et al., 2008)

Another concern is the development of proper formulation of the therapeutic drugs and functional food to enhance the mechanism of drug delivery systems to protect them from degradation and provide controlled release. In spite of the fact that pleuran was applied to different experimental animals as powder, hydrogel drink or adjuvant compounds but few data are available regarding the delivery mechanisms and comparison between different drug delivery systems. It is believed that β-glucan activity is dependent on the pH of the site where they interact with immune system compound. In acidic environment such as stomach the biological efficiency will decline through disruption of triple helix structure (Rop et al., 2009). These clearly indicate the importance of studying of drug delivery systems for pleuran.

Animal experiment and clinical trials is one the basic requirement steps which involve in releasing a drug to the market. They not only provide us the precise determination of therapeutic effect of pleuran, but also assist in gathering the basic information of pleuran, mechanism of action. Unfortunately, less research target animal experiment and clinical trials for evaluating pleuran medicinal properties compare to other compounds of therapeutic mushrooms.

CONCLUSION

Medicinal mushroom compounds are promising natural and safe resourse in therapeutic application of cancers and autoimmune disease. Pleuran is one of these compounds which derived mainly from *P. ostreatus* and research until now considered it as one of the most potent anti-tumor and immunomodulators. Until now, no sufficient data in term of he mechanism of therapeutic action of this β-glucan are available and further studies are required in this research area.

REFERENCES

[1] Aida, F.M.N.A., Shuhaimi, M., Yazid, M., Maaruf, A.G (2009). Mushroom as a potential source of prebiotics: a review. *Trends in Food Science and Technology*. 20: 567-575.

[2] Akramiene, D., Kondrotas, A., Didziapetriene, J., Kevelaitis, E (2007). Effects of β-glucans on the immune system. *Medicina (Kaunas)*. 43: 597-606.

[3] Bae, J., Sinha, J., Park, J., Song, C., Yun, J (2000). Optimization of submerged culture conditions for exo-biopolymer production by *Paecilomyces Japo*nica. *Journal of Microbiology and Biotechnology*. 10: 482-487.

[4] Bobek, P and Galbary, S (2001). Effect of pleuran (beta-glucan from *Pleurotus ostreatus*) on the antioxidant status of the organism and an dimethylhydrazine induced precancerous lesions in rat colon. *Br J Biomed Sci*. 58:164-168.

[5] Bobek, P., Nosalova, V., Cerna, S (2001). Effect of pleuran (beta-glucan *Pleurotus ostreatus*) in diet or drinking fluid on colitis rats. *Mol Nutr fod Rev*. 45: 360-363.

[6] Bobek, P., Ozdin, Lubomir., Kuniak, L (1997). Effect of Osyter mushroom and isolated β-glucan on lipid peroxidation and the activities of antioxidative enzymes in rats fed the cholesterol diet . *Nutritional Biochemistry*. 8: 469-471.

[7] Bobovcak, M., Kuniakova, R., Gabriz, J., Majtan, J (2010). Effect of pleuran (β-glucan from *Pleurotus ostreatus*) supplementation on cellular immune response after intensive exercise in elite athletes. *Appl Physiol Nutr Metab. 35:755-62.*

[8] Bohn, J.A and BeMiller, N (1995). (1→3)-β-D-Glucans as biological response modifiers: a review of structure- functional activity relationships. *Carbohydrate Polymers*. 28: 3-14.

[9] Borchers, A.T., Keen, C.L., Gershwin, M.E (2004). Mushrooms, Tumors, and Immunity: an update. *Exp Biol Med (Maywood)*.229:393-406.

[10] Burns, P.J., Yeo, P., Keshavarz, T., Roller, S., Evans, C.S (1994). Physiological studies of exopolysaccharide production from Basidiomycetes *Pleurotus sp. Florida*; effect of C and N source on polysaccharide production for potential as a hypocholesterolemic, antitumor and a fat mimetic. *Enzyme Microbial. Technol*. 34, 566–572.

[11] Carbonero, E.R., Gracher, A.H.P., Smiderle, F.R., Rosado, F.R., Sassaki, G.L., Gorin, P.A.J., Iacomini, M (2006). A β-glucan from the fruit bodies of edible mushrooms *Pleurotus eryngii* and *Pleurotus ostreatoroseus*. *Carbohydrate Polymers*. 66: 252-257.

[12] Chihara, G (1993). Medical aspects of *Lentinan* isolated from *Lentinus Edodes* (Berk) Sing. In : Chang, S.T., Buswell, J.A., Chiu, S.W, Editors, *Mushroom Biology & Mushroom Products*, the Chinese University Press, Hong Kong. pp. 261-266.

[13] Daba, A.S (2006). Production of anticancer compound from oyster mushroom fruit bodies. WIPO Patent Application WO/2006/119783.

[14] Daba, A.S and Ezeronye, O.U (2003). Anti-cancer effect of polysaccharide isolated from higher *basidiomycetes* mushrooms. *African Journal of Biotechnology*. 2: 672-678.

[15] Daba, A.S., Kabeil, S.S., Botros, W.A., El-Saadani, M.A (2008). Production of mushrooms (*Pleurotus ostreatus*) in Egypt as a source of nutritional and medicinal food. *World Journal of Agricultural Sciences*. 4: 630-634.

[16] El Enshasy, H. (2007). Filamentous Fungal Cultures-Process Characteristics, Products, and Applications. *In* Bioprocessing for value-added products from renewable resources. (S.T. Yang, Ed.). pp. 225-261. Elsevier Press

[17] El-Enshasy, H., Daba, A., El-Demellawy, M., Ibrahim, A., El Sayed, S., El Badry, I (2010). Bioprocess development for large scale production of anticancer exo-polysaccharide by *Pleurotus ostreatus* in submerged culture. *Journal of applied science*. 10: 2523-2529

[18] Frederick ,R.V., Michael, F. P, and Carl, R. A (1995). A Compendium of Vaccine Adjuvants and Excipients (2nd Edition).In: Vaccine Design The Subunit and Aojuvant Approach (Eds Powell, M.F and Newman, M.J) Plenum Press, New York. Pp 141-228.

[19] Gao, Y., Chan, E., Zhou, S (2004). Immunodulating activities of *Ganoderma*, a mushrooms with medicinal properties. *Food Review International*. 20: 123-161.

[20] Gern, R.M.M., Wisbeck, E., Rampinelli, J.R., Ninow, J.L., Furlan, S.A (2008). A alternative medium for production of *Pleurotus ostreatus* biomass and potential antitumor polysaccharides *Bioresource Technology*. 99:76-82

[21] Gibriel, A.Y., Ahmed, M., Rasmy, N., Rizk, I., Abdel-Rehem, N.S (1996). Cultivation of Oyster mushrooms (*Pleurotus spp.*): Evaluation of Different Media and Organic Substrates. *Mushroom Biology & Mushroom Products*. Royse (ed). Penn State Univ. ISBN 1-883956-01-3.

[22] Gregori, A., Svagelj, M., Pohleven, J (2007). Cultivation techniques and medicinal properties of *Pleurotus* spp. *Food Technol Biotechnol*. 45: 238-249.

[23] Gutierrez, A., Prieto, A., Martinez, A.T (1996). Structural characterization of extracellular polysaccharides produced by fungi from the genus *Pleurotus*. *Carbohydrate Resear*ch. 281: 143-154.

[24] Hearst, R., Nelson, D., Mccollum, G., Millar, B.C., Maeda, Y., Godsmith, C.E., Rooney, P.J., Loughrey, A., Rao, J.R., Moore, J. E (2009). An examination of antibacterial and antifungal properties of constituents of Shiitake (*Lentinula edodes*) and Oyster (*Pleurotus ostreatus*) mushrooms. *Complementary therapies in clinical practice*. 15: 5-7.

[25] Jayakumar, T., Sakthivel, M., Thomas, P.A., Geraldine, P (2008). *Pleurotus ostreatus*, an oyster mushroom, decreases the oxidative stress induced by carbon tetrachloride in rat kidneys, heart and brain. *Chemico-Biological Interactions*. 176: 108-120.

[26] Jedinak, A and Sliva, D (2008*). Pleurotus ostreatus* inhibits proliferation of human breast and colon cancer cells through p53-dependent as well as p53-independent pathway (2008). *Int J Oncol*. 33: 1307-1313.

[27] Jung, G. T., Ju, I. O., Yu, Y. Z., Ryu, J., Choi, J. S., Choi, Y. G (2003). Mycelial yield of *Pleurotus ostreatus* using thinned apple, pear and peach on submerged culture. *Biotechnology and Bioprocess Engineering*. 8:286-290

[28] Kalmis, E., Nuri, A., Hassan, Y., Fatih, K (2008). Feasibility of using olive mill effluent (OME) as a wetting agent during the cultivation of oyster mushroom, *Pleurotus ostreatus*, on wheat straw. *Bioresource Technology*. 99: 164-169.

[29] Karacsonyi, S and Kuniak, K (1994). Polysaccharides of *Pleurotus ostreatus*: isolation and structure of pleuran, an alkali-insoluble β-D-glucan. *Carbohydrate Polymers*. 24: 107-111.

[30] Kavanagh, K (2005). *Fungi Biology and Applications*. Ireland: John Wiley & Sons.

[31] Kim, S.W., Hwang, H.J., Park, J.P., Cho, Y.J., Song, C.H., Yun, J.W (2002). Mycelial growth and exo-biopolymer production by submerged culture of various edible mushrooms under different media. *Letters in Applied Microbiology*. 34: 56-61.

[32] Kogan, G (2000). (1→3,1→6)-β-D-Glucans of yeast and fungi and their biological activity. *Studies in Natural Products Chemistry*. 23: 107-

[33] Lavi, I., Friesem, D., Geresh, S., Hadar, Y., Schwartz, B (2006). An aqueous polysaccharide extract from the edible mushroom *Pleurotus ostreatus* induces anti-proliferative and pro-apoptoxic effects on HT-29 colon cancer cells. *Cancer Letters*. 244: 61-70.

[34] Lin, S. E. and Chen, H. Y (2007). Factors affecting mycelial biomass and exopolysaccharide production in submerged cultivation of *Antrodia cinnamomea* using complex media. *Bioresource Technology*. 98: 2511–2517.44

[35] Lindequist, U., Niedermeyer, T.H.J., Dieter Julich, W (2005). The pharmacological potential of mushrooms. *eCAM*. 2: 285-299.

[36] Lomberh, M.L., Solomko, E.F., Buchalo, A.S., Kirchhoff, B (2002). Studies of medicinal mushrooms in submerged cultures. *Mushroom Biology and Mushroom Products. Sánchez et al. (eds).UAEM. ISBN 968-878-105-3.*pp 367-377.

[37] Maity, K.K., Patra, S., Dey, B., Bhunia, S.K., Mandal, S., Das, D., Majumdar, D.K., Maiti, S., Maiti, T.K., Islam, S.S (2010). Heteropolysaccharide from aqueous extract of an edible mushroom, *Pleurotus ostreatus* cultivar: structural and biological studies. *Carbohydrate Research.*

[38] Mantovani, M.S., Bellini, M.F., Angeli, J.P.F., Oliveira, R.J., Silva, A.F., Ribeiro, L.R (2008). β-glucans in promoting health: Prevention against mutation and cancer. *Mutation research.* 658: 154-161.

[39] Manu-Tawiah, W and Martin, A. M (1986). Study of operational variables in the submerged growth of *Pleurotus ostreatus* mushroom mycelium. *Applied Biochemistry and biotechnology.* 14: 221-229.

[40] Manzi, P and Pizzoferrato, L (2000). Beta glucans in edible muhrooms. *Food Chemistry.* 68:315-318.

[41] Morris, H.J., Marcos, J., LLaurado, G., Fontaine, R., Tamayo, V., Garcia, N., Bermudez, R.C (2003). Immunodulating effects of hot-water extract from *Pleurotus ostreatus* mycelium on cyclophosphamide treated mice. *Micologia Aplicada International.* 15: 7-13.

[42] Nehad, E.A and El-Shamy, A.R (2010). Physiological studies on the production of exopolysaccharide by Fungi. *Agriculture and Biology Journal of North Africa.*

[43] Nosalova, V., Bobek, P., Cerna S, Galbavy, Stvrtina, S (2001). Effects of pleuran (beta-glucan isolated from *Pleurotus ostreatus* on experimental colitis in rats. *Physiol Res.* 50:575-581.

[44] Novak, M and Vetvicha, V (2008). B-glucans, history, and the present: Immunomodulatory aspects and mechanisms of action. *Journal of Immunotoxicology.* 5: 47-57.

[45] Novak, M and Vetvicha, V (2009). Glucans as biological response modifiers. Endocrine, Metabolic & Immune Disorders-Drug Target. 9:67-75.

[46] Pramanik, M., Mondal, S., Chakraborty, I., Rout, D., Islam, S.S (2005). Structural investigation of a polysaccharide (Fr.II) isolated from the aqueous extract of an edible mushroom, *Pleurotus sajor-caju. Carbohydrate Resear*ch. 340: 629-636.

[47] Rop, O., Micek, J., Jurikova, T (2009). Beta-glucans in higher fungi and their health effects. *Nutrition Reviews.* 67:624-631.

[48] Rosado, F. R., Germano, S., Carbonero, Da Costa, S.M.G., Iacomini, M., Kemmelmeier, C (2003). Biomass and exopolysaccharide production in submerged cultures of *Pleurotus ostreatoroseus* Sing and *Pleurotus ostreatus* "florida". *J. Basic Microbiol.* 43: 230–237.

[49] Rout, D., Mondal, S., Chakraborty, I., Islam, S.S (2006). The structure of a polysaccharide from Fraction-II of an edible mushroom, *Pleurotus florida. Carbohydrate Research.* 341: 995-1002.

[50] Roy, S.K., Das, D., Mondal, S., Maiti, D., Bhunia, B., Maiti, T.K., Islam, S.S (2009). Structural studies of an immunoenhancing water-soluble glucan isolated from hot water extract of an edible mushroom, *Pleurotus florida*, cultivar *Assam florida. Carbohydrate Research.* 344: 2596-2601.

[51] Sainos, E., Diaz-Godinez, G., Loera, O., Montiel-Gonzalez, A.M., Sanchez, C (2006). Growth of *Pleurotus ostreatus* on wheat straw and wheat grain based media: biochemical aspects and preparation of mushroom inoculum. *Appl. Microbiol. Biotechnol.* 72: 812-815.

[52] Sarangi, I., Ghosh, D., Bhutia, S.K., Mallick, S.K., Maiti, T.K (2006). Anti-tumor and immunomodulating effects of *Pleurotus ostreatus* mycelia-derived proteoglycans. *International Immunopharmacology*. 6: 1287-1297.

[53] Shamtsyan, M., Konusova, V., Maksimova, Y., Goloshchev, A., Panchenko, A., Simbirtsev, A., Petrishchev, N., Denisova, N (2004). Immunomodulating and anti-tumor action of extracts of several mushrooms. *Journal of Biotechnology*. 113: 77-83.

[54] Silveira, M.L.L., Furlan, S.A., Ninow, J.L (2008). Development of an alternative technology for the oyster mushroom production using liquid inoculum. *Cienc. Technol. Aliment.,Campinas*. 28: 858-862.

[55] Smith, J.E., Rowan, N.J., Sullivan, R (2002). Medicinal mushrooms: a rapidly developing area of biotechnology for cancer therapy and other bioactivities. *Biotechnology Letters*. 24: 1839-1845.

[56] Stajic, M., Persky, L., Friesem, D., Hadar, Y., Wasser, S.P., Nevo, E., Vukojevic, J (2006). Effect of different carbon and nitrogen sources on laccase and peroxidases production by selected *Pleurotus* species. *Enzyme and Microbial Technolog* . 38: 65–73

[57] Sun, Y and Liu, J (2009). Purification, structure and immunobiological activity of a water-soluble polysaccharide from the fruiting body of *Pleurotus ostreatus*. *Bioresource Technology*. 100: 983-986.

[58] Sun, Y., Li, T., Yan, J., Liu, J (2010). Technology optimization for polysaccharide (POP) extraction from the fruiting bodies of *Pleurotus ostre*atus by Box-Behnken statistical design. *Carbohydrate polymers*. 80: 242-247.

[59] Surenjav, U., Zhang, L., Xu, X., Zhang, X., Zeng, F (2006) .Effects of molecular structure on antitumor activities of (1-3) –β-D-glucans from different *Lentinus edodes*. *Carbohydrate Polymer*. 63: 97-104.

[60] Synytsya, A., Mickova, K., Synytsya, A., Jablonsky, I., Spevacek, J., Erban, V., Kovarikova, E., Copikova, J (2009). *Carbohydrate Polymers*. 76: 548-556.

[61] Tang, J. Y and Zhong, J. J (2003). Role of oxygen supply in submerged fermentation of *Ganoderma lucidum* for production of *Ganoderma* polysaccharide and ganoderic acid. Enzyme *and microbial technology*.32:478-484

[62] Téllez-Téllez, M., Fernández. F. J., Montiel-González, A. M., Sánchez, C. and Díaz-Godínez, G (2008). Growth and laccase production by *Pleurotus ostrea*tus in submerged and solid-state fermentation. *Appl Microbiol Biotechno*l. 81:675–679.

[63] Tong, H., Xia, F., Feng, K., Sun, G., Gao, X., Sun, L., Jiang, R., Tian, D., Sun, X (2009). Structural characterization and in vitro antitumor activity of a novel polysaccharide isolated from the fruiting bodies of *Pleurotus ostreatus*. *Bioresource Technology*. 100: 1682-1686.

[64] Wang, J.C., Hu, H. S., Chin Liang, Z (2005).Optimization for the production of water-soluble polysaccharide from *Pleurotus citrinopi*leatus in submerged culture and its antitumor effect. *Appl Microbial Biotechnol*. 67: 759-766.

[65] Wasser, S.P (2002). Medicinal mushrooms as a source of antitumor and immunomodulating polysaccharides. *Applied Microbiology and Biotechnology*. 60:258-274.

[66] Wasser, S.P and Wies, A.L (1999). Medicinal properties of substances occurring in higher basidiomycetes mushrooms: Current perspectives (Review). *Intl. J. Med. Mushrooms*. 1;31-62.

[67] Wu, J. Z., Cheung, P., Wong, K., Huang, N (2004). Studies on submerged fermentation of *Pleurotus tuber-regium* (Fr.) Singer. Part 2: effect of carbon-to-nitrogen ratio of the culture medium on the content and composition of the mycelial dietary fibre. *Food Chemistry*. 85: 101–105.

[68] Wu, Y. C., Liang, C. Z., Lu, P. C., Wu.,H. S (2008). Effect of carbon and nitrogen sources on the production and carbohydrate composition of exopolysaccharide by submerged culture of *Pleurotus citrinopileatus*. *Journal of Food and Drug Analy*sis. 16: 61-67.

[69] Yildiz, A., Karakaplan, M., Aydin, F (1998). Studies *on Pleurotus ostreatus* (Jacq.ex Fr) Kum. Var. salignus (Pers. Ex Fr) Konr. Et Maubl.: cultivation, proximate composition, organic and mineral composition of carpophores. *Food Chemistry*. 61: 127-130.

[70] Xiao, H. J., Min-Xiao, D., Xiong, Q., Qi-Liang, Z., Zhong, J. J (2010). Nutritional requirements for the hyperproduction of bioactive exopolysaccharide by submerged fermentation of the edible medicinal fungus *Cordyceps taii. Biochemical Engineering journal*. 49:241-249.

[71] Yildiz, S., Gezer, E.D., Yildiz, U.C., Temiz, A., Dizman, E (2006). Effects of different substrate combinations on the mycelial growth of *Pleurotus ostreatus*. *Proceedings of the Fifth International Conference on Mushroom Biology and Mushroom Products, Shanghai, PR China*. P.551

[72] Yoshioka, Y., Tabeta, R., Saito, H., Uehara, N., Fukuoka, F (1985). Antitumor polysaccharides from *P. ostreatus* (FR) Quel: Isolation and structure of a β-glucan. *Carbohydrate Research*. 140: 93-100.

[73] Zhang, R.H., Li, X.J., Fadel, J.G (2002). Oyster mushroom cultivation with rice and wheat straw. *Bioresource .Technology*. *82*: 277-284.

[74] Zhang, M., Cui, S.W., Cheung, P.C.K., Wang, Q (2007). Antitumor polysaccharides from mushrooms: a review on their isolation process, structural characteristics and antitumor activity. *Trends in Food Science &Technology*. 18: 4-19.

[75] Zhang, R., Li, X., Fadel, J.G (2002) Osyter mushroom cultivation with rice and wheat straw. *Bioresource Technology*. 82: 277-284.

[76] Zhong, J. J and Tang, J. Y (2004). Submerged cultivation of medicinal mushrooms for production of valuable bioactive metabolites. *Advanced biochem Engineering/Biotechnology*.87: 25-59.

In: Mushrooms: Types, Properties and Nutrition
Editor: Sophie Andres and Niko Baumann

ISBN 978-1-61470-110-1
2012 Nova Science Publishers, Inc.

Chapter 8

ENHANCING THE MEDICINAL PROPERTIES OF *AGARICUS SUBRUFESCENS* BY GROWING PRACTICES

Diego Cunha Zied[1*], Francisco José Gea Alegría[2] and Arturo Pardo Giménez[2]

[1] Faculdade de Ciências Agronômicas, Universidade Estadual Paulista, Módulo de Cogumelos, Departamento de Produção Vegetal (Defesa Fitossanitária), Brazil

[2] Centro de Investigación, Experimentación y Servicios del Champiñón (CIES), Quintanar del Rey, Cuenca, Spain

ABSTRACT

The *Agaricus subrefescens* Peck [*A. blazei* (Murrill) ss. Heinemann] mushroom has been widely studied in various parts of the world, due to its medical and pharmacological properties. As substances of interest, the cell wall of the fungi contains polysaccharides called β-glucans, which have a structural function. During the 1980s, the *A. subrufescens* mushroom was imported to Japan due to its alleged health effects and is widely used today in Oriental countries both as an edible mushroom, considered a functional food, and as a natural therapy in the form of a medicinal extract, used mostly for the prevention and treatment of cancer. In accordance with Brazilian tradition, it could be useful against a variety of diseases, such as diabetes, atherosclerosis, hepatitis, hypercholesterolemia and heart disease, among others. The mushroom is commercialized in several countries as a nutraceutical product, which is a novel class of dietary supplements, including partially refined extract or dried biomass from the mushroom made into a capsule or tablet. In general, the compositional analysis of *A. subrufescens* mushrooms is water (84-87%), protein (30-33%), fat (0.82-1.3%), fiber (5.6-6.8%) and minerals (5.9-7.1%), with β-glucan contents between 4.4 and 6.9 g 100 g^{-1} of mushroom. Active metabolites can be isolated from basidiomes; pure culture mycelia, culture filtrate, and currently numerous attempts are in progress to obtain active metabolites from the mycelia through submerged fermentation culture due the cheaper preparations. The first separation of active anticancer compounds purified from the sodium hydroxide extract of the fruit body

* Faculdade de Ciências Agronômicas, Universidade Estadual Paulista, FCA/UNESP. Módulo de Cogumelos, Departamento de Produção Vegetal (Defesa Fitossanitária), Brazil. E-mail: dczied@gmail.com

occurred in 1989. Following this achievement, several controversies have arisen, mainly regarding the nutraceutical quality of the extracts obtained and in relation to the tests conducted, emphasizing that the chemical composition and the stage of maturity of the mushrooms, as well as the cultivation practices used in growing them, are crucial to obtaining quality basidiomes with high concentrations of β-glucans. Thus, in this book chapter, the problems arising from the variability of the content of β-glucans of mushrooms due to different practices and growing techniques adopted for the *A. subrufescens* production are discussed. To achieve this, five strains (ABL 99/28, ABL 99/30, ABL 03/44, ABL 04/49 and ABL 06/59), cultivated in three composts (Massai straw + sugar cane bagasse, oat straw + sugar cane bagasse and Aruana straw + sugar cane bagasse), with four casing layers (soil + charcoal, soil + coconut fiber, soil + peat moss and soil + composted pine bark) in four growing environments (climatized chamber, greenhouse with a transparent plastic film, greenhouse with a milky-white plastic film and greenhouse with a Duplalon® plastic film) were studied, with the aim of analyzing the variability of β-glucan content in the harvested mushroom and the agronomic performance of the growing practices. The ABL 04/49 strain cultivated with Massai straw and sugar cane bagasse showed the highest concentrations of β-glucans, but not a best yield, which was achieved by the ABL 99/30 strain cultivated with oat straw + sugar cane bagasse and obtained good β-glucan content and high yield. For agronomic performance, soil + composted pine bark was highlighted, influenced by the cultivation environment, to achieve high yield. Finally, following the presentation of a series of results, the creation of a growing protocol is suggested, which seeks to adopt cultivation practices that enhance the presence of β-glucans in the mushrooms.

Keywords: *Agaricus subrufescens*, β-glucans, growing practices, agronomic performance.

INTRODUCTION

The *Agaricus subrefescens* Peck [*A. blazei* (Murrill) ss. Heinemann] mushroom has been widely studied in various parts of the world, due to its medical and pharmacological properties. As substances of interest, the cell wall of fungi contains polysaccharides called β-glucans, which have a structural function. The main commercial target of the basidiomes is the pharmaceutical industry, but dried basidiomes are also sold as a complementary therapeutic treatment, in order to enhance the immunological system (Eira, 2003).

In Brazil, *A. subrufescens* is known as "Cogumelo do Sol®", "Cogumelo Medicinal", "Cogumelo de Piedade" or simply as "blazei", while in the USA, it is known as the "Almond mushroom" and as "Royal Sun Agaricus", in Japan as "Himematsutake", "Agarikusutake" and "Kawarihiratake" and in China as "Ji Rong Song" (Kopytowski Filho et al., 2006; Firenzuoli et al., 2007). According to Zied et al. (2009a), the basidiomes of *A. subrufescens* possess a pileus of 2-7 cm in diameter, on average; they are semiglobose, light brown to cream, with small white flakes on the upper part, which protrudes from the pileus (2-3mm). The stem narrows at the junction with the pileus (4-13 cm long by 1-3 cm in diameter), is uniform in thickness or has a bulbous white base (Figure 1).

The lamellae are arranged very close together, are not connected to the stem, measure 0.8-1.0 cm in width. The spores are whitish to gray-brown and dark brown at senescence, elliptical (4.9-6.4 by 4.0-4.7 μm), thick walled that have no pores, possess predominantly tetrasporic basidia and cystidia that are claviform, spear-like (18-27 by 6-10 μm) or catenulated, with globose elements (5-7 μm in diameter) that are hyaline or brown (Heinemann, 1993).

Zied et al., 2009a.

Figure 1. A: Morphological aspects of the fungus, B: Fungus spores in senescence and C: Elliptical spores.

Regarding its nomenclature, several controversies and references can be found in the literature, which describes *Agaricus subrufescens* as: *Agaricus silvaticus*, *Agaricus brasiliensis* and *Agaricus blazei*, among others (Kerrigan, 2005; González Matute et al., 2010). Given this fact, in early 2008, a project headed by Philippe Callac[1], a French researcher, and collaborators was conducted to study the biodiversity of this species in different parts of the world to verify their molecular characteristics.

For Almond mushroom production, a lingo-cellulosic substrate is used with a carbon: nitrogen (C:N) ratio at the end of Phase II of 25-27:1 and N content from 1.5-1.9% (Kopytowski Filho, 2002). Regarding the mix of the casing layer, sand clay soils (55% sand, 10% silt and 35% clay), with bulk density of 1.0-1.1 gcm^{-3} and water holding capacity of 40-50%, to which 25% volume of organic material (charcoal, peat, coconut fiber, etc..) is added, have presented good results in relation to yield and the quality of fruiting bodies or basidiomes (Zied et al., 2011).

The cultivation temperatures used for the Brazilian strains are as follows: 28°C during incubation, between 25-28°C during fruiting and at pinning, a decrease in temperature is recommended to a value close to 20°C. The initial CO_2 levels following the addition of casing can be high ($\geq$ 3,000 ppm), while after colonization of the casing, the CO_2 content diminishes to below 800 ppm. The values of relative humidity during the vegetative growth vary from 85 to 90% and during the reproductive growth from 80 to 85% (Zied & Minhoni, 2009b).

Currently, only a few countries produce *A. subrufescens* on a commercial scale, including Brazil, China, USA and Taiwan (Zied et al., 2010; Wang et al., 2010), but on a experimental scale, countries like Spain, France, Korea, Canada, Argentina and Japan (Ohno et al., 2001; Wang et al., 2010; González Matute et al., 2010) can also be included in this list.

MEDICINAL PROPRIETIES

Glucans are side branches of a (1-6)-β-backbone, as described by Ohno et al. (2001), who reported that the active fraction of β-glucans of *A. blazei* basidiomes had a (1-6)-β-backbone structure (or functional center) with (1-3)-β-side branches, in the ratio of 1:2; while the linear (1,6)- β-glucan seems to be inactive (Figure 2).

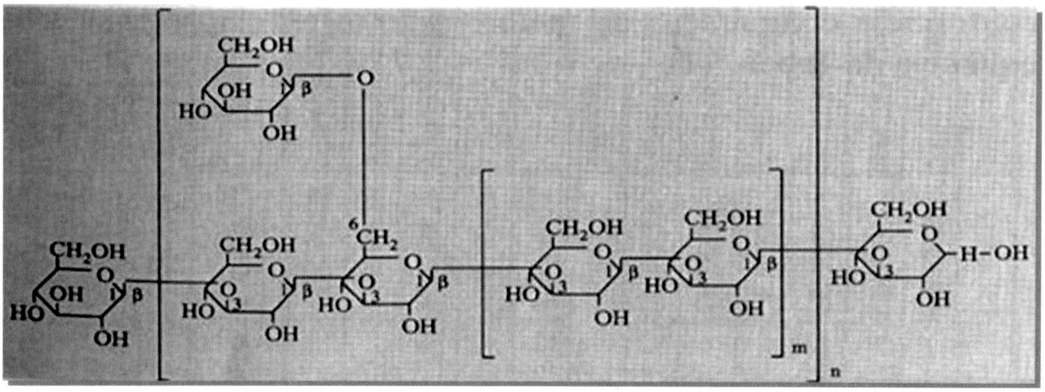

Figure 2. (1–6) -β-backbone structure (or functional center) with (1–3) -β-side branches.

The biochemical importance of the (1-3)-β-side branches has been confirmed and has shown enhancement of the immunomodulatory activity of polysaccharides (Dong et al., 2002); moreover, Mizuno et al. (1990) reported an important antitumor activity linked to the water-soluble (1-6)-(1-3)-β-D-glucan. However, a significant increase in water-soluble (1-4)-α-glucan with apparent antitumor activity occurs during maturation (Mizuno et al., 1990); thus, cap-opened, more fragile mature fruiting bodies of *A. blazei* should be selected over immature basidiomes for the production of nutraceuticals, because these contain the most useful glucans (Camelini et al., 2005). In addition, an α-1,6 and ∞-1,4 glucan complex (Mizuno et al., 1998) and a glucomannan with a main chain of β-1,2-linked D-mannopyranosyl residues have been isolated from this mushroom and shown to inhibit tumorigenesis (Mizuno et al., 1999).

From 1999 to 2003, at the Faculty of Agricultural Sciences of São Paulo State University (*Faculdade de Ciências Agronômicas, Universidade Estadual Paulista*, FCA/UNESP), a Thematic Project was developed concerning edible and medicinal mushrooms, their cultivation technology, biochemical characterization and protective effects of *Agaricus blazei* Murrill (Almond mushroom) and *Lentinula edodes* (Berk.) Pegler (Shiitake), under protocol no. 98/07726/5. The project involved analyses and studies regarding the quantity of polysaccharides (β-glucans) in the mushroom *A. blazei* (Table 1).

At the time, Eira affirmed that the quantity of β-glucans present in the mushroom could vary depending on growing conditions, substrate composition and type of strain (Eira, 2003). However, several other factors also influence the concentration of β-glucans from *A. blazei* mushroom, such as local of growing (field or greenhouse), compost quality, rainfall frequency, sudden change in temperature and amount of pests in the cycle of cultivation. These factors also affect the content of β-glucans in cereal crops (Park et al., 2003).

Table 1. Enzymatic analysis of β-glucan in samples of *A. blazei* (ABL), blinded to strain

Sample	Strains	Powder color	mg of β-glucans
1	ABL – 99/28	yellowish brown	22.37
2	ABL – 99/28	yellowish brown	44.74
3	ABL – 99/28	yellowish brown	52.96
Mean	ABL – 99/28		40.02
Standard deviation	ABL – 99/28		15.83

Source: Eira, 2003.

The amount of light in the cultivation environment (Lux), the type of casing layer, the duration of the cycle cultivation, the number of mushrooms harvested per mass of compost and storage time of dried mushroom post-harvest, are also possible factors that influence the quality of the final product for use as medicine. Thus in this book chapter, the problems arising from the variability of the concentrations of β-glucans of the mushrooms due to different practices and growing techniques adopted for *A. subrufescens* production are discussed. For this purpose, two experiments were conducted and compared regarding the medicinal potential of harvested mushrooms. Finally, following the presentation of a series of results, the creation of a growing protocol is suggested, which seeks to adopt cultivation practices that enhance the presence of β-glucans in the mushrooms.

Variability Study of β-Glucan and the Agronomic Performance of Agaricus Subrufescens According to Different Strains and Types of Compost

A variety of edible mushrooms have gained increasing importance in modern medicine due to their high medicinal values. *Agaricus campestris*, *Flamulina mellae* and *Flamulina odilpis* are known to possess antibacterial activity against *Staphylococcus aureus*, *Salmonella typhii* and *Escherichia coli* (Bose, 1953). A variety of antifungal activities were determined for *L. edodes*, the *Cortinellus* Shiitake and for *Coprinus comatus* (Bohus et al., 1961; Herman, 1962). From the giant puffball mushrooms, *Calvatia giantea*, an antitumor substance called "Calvacin" has already been obtained (Beneke, 1963).

Used for the same purpose, *A. blazei* is marketed in many countries as a nutraceutical product, a novel class of dietary supplements including partially refined extract or dried biomass from the mushroom made into a capsule or tablet (Chamg & Buswell, 1996). However, much controversy exists in the literature concerning the medicinal substances identified in *A. blazei*, which have been highlighted by several authors (Camelini et al., 2005; Kawagishi et al., 1989; Mizuno et al., 1990; Osaki et al., 1994; Takatu et al., 2001; Kawagishi et al., 1988; Ohno et al., 2001).

Thus there is considerable concern within the scientific community regarding which active substances in mushrooms are genuinely valuable, which have healing, pharmacological and medicinal properties, quite aside from any important agronomic and physiological changes involved in the process of cultivation, which might influence the chemical characteristics of such mushrooms.

Firenzuola et al. (2007) emphasized the absence of scientific papers presenting concrete data that addressed the characteristics of the crop and the agronomic performance and physiological behavior of the varieties of *A. blazei* used in pharmacological, biochemical and medical research. However, the cultivation of *A. blazei* in Brazil is driven by adjustments to the practices adopted in the production of *A. bisporus* and the adoption of management systems and technical research results published recently (Kopytowski & Minhoni, 2004; Zied et al., 2009c; Zied et al., 2010; Colauto et al., 2010).

In Brazil, no definitive protocol exists regarding cultivation; there are no guidelines that clarify which cultivation practices influence the presence of active substances in the mushrooms harvested. Researchers should also bear in mind the agronomic performance of *A. subrufescens*, which should include the feasibility of such protocols for the grower and the sustainability of the activity. To this end, this study aimed to compare the medicinal potential (β-glucans) and agronomic performance (yield, number and weight of the mushrooms) of *A. subrufescens* in relation to the use of different strains and composts.

Methodology

Five commercial strains of *A. subrufescens* were used, which had been conserved in the culture collection of the Mushroom Research Center (*Banco de Matrizes do Módulo de Cogumelos, Departamento de Produção Vegetal, Defesa Fitossanitária*) of the FCA/UNESP, and their main features are described in Table 1.

The culture medium used for the production of subculturing was prepared by boiling 60 g of substrate that was added to 900 ml of distilled water and 15 g of agar, then autoclaved twice for 30 minutes at 121°C, following by an interval of 24 hours. After cooling, the inoculation of *A. subrufescens* was initiated.

The substrate used for production of the inoculum was based on grains of triticale (*Triticum secale*), gypsum and limestone. The grains were first boiled and then 20 g kg^{-1} of limestone and 160 g kg^{-1} of gypsum were added. Following homogenization, the substrate was autoclaved for 3 hours at 121°C and then the inoculation of *A. subrufescens* was initiated. The methodology followed the procedures adopted by Minhoni et al. (2005).

The method used for the production of the compost was "traditional", with duration of 26 days for Phase I, a process involving pre-wetting for 7 days and a fermentation process for 19 days that included six turnings. Phase II composting lasted 9 days: 8 hours "pasteurization" at 59 ± 1°C and 8 days "conditioning" at 47 ± 2°C. The three different composts were prepared at the same time and followed the same production process. Table 2 presents the materials used in each compost and their chemical characteristics at the end of Phase II of composting. The C:N ratio was initially calculated on the basis of obtaining composts with predetermined values of 33-37:1.

Following pasteurization and conditioning, the compost was mixed with the inoculum and placed in plastic boxes containing 12 kg of compost and 120 g of inoculum. Then the three composts were inoculated with the five strains of *A. subrufescens* they were incubated in a dark chamber at 28 ± 2°C, relative humidity of 50 ± 10% and no ventilation for 21 days. During this period, inspections were periodically made to check the growth of mushroom mycelium and for possible contamination of the composts by pests and diseases.

The casing used was a mixture of soil + charcoal (4:1, v/v), to which limestone was added to correct pH, which was then moistened to approximately 35% and pasteurized for 12 hours at 58ºC. The soil used in this study followed the quality guidelines described by Zied (2008), which are: sand content of 427 g kg^{-1}, silt of 395 g kg^{-1}, clay of 178 g kg^{-1}, amount of Al + H of 9.6 $mmol_c$ dm^{-3}, V% of 92% and finally Mg^{+2} of 3 $mmol_c$ dm^{-3}.

When the mycelia were fully developed in the compost, an additional layer of casing was pressed over the top, until it reached 5 cm in depth. The boxes with compost and the casing layer were taken to a greenhouse with milky-white plastic film and partial control of temperature, humidity and aeration. During cultivation, the air temperature ranged from 10-40ºC, compost temperature ranged from 18-32ºC and relative humidity ranged from 35-100%.

The induction of mushrooms occurred naturally, according to the weather conditions of the day or week. The total production time was 120 days and at 21 days, the presences of primordia were observed. According to the strain and compost used, the harvest flush ranged from 2 (ABL 99/30 strain inoculated in compost 1) to 7 (ABL 03/44 strain inoculated in compost 3) during cultivation.

Collection was performed manually, followed by pre-cleaning of the mushrooms by scraping the base of the stalks to remove the residue of the casing layer material. Next, the mushrooms were identified and assessed regarding yield, weight and number of mushrooms and the amount of β-glucan. The statistical software Sisvar 3.2 was used to analyze the means obtained for each of the treatments (strain + compost) and compare them using the Tukey test (≤ 0.05).

Results

Analysis of the results presented in Table 3 verified that the strain ABL 99/30 showed the highest yield and number of mushrooms. As expected, the strain ABL 06/53 presented the greatest basidiome weight, resulting from low yield. In relation to the β-glucan values, the variety ABL 04/49 showed the highest value (6.89 g g^{-1}), approximately 35% more than the strain ABL 99/28, which showed the lowest value (4.42 g g^{-1}).

According to the variables analyzed, the compost used showed no statistical differences when evaluated alone (single factor).

Analysis of the results obtained for the strains ABL 99/30 and 04/49, grown in three composts, verified that the highest yield was obtained with variety ABL 99/30 and compost 2 (20.17%), which provided a large number of mushrooms harvested (142.8 units), with an mean of 99 units; however, these presented reduced weight of 17.37 g, compared to the mean of 18.38 g for all 3 composts, and β-glucan content of 5.16 g g^{-1}, mean of 5.99 g g^{-1} (Table 4). Another possible option was the strain ABL 99/30 with compost 3, which resulted in high yield and number of mushrooms harvested than the mean, but still presented weight and β-glucan below the mean of the results obtained for this double factorial (strain *vs* compound).

Table 2. Codes of the strains, Collection year, City, State, Country and the main characteristics

Encoding	Collection year	City of the collection	State	Country	General characteristics
ABL 99/28	1999	Botucatu	São Paulo	Brazil	Strain maintained in the culture collection of the Mushroom Research Center, characterized by its medium size, strong texture and regular yield.
ABL 99/30	1999	Piedade	São Paulo	Brazil	Mushroom isolated from a commercial cultivation of the Atushi group, characterized by its medium size, strong texture, high yield and slightly lower fruiting temperature (± 25°C).
ABL 03/44	2003	Lençóis Paulista	São Paulo	Brazil	Mushroom isolated from a commercial cultivation of Santa Fé Farm, characterized by its large size, strong texture, regular yield.
ABL 04/49	2004	São José do Rio Preto	São Paulo	Brazil	Mushroom isolated from a commercial cultivation of the grower Goto, characterized by its large size, strong texture, regular yield and higher fruiting temperature (± 28°C),
ABL 06/53	2006	Brasília	DF	Brazil	Mushroom isolated from a commercial cultivation of the company Brasmicel, characterized by its large size, moderate texture, low yield and high precocity.

Observations: the variety ABL 06/53 presented high susceptibility to attack by *Verticillium fungicola* and *Micogone perniciosa*.

Table 3. Materials used for each compost and their respective chemical characteristics at the end of Phase II of the composting process

Formulation	Dry weight of the materials (kg)									
	Sugar cane bagasse	Oat straw	Massai straw	Aruana straw	Soybean meal	Urea	Ammonium Sulfate	Superphosphate	Gypsum	Limestone
Compost 1	141	-	130	-	9.2	2	1.5	4	15	25
Compost 2	141	130	-	-	9.2	2.5	2.4	4	15	25
Compost 3	141	-	-	131	9.2	1	1	4	15	25

	Chemical elements														
	N	P_2O_5	K_2O	Ca	Mg	S	MO	C	Na	Cu	Fe	Mn	Zn	C/N	pH
	%								mg kg^{-1} dry material						
Compost 1	1.81	1	1.09	5.15	0.48	1.36	64	35.6	440	0	500	260	20	20/1	6.25
Compost 2	1.76	0.66	1.04	5.44	0.48	1.26	68	37.8	440	0	400	160	18	21/1	7.04
Compost 3	1.6	0.64	1.31	5.20	1.68	1.40	65	36.2	1160	172	400	306	212	23/1	6.94

Compost 1: Substrate prepared with sugar cane bagasse (*Saccharum officinarum*) and Massai (*Panicum maximum* cv. Massai) straw.
Compost 2: Substrate prepared with sugar cane bagasse (*Saccharum officinarum*) and oat (*Avena sativa*) straw.
Compost 3: Substrate prepared with sugar cane bagasse (*Saccharum officinarum*) and Aruana (*Panicum maximum* cv. Aruana) straw.

Table 4. Agronomic performance and quantity of β-glucan obtained per treatment

Variables analyzed	Yield (%)	No. of mushrooms (u)	Mushroom weight (g)	β-glucan (g $100g^{-1}$)
ABL 99/28	8.75 b	61.23 b	17.64 b	4.42 c
ABL 99/30	17.08 a	124.25 a	17.75 b	5.10 b
ABL 03/44	10.69 b	75.70 b	18.43 b	5.40 b
ABL 04/49	11.33 b	75.62 b	19.02 b	6.89 a
ABL 06/53	3.19 c	22.48 c	22.55 a	4.90 bc
Compost 1	10.77 A	77.98 A	18.37 A	5.19 A
Compost 2	10.72 A	70.28 A	19.99 A	5.44 A
Compost 3	9.13 A	67.31 A	18.88 A	5.40 A
Mean	10.21	71.86	19.08	5.34

(*) Values followed by a different letter within a column are significantly different at 5% level according to Tukey's test.

The medicinal potential of ABL 04/49 was evident due to the high values of β-glucan obtained. However, the yield and number of mushrooms was below the mean (Table 4), considering that the harvested mushrooms presented high weight. In contrast, when comparing the yield of this strain grown in three different composts (11.5, 11.2 and 11.2%) with the mean yield obtained of 10.2% (Table 3), these values were greater than the mean for the factors of strain and compost analyzed alone.

Both factors, strain and compost, affect the amount of β-glucan present in the basidiomes when the values for their interaction were analyzed; e.g., compost 3 cultivated with strain ABL 04/49 showed low amounts of β-glucan (5.05 g $100g^{-1}$), whereas strain ABL 04/49 (8.11 g $100g^{-1}$) showed approximately 40% greater quantities of β-glucan than the strain ABL 99/30 (8.11 g $100g^{-1}$) cultivated in the same Massai-based compost.

The factors of yield, number and weight of the basidiomes seem to influence the quantity of β-glucan in an indirect way (Table 3), in which the strain ABL 06/53 showed low yield and less than the mean number of mushrooms and, consequently, a low quantity of β-glucan (less than the mean). Wang et al. (2010) tested six different substrate formulations in the production of *A. blazei*, with the final N content of the compost varying from 1.2 to 1.55 mg kg^{-1} of dry material, and determined no significant difference in the final value of polysaccharides of the basidiomes harvested. The same authors used the following raw materials to produce the composts: asparagus straw, cottonseed hull, cow manure, soybean cake and gypsum (Wang et al., 2010).

Conclusion

The factors strain and compost affected all the variables analyzed. The analysis highlighted the variety ABL 99/30 due to its high yield and the strain ABL 04/49 due to its high quantity of β-glucan. Correlating the results based on compost, the oat-based compost showed high yield when cultivated with variety ABL 99/30 and Massai-based compost showed high β-glucan when cultivated with the variety ABL 04/49.

Table 5. Values for agronomic performance and quantity of β-glucan according to the varieties ABL 99/30 and ABL 04/49

Variables analyzed	Yield (%)		No. of mushrooms (u)		Mushroom weight (g)		β-glucan (g $100g^{-1}$)	
	ABL 99/30	**ABL 04/49**	**ABL 99/30**	**ABL 04/49**	**ABL 99/30**	**ABL 04/49**	**ABL 99/30**	**ABL 04/49**
Compost 1	14.78 A a	11.58 A a	100.87 A a	82.48 A a	18.72 A a	17.33 A b	4.89 B a	8.11 A a
Compost 2	20.17 A a	11.20 B a	142.87 A a	68.25 A b	17.37 B a	21.04 A a	5.16 A a	7.52 B a
Compost 3	16.29 A a	11.22 A a	129.00 A a	76.12 A b	17.16 A a	18.67 A ab	5.23 A a	5.05 A b
Media	14.20		99.00		18.38		5.99	

Capital letters compare the results along the line and lowercase letters down the columns (Tukey ≤ 0.05).

VARIABILITY STUDY OF AGARICUS SUBRUFESCENS β-GLUCAN AND AGRONOMIC PERFORMANCE ACCORDING TO THE CASING LAYER AND CULTIVATION ENVIRONMENT

The incidence of cancer is gradually increasing and the diversity of diseases in various human organs changes every year (Ohno et al., 2001). Besides the fight against cancer with traditional surgery, radiation and chemotherapy, there is strong belief that immunotherapy could be an important tool for cancer treatment (Sone, 1999). Immunotherapy and biotherapy include various approaches, such as biological response modifiers, cytokines, lymphocyte transplantation, gene therapy, use of medicinal plants and alternative medications, such as mushrooms (Suzuki et al., 1998).

Clinical trials of these therapies are being conducted in several parts of the world and the evidence available has demonstrated their effectiveness, although the precise mechanisms of action remain difficult to understand (Rios-Hernandez et al., 1994; Mohagheghpour et al., 1995; Di Luiz, 1999; Ohno et al., 2001). As part of the trials, some studies have focused on *A. blazei* as a potential agent that possesses medicinal and pharmaceutical properties for the treatment of diseases like AIDS, diabetes, hypertension and hepatitis (Mizumo, 2002).

It should be emphasized that some of this information, although valuable, is being misused in misleading advertisements by the part of the media, which lack a scientific basis and end up manipulating data obtained using purified substances extracted from the mushrooms, which, in the purest and most concentrated form, are conceptually drugs; i.e., their use is clearly different from that of nutraceutical whole mushrooms ingested during food consumption (Eira, 2003).

A certain degree of controversy and discussion has been generated concerning the use of mushrooms in immunotherapy or biotherapy. The first hypothesis raised, and later experimentally evaluated, was the influence of the stage of growth and maturation of *A. blazei* mushrooms on the quality of the extracts used in experiments, particularly regarding the presence and structure of the biochemical compounds contained in the extracts (Mol & Wessels, 1990; Camelini et al, 2005).

The second refers to growing practices that also influence the medicinal properties of mushrooms. Previous studies have verified that the genetic strain and the materials used in the compost also influence the quantity of β-glucan in the basidiomes, such that the ABL strain 04/49 showed high amount of β-glucans, around 8.11 g g^{-1}, while strain ABL 99/30 showed lower amounts of β-glucan, 4.89 g g^{-1}. Both strains were grown in the same compost based on sugar cane bagasse and Massai straw (*Panicum maximum* CV, MASSAI).

Thus, strain ABL 99/30 and the oat-based compost were selected for a second experiment, the purpose of which was to evaluate the influence of the cultivation environment (3 greenhouses with different plastic films and a climatized chamber with no sunlight) and the casing layer (4 soil-based casing layers mixed with charcoal, coconut fiber, composted pine bark and peat moss) on the quantity of β-glucan in the basidiomes.

Methodology

The strain ABL 99/30, stored in the culture collection of the Mushroom Research Center of the FCA/UNESP, was isolated in 1999 in the town of Piedade from a commercial cultivation of the Atushi Group in the State of São Paulo, Brazil. It is characterized as a medium sized mushroom, with a strong texture and high yield (10-18%), a slightly lower fruiting temperature (± 25°C) and contents of β-glucan varying from 4.89 to 5.23 g $100g^{-1}$, protein from 30.35 to 33.33%, ash from 5.87 to 7.06%, fiber from 5.63 to 6.84% and finally, fat varying from 0.86 to 1.13%. The entire inoculum production process was performed as described by Zied et al. (2010).

The compost used consisted of a mixture of sugar cane bagasse (*Saccharum officinarum*), oat straw (*Avena sativa*), soybean meal, urea, ammonium sulfate, superphosphate, gypsum and limestone (Table 2, Compost 2 "dry weight of the materials"), which also followed the same methods and procedures used in Phase I and II during the substrate production process. At the end of Phase II of composting, the compost presented the following chemical content: 1.66% of N; 0.45% of P_2O_5; 0.26% of K_2O; 5.53% of Ca; 0.23% of Mg; 0.98% of S; 65% of MO; 36.2% of C; 300 mg kg^{-1} of Na; 10 mg kg^{-1} of Cu; 3300 mg kg^{-1} of Fe; 128 mg kg^{-1} of Mn; 28 mg kg^{-1} of Zn; a C:N of 22:1; and pH of 6.83.

The compost was inoculated with 1% ABL 99/30 mycelia in relation to compost wet weight. Thus, 12 kg of compost were mixed homogeneously with 120 g of inoculum, in a plastic box previously lined with plastic film. The composts were incubated in a climatized chamber in the absence of light, at 28 ± 2°C, relative humidity of 50 ± 10% and in the absence of ventilation for 16 days.

Four casing layers were prepared: soil + charcoal (4:1, v/v); soil + coconut fiber (4:1, v/v); soil + composted pine bark (4:1, v/v); and soil + peat moss "*Sphagnum*", using the same soil as the previous experiment. First, limestone was added to the soil to raise the pH to between 7 and 7.5, then it was moistened and mixed with the organic materials. Table 5 shows the main physicochemical materials used in the experiment.

After mixing, the casing layers were pasteurized for 8 hours at 58°C, which was reduced to 24°C. Next, a 5 inch thick casing layer was overlaid on the inoculated compost in a plastic box containing 12 kg of compost, to which 12.5 liters of casing were added. The boxes containing the compost and casing layer were placed under four different environmental conditions (Figure 3): a climatized chamber (12 x 2.3 x 2.3 m, length x width x height); an experimental greenhouse with Duplalon® plastic film (8 x 4 x 2.5 m, LxWxH); an experimental greenhouse with milky-white plastic film (8 x 4 x 2.5 m, LxWxH); and an experimental greenhouse with transparent plastic film (8 x 4 x 2.5 m, LxWxH).

In each cultivation environment, variations occurred in the environmental variables, compost temperature, ambient temperature, relative humidity and quantity of light. Table 6 shows how the environmental variables behaved in each cultivation environment during production.

Thus the induction of primordia occurred naturally, according to the daily or weekly weather conditions, for crops cultivated in greenhouses with Duplalon®, milky-white and transparent plastic films. However, for cultivation in the climatized chamber, the flushes were controlled and induced to obtain five production flushes during the cultivation period. To achieve this, the environmental variables of temperature, relative humidity and ventilation were maintained according to the methodology described by Zied (2008).

Figure 3. Environments used in this research, A: climatized chamber, B: greenhouse with Duplalon® plastic film, C: greenhouse with milky-white plastic film, C: greenhouse with transparent plastic film.

The total production time was 120 days and at 21 days, the presence of primordia was observed in the climatized chamber, greenhouses with Duplalon® and milky-white film. Depending on the casing layer and the cultivation environment used, the harvest flush ranged from 2 (charcoal casing layer grown in the greenhouse with Duplalon® film) to 5 (all casing layer grown in the climatized chamber) during the cultivation period.

Collection was performed manually, followed by pre-cleaning of the mushrooms by scraping the base of the stalks to remove the residue of the casing layer material. Next, the mushrooms were identified and assessed regarding yield, weight and number of mushrooms and the amount of β-glucan. The statistical software Sisvar 3.2 was used to analyze the means obtained for each of the treatments (strain + compost) and compare them using the Tukey test (≤ 0.05).

Results

According to the results in Table 8, all the variables analyzed (basidiome yield, number and weight) presented significant differences, except for the quantity of β-glucan present in the mushroom.

The highest yield was verified in cultivation conducted in the climatized chamber and under these controlled conditions. No statistically significant differences in the values were observed for the different casing layers. The casing layers with soil + charcoal and soil + composted pine bark showed the highest yield when grown in semi-conditioned environments (greenhouses with Duplalon®, milky-white and transparent film). The low yields obtained in the semi-conditioned environments can be explained by the high temperatures verified the ambient and the compost (Table 7).

Basidiome weight was directly influenced by the type of casing layer and cultivation environment, such that the use of soil + composted pine bark resulted in a harvest with heavy mushrooms. To achieve high basidiome weight in cultivations conducted in a greenhouse, the use of casing layers presenting lower density and higher water retention capacity (soil + composted pine bark and soil + peat moss) is recommended.

The largest number of basidiomes was harvested in the climatized chamber using casing layers of soil + charcoal and soil + coconut fiber. These casing layers also presented larger quantities of harvested mushrooms in the greenhouses with milky-white and transparent film, environments that had a high rate of casing layer water evaporation, since they possess high density and low water retention capacity that favores a higher mushroom yield, though with reduced weight of mushroom.

Similar results were obtained by Zied et al. (2011), who emphasized that a negative correlation was verified between water-holding capacity and the number of mushrooms, while positive correlations were verified between bulk density and porosity and the number of mushrooms.

The quantity of β-glucan showed no significant differences in relation to the treatments analyzed, in disagreement with the results obtained by Park et al. (2003), who compared the quantity of β-glucan in mushrooms cultivated in Brazil and Japan. In the aforementioned study, the authors used mushrooms produced in greenhouses and the field in Brazil and only mushrooms produced in greenhouses in Japan and concluded that *A. blazei* cultivated in greenhouses presents a lower quantity of β-glucan (7.6 ± 2.8 g $100g^{-1}$ of mushrooms produced in Japan and 8.4 ± 0.9 g $100g^{-1}$ of mushrooms produced in Brazil) than those cultivated in the field (10.1 ± 2.1 g $100g^{-1}$ of mushrooms) (Park et al., 2003). It should be emphasized that the strain and compost factors used by the authors may have influenced the results of β-glucan present in the basidiomes.

Conclusion

Although not influence statistically the values of β-glucan, the variables casing and environment of growth varied widely according to the treatment used. The choice of the casing layer is important to get the mushrooms with desired physical characteristics. The cultivation in climatized chamber has great feasibility for the sustainability of the sector fungicola, allowing cultivation in all seasons.

Table 6. Physicochemical characteristics of the casing layers used in the experiment

	pH	OM	P	H + Al	K	Ca	Mg	SB	PBS	CEC	Fe	Mn	CE	BD	WHC	
					Chemical elements											
					%					mg kg^{-1} dry material			µs/cm	g cm^{-3}	S. 0,006MPa	
S + CH	6.8	9	4	10	1.1	64	3	68	88	78	5	0.6	190.7	1.12	0.48	0.20
S + CF	6.7	52	19	11	1.7	135	17	154	94	164	13	2.5	492.0	0.86	0.58	0.22
S + CP	7.1	31	6	9	2.1	163	8	173	95	182	5	0.6	104.5	0.84	0.62	0.25
S + PM	7.3	45	22	10	3.2	183	10	196	96	206	10	0.8	367.8	0.81	0.72	0.26

S = soil; CH = charcoal; CF = coconut fiber; CP = composted pine bark; PM = peat moss.

OM: Organic Matter, SB: Sum of bases: CEC: Cation exchange capacity, PBS: percentage of base saturation, BD: bulk density, WHC: water holding capacity.

Table 7. Values of the environmental variables according to the cultivation environment

Environmental variables	Ambient temperature (°C)			Compost temperature (°C)			Relative humidity (%)			Luminosity (20,000 lux)		
	Max	Min	Mean	Max	Min	Mean	Max	Min	Mean	Max	Min	Mean
CC	30.1	16.3	25.9	29.0	17.5	25.0	100	90	97	35	35	35
GDF	41.2	16.7	24.2	30.5	17.7	21.0	100	35	60.8	23	0	11.5
GMF	46.9	17.2	27.6	33.4	18.2	24.3	100	37	54.3	1390	0	695
GTF	55.3	17.0	30.6	38.3	17.9	29.6	100	32	48.7	2990	0	1495
Mean	43.3	16.8	27.0	32.8	17.8	24.9	100	48.5	65.2	1110	0	560

CC = climatized chamber; GDF = greenhouse with Duplalon® film; GMF = = greenhouse with milky-white film; GTF = greenhouse with transparent film.

Table 8. Agronomic performance obtained for each treatment

Casing layer (4:1, v/v)	Cultivation environment			
	Climatized chamber	Greenhouse + Duplalon® film	Greenhouse + milky-white film	Greenhouse + transparent film
	Yield (%)			
Soil + charcoal	14.72 A a	3.63 A b	6. 10 A b	5.03 A b
Soil + coconut fiber	12.71 A a	0.35 B b	0.87 C b	1.45 B b
Soil + pine bark	15.37 A a	3.94 A b	4.32 BA b	5.53 A b
Soil + peat moss	13.60 A a	2.00 AB b	2.11 CB b	1.04 B b
	Mushroom weight (g)			
Soil + charcoal	13.71 A a	13.62 B a	11.61 B a	15.70 A a
Soil + coconut fiber	11.77 A a	13.33 B a	16.02 B a	15.23 A a
Soil + pine bark	15.96 A b	27.85 A a	33.52 A a	24.25 A ab
Soil + peat moss	14.11 A a	17.38 AB a	19.91 B a	21.74 A a
	No. of basidiomes (u)			
Soil + charcoal	137.28 AB a	34.14 A b	68. 57 A b	43.66 AB b
Soil + coconut fiber	174.54 A a	37.85 A b	35.85 AB b	52.14 A b
Soil + pine bark	97.85 B a	1.75 A b	4.15 B b	8.65 B b
Soil + peat moss	120.14 B a	21.14 A b	13.85 B b	9.20 B b
	β-glucan (g $100g^{-1}$)			
Soil + charcoal	5.17 A a	3.97 A a	3.75 A a	6.62 A a
Soil + coconut fiber	2.51 A a	6.01 A a	6.65 A a	2.03 A a
Soil + pine bark	4.86 A a	4.20 A a	3.68 A a	4.71 A a
Soil + peat moss	9.07 A a	7.01 A a	5.37 A a	4.38 A a

Capital letters compare results between columns and lowercase letters between the rows, for each variable analyzed, yield, number and weight of the mushrooms (Tukey ≤ 0.05).

THE CREATION OF A GROWING PROTOCOL

The spatial variability rate of the quantity of β-glucan and the yield according to the variables analyzed (strain, compost, casing layer and cultivation environment) is presented in Figure 4. The variations in β-glucan values within each individual factor were greatest in the strain (35.8%), followed by the casing layer (34.5%), the cultivation environment (15.7%) and compost (9.9%), while for the yields, the greatest was the cultivation environment (82.1%), followed by the strain (81.3%), casing layer (49.1%) and compost (15.2%).

Thus, we propose minimum values for quantity of β-glucan of 5 g $100g^{-1}$ and for yield of 10% to adopt the following cultivation practices:

- Use of varieties ABL 99/30, ABL 03/44 and ABL 04/49, because these demonstrated good agronomic performance and the presence of β-glucan under the experimental conditions evaluated;
- The compost can be prepared using sugar cane bagasse and oat, Massai or Aruana straw as raw materials. Adequate formulation of the substrate is of paramount importance to obtain a good yield, since a balanced compost shows minimal variation in yield and quantity of β-glucan;
- The casing layer is a determining factor in the presence of β-glucan in the basidiomes and for a successful yield, since this factor presented great variation between the experimental treatments. This can be chosen according to the desired final physical characteristics of the harvested basidiomes; however, a good suggestion is the use of peat moss and charcoal added to the soil (1:4, v/v);
- The cultivation environment is a conclusive factor together with the yield values and a major source of the abandonment of mushroom production as a commercial crop by growers due to the variation in yields during the cultivation cycles throughout the year. Thus, cultivation in climatized chambers is recommended since this does not directly influence the values of β-glucan in the harvested basidiomes.

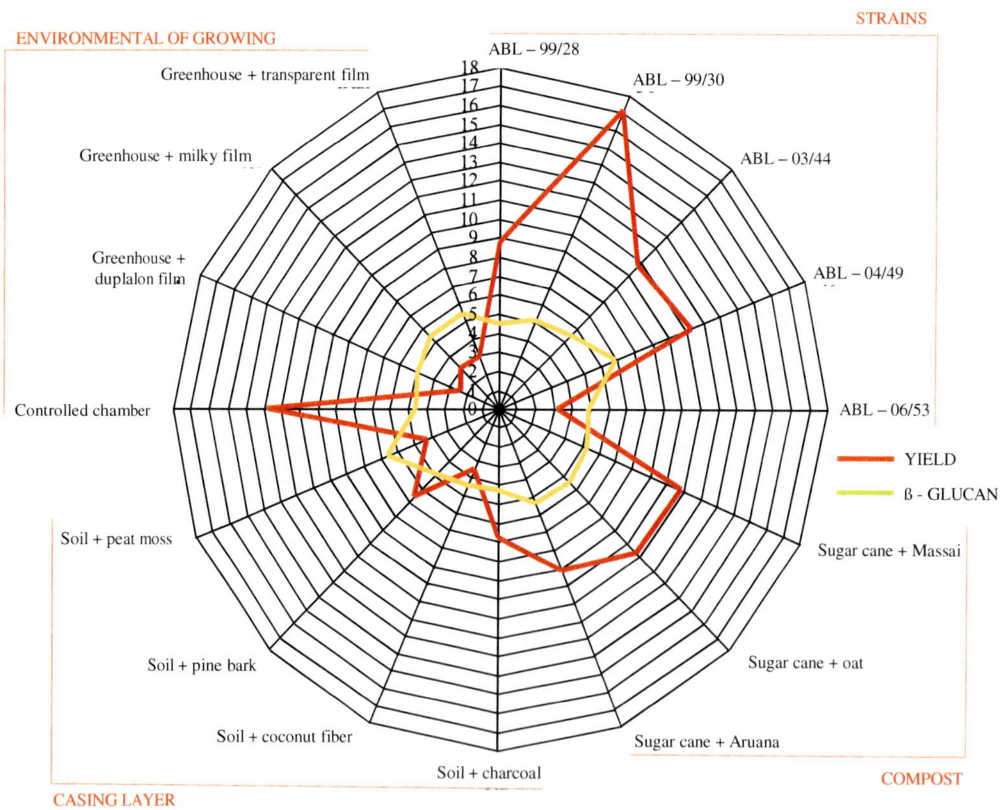

Figure 4. Variability in the data of β-glucan (g 100g-1) and yield (%) depending on cultivation practices studied.

CONCLUSION

To valorize, certify and potentiate the international market of *Agaricus subrufescens*, local, regional and federal research should be conducted in accordance with the climatic characteristics and cultivation environments, type of strain, casing layer and raw materials available in each location, remembering that these basic requirements are determining factors to enhance the quantity of β-glucan in the basidiomes.

REFERENCES

Beneke, E.A. (1963). Clavatia, Clavacin and cancet. *Mycologia, 55*, 227-270.

Bose, S.R. (1953). Antibacterial substances from some higher fungi in India. *Indian Journal of Pharm., 15*, 179-291.

Bohus, G., Galz, C., & Sehelber, E. (1961). The antibiotic action of higher fungi on resistant bacteria and fungi. *Acta Biol. Acad. Sci. Hung., 12*, 1-12.

Chang, S.T., & Buswell, J.A. (1996). Mushroom nitriceuticals. *World J. Microb. Biotech., 12*, 473-476.

Camelini, C.M., Maraschin, M., Mendonça, M.M., Zucco, C., Ferreira, A.G., & Tavares, L.A. (2005). Structural characterization of b-glucans of *Agaricus brasiliensis* in different stages of fruiting body maturity and their use in nutraceutical products. *Biotechnology Letters, 27*, 1295–1299.

Colauto, N.B., Silveira, A.R., Eira, A.F., & Linde, G.A. (2010). Alternative to peat for *Agaricus brasiliensis* yield. *Bioresource Technology, 101*, 712-716.

Di Luzio, N.R. (1999). Springer Semin. *Immunopathol, 8*, 387-400.

Dong, Q., Yao, J., & Yang, X. (2002). Structural characterization of watersoluble of b-D glucan from fruiting bodies of *Agaricus blazei* Murr. *Carbohyd Res., 337*, 1417–1421.

Eira, A.F. (2003). *Cultivo do cogumelo medicinal Agaricus blazei (Murrill) ss. Heinemann.* Viçosa: Editora Aprenda Fácil, pp. 308.

Firenzuoli, F., Gori, L., & Lombardo, G, (2007). The Medicinal Mushroom *Agaricus blazei* Murrill: Review of Literature and Pharmaco-Toxicological Problems. *eCAM, 5(1)*, 3–15.

González Matute, R., Fligas, D., & Curvetto, N. (2010). *Agaricus blazei* production on non-composted substrates based on sunflower seed hulls and spent oyster mushroom substrate. *World J. Microbiol. Biotechnol.*, DOI 10.1007/s11274-010-0582-5.

Heinemann, P. (1993). Agarici Austroamericani. VIII. Agariceae des regions intertropicales d'Amérique du Sud. *Bulletin du Jardin Botanique National de Belgique, 62*, 355-384.

Herman, H. (1962). Cortinellin, Aine antibiotisch Wrksam Substanz aus Cortinellus Shitake. *Naturwissenschften, 49*, 542.

Kawagishi, H., Nomura, A., Yumen, T., Mizuno, T., Hagiwara, T., & Nakamura, T. (1988) *Carbohydr. Res.*, 183, 150-154.

Kawagishi, H., Ryuichi, R.I., Kanao, T., Keishiro, T.M., Hitoshi, S., & Hagiwara, I.T. (1989). Fractionation and antitumor activity of the water-insoluble residue of *Agaricus blazei* fruiting bodies. *Carbohyd Res,186*, 267–273.

Kerrigan, R. W. (2005). Agaricus subrufescens, a cultivated edible and medicinal mushroom, and its sinonyms. *Mycologia, 97*, 12-24.

Kopytowski Filho, J. (2002). Relação C/N e proporção de fontes nitrogenadas na produtividade de *Agaricus blazei* Murrill e poder calorífico do composto. 101 f. *Dissertação* (Mestrado em Agronomia) - Faculdade de Ciências Agronômicas, Universidade Estadual Paulista, Botucatu, 2002.

Kopytowski Filho, J., & Minhoni, M.T.A. (2004). C/N ratio on yield of *Agaricus blazei* Murrill ss. Heinemann. *Mushroom Science, 16,* 213-220.

Kopytowski Filho, J., Minhoni, M.T.A., & Rodriguez Estrada, A. (2006). *Agaricus blazei* – "The Almond Portobello": cultivation and commercialization. American Mushroom Institute. *Mushroom News, 54 (2),* 22-28.

Minhoni, M. T. A., Kopytowski Filho, J., & Andrade, M. C. N. *Cultivo de Agaricus blazei Murrill ss. Heinemann.* 3. ed Botucatu: Fundação de Estudos e Pesquisas Agrícolas e Florestais, 2005. p 101.

Mizuno, T., Hagiwara, T., Nakamura, T., Ito, H., Shimura, K., & Sumiya, T. (1990). Antitumor activity and some properties of water-soluble polysaccharides from ''Himematsutake'', the fruiting body of *Agaricus blazei* Murril. *Agric. Biol. Chem., 54,* 2889-96.

Mizuno, M., Morimoto, M., Minato, K., & Tsuchida, H. (1998). *Biosci. Biotechnol. Biochem., 62,* 434-437.

Mizuno, M.., Minato, K.., Ito, H., Kawade, M., Terai, H., & Tsuchida, H. (1999). *Biochem. Mol. Biol. Int., 47,* 707-714.

Mizumo, M. (2002). Medicinal Properties and Clinical Effects of Culinary-Medicinal Mushroom *Agaricus blazei* Murrill (Agaricomycetideae) (Review). *International J. Medicinal Mushroom, 4,* 299-312.

Mohagheghpour, N., Dawson, M., Hobbs, P., Judd, A., Winant, R., Dousman, I., Waldeck, N., Hokama, L., Tuse, D., & Kos, F. (1995). *Adv. Exp. Med. Biol., 383,* 13-22.

Mol, P.C., & Wessels, J.G.H. (1990). Differences in wall structure between substrate hyphae and hyphae of fruit-body stipes in *Agaricus bisporus. Mycol. Res., 94,* 472–479.

Ohno, N., Furukama, M., Miura, N.N., Adachi, Y., Motoi, M., & Yadomae, T. (2001). Antitumor β -Glucan from the Cultured Fruit Body of *Agaricus blazei.* Biol. Pharm. Bull., 24(7), 820-828.

Osaki, Y., Kato, T., Yamamoto, K., Okubo, J., & Miyazaki, T. (1994). Antimutagenic and bactericidal substances in the fruit body if a basidiomycete *Agaricus blazei*, Jun-17. *Yakugaku Zasshi, 114,* 342-345.

Park, Y. K., Ikegaki, M., Alencar, S. M., & Aguiar, C. L. Determinação da concentração de β-glucano em cogumelos *Agaricus blazei* Murril por método enzimático. Ciência Tecnologia de Alimento, Campinas, 23 (3): p. 312-316, 2003.

Rios-Hernandez, M., Dos-Santos, N. J., Silvia-Cardoso., Bello-Garciga, J. L., & Pedroso M., (1994). *Arch. Med. Res., 25,* 179-180.

Sone, S. (1999). "Clinical Oncology," 2nd ed., *Jpn. Soc. Med. Oncology.* pp. 475-493.

Suzuki, T., Ohno, N., Ohshima, Y., & Yadomae, T. (1998). FEMS Immunol. Med. Microb., 21, 223-230.

Takaku, T., Kimura, Y., Okuda, H. (2001). Isolation of an antitumor compound from *Agaricus blazei* Murril and its mechanism of action. *J. Nutr, 131,*1409-1413.

Wang, Q., Li, B.B., Li, H., Han, J.R. (2010). Yield, dry matter and polysaccharides content of the mushroom *Agaricus blazei* produced on asparagus straw substrate. *Scientia Hoticulturae, 125,* 16-18.

Zied, D.C. (2008). Camadas de cobertura com diferentes combinações de solos e ambientes de cultivo na produção do cogumelo *Agaricus blazei* (Murrill) ss. Heinemann. Tese. Universidade Estadual Paulista, Brasil. 120 pp.

Zied, D.C., Pardo, J.E., Pardo, A. & Minhoni, M.T.A. (2009a). Características generales, producción y comercialización de *Agaricus blazei* (Murril) ss. Heinemann (*A. brasiliensis*): Una nueva alternativa de cultivo de hongo en España. In: Consejería de Agricultura y Medio Ambiente y Diputación Provincial de Cuenca (Eds.), *V Jornadas Técnicas del Champiñón y otros Hongos Comestibles en Castilla-La Mancha* (pp. 1-20). Cuenca (in press).

Zied, D.C., & Minhoni, M.T.A. (2009b). Influence of the environment of production in yield of mushroom *Agaricus blazei* ss. Heinemann (*A. brasiliensis*). *Energia na Agricultura, 24*, 17–36.

Zied, D.C., Minhoni, M.T.A., Kopytowski Filho, J., Arruda, D.P., & Andrade, M.C.N. (2009c). Production of *Agaricus blazei* ss. Heinemann (*A. brasiliensis*) in function of different casing layers and composts. *Inci, 34*, 437–442.

Zied, D.C., Minhoni, M.T.A., Kopytowski Filho, J., & Andrade, M.C.N. (2010). Production of *Agaricus blazei* ss. Heinemann (*A. brasiliensis*) on different casing layers and environments. World J Microbiol Biotechnol. doi: 10.1007/s11274-010-0367-x.

Zied, D.C., Minhoni, M. T. A., Kopytowski-Filho, J., Barbosa, L. & Andrade, M. C. N. (2011). Medicinal mushroom growth as affected by non-axenic casing soil. *Pedosphere, 21(2)*, 146-153.

In: Mushrooms: Types, Properties and Nutrition
Editor: Sophie Andres and Niko Baumann

ISBN 978-1-61470-110-1
2012 Nova Science Publishers, Inc.

Chapter 9

HEMOLYTIC LECTINS OF HIGHER MUSHROOMS

V. O. Antonyuk and R. S. Stoika
Institute of CellBiologyNationalAcademy of Sciences of Ukraine.
Drahomanov street 14/16, L'viv, 79005, Ukraine

ABSTRACT

The results of study of 4 specific lectins isolated from higher *Amanita* genus mushrooms and possessing hemolytic activity are presented. Two of them – *A. phalloides and A. virosa*, belong to *Amanitaceae* family, while two others - *Mycena pura* and *Laetiporus sulphureus*, belong to *Tricholomataceae* and *Corticiaceae* families, correspondingly. The purification methods used for their isolation are described, and the comparative analysis of physico-chemical and biological properties of these lectins have been done. It was shown that according to the peculiarities of the action towards erythrocytes, these lectins can be divided into two groups. While in *Mycena pura* and *Laetiporus sulphureus* lectins, the hemagglutinating activity is higher than the hemolytic activity, in *A. virosa* and *A. phalloides* lectins, situation is opposite and the hemolytic activity is higher than the hemagglutinating activity. The hemolytic activity of the lectins under study was inhibited by the polyethylene glycol of high molecular mass. At the same time, the presence of polyethylene glycol did not influence the hemagglutinating ability of the lectins. This allowed studying their carbohydrate specificity, and it was found that the lectins of *Amanita* genus mushrooms considerably differed in this specificity. The hemolytic lectins also possessed antimicrobial action and inhibited growth of Gram-positive microorganisms and *Proteus*. Besides, they demonstrated cytotoxic action towards cultured mammalian leukemia cells. In general, these mushroom lectins resemble cytolytic lectins of lower invertebrates. Their potential biological role is considered.

INTRODUCTION

Lectins is a group of proteins of the non-immune origin, notable for their reversible and selective binding to carbohydrates and carbohydrate determinants of biopolymers without changing their covalent structure. Except of classic lectins, there are so-called bi-functional lectins which possess an enzymatic activity. This activity can be responsible for their high toxicity towards the eukaryotic cells. As an example of the cytotoxic heterodimeric lectins could be ricin, viscumin, modeccin, abrin, volcensin, and some other lectins which possess RNA N-glycosidase activity. Its expression leads to inhibiting protein synthesis by the irreversible ribosome inactivation via blocking functioning of the elongation factors EF-1 and EF-2 [1,2]. Other enzymatic activities were also described in lectins - α-galactosidase activity was detected in the mung bean *Phaseolus aureus* lectin [3], and the lichen *Xanthoria parietina*lectin was shown to possess arginase activity [4].

Recently, the lectins demonstrating both hemagglutinating and hemolytic activities were found in lower invertebrates, such as sea cucumber (*Cucumaria echinata*) [5], American horseshoe crab(*Limulus polyphemus*) and Japanese crab (*Tachypleus tridentatus*) [6]. Earlier, the hemolytic lectin was discovered in the earth worm *Eisenia fetida* [7]. It was considered that such lectins can exist only in animals, however, the hemolytic lectin was also isolated by Konska et al. [8] and Tateno, Goldstein [9] from *Laetiporus sulfureus* mushroom. There were no hemolytic lectins found in plants. It is known that the hemolytic substances in plants are of non-protein nature and mostly represented by different saponins.

We have applied the method of osmotic protection of erythrocytes by the polyethylene glycols of various molecular mass, and discovered that a toxic protein isolated from *Amanita phalloides*, known as "phallolysin", and the analogical protein isolated from *Amanita virosa*, are also lectins [10,11]. Another hemolytic lectin was discovered by us in the fruiting bodies of *Mycena pura* (family *Tricholomataceae*) mushroom [12].

Thus, it was suggested that the hemolytic lectins are of wider spreading among the basidial mushrooms. They are very hard to isolate, like some high molecular hemolysins, in particular, hemolysin of *Polyporus squamosus* mushroom.

The main goal of this workwas to conduct a comparative analysis of hemolytic lectins isolated from the highest fungi, and to consider their potential biological value. Here were presented the results of our studies devoted to 4 hemolytic lectins purified from *Bacidio-mycetes* mushrooms: lectins from *A. phalloides* and *A. virosa*which belong to *Amanita* genus,*Amanitaceae* family, and two lectins isolated from mushroom species - *Mycena pura* and *Laetiporus sulphureus* which belong to other families - *Tricholomataceae* and *Corticiaceae,* correspondingly.

MATERIALS AND METHODS

Fruit bodies of the *Amanitaphalloides*, *Amanitavirosa*and*Mycenapura*used for isolating the lectins, were collected in Skole district of the Lviv region (western part of Ukraine). They were transported to the laboratory at the same day when being collected. The fruit bodies of *Laetiporussulphureus*were collected in the outskirts of the city of Lviv.

As it was shown earlier [13], in spite of differences in their carbohydrate specificity, these lectins can be affinity absorbed by the glutaraldehyde cross-linked ovomucin that is important for their purification. However, concrete methods of purification of the lectins from different mushroom species have to be developed.

For purification of the lectin from *Laetiporus sulphureus* mushroom, fresh fruit bodies (2.0 kg) were grinded, and 3 liters of 1 % NaCl were added to the resulting mass followed by homogenization in blender. Extract (4.2 liters) was released through dense textile and cleared up by centrifugation at 1,500-2,000*g* for 10 min. The supernatant was slowly acidified to pH 5.0, and the precipitate was removed by centrifugation. The obtained supernatant was alkalinized to pH 7.5–8.0, and the precipitate was removed again by centrifugation at 1,500-2,000*g* for 10 min. Mucous polysaccharides present in this extract and making problems at further purification steps, we removed by salting-out with ammonium sulphate at 40 % saturation, while the lectin bulk salting-out was achieved at 40–90 % saturation. Thus, the supernatant cleaned off the polysaccharides was supplemented with ammonium sulphate to 90 % saturation, and the sediment was removed by centrifugation (5,000 g, 30 min) and dissolved in distilled water (40–60 ml). After dialysis against water, liquid's density should be in range of 1.03–1.06 g. Further purification was carried outby the affinity chromatography. A clean transparent liquid was applied onto a column (V = 120 ml) of affinity sorbent (see: Materials and Methods) that was previously equilibrated with 0.05 M phosphate buffer, pH 7.0, supplemented with 0.5 M NaCl. After passing through the column of the prepared extract from *L. pergamenus*, the column was washed with the same buffer until the absorbance (280 nm) in the eluate was less than $\lambda_{280} < 0.1$. The lectin adsorbed on the column was eluted with 3 % lactose solution in phosphate buffer. The lectin yield from the column was monitored by measuring protein content, and, in parallel, by the titer of hemagglutination of rabbit erythrocytes. The lectin containing fraction was precipitated with ammonium sulphate at 90% saturation, and after centrifugation, the precipitate was dissolved in a small volume (20 – 40 ml) of distilled water and dialyzed against 0.05 M phosphate buffer. After thrice-repeated replacement of the dialysis solution, the resulting lectin was applied onto a column filled with DEAE-Toyopearl (5.0 × 0.8 cm) equilibrated with the same buffer solution. The lectin passed through the column without retention, while most other materials were retarded on the column. At an increase of ionic strength of the buffer solution to 0.1 M, most of the lectin passed through the column, however, it was still contaminated with other non-lectin materials. After the ion exchange chromatography, the purified lectin was dialyzed against distilled water and lyophilized.

In general, purification procedure of the lectin from fruit bodies of *Mycena pura* mushroom is similar to procedure described above. However, there was no necessity in the 2-step precipitation by ammonium sulphate, since a viscous polysaccharide is not present in the *Mycena* extract. Besides, the lectin elution from the affinity sorbent was carried out with 1% acetic acid. This lectin is not destroyed by the ethanol. Thus, it was possible to use at the intermediate stages of purification lectin precipitation by two volumes of cooled ethanol.

The purification of the hemolytic lectin from the fruit bodies of *A. phalloides* and *A. virosa* was carried out by a method described earlier [10,14] with minor modifications. Fresh fruit bodies of *A.virosa* were pressed out, and then thiourea was added to the obtained juice in final concentration 0.1%. pH of the juice was adjusted to 4.5, and the pellet was removed by centrifugation (see above). The pH of the juice was adjusted to 8.0, and again, the pellet was removed by centrifugation. Then, pH was reduced to 7.0, and the extract was dialyzed against 0.1 M phosphate buffer, pH 7.0. After three replacements of the dialysis solution, the resulting material was applied onto the ion-exchange DEAE-Toyopearl column (15 x 2 cm) equilibrated with the same buffer solution. The column elution was performed at room temperature with flow rate 2 ml/min. The unbound material was pooled and concentrated by 90% saturation of the ammonium sulphate. Highly pigmented material was mostly retained by the ion-exchanger, while the toxovirin and toxic lectin passed through the column without retention (Figure 1).

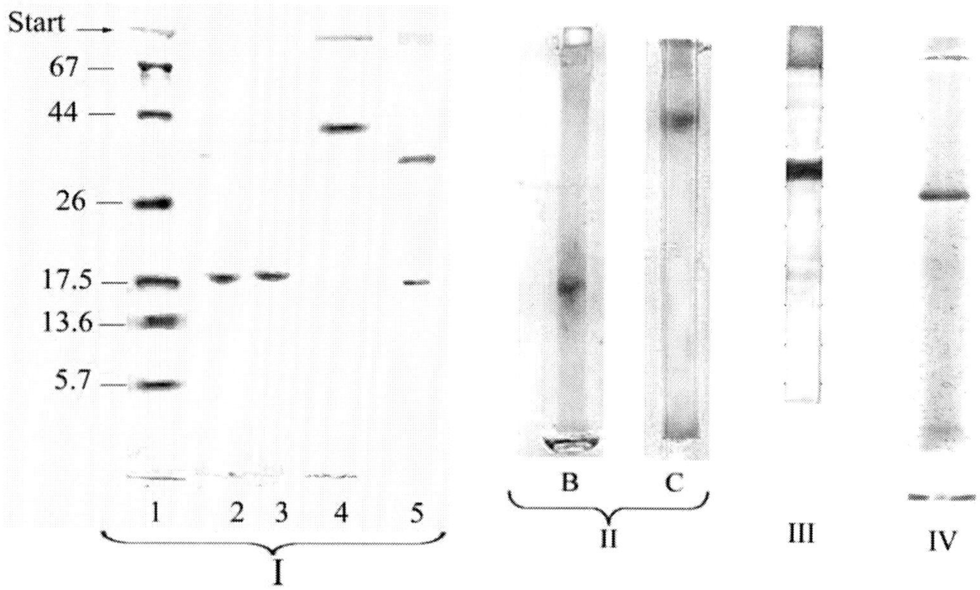

Figure 1. **A** - Electrophoresis in PAAG gradient (10–20 %) with 0.1% SDS-Na:

1 – Mol. mass markers;

2 – *A. virosa* lectin;

3 – *A. virosa* lectin + 2-mercaptoethanol;

4 – *Mycenapura* lectin;

5 – *Laetiporus sulphureus*lectin;

II - Disc electrophoresis in 7.5% PAAG:1 – *A.virosa* lectin, pH 8.6;2 – *A.virosa* lectin, pH 4.3;

III - Disc electrophoresis in 10 % PAAG,pH 8.6: *Mycena pura* lectin;

IV - Disc electrophoresis in 12.5% PAAG, pH 8.6 : *Laetiporus sulphureus* lectin

Further isolation of the hemolytic lectin was carried out on the column filled up with the affinity sorbent (cross-linked ovomycin), as described [13]. The column was washed up by the phosphate buffered saline (PBS), pH 7.4, until the eluent absorbance showed $E_{280}<0.1$. The lectin was retained by the sorbent, and then it was eluted by 1% acetic acid (Figure 2).

Collected fractions were combined and protein was concentrated by dialysis against saturated (100%) solution of the ammonium sulfate.

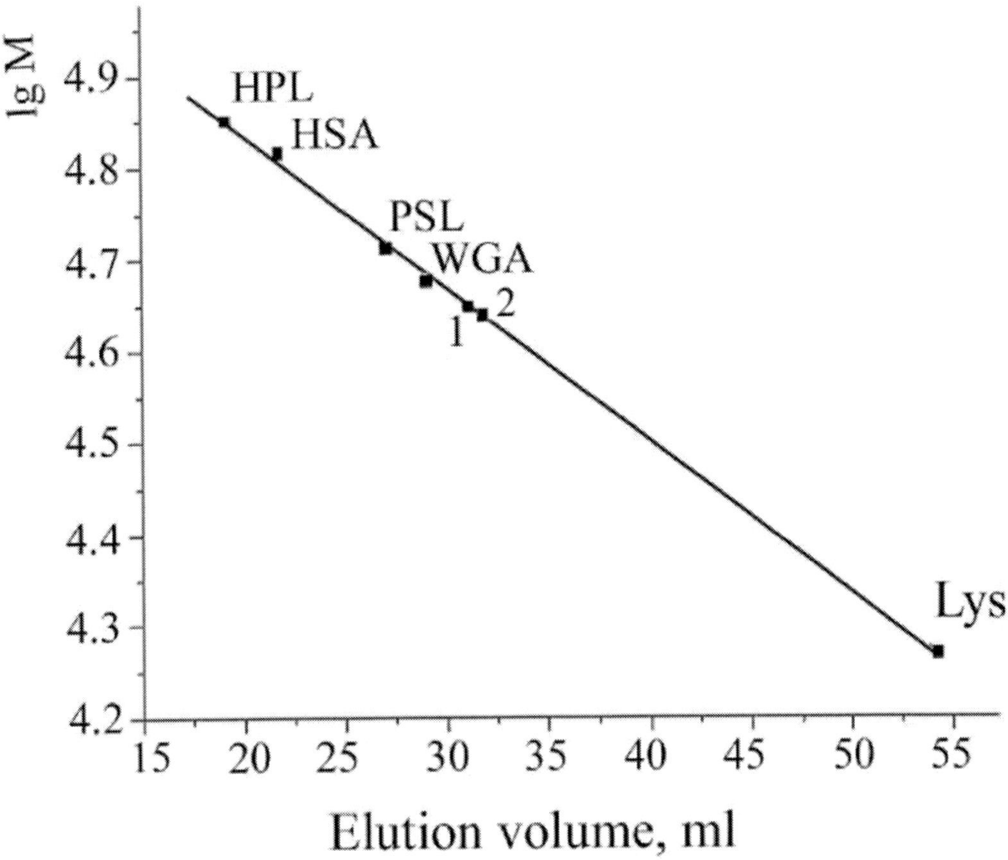

Fig2. Gel-filtration (for determination of the molecular mass) of *A. virosa* (1) and *A. phalloides* (2) lectins was performed on the Toyopearl-HW-55 column (39 x 1.5 cm) eluted with 0.1 M acetate buffer solution, pH 6.4, supplemented with 0.5 M NaCI (the elution rate equaled 0.3 ml/min).

Electrophoretic Studies

Purity of the obtained lectin preparations was studied by disk-electrophoresis in 7.5–12.5 % polyacrylamide gel (PAAG) in the alkaline buffer system (pH 8.6), and by electrophoresis in 7.5% PAAG in the acidic buffer system (pH 4.3). The molecular mass of polypeptide chains of *A. virosa* toxic lectin and of toxovirin was determined by electrophoresis in gradient 10-15% PAAG supplemented with 0.1% sodium dodecyl-sulphate (SDS) [15]. Proteins used as molecular mass standarts were: chicken egg lysozyme (14.3 kDa), *Lens culinaris* lectin (5.7 + 17.5 kDa), *Helix pomatia* lectin (26 kDa), and human blood serum albumin (69 kDa).

Determination of the molecular mass of *A. phalloides* and *A. virosa* lectins was carried out by gel-chromatography on the Toyopearl HW-55 column (39 x 1.5 cm), eluted with0.1 M acetate buffer solution, pH 6.4, supplemented with 0.5 M NaCI (elution rate equaled 0.3 ml/min). The molecular mass markers were: chicken egg lysozyme (14.3 kDa), pea seed lectin (48 kDa), wheat germ agglutinin (36 kDa), *Helix pomatia* lectin (78 kDa), and human blood serum albumin (69 kDa).

Study of the Hemolytic Activity of Purified Lectins

Citrate-supplemented blood of human donors and some animals (see table 2, 3) was used for preparing erythrocytes. Blood was centrifuged (500 g, 5 min, room temperature) to pellet the erythrocytes which were washed three times with PBS. Erythrocytes were used for measuring hemolytic activity of studied lectins. For that purpose, basic solution of the purified lectin was prepared (1 mg of the lectin was dissolved in 10 ml PBS) and a series of its double dilutions in the PBS (0.05 ml) was prepared: 0.5 mg/ml, 0.25 mg/ml, 0.125 mg/ml, and so on. Then, 0.05 ml of 2 % suspension of erythrocytes of selected animal species or human was added to each test tube, and minimal time during which complete hemolysis of erythrocytes occurred in the test tubes was defined. This time was determined by visual observation of the reaction mixture after centrifugation of the microtubes for 30 sec at 1,000 g. Completed hemolysis was achieved when erythrocyte pellet was absent and the supernatant got red color. The diagram of dependence of hemolysistime upon the lectin concentration was drawn.

Experiments on the osmotic protection against hemolysis were executed by using rabbit erythrocyte, as described earlier [10]. For that purpose, 12% solutions of polyethylene glycol (PEG) of different molecular mass (400, 600, 1,350, 1,500, 3,000 and 4,000 Da) were prepared by using PBS, pH 7.4. Equal volume *A. virosa* lectin solution in 0.125 mg/ml concentration and 0.05 ml of 2% suspension of rabbit erythrocytes were added to the microtest-tube which contained 0.05 ml of specific PEG solution. In the control test-tube, corresponding PEG was replaced by the PBS. The mixture was shaken up, and the hemolysis was observed. PEGs in whose presence the hemolysis did not occur, were selected for further study. They were utilized for determination of minimal PEG concentration which inhibited red cell hemolysis. For that purpose, a series of double dilutions of PEG was prepared, as described above, and equal volume (0.05 ml) of *A. virosa* lectin solution and 0.05 ml of 2% suspension of erythrocytes were added. The least concentration of PEG which completely blocked the lectin induced hemolysis, was defined (in control, 100% hemolysis was detected).

Study of Carbohydrate Specificity of the Lectin

Carbohydrate specificity of the lectin was determined, as described [10] with minor modification that the buffer (PBS, pH 7.4) contained PEG-1,500 in final concentration 6%. We have calculated minimal concentration of specific carbohydrate which completely inhibited hemagglutinating activity of the lectin used in the titer 1:4.

The following substances were used to characterize the carbohydrate specificity of the lectin: D-glucose, D-fructose, D-galactose, sucrose, maltose, lactose (*Soyuzchimreactiv*, RF), raffinose (*Fluka*, Switzerland), α- andβ-methyl-D-galactosides, L-rhamnose, cellobiose, N-acetyl-D-galactopyranosideandN-acetyl-D-glucopyranoside (*Chemapol*, ChechRepublic), D-mannose, D-turanose, L-ribose (BratislavaChemicalInstitute, Slovakia), melibiose, α-methyl-D-mannoside, L-fucose (*KochLight*, UK), phenyl-α-2-acetamido-D-glucopyranoside, 4-nitrophenyl-β-D-gluco- and -galactopyranoside – from ("Serva" (Germany), 4-nitrophenyl-β-D-glucopyranoside. They were prepared, as described by Kabat &Mayer [16].

The interaction of the lectins with glycoproteins and polysaccharides was defined by using glycogen of porcine liver, ovomucoid, thrice re-crystallized ovalbumin (*Biolar*, Latvia), and yeast mannan [17]. Group-specific substances H, A, and B were isolated from the cyst liquid obtained after ovary surgery in patients with the corresponding blood groups. The above mentioned substances were purified by the described method [18].

Study of the Antimicrobial Action of the Lectins

The hemolytic lectins of *A. virosa* and *M. pura* mushrooms and the lectins isolated from other sources were studied. They exhibited distinct carbohydrate specifity (Scientific Manufacturing Cooperative "Lectinotest", Ukraine). These were the lectins obtained from *Canavalia ensiformis* (Con A), *Helix pomatia* (HPA), *Sambucus nigra* bark (SNA), wheat germ (WGA), pea (PSA), and *Perca fluviatilis*roe (PFA). To compare their inhibiting action towards growing bacteria, discs saturated with widely used antibiotics such as gentamycin and nistatin were applied.

Daily suspended cultures of Escherichia coli, Pseudomonas aeruginosa, Candida albicans, Klebsiella pneumoniae, Staphylococcus epidermidis, Staphylococcus aureus, Bacillus subtilis, Proteus vulgaris, Corynebacterium xerosis, Streptococcus faecalis were sowed out as an entire lawn in Petri dishes. Suspension of these microbes was prepared by using the turbidity standard of 5 units. Each dish was divided into eight sectors, and 50 μg of solution of various lectins were dropped in the centre of each sector. Five trial series were carried out. The obtained cultures were incubated in the thermostat at 37 ^{0}C for 24 h.

Mammalian Cells and their Culturing

Murine leukemia L1210 cells and human leukemia T-cells of CEM T4 and Jurkat lines were obtained from the collection at R. E. Kavetsky Institute of Experimental Pathology, Oncology and Radiobiology, National Academy of Sciences of Ukraine (Kyiv). Cells were cultured in the Dulbecco's modified Eagle's medium (DMEM, Sigma, USA) supplemented with 10% fetal bovine serum (FBS) and 100 units/ml of gentamycin (Sigma, USA). For measuring cytotoxic effect of *A.virosa* and *A. phalloides* lectins, cells were seeded in 24-well plastic dishes in DMEM in the presence of 10% FBS. Approximately 300 000 cells of L1210 line per well in 1 ml, 200 000 cells of CEM-T4 per well in 1 ml, and 500 000 cells of Jurkat line per well in 1 ml were present at the beginning of cytotoxicological experiments. After 24 hours, tested substances were added. Number of viable cells was counted in the hemocytometric chamber at various time intervals. The ratio of dead cells was defined after their staining with trypan blue (0.1%, w/v) and counting under the light microscope (MikMed, LOMO, Russian Federation).

Results and Discussion

For purification of lectins from the fruit bodies of studied mushrooms, the described methods based on applying cross-linked ovomucin as affinity sorbent (see «Materials and Methods») were used. The highest yield of the lectin was obtained in case of utilizing fruit bodies of *Amanita phalloides*, while there was 5–6 times less lectin obtained from other mushroom species (Table 1). For fresh fruit bodies of *Mycena pura,* it made 59 % of theoretically possible output calculated according to the lectin activity.

Table 1. Yield of the hemolytic lectins (per 1 kg of fresh material) from the fruit bodies of mushrooms at their purification with the use of cross-linked ovomucin as affinity sorbent [13]

	Lectinsource	Lectin yield (per 1 kg of fresh material)
1	*Mycena pura*	9.0 mg
2	*Laetiporus sulphureus*	9.1 mg
3	*Amanita virosa*	10.8 mg
4	*Amanita phalloides*	55.0 mg

The activity of lectin in the fruit bodies of *Mycena pura* was preserved at the rapid drying (+56°C), freezing (-18°C), ammonium sulphate precipitation, during keeping extract (in the presence of 0.1 % sodium azide) for 72 h at room temperature, however, it was diminished by ethanol or acetone precipitation. The lectin is stable in broad range of pH 4.0–10.5.

All purified lectins were relatively thermolabile, since their heating for 30 min at +65°C fully deprived both the hemolytical and hymagglutinating activities. At the same time, dialysis for 8 h of the lectin solution against 1% EDTA (disodium salt) did not affect these activities of the lectin that suggests their independence upon calcium or magnesium ions.

A. virosa and *A. phalloides* lectins possess molecular mass 17-18 kDa (SDS-PAAG) which was not changed by the treatment with 2-mercaptoethanol (Figure 1). However, the molecular mass determination by means of gel filtration on the ToyopearlHW-55 (Toyosoda, Japan) column showed the value of 36 kDa (Figure 2). This discrepancy suggests that the lectins consist of two polypeptide chains of identical or very similar molecular mass.

The lectins isolated from *Mycena pura* and *Laetiporus sulphureus* differ more significantly. At the electrophoresis in gradient 10-20 % PAAG supplemented with 0.1% sodium dodecyl-sulphate without of β-mercaptoethanol the *Mycena pura* lectin showed a band corresponding to a molecular mass ~40 kDa. While the*Laetiporus sulphureus* lectin has a tetrameric structure and consists of 4 subunits of two size types of 35 kDa and 17 kDa (Figure 1, I-5). After freeze drying, *Mycena pura* and *Laetiporus sulphureus* lectins appear as white powders that are soluble in water-salts solutions. At the same time, in the case of *A. virosa* and *A. phalloides* lectins, freeze drying is undesirable, since it diminishes solubility of the lectin preparations.

When the action towards erythrocytes was taken into account, these mushroom lectins can be divided into two groups. For *Mycena pura* and *Laetiporus sulphureus* lectins, the hemagglutinating activity towards erythrocytes of most species tested is much higher than the hemolytic activity. Opposite, for *Amanita virosa* and *A. phalloides* lectins, the hemolytic activity is higher than the hemagglutinating activity (Table 2 and 3).

Table 2. Minimal hemolyzing and hemagglutinating concentration of *A. phalloides* and *A. virosa* lectinswhen acting towards erythrocytes of different species

	Origin of erythrocytes	**Minimal lectin concentration (µg/ml)**			
		Amanita phalloides lectin		*Amanitavirosa*lectin	
		hemagglutinating	hemolyzing	hemagglutinating	hemolyzing
1	Cow	> 1,000	> 1,000	> 1,300	> 1,000
2	Sheep	> 1,000	> 1,000	900	> 1,000
3	Carp	> 1,000	> 1,000	> 1,000	> 1,000
4	Horse	n.d.	n.d.	> 1300	600
5	Dog	4	16	40	60
6	Rat	1,000	6	600	1.5
7	Rabbit	0.08	0.16	0.02	0.1
8	Human I(O)	256	16	80	7.5
9	Human, II (A)	256	16	80	3.8
10	Human, III (B)	256	16	40	7.5

n.d. – not determined

Table 3. Minimal hemolyzing and hemagglutinating concentration of *Mycena pura*and *Laetiporus sulphureus* lectinswhen acting towards erythrocytes of different species

	Origin of erythrocytes	**Minimal lectin concentration (µg/ml)**			
		Mycena pura lectin		*Laetiporus sulphureus* lectin	
		hemagglutinating	hemolyzing	hemagglutinating	hemolyzing
1	Cow	125	> 1,000	2,500	> 1,000
2	Sheep	16	> 1,000	> 1,000	> 1,000
3	Horse	8	> 1,000	n.d.	n.d.
4	Dog	8	250	n.d.	n.d.
5	Rat	2	12	1	32
6	Rabbit	0.06	4	2	32
7	Human I(O)	4	30	16	128
8	Human, II (A)	1	20	16	128
9	Human, III (B)	2	30	16	128

n.d. – not determined

A. virosa lectin, similarly to *A. phalloides*lectin (see table 2), caused hemolysis of rabbit, human, dog and rat erythocytes, although specific hemolytic concentrations of these lectins were different. The hemolysis of sheep, cow and carp erythocytes was not observed at application of the lectin in concentration up to 1 mg/ml. Dependence of the hemolysis time upon the concentration of *A. virosa* lectin is shown in Figure 3. These data evidence that rabbit erythrocytes are the most sensitive to the hemolytic effect of *A. virosa*, and they are followed by rat and human erythrocytes, while dog erythrocytes were found to be the most resistant to the lectin–induced hemolysis. The diagram also demonstrates that for each type of erythrocytes, there is a threshold lectin concentration below which the hemolysis is not observed. For dog erythrocytes, such concentration is about 60 μg/ml, for human erythrocytes – 7.5 μg/ml, for rat erythrocytes – 1.5 μg/ml, and for rabbit erythrocytes – 0.1 μg/ml. Summarizing, the hemolytic capability of *A. virosa* lectin is 6–8 times higher than the hemolytic capability of similar *A. phalloides*lectin [11].

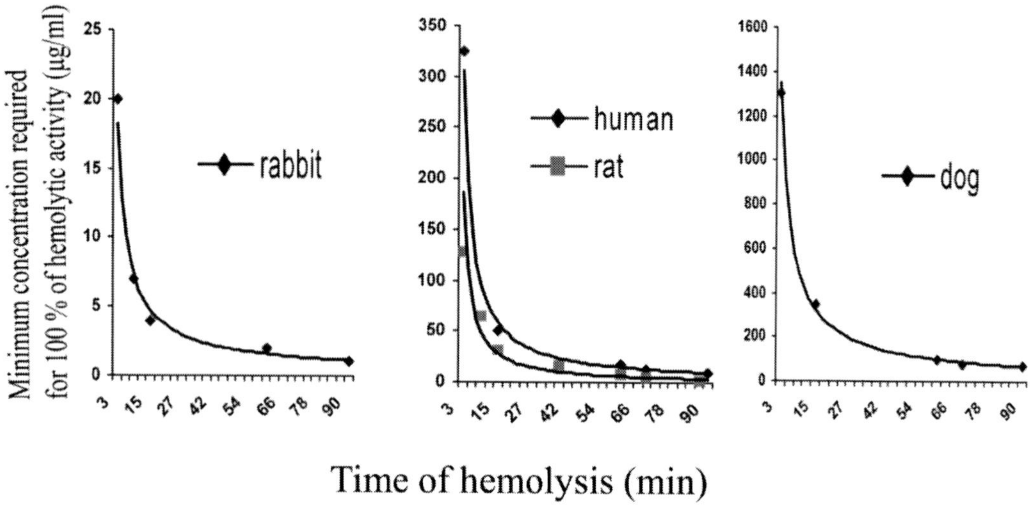

Figure 3. Dependence of the hemolysis speed of human, rat, rabbit and dog erythrocytes upon the concentration of *A. virosa* lectin (see "Materials and Methods").

Graphical dependence of hemolysis time on concentration of other hemolytic lectins has the same general view, differing only in numerical values.

Hemolysis induced by the *A. virosa* lectin, similarly to hemolysis induced by the death cap (*A. phalloides*) lectin (phallolysin) [10], was inhibited by PEG-1,350-6,000, and was not inhibited by PEG-400 or PEG-600. Hence, the size of pores created by the *A. phalloides*lectin in plasma membrane of rabbit erythrocytes, is larger than the hydrodynamic diameter of PEG-600, and less than the hydrodynamic diameter of PEG-1,350. According to the calculations performed by Scherrer and Gerhardt [20], the hydrodynamic diameter of PEG-600 molecule equals 1.6 nm, and that of PEG-1,350 – 2.3 nm. These data allow one to suggestthat *A. virosa* lectin forms ion-permeable pores in plasma membrane of rabbit erythrocytes with the hydrodynamic diametersmaller than 2.3 nm and larger than 1.6 nm. These ion-permeable pores are less than the pores (3.8 nm) formed by the hemolytic lectin isolated from the *Laetiporussulphurous (Bull. et Fr.)* mushroom [9], and larger than the pores (1.6 nm) formed by the beta-toxin isolated from *Clostridium perfringens* [21].

The detection of the hemagglutinating activity in the above mentioned lectins is complicated because of the presence of the hemolytic activity. If the hemagglutinating activity is well expressed and considerably prevails the hemolytic activity (for example, in *Mycena pura* and *Laetiporus sulphureus* lectins), then it is no so problematic to estimate the first one. However, in the *A. virosa* and *A. phalloides* lectins, the hemolytic activity is much higher comparing the hemagglutinating one, and, thus, significant complications appear at evaluating the agglutinating activity. Since, as noted above, at high concentration of PEG the hemolysis is suppressed, that enables measuring the hemagglutination. This fact is also helpful when evaluating inhibition of hemagglutination by the carbohydrates that are utilized for determination of lectin carbohydrate specificity. Application of PEG enabled to define carbohydrate specificity of *A. virosa* and *A. phalloides* lectins, which was not possible to carry out without hemolysis inhibition.

Carbohydrate specificity is major functional characteristics of lectins that often determines potentials of their subsequent bio-applications and explains specific lectin binding to histological structures. Carbohydrate specificity of lectins differs considerably which could be a reason of some differences in their hemolytic properties. Similar index of agglutination titers of erythrocytes of various animal species and human can evidence to similarity of carbohydrate moiety of lectin receptors.

A. virosa and *A. phalloides* mushrooms are known to be morphologically related, belong to one subfamily (*Lepidella*) and one section (*Phalloides*) which assumes likeness of their chemical composition. However, agglutination ability of *A. virosa* lectin is somewhat higher than such ability of the *A. phalloides* lectin. This testifies to possible differences in the carbohydrate specificity of these lectins. The results of carbohydrate specificity study of these lectins are presented in table 4.

The results presented in Table 4, show that *A. phalloides* and *A. virosa* toxic lectins differ considerably in their carbohydrate specificity. *A. virosa* lectin does not interact with monosaccharides, and only their 4-nitrophenyl derivates were strong inhibitors of its action. It can be suggested that active sites of *A. virosa* lectin recognize relatively large oligosaccharide structures, since simple sugars and saccharides had no effect on the activity of this lectin.

Regarding glycoproteins, *A. virosa* lectin interacted best of all with ovine submaxillary mucin, α_2-macroglobulin, and group specific substances of human erythrocytes. Immunoglobulin G, bovine thyreoglobulin, ovomucoid, transferrin, and bovine submaxillary mucin were weaker inhibitors of *A. virosa* lectin. Thus, similarly to *A.phalloides* lectin and unlike related *A. muscaria* and *A. rubescens* lectins [22], *A. virosa* lectin does not prefer O-type glycoproteins over the N-type glycoproteins.

The *Laetiporus sulphureus*lectin interacted only with lactose among all other tested carbohydrates (Table 5).

Table. 4. Carbohydrate specificity of *A. phalloides* and *A. virosa* lectins

	Carbohydrate or glycoprotein	Minimal hapten concentration (mM or %) needed for inhibition of 4 hemagglutinating units of lectin activity	
		A. phalloides [10]	*A. virosa*
1	D-galactose	0.78	—
2	N-acethyl-D-galactosamine	6.25	—
3	N-acethyl-D-glucosamine	25	—
4	Lactose	0.78	—
5	α-methyl-D-galactopyranoside	25	—
6	α-methyl-D-mannopyranoside	—	—
7	β-methyl-D-galactopyranoside	0.78	—
8	Blood group specific substance B	0.06 %	0.06 %
9	Blood group specific substance A	0.03 %	0.03 %
10	Blood group specific substance H	0.03 %	0.03 %

Table 4. (Continued)

	Carbohydrate or glycoprotein	Minimal hapten concentration (mM or %) needed for inhibition of 4 hemagglutinating units of lectin activity	
		A. phalloides [10]	*A. virosa*
11	L-rhamnose	12.5	—
12	Mellibiose (Galα1,6Glc)	3.12	
13	Raffinose (Galα1,6Glcβ1,2DFrucf)	3.12	—
14	Gentiobiose (Glcβ1,6Glc)	100	—
15	Trehalose (Glcβ1,1Glc)	100	—
16	Maltose (Glcα1,4Glc)	50	—
17	Bovine thyreoglobulin	0.5 %	0.125 %
18	Ovomucoid	0.06 %	0.125 %
19	Asialo-ovomucoid	0.015 %	0.25 %
20	Transferrin	—	0.25 %
21	Ovine submaxillary mucin	0.125 %	0.03 %
22	α_2-macroglobulin	—	0.03 %
23	Immunoglobulin G	0.25 %	0.08 %
24	Ovalbumin	1.0 %	—
25	α-phenyl-N-acethyl-D-glucosamine	20	—
26	4-nitrophenyl-α-D-mannopyranoside	—(40 mM)*	20
27	4-nitrophenyl-β-D-glucopyranoside	5	2.5
28	4-nitrophenyl-β-D-galactopyranoside	0.3	20

Notes: D-glucose, D-mannose, L-fucose, D-ribose, fructose, cellobiose (Glcβ1,4Glc), sucrose, β-methyl-D-xylopyranoside, 4-nitrophenyl-α-L-fucopyranoside are not included the list, since they do not interact with studied lectins.

* For 40 mMhapten concentration

Table 5. The interaction of *Laetiporus sulphureus* lectin with carbohydrates and glycoproteins

	Carbohydrate or glycoprotein	**Minimal hapten concentration (mM or %) needed for inhibition of 4 hemagglutinating units of lectin activity**
	Lactose (Galβ1,4Glc)	12.5
	Orozomucoid (alpha-GP)	0.5 %
	Alkaline phosphatase from calf small intestine	0.03 %
	Transferrin	0.5 %
	Sheep submaxillary mucin	0.25 %
	Bovine thyreoglobulin	0.25 %
	Blood group specific substance H	0.25 %
	Blood group specific substance A	0.125 %
	Blood group specific substance B	0.125 %

Note: carbonhydrates and glycoproteins (see "Materials and Methods") which did not interact with this lectin are not presented in the table.

Our data are in agreement with the results of *Tateno and Goldstein* [9] who showed that lactose (Galβ1-4Glc) was 4-10 times more active inhibitor of lectin activity than related disaccharides, such as Galβ1-4 Man, Galβ1-4Fru (lactulose), or Galβ1-4 Glucitol (lactitol).

Lectin of Lilac Bell Cap (*Mycena pura* fungus agglutinin*,* MPFA) poorly interacted with α-methyl-D-mannopyranoside, D-glucose and D-mannose and some disaccharides which contain glucose. Therefore, we decided to compare its carbohydrate specificity with that of the lectin of pea (*Pisum sativum* agglutinin, PSA, *Fabaceae* family), and the lectin of a snowflake (*Leucojus vernus* agglutinin, LVA, *Amaryllidaceae* family), which belong to two different subgroups of D-mannose-specific lectins. The results of this study are presented in Tables 6 and 7.

All three studied lectins (PSA, LVA and MPFA) bind better the compounds of oligosaccharide structures of the glycoconjugate components, while their interaction with other substances differ considerably. For instance, LVA binds strongly yeast mannan, PSA does this with medium strength, and *Mycena* lectin does not bind it at all. PSA binds strongly bovine thyroglobulin and sheep submaxillary mucin, while the interaction of LVA and MPFA with the mucin is medium. The *Mycena* lectin strongly interacts with calf intestinal alkaline phosphatase, while PSA poorly interacts with this enzyme, and LVA does not bind it at all. This testifies to higher complementation of centres of carbohydrate interaction of the *Mycena* lectin, PSA and LVA towards the carbohydrates of complex and branched structure (Table 7).

Table 6. The interaction of *Mycena* lectin with mono- and disaccharides

	Carbohydrate	**Minimum hapten concentration for inhibition of hemagglutination (mM)**		
		MPFA	**PSA**	**LVA**
1	D-glucose	100	25	-
2	α-methyl-D-glucopyranoside	-	3.12	-
3	D- mannose	50	6.25	100
4	α-methyl-D-mannopyranoside	25	3.12	50
5	N-acethyl-D-glucosamine	-	3.12	-
6	Melibiose (Galα1,6Glc)	-	-	100
7	Maltose (Glcα1,4Glc)	-	25	-
8	Trehalose (Glcβ1,1Glc)	-	25	-
9	Gentiobiose (Glcβ1,6Glc)	100	-	-
10	Saccharose (Glcβ1,2Fru)	-	100	-
11	Turanose (Glcβ1,3Fru)	100	100	100
12	Glucose-6-phosphate dipotassium salt	100	-	-

Table 7. Interaction of lectins with polysaccharides and glycoproteins

Glycoprotein or polysaccharide	Minimal concentration of glycoprotein, inhibiting hemagglutination of lectins, %		
	MPFA	PSA	LVA
Transferrin	0.25	0.125	-
Bovine thyreoglobulin	0.125	0.03	0.5
Alkaline phosphatase from calf small intestine	0.002	0.5	-
Orozomucoid (alpha-GP)	0.125	-	-
Ovomucoid	0.06	0.5	1
Asialo-ovomucoid	0.015	-	-
Sheep submaxillary mucin	0.125	0.06	0.25
Bovine submaxillary mucin	1	0.125	-
Decialized bovine submaxillary mucin	0.5	0.06	-
Human G immunoglobulin	-	0.125	-
α_2-macroglobulin	1	-	-
Group-specific substance H, 1%	0.125	0.5	-
Group-specific substance A, 1%	0.125	0.25	-
Group-specific substance B, 1%	0.25	0.5	-
Ovalbumin	0.125	-	-
Pig liver glycogen (1%)	0.5	1	-
Yeast mannan (1%)	-	0.125	0.03
Starch (1%)	1	1	1

Note: Dash means the absence of interaction at 1% concentration.

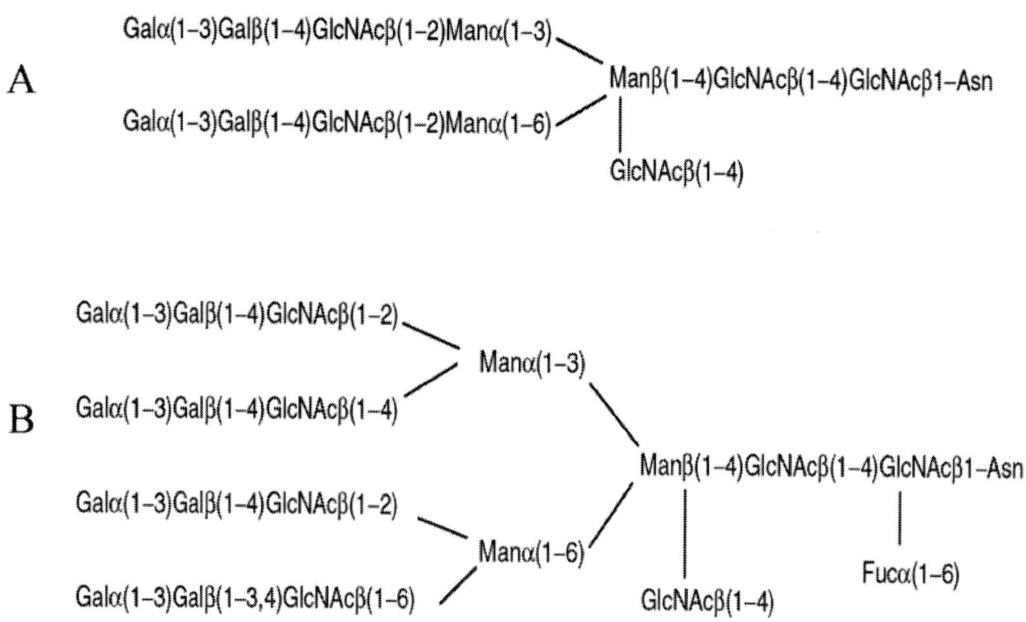

Figure 4. Glycopeptide structure attached to Asp249 (A) and Asp410 (B) of calf small intestinal alkaline phosphatase [24].

There are data in literature demonstrating presence of alkaline phosphatase (E.C.3.1.3.1) in all tissues and liquids of the organism except blood vessel walls and hyaline cartilage. This enzyme is abundant in the epithelium wall of small intestine, flexuous renal tubules, bone tissue, breast gland during lactation, and leukocytes [12]. Alkaline phosphatase of the calf small intestine is zinc-containing two-chain glycoprotein with $M_r = 65$ kDa, that has two sites of glycosylation– near asparagine-249 containing a basic glycosylresidues. It possesses two carbohydrate antennas containing D-galactose, D-glucosamine, and D-mannose, while the glycosylated moiety at asparagine-410 has four antennas and contains L-fucose (Figure 4). Total amount of carbohydrates in the alkaline phosphatase ranges from 8 to 17% depending on type and localization of this enzyme [24]. Structures of the alkaline phosphatase in human and other mammalians differ not considerablly. Due to high amount and variety of carbohydrates in the alkaline phosphatise, one might expect its interaction with many lectins of various carbohydrate specificity.

Table 8. Interaction of lectins with alkaline phosphatase (E.C. 3.1.3.1) of calf small intestine

Lectin	Specificity to mono- or oligosaccharides	Minimal concentration of the enzyme, inhibiting agglutination	
		%	mM
LABA (*Laburnum anagyroides* bark agglutinin)	L-Fuc	0.25	0.04
LTA (*Tetragonolobus purpureus* agglutinin)	L-Fuc	1	0.16
LLA (*Lucioperca lucioperca* agglutinin)	L-Fuc	0.03	0.005
PFA (*Perca fluviatilis* roe agglutinin)	L-Fuc	0.06	0.01
STA (*Solanum tuberosus* agglutinin)	$(NAcDGlc)_4$	0.125	0.02
WGA (Wheat germs agglutinin)	$(NAcDGlc)_3$	0.125	0.02
PNA (Peanut agglutinin)	DGal > GalNAc	0.125	0.02
GNA (*Galanthus nivalis* agglutinin)	Manα(1-3)Man	-	>0.16
LVA (*Leucojum vernum* agglutinin)	Manα(1-3)Man (?)	-	>0.16
PHA-P (Phytohemagglutinin-P)	Complex oligosaccharide	0.5	0.08
SBA (Soy been agglutinin)	GalNAc >> Gal	0.015	0.0025
HPA (*Helix pomatia* agglutinin)	A-GalNAc	0.015	0.0025
LCA (*Lens culinaris* agglutinin)	D-Man > DGlc	0.5	0.08
PSL (*Pisum sativum* agglutinin)	D-Man > DGlc	0.5	0.08
SRA (*Sambucus racemosa* agglutinin)	Neu5Aca(2-6)Gal >DGal	0.03	0.005
VAA (*Viscum album* agglutinin)	DGal	-	>0.16
LSFA (*Laetiporus sulphureus* fungus agglutinin)	Lactosamin	0.03	0.005
LTFA (*Lactarius torminosus* fungus agglutinin)	Complex oligosaccharide	0.125	0.02
LPFA (*Lactarius pergamenus* fungus agglutinin)	Complex oligosaccharide	0.06	0.01
PAFA (*Paxillus atrotomentosus* agglutinin)	Complex oligosaccharide	-	0.16
MPFA (*Mycena pura* fungus agglutinin)	D-Man, DGlc	0.002	0.0003

Note: Dash means the absence of interaction at 1% concentration

Taking into consideration the presented data, we have also studied the interaction of alkaline phosphatase with the lectins that are most frequently used in the histochemical investigations. The results presented in Table 8, testify to the fact that many lectins bind to the alkaline phosphatase. In particular, strong interaction of the alkaline phosphatase was demonstrated for the lectins of soy bean and edible snail, and this enzyme was the strongest inhibitor of activity of *Mycena pura* lectin. The lectins of *Amaryllidaceae* family (snowdrop and snowflake), and, unexpectedly, the lectin of the mistletoe did not interact with the alkaline phosphatase.

The results of study on the binding of above mentioned lectin to various glycoproteins indicates that a big part of them belongs to such disaccharide links as GlcNAcβ(1-2)Manα(1-6) or GlcNAcβ(1-2)Manα(1-2), while terminal residues of disaccharides Galα(1-3) Galβ(1-4) do not abolish interaction of the lectin with the carbohydrates. This suggests that these links are not essential for such interaction.

THE EFFECT OF HEMOLYTIC LECTINS TOWARDSOTHER TYPES OF CELLS

Action of hemolytic lectins towards cultured leukemia cells. *A. virosa* lectin was found to possess a cytotoxic action towards the mammalian leukemia cells of different lines: human CEMT4 and Jurkat cells and murine L1210 cells. This lectin applied in concentration $\geq$ 5 μg/ml expressed a cytotoxic action towards to L1210 cells as soon as in one hour after its adding to the cultural medium. CEM T4 and Jurkat cell lines are more sensitive to a cytotoxic action of the lectin whose minimum toxic action appeared at considerably lower concentrations (Table 9).

Table 9. Cytotoxicity characteristics *A. virosa* lectin action (24 h)

Cell line	Minimal active dose μg/ml	Cytotoxicity characteristics			
		ID 50	ID 100	LD 50	LD 100
L 1210	0.5	2.5-5	5	2.5-5	10
CEM T4	~0.01-0.05	0.5	2.5	0.5	2.5
Jurkat	~0.005-0.01	1	2.5	1	2.5

Antimicrobial activity of studied lectins was revealed only in the hemolytic lectins (*Mycena pura* and *Amanita virosa*), whereas the non-hemolytic lectins (ConA, HPA, SNA, WGA, PSA and *Perca fluviatilis* roe agglutinin) did not affect growth of microorganisms. Mycena lectin mainly inhibited growth of gram-positive microorganisms as well as Proteus (gram-negative microflora). In particular, growth delay area for *C. xerosis*appeared to be 33 mm, *S. aureus* – 28 mm, *B. subtilis* – 23 mm. The studied lectin weakly influenced *C. albicans* and did not affect growth of *E. coli* and *P. aeruginosa* (Figure 5).

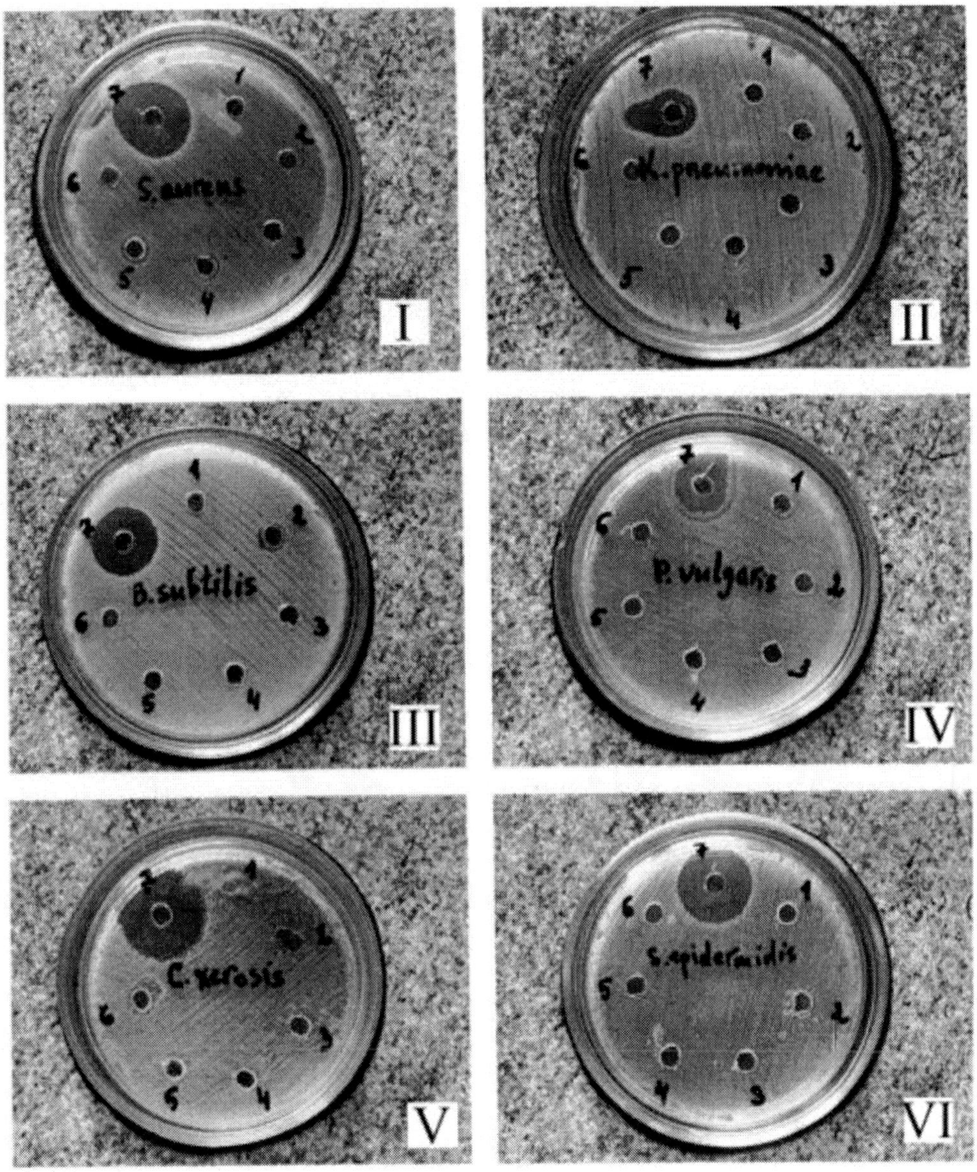

Picture was kindly provided by Nemchenko I. I., PhD.

Figure 5. Petri dishes with cultures of microorganisms: I - *Streptococcus aureus;* II –***Klebsiella*** *pneumoniae;* III - *Bacillus subtilis;* IV -*Proteus vulgaris;* V - *Corynebacterium xerosis;* VI - *Staphylococcus epidermidis.*

50 μg of various lectins in solution were dropped in the centre of each sector: 1 - *Canavalia ensiformis* (Con A), 2 - *Helix pomatia* (HPA), 3 - *Sambucus nigra* bark (SNA), 4 - wheat germ (WGA), 5 - pea (PSA), 6 - *Perca fluviatilis*roe (PFA), 7 - *Mycena pura* lectin.

A. virosa lectin inhibited growth mainly of gram-positive microorganisms, such as *C. xerosis, S. aureus, S. epidermides, K. pneumoniae, P. vulgaris, B. subtilis*, however, it did not have a baceriostatic effect towards gram-negative microflora, namely *E. coli* and *P.*

aeruginosa. Growth delay areas at the application of these lectins were found to be smaller than those at the applicaion of the *Mycena pura* lectin (with the exception of *Str. Faecalis* lectin) (Table 10). The antimicrobial activity of lectins was compared with the effects of known antibiotics, such as gentamycin, an inhibitor of protein synthesis, and nystatin which disrupts integrity of cell membrane. These data are also presented in Table 10.

Different action of the hemolytic lectins of mushrooms towards gram-positive and gram-negative microorganisms could be explained by differences in their cell wall structure. Gram-positive bacteria suppressed by the lectins, have thick wall that does not contain lipids, whereas gram-negative microbe wall is thinner and contains large amount of glycolipids [25]. Hence, lectin action most likely depends on presence of specific carbohydrates in cellular wall, but not on cell wall thickness.

Table 10. Size of growth inhibition zones in various species of microorganisms under the action of lectins of *Amanita virosa* and *Mycena pura* (50 μg of lectin per hole) and some antibiotics

Microorganism	Diameter of growth delay area, mm		
	***M. pura* lectin, 50 μl**	***A. virosa* lectin, 50 μl**	**Gentamycin**
E. coli	-	-	*24*
P. aeruginosa	-	-	*31*
C. albicans	*9*	-	*25*
			Nystatin, 80 Units
K. pneumoniae	*20*	*14*	*7*
S. epidermidis	*25*	*18*	*35*
S. aureus	*28*	*17*	*26*
B. subtilis	*23*	*11*	*30*
P. vulgaris	*21*	*13*	*26*
C. xeroxis	*33*	*22*	*36*
Str. faecalis	*22*	*24*	*32*

Note: Through-line means absence of growth delay area.

Conclusion

Detection of cytotoxic and antimicrobial action in the hemolytic lectins of higher mushrooms allows considering them as cytolytic lectins, because such action is not limited by the erythrocytes. The mechanism of this action is based on the formation of pores in plasma membrane of targeted cells. Although these lectins are capable of destroying cell wall, their toxicity is much lower comparing with the toxicity of such ribosome-inactivating lectins as ricin, abrin, or viscumin.

Up to date, the cytolytic lectins were discovered mainly in low vertebrates with primitive immune system. In these animals, recognition and lysis of foreign cells is performed without involving antibodies of immunoglobulin nature that are absent there. In contrast to the invertebrates whose cytolysis is mediated by complement system and includes a complex consisting of a series of activated proteases, effector and regulator proteins, in the invertebrates, the lysis of foreign cells, first of all microorganisms, could be conducted, in particular, by the cytolytic lectins. The last ones were not discovered in plants, and before this work, they were described only in one species of fungi - *Laetiporussulphureus*. Discovery of similar type of lectins in three other species of higher fungi points out to possible relations of some groups of fungi lectins with the lectins of lower invertebrates. This evidence can be an additional argument in favour of the fact that the sub-kingdom of *Eumycota* has separated from the general evolutionary mainstream somewhat later in comparison with separation of plant sub-kingdom.

Potential toxicological value of the hemolytic lectins of *A. virosa* and *A. phalloides* cannot be excluded. These poisonous mushrooms can be collected by mistake instead of the russule.

When the poisonous mushrooms of *Amanita* genus are eaten by mistake in raw form, such lectins can endow a certain contribution to general toxic effect towards the mammalian organism.

REFERENCES

[1] Endo Y., Tsurugi K. RNA N-glycosidase activity of ricin A-chain. Mechanism of action of the toxic lectin ricin on eukaryotic ribosomes.*J. Biol. Chem.* - 1987.-V. 262.- P.8128-8130.

[2] Endo Y., Mitsui K., Motizuki M., Tsurugi K. Mechanism of action of ricin and related toxic lectins on eucariotic ribosomes: the site and characteristics of the modification in 28s rRNA caused by the toxins. *J. Biol. Chem.*-1987 - V.262.- P.5908-5912.

[3] Hankins C. N., Shannon L. M. The physical and enzymatic properties of phytohemagglutinin of Mung bean *J. Biol. Chem.* – 1978. –V. 253, No 21. – P. 7791-7797.

[4] MolinaM. C., Vicente C. Purification and characterization of two isolectins with arginase activity from the lichen Xanthoria parietina *J. Biochem. Mol. Biol.* – 2000.- V.33, No 4. P. 300 -307.

[5] Hatakeyama T., Kohzaki H., Nagatomo H., Yamasaki N. Purification and characterization of four Ca^{2+}-dependent lectins from the marine invertebrate *Cucumaria echinata. J. Biochem.*-1994.-V. 116.-P. 209-214.

[6] Armstrong P.B, Swarnakar S, Srimal S, Misquith S, Hahn E.A, Aimes R.T, Quigley J.P. A cytolytic function for a sialic acid-binding lectin that is a member of the pentraxin family of proteins. *J Biol Chem.* –1996.- V.271, No 25.- P.14717-14721.

[7] Roch, P., Canicatti, C., Valembois, P. Interaction between earthworm hemolysins and sheep red blood cell membranes. *Biochim. Biophys. Acta.*-1989.-V.2.-P. 193– 198.

[8] Konska G, Guillot J, Dusser M, Damez M, Botton B.Isolation and characterization of an N-acetyllactosamine-binding lectin from the mushroom *Laetiporus sulfureus.J Biochem (Tokyo).*–1994.-V.116, No 3.-P.519-523.

[9] Tateno H, Goldstein I.J.Molecular cloning, expression, and characterization of novel hemolytic lectins from the mushroom *Laetiporus sulphureus*, which show homology to bacterial toxins.*J Biol Chem.* 2003.-V. 278, No 42. - P. 40455-40463.

[10] Antonyuk V.O. Study of carbohydrate specificity of hemolytic lectin from death-cap mushroom (*Amanita phalloides (Vaill. Fr.) Secr.*). *Biopolymers and Cell*.- 2005.-V. 21, No 4.P. 1-7 (In Ukrainian).

[11] Antonyuk V.O., Klyuchivska O.Yu, Stoika R.S. Cytotoxic proteins of Amanita virosa Secr. mushroom: purification, characteristics and action towards mammalian cells *Toxicon*.- 2010.- V.55. P. 1297 -1305

[12] Antonyuk V. O., Nemchenko I. I., Tymchuk I. V., . Danileuchenko V. V, StoikaR. S. Study of Hemolytical and Antimicrobial Action of Poisonous Mushrooms Lectins of *Amanita virosa* Secr. and *Mycena pura* /Fr./ Kumm. *Biopolymers and Cell*.- 2010. - V.26, No 1. – P.29-35 (In Ukrainian).

[13] AntonyukV.O. 1990. Methodforobtaining of affinity sorbent for purification of lectin. Patent (Author's Certificate) No 1554961, *Bull Izobret. USSR* No 13 (In Russian).

[14] Lutsik-Kordovsky M. D., Antonyuk V.A., Stasyk T.V., Yakymovych M.Ya, Yakymovych L.A., Hellman U., Soushelriytsky S., Stoika R.S., Lutsik A.D. Hemagglutinating lectin from fruit bodies of Amanita phalloides: isolation, properties and biological activity *Abstr. Interlec 18*, 27-31 July 1999 Univ. Portsmouth, P.77.

[15] Maurer G, Disk-electrophoresis. Mir,:Moscow, pp. 55-102 .- 1971 (In Russian translation).

[16] Kabat E.A., Mayer M.M. Experimental Immunochemistry.- Illinois, USA: Charles Thomas Publisher, - 690 P. -1964.

[17] Lazur'evskyi G.V., Terent'eva I. V.,Shamshurin A.A. Practical Works in Chemistry of Natural Compounds,High school, Moscow – 336 P. -1966 (In Russian).

[18] 18. Methods of Carbohydrate Chemistry /Ed. by Kochetkov N.K., Mir,: Moscow, 512 p. - 1967. (In Russian)

[19] Lutsik AD, Detyuk ES, Lutsik MD, Lectins in Histochemistry. Lvov University Publishing House, Lvov - 1989.(In Russian).

[20] Scherrer R. Gerhardt P. Molecular Sieving by the Bacillus megaterium cell wall and protoplast *J Bacteriol.*1971.-V. 107, No 3.-P. 718–735.

[21] Nagahama M., Hayashi S., Morimitsu S., Sakurai J. Biological activities and pore formation of *Clostridium perfringens* beta toxin in HL 60 cells. *J Biol. Chem.* 2003.-V. 278, N 38.-P. 36934-36941.

[22] Lutsik-Kordovsky M.D., Stasyk T.V., Stoika R.S. Analysis of cytotoxicity of lectins and non-lectin proteins from Amanita mushrooms. *Experimental oncology* (Kiev).- 2001.-V. 23 №1. P. 43-45.

[23] MillánJ. L. Alkaline Phosphatases Structure, substrate specificity and functional relatedness to other members of a large superfamily of enzymes *Purinergic Signal.* 2006 V. 2, No 2.- P. 335–341

[24] Bublitz R., Hoppe H., Cumme G. A., Thiele M., Attey A., Horn A. Structural study on the carbohydrate moiety of calf intestinal alkaline phosphatase.*J. Mass Spectrom.* - 2001; V. 36.- P. 960-972.

[25] Kauzer F.H., Bienz K.A., Eskert J., Zinkernagel R.M. Medical Microbiology.-Stuttgart-New York: Thieme.- 698 P.- 2005.

In: Mushrooms: Types, Properties and Nutrition
Editors: S. Andres and N. Baumann
ISBN 978-1-61470-110-1

Chapter 10

RADIOACTIVITY OF EUROPEAN WILD GROWING EDIBLE MUSHROOMS

Pavel Kalač*
Department of Applied Chemistry, Faculty of Agriculture, University of South Bohemia, CZ-37005 České Budějovice, Czech Republic

ABSTRACT

Wild growing mushrooms are widely consumed as a delicacy in several European countries, at level up to several kg per year per capita. Activity concentrations of the natural isotope ^{40}K are usually 800-1,500 Bq kg^{-1} dry matter (DM). Other natural radionuclides with leading ^{210}Pb and ^{210}Po are of lower importance. Activities of ^{137}Cs from nuclear weapons testing below 1,000 Bq kg^{-1} DM were commonly reported until 1986. The situation changed dramatically after the accident of Chernobyl nuclear power station in 1986. Activities up to over 100,000 Bq kg^{-1} DM of ^{137}Cs and to a lesser extent of ^{134}Cs were observed in some edible species in the following years. Commonly, mycorrhizal species accumulate radiocesium more than species with saprotrophic or parasitic nutritional strategy. *Xerocomus badius, X. chrysenteron, Suillus variegatus, Rozites caperata, Laccaria amethystina* and *Hydnum repandum* belong among the radiocesium highly accumulating and widely consumed species. Activity concentrations have been affected by several environmental factors, such as rate of soil contamination by the Chernobyl fallout, the depth from which mycelium takes nutrients and time since the accident. Most of the ^{137}Cs in forest soils appear to be available for uptake by mushrooms until now. A considerable consumption of accumulating species collected from the sites heavily contaminated in 1986 can be still of a health concern. The contamination can be reduced by soaking or cooking of dried or frozen mushroom slices. Until now, meat of wild boars eating some mushroom species from heavily contaminated areas can highly surpass statutory limit for ^{137}Cs.

Keywords: edible mushrooms; radioactivity; radionuclides; radiocesium; health concern; effective dose; game meat.

---

* E-mail: kalac@zf.jcu.cz; Phone: +420 387 772 657; Fax: +420 385 310405

1. Introduction

The collection of wild growing edible mushrooms (higher fungi, macrofungi) as a delicacy has historical tradition in various parts of the world. This recreational and sometimes economic activity has been widespread e.g. in Slavonic countries of central and east Europe. Thus, a possibility of its deprivation or limitation due to the fears of mushroom contamination is considered as an abatement of lifestyle. Such situation arose in years following disaster of Chernobyl nuclear power station in 1986.

Radionuclide contamination of the environment has originated from the fallout following nuclear weapons testing, operation of nuclear energy-generating industries including their accidents and from medical uses of radioisotopes. Fungi (both filamentous and fruit-forming) are very efficient in absorbing radionuclides and are an important component of long-term accumulation of radionuclides owing to the long-living and huge hyphal network and biomass in upper horizons of forest soils.

High radioactivity levels of some wild growing mushroom species were reported in the 1960′s (Grüter, 1964). The contamination of most of Europe with radioactive fallout from the Chernobyl accident triggered the extensive research and monitoring of the environment including mushrooms. Numerous original papers on mushroom radioactivity were published during the last 25 years. Moreover, several reviews of the literature are available. They focus either on environmental aspects and transfer particularly of radiocesium to fungi (both micro-fungi and mushrooms) (Gillett and Crout, 2000; Duff and Ramsey, 2008; Dighton et al., 2008) or on aspects of radionuclides intake from the consumption of various mushroom species (Kalač, 2001).

This chapter summarizes recent information on the radioactivity of European edible mushroom species. Only the weightiest articles until 2000 will be cited, further papers from that period are referred in the above mentioned reviews.

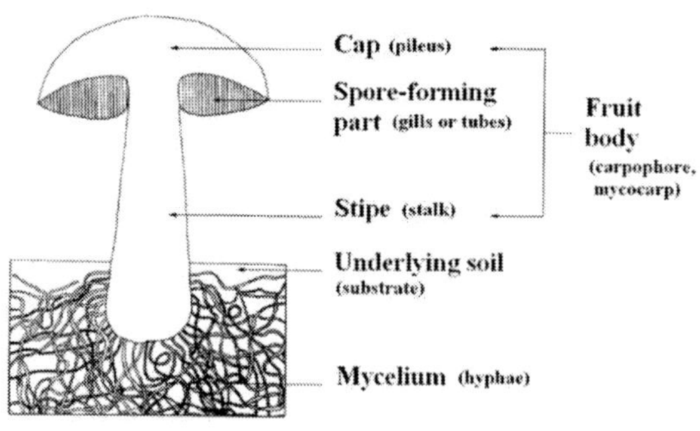

Figure 1. A sketch of a typical mushroom.

2. Mycological Terms

Some mycological terms used in this chapter are given in Figure 1. The term mushroom will be used for the fruit body, mostly aboveground, of higher fungi. A fruit body is formed from spacious underground mycelium by the process of fructification. The lifetime of the bulk of fruit bodies is only 10-14 days.

According to their nutritional strategy, mushrooms can be divided into three groups. Mycorrhizal (symbiotic) species form a close, mutually favorable relationship with their host plant, mostly a tree. Saprotrophic species (or saprophytes) live on and metabolically consume organic matter. Parasitic species live on other species in a non-symbiotic relationship.

3. Radioactivity Units and Legislation

Published data on radioactivity levels are usually expressed per dry matter (DM) content. For unification, the commonly accepted content of 10% of dry matter in mushroom fruit bodies was used for the calculation in this chapter, if published data were given per fresh matter.

One Bq (becquerel) has been the unit for the activity of a radioactive source in which, on average, one atom decays per second. Activity concentration, i.e. activity per dry matter unit, will be used in this chapter. Within the European Union (EU), food statutory limits for radiocesium are 600 and 370 Bq per kg of fresh matter for adults and children, respectively. Thus, the limit for adults would be 6,000 Bq per kg of dry matter for mushrooms. Codex Alimentarius (FAO/WHO, 1991) defines explicitly the intervention levels for foodstuff being 1,000 Bq kg^{-1} (fresh matter) for ^{137}Cs.

Nevertheless, the considerable increase of foodstuff radioactivity following the Chernobyl disaster induced within the EU (formerly the EC) series of legislative provisions and recommendations (e.g. 2003/274/EC for mushrooms). Under the regulation from 1987, the maximum permitted level of ^{137}Cs was 1,250 Bq kg^{-1} fresh matter (i.e. 12,500 Bq kg^{-1} DM) for mushrooms. A similar limit of 1,000 Bq kg^{-1} fresh matter (i.e. 10,000 Bq kg^{-1} DM) was recommended by the International Atomic Energy Agency (IAEA) in 1994.

A possible risk of radioactivity for human health is expressed as the effective dose (E) given in mSv (millisievert) per year. The acceptable yearly burden for an adult, of the public, recommended by the IAEA (1996) has been 1 mSv. A contribution to the yearly effective dose to an adult from mushroom consumption can be calculated as follows:

$E = Y \times Z \times d_c$,

where Y = annual intake of mushrooms (kg DM per person),

Z = activity concentration (Bq kg^{-1} DM),

d_c = dose coefficient (conversion factor) defined as the dose received by an adult per unit intake of radioactivity. The values are 1.3×10^{-8}, 1.9×10^{-8}, 6.2×10^{-9}, 6.9×10^{-7} and 1.2×10^{-6} Sv per Bq for ^{137}Cs, ^{134}Cs, ^{40}K, ^{210}Pb and ^{210}Po, respectively.

4. Natural Radionuclides in Mushrooms

Mushrooms contain considerably higher levels of potassium than foods of plant origin (see chapter 6 of this book). Potassium is not distributed evenly in a fruit body. The content usually decreases in order cap > stipe > gills or tubes in spore-forming part > spores. The accumulation factor (ratio of the contents in fruit body DM and in DM of underlying soil) is usually between 20 and 40. The radionuclide ^{40}K is present in the mixture of potassium isotopes at the constant level of 0.017%. Its half-life is extremely long.

Usual activity concentration of ^{40}K in numerous species of wild growing mushrooms, mostly from central Europe, were 800-1,500 Bq kg^{-1} DM between 1984 and 1992 (Kalač, 2001 and references therein). Transfer factors for ^{40}K from underlying soil to fruit bodies exceeding value of 10 were reported for *Xerocomus badius*, *Amanita rubescens* and *Lycoperdon perlatum* (Eckl et al., 1986).

Data on ^{40}K published during the last decade are given in Table 1. The values are very similar to those reported until 2000. Thus, it seems that the incorporation of the stable potassium isotopes ^{39}K and ^{41}K and hence also ^{40}K is self-regulated by physiological requirements of a mushroom. The extremely high activity concentration of ^{40}K up to 12,000 Bq kg^{-1} DM was reported in *Laccaria laccata* (Mietelski et al., 2010).

Unlike ^{137}Cs from radioactive fallout, natural ^{40}K is distributed in vertical profile of forest soils evenly (Krolak et al., 2010).

Isotope ^{210}Pb (half-life 22 years), a decay product of naturally occurring ^{238}U, ^{226}Ra or ^{222}Rn, is present in mushroom fruit bodies in activity concentration usually for two orders of magnitude lower than ^{40}K (Table 2). Activity concentrations of ^{210}Pb are consistently higher – often by one order of magnitude – than those of its long-lived precursor ^{226}Ra (see below). This indicates that a very small ^{210}Pb proportion, if any, originates from the decay of ^{226}Ra taken up from the soil (Kirchner and Daillant, 1998). Dose coefficient of ^{210}Pb is over 100 times higher than that of ^{40}K (see the previous section) and the effective dose from its intake from mushrooms should not be overlooked. For instance, such situation was observed in northern Finland with low contamination with the Chernobyl radiocesium (Vaaramaa et al., 2009).

A further natural radionuclide of ^{238}U decay, ^{210}Po (half-life 138 days), was determined in Polish mushrooms by Skwarzec and Jakusik (2003). Among 20 tested both edible and inedible species, the highest mean activity concentration of about 40 Bq kg^{-1} DM was observed in *Boletus edulis* and *Leccinum scabrum*, while the lowest one, 2.1- 4.3 Bq kg^{-1} DM, in *Xerocomus badius* and *X. subtomentosus*, both being typical radiocesium accumulators (see below). This is interesting because all these species are taxonomically closely related. Higher ^{210}Po levels were reported for caps than for stipes. Despite the high dose coefficient (1.2×10^{-6} Sv per Bq), even at high *B. edulis* consumption of 5 kg (fresh matter) per year, an effective dose would be 37 μSv. Levels of ^{210}Po in mushrooms are considerably higher than those in foods of both plant and animal origin.

Activity concentrations of ^{210}Po higher for one order of magnitude reported Vaaramaa et al. (2009) in Finnish mushrooms. They also observed the highest levels in Boletaceae family, namely in *Leccinum vulpinum*, while the lowest in Russulaceae family. A higher accumulation was proved in caps than in stipes. Yearly radiation doses from ^{210}Po ingested from mushrooms in Finnish conditions were reckoned up to about 4 μSv.

Rarely reported data on ^{226}Ra in mushrooms indicate low activity concentrations in France (Kirchner and Daillant, 1998) and even in a Romanian uranium mining area (Popa et al., 2010).

Radionuclide ^{7}Be was detectable only in a few of 72 samples of various Spanish mushroom species (Baeza et al., 2004).

Overall, ^{40}K is the main natural radionuclide of edible mushrooms.

Table 1. Activity concentrations of natural isotope ^{40}K (kBq kg^{-1} DM) in wild growing mushrooms

Species	Activity	Year(s) of collection	Country	Reference
Xerocomus badius (28 pooled samples)	0.50 – 1.48	1996-1998	Poland	Malinowska et al., 2006
12 species	0.59 – 2.02	2002	Turkey	Karadeniz and Yaprak, 2010
11 species	0.72 – 1.78		Turkey	Turhan et al., 2007
10 species	0.8 – 1.8	2004	8 European countries	Szántó et al., 2007
70 species (including inedible)	mostly 1.0 – 2.5	2006, 2007	Southern Poland (site of a hot spot of Chernobyl fallout)	Mietelski et al., 2010

Table 2. Activity concentrations of natural isotope ^{210}Pb (Bq kg^{-1} DM) in wild growing mushrooms

Species	Activity	Year(s) of collection	Country	Reference
18 species	<1.8 – 36.5 mean 12.0	1991-1997	France	Kirchner and Daillant, 1998
Xerocomu badius (80 pooled samples)		1996-1998	Poland	Malinowska et al., 2006
caps	1.8 – 36.4			
stipes	1.1 – 26.2			
32 species (including inedible)	0.8 – 202 median 16.6	1999-2001	Spain	Guillén et al., 2009
17 species (including inedible)		2006, 2007	Finland	Vaaramaa et al., 2009
caps	1.6 – 16.2			
stipes	1.0 – 15.5			

5. Artificial Radionuclides in Mushrooms Until 1985

Several radionuclides were discharged into the global environment through nuclear weapons testing until 1963. The total release of the most important radionuclide, ^{137}Cs, was estimated as 9.6 x 10^{17} Bq (for this section, see Kalač, 2001 and references therein).

Most mushroom species have limited abilities to accumulate non-radioactive isotopes of cesium. The reported bioaccumulation factors fruit body/underlying soil of the non-radioactive Cs are not significantly different from those for vascular plants. However, in radiocesium from the fallout, the observed values were higher for at least one order of magnitude.

Limited data were reported for mushroom radioactivity until 1985 in central Europe. Activity concentrations of ^{137}Cs below 1,000 Bq kg^{-1} DM were usual. *Xerocomus badius* and *X. chrysenteron* were already then identified as the accumulating species.

Mushrooms did not accumulate ^{90}Sr or radioisotopes of plutonium at toxicologically significant levels.

6. Mushroom Radioactivity After the Chernobyl Accident

6.1. Chernobyl Accident

The disaster of the Chernobyl (Ukraine) nuclear power station on 26 April 1986 released into the environment about 3.8 x 10^{16} Bq from ^{137}Cs decay. The ratio ^{137}Cs to ^{134}Cs was approximately 2 : 1. The main radionuclides produced during an explosive fission reaction are ^{137}Cs and ^{90}Sr with long half-lives 30.17 and 28.8 years, respectively. Radiocesium ^{134}Cs (half-life 2.06 years), also important for mushroom contamination, has been produced in reactors during long-term fission. Moreover, ^{144}Ce, ^{131}I, ^{95}Nb, ^{239} Pu, ^{240}Pu, ^{103}Ru, ^{106}Ru, ^{230}Th, ^{232}Th and ^{95}Zr were detected in mushrooms early after the accident (Mietelski et al., 2002; Zarubina, 2004). Nevertheless, their toxicological risk from mushroom consumption was limited.

The contamination levels of an area were affected by direction and speed of radioactive cloud, distance from Chernobyl (here and hereafter is meant the power station), and particularly by rainfall or only dry depositions during passage of the clouds. Fallout levels were thus very different even in relatively adjacent sites. For instance, Bakken and Olsen (1990) observed in Norway up to fivefold differences even within squares 20 x 20 cm. Contamination levels varied very widely in the orders of 10^1-10^5 Bq m^{-2}. In the heavily polluted areas adjacent to Chernobyl, ^{137}Cs radioactivity may be of a health concern for up to next three centuries.

During the initial years after the accident, both radiocesium isotopes participated in mushroom radioactivity. Later, since about mid-1990's, ^{137}Cs has remained the crucial radionuclide.

6.2. Radiocesium Transfer From Soils to Mushrooms

Edible mushrooms have been collected mostly in forests, which present semi-natural ecosystems. Forest soils differ from agricultural cultivated lands. Temperate forest soils are multi-layer with forest floor, hemi-organic and mineral layers. Fungal and microbial activities likely contribute considerably to the long-term retention of radionuclides in organic layers of forest soils. A high variability is thus commonly observed in radionuclide transfers and redistribution patterns in contaminated forests. Soil compartments represent the major reservoir of radionuclides, which can cause long-term contamination of the food chain. Organic matter of forest soils has a lower affinity for cesium than mineral components. Cesium is thus easily available for mushroom mycelium located in organic layers. Long lasting availability of some radionuclides was shown to be the source of much higher transfer in forest ecosystems than in agricultural lands (Calmon et al., 2009).

Contamination levels of ^{137}Cs observed in years following the Chernobyl accident even within the same species show both high spatial and temporal variability. Several factors have been implicated: mycelium habitat and depth, forest type – fruit body location, sampling strategy, soil clay content and soil moisture and/or microclimate (Gillett and Crout, 2000 and references therein).

Depth of mycelium is a very weighty factor affecting initiation and level of mushroom radioactivity. In species with superficial mycelium (e.g. of genera *Collybia* and *Clitocybe*), the contamination attained within a few months of fallout whilst other deeper penetrating species (e.g. *Boletus edulis*) achieved contamination peaks several years after the deposition. Therefore, ecological half-lives can be estimated with difficulty (Gillett and Crout, 2000).

Transfer of a radionuclide from soil to mushroom fruit body has been usually expressed as "aggregated transfer coefficient" (T_{ag}), defined as the ratio between mushroom activity at time t and the initial deposit of the radionuclide at time $t = 0$ (for post-Chernobyl ^{137}Cs assumed to 1st May 1986). The parameter must be decay-corrected to the time of a mushroom sampling. A simplified parameter has been transfer factor (TF; or transfer ratio) defined as the ratio of mushroom activity (Bq kg^{-1}) to underlying soil deposit (Bq m^{-2}), both at time t. Practical differences between TF and T_{ag} are relatively small compared to the effects of other factors. Both the parameters are usually expressed as m^2 per kg DM of a mushroom.

Nevertheless, estimation of credible TF values requires consideration of the localization of the mycelia within the soil column. Sampling the whole organic layer can cause overestimation or underestimation of the transfer ratios (Karadeniz and Yaprak, 2007). As reported Baeza et al. (2005) for Spanish pine forests, mushroom mycelium is generally localized in the surface layer of soil (0-5 cm). However, different soil horizons were sampled and analyzed (e.g. 0-2, 0-5, 0-10 cm) by various authors and the reported transfer factors are thus hardly comparable.

Transfer factors for ^{137}Cs and stable ^{133}Cs determined in 1998 and 2002 in a scale of mushroom species were fairly constant both in Belorussian (Yoshida et al., 2004) and Turkish forests (Karadeniz and Yaprak, 2007), respectively, with different level of ^{137}Cs contamination. Thus, some 15 years after the contamination, the studied mushroom species did not distinguish between stable ^{133}Cs and ^{137}Cs from Chernobyl. ^{133}Cs could be thus partially used for the evaluation of ^{137}Cs behavior.

There exists a consensus of various authors, that mushrooms of different nutritional strategy accumulate ^{137}Cs in the order mycorrhizal > saprotrophic > or ≈ parasitic (overall in Gillett and Crout, 2000).

Mushrooms showed a greater preference for rubidium and potassium than for cesium. The uptake of ^{137}Cs could be thus prevented by providing additional K or Rb at contaminated sites (Vinichuk et al., 2010).

Transfer factors of ^{137}Cs for mushrooms from coniferous forests usually exceeds the factors for mushrooms in deciduous forests (Vinichuk and Johanson, 2003 and references therein).

Until now, most of the ^{137}Cs in forest soils appears to be available for uptake and radioactive decay likely remains the main factor for a reduction of the radionuclide in forest ecosystems. Several models for ^{137}Cs migration in forest ecosystems after the Chernobyl accident were developed, however, they have a very high degree of divergence of predictions, particularly when forecasting edible mushrooms contamination.

Only a limited information is available on transfer of other radionuclides to mushroom fruit bodies. The transfer from soil to mushrooms determined in Spanish forests was low for detected both man-made and natural radionuclides. The efficiencies were ranked as follows: $^{228,\ 230,\ 232}Th \approx {}^{40}K \geq {}^{137}Cs \geq {}^{234,238}U \approx {}^{226}Ra \geq {}^{90}Sr \geq {}^{239+240}Pu \approx {}^{241}Am$. *Hebeloma cylindrosporum*, known by its high transfer factor for ^{137}Cs, was found as an accumulator of other studied radionuclides and can be thus used as a good bioindicator of the radioactivity of an ecosystem (Baeza et al., 2006a; Baeza and Guillén, 2006).

6.3. Mushrooms Radioactivity Since 1986

Skin and flesh had very similar radiocesium concentrations, while the spore-bearing parts (gills or tubes) had for about 50-100% higher levels (Bakken and Olsen, 1990), in a work of Heinrich (1993) even for about 250% higher. Such information was proved by Baeza et al. (2006b) who separated fruit bodies to cap + gills and stipes. Maximum level of ^{134}Cs was found in mature fruit bodies. Also Malinowska et al. (2006) reported higher activity concentrations in caps than stipes of *Xerocomus badius* and similar phenomenon was observed in various species of several families (Bazała et al., 2005).

High accumulation of radiocesium particularly in caps of *X. badius* has been ascribed to natural pigments badion and norbadion A causing chocolate brown or golden yellow coloration of this and related species (Aumann et al., 1989). These pigments have numerous acid-base functional groups, which are able to bind monovalent cations including Cs^{+}. Nevertheless, only about 12 % of the total radiocesium was detected in cap skin (epicutis) of *X. badius*, in proportion to its weight (Neukom and Gisler, 1991).

6.3.1. Mushrooms Radioactivity between 1986 and 2000

Data on mushroom radioactivity during the years following the Chernobyl accident were reviewed by Gillett and Crout (2000), Kalač (2001) and particularly by a comprehensive overview of Duff and Ramsey (2008). The latter review collates data until 2000. Numerous references are available therein and are not thus given here.

Overall, activity concentrations varied very widely. In some mycorrhizal species from several hundreds to above 100,000 Bq kg^{-1} DM, in saprotrophic and parasitic species between a few hundreds and a few thousands Bq kg^{-1} DM. The EU statutory limit of 12,500 Bq kg^{-1} DM was often substantially surpassed. Commonly consumed species with different rates of radiocesium accumulation are given in Table 3.

Table 3. Selected edible mushroom species with different rates of radiocesium accumulation

High	Medium	Low
Cantharellus lutescens (M)	*Agaricus silvaticus* (S)	*Amanita rubescens* (M)
Cantharellus tubaeformis (M)	*Boletus edulis* (M)	*Armillariella mellea* *
Hydnum repandum (M)	*Cantharellus cibarius* (M)	*Calocybe gambosa* (S)
Laccaria amethystina (M)	*Leccinum aurantiacum* (M)	*Laccaria laccata* (M)
Rozites caperata (M)	*Leccinum scabrum* (M)	*Lepista nuda* (S)
Russula cyanoxantha (M)	*Russula xerampelina* (M)	*Lycoperdon perlatum* (S)
Suillus variegatus (M)		*Macrolepiota procera* (S)
Xerocomus badius (M)		
Xerocomus chrysenteron (M)		

Nutritional strategy: M ... mycorrhizal, S ... saprotrophic, * mostly parasitic, but with an adaptable strategy

Very wide differences exist within a species. A unique study was carried out in Poland with 278 samples of highly accumulating and widely consumed *Xerocomus badius*, to cover systematically the whole country area, in 1991. Maps of ^{137}Cs, ^{134}Cs and ^{40}K levels in this species were prepared. The most frequent activity concentrations were 2,000-10,000 and 200-600 Bq kg^{-1} DM for ^{137}Cs and ^{134}Cs, respectively (Mietelski et al., 1994).

An extremely high, probably record value of 862,100 Bq kg^{-1} DM was determined in inedible *Paxillus involutus* from a site about 70 km west of Chernobyl (Vinichuk and Johanson, 2003).

Data on ^{134}Cs were reported only until mid-1990's due to the short half-life of the radioisotope. Nevertheless, traces were detected in some species in Poland even in 2006 and 2007 (Mietelski et al., 2010).

Maximum level of mushroom radioactivity was observed in 1987 and mainly in 1988. This was caused by a subsequent supply of the radionuclides to the upper soil horizon through dropped needles and leaching from needles and bark.

6.3.2. Mushrooms Radioactivity Since 2001

The frequency of original papers with information on radioactivity of mushrooms collected after 2000 markedly decreased as compared with the previous period. All following data deal with ^{137}Cs activity.

Interesting long-term data reported Mascanzoni (2009) in series of activity concentrations determined between 1986 and 2007 in a site of southern Sweden with high Chernobyl fallout contamination of 35,000 Bq m^{-2}. In accumulating *Suillus luteus*, values of the activity, decay-corrected to 1st May 1986 to avoid a systematic bias of ^{137}Cs decay, remained stable at level about 120,000 Bq kg^{-1} DM, while in a *Cantharellus* spp. the activity increased about three times during the period. Reasons of the difference were not convincingly explained.

Activity concentrations mostly below 100 Bq kg^{-1} DM were determined in 33 samples of 15 species gathered from several sites in the Czech Republic and the Slovak Republic with relatively low Chernobyl fallout, during 2000-2004. *S. luteus* and particularly *X. badius* were out of the common run. The maximum measured level in *X. badius* was 6,260 Bq kg^{-1} DM (Dvořák et al., 2006). Similarly, caps of *X. badius*, *S. luteus* and *Tricholoma equestre* collected in northern Poland in 2004, had activities 2,700, 1,880 and 1,180 Bq kg^{-1} DM (Bazała et al., 2005).

Mean activity concentration of 18 samples belonging to 10 species, collected in eight European countries from Ukraine to Belgium during 2004, was 834 ± 1,290 Bq kg^{-1} DM. Also in this work, *X. badius* samples from Poland and Ukraine had the highest determined activities (Szántó et al., 2007).

A specific situation was reported from the most polluted area of Poland, "Opole region hot spot", where the Chernobyl contamination was up to >60,000 Bq m^{-2} in 1993. In over 70 species of both edible and inedible mushroom species, collected during 2006 and 2007, a very wide range between 30 and 54,070 Bq kg^{-1} DM was determined. Two families showed extraordinary high means, Suillaceae (nearly 12,000 Bq kg^{-1} DM) and mycorrhizal species of Tricholomataceae (about 10,000 Bq kg^{-1} DM). Commonly, mycorrhizal species had higher activity concentrations than saprotrophic ones (Mietelski et al., 2010).

Quite different situation was reported from Spain with minimum Chernobyl fallout (Baeza et al., 2004). Median of 72 samples belonging to 49 species was only 3.44 Bq kg^{-1} DM with accumulating genera *Hebeloma*, *Tricholoma* and *Lactarius*. Similarly as in other reports, mycorrhizal species accumulated more ^{137}Cs than saprotrophic ones. Unfortunately, period of mushroom sampling was not given.

In conclusion, mushroom contamination with ^{137}Cs during the decade since 2001 remained relatively stable. The two main recognized factors, level of a site contamination by the Chernobyl fallout and mushroom species, have to be taken under consideration.

7. Decrease of Mushroom Radioactivity by Culinary Treatments

Data on the radioactivity changes during various courses of mushroom preservation have been rare. Only limited information has been available for a decrease of ^{137}Cs during mushroom soaking and cooking.

The extraction of radiocesium from intact fruit bodies into soaking water or table salt solution is low due to mushroom lipophilic gel-like surface layer. On the contrary, shrinkage between 36-87% of the initial values was reported from boiled slices of several mushroom species. The decrease was higher from dried or frozen slices with disrupted tissue structure than from fresh ones (Steger et al., 1987; Klán et al., 1988; Neukom and Gisler, 1991; Skibniewska and Smoczyński, 1999; Butkus and Dimaviciene, 2009). Naturally, the extracts must be discarded. Such decontamination removes a lot of compounds affecting attractive flavor.

8. Radioactivity Burden Due to Mushroom Consumption

As results from the previous sections, the calculations of credible radioactivity dose from ingested mushrooms have been rather complicated. Very wide variability of activity levels both among and within species, limited knowledge of radioisotopes bioavailability, likewise fragmentary information on wild mushroom consumption within the population are the main imperfect input data. The reported calculations were carried on for fresh or dried mushrooms and did not take into consideration possible losses of radioisotopes during culinary treatments. Thus, the following values should be appreciated as the qualified estimations.

Using cultivated *Pleurotus eryngii,* Baeza and Guillén (2004) determined very high availability of about 97% of Cs, Sr and Pu, as only about 3% of radioactivity was associated with non-digestible neutral detergent fibre fraction.

As mentioned above, permissible dose limit has been 1,000 μSv $year^{-1}$ for adults.

In Bryansk region, the most contaminated area of Russia about 200 km apart of Chernobyl, dose of radiocesium ingested by the rural population from mushrooms was determined to be about 500 μSv per year during 1994-1995 with peaks in fall. Mean annual consumption 5.5 and 9.3 kg of wild mushrooms was used for females and males, respectively (Skuterud et al., 1997). Mushrooms could contribute up to 60-70% of dietary ^{137}Cs intake of adults during the period 1986-1994 in the region (Shutov et al., 1996).

For mushrooms collected in Poland between 1996 and 1998, the annual incurred effective dose was 80 and 10 μSv for ^{137}Cs and ^{210}Pb, respectively. These values were calculated for annual consumption of 10 kg of fresh mushrooms with mean activity concentrations of 5,800 and 13 Bq kg^{-1} for ^{137}Cs and ^{210}Pb, respectively (Malinowska et al., 2006).

The contribution of 200 μSv to the effective dose was calculated for extreme mushroom consumers in former Czechoslovakia by Klαn et al. (1988) for the period immediately after the Chernobyl accident. The input data were 10 kg per year of *X. badius* and *X. chrysenteron* with high mean levels of 10,000 and 2,200 Bq kg^{-1} DM of ^{137}Cs and ^{134}Cs, respectively. Even though the Chernobyl fallout was relatively low on the territory of Czechoslovakia, mean yearly intake of 1,550 Bq from mushrooms was calculated for 1988. Mushrooms thus became the most significant source of internal contamination (Horyna, 1991).

Markedly lower burden has been in countries with very low Chernobyl fallout together with low consumption of wild mushrooms. This was reported from the United Kingdom for the period 1994-1996 (Barnett et al., 1999), Spain (Baeza et al., 2004) or Turkey (Karadeniz and Yaprak, 2010). The annual effective doses were extremely low, ranging from <1 to 2 μSv. Thus, natural radioisotopes have been the main sources of the burden under such conditions.

9. Radiocesium in Meat of Game Eating Mushrooms

Considerable increase of radioactivity in tissues of animals eating mushrooms was reported particularly from Scandinavian and central European countries in years following Chernobyl accident. These regions were stricken by extensive fallout and had a high density of mushrooms. The contamination was observed both in wild animals, such as roe deer (*Capreolus capreolus*), red deer (*Cervus elaphus*), reindeer (*Rangifer tarandus*) and wild boar

(*Sus scrofa*), and in grazing domestic goats and sheep. Commonly, meat of red deer was less contaminated than that of roe deer or wild boar. Radioactivity levels have often and highly surpassed the limit of 600 Bq kg^{-1}.

Time trends of meat radioactivity in following years have been different for ruminants and wild boars. In roe deer and red deer, activity concentrations of ^{137}Cs are rather low, except for the mushroom season in fall, and continually decreasing with time from the Chernobyl fallout. On the contrary, wild boar meat has retained high level of radioactivity, even increasing in some periods (Strebl and Tataruch, 2007). This is due to different habitat and dietary habits. Roe deer lives usually in a relatively small territory and feeds mostly on less contaminated herbal plants, apart from the mushroom season. Wild boar belongs to omnivores accepting a variety of food and is capable to move over long distances (Semizhon et al., 2009).

Time-dependency of ^{137}Cs activity concentrations in roe deer meat has steadily decreased since 1987. It mostly declined below the hygienic limit during the mid-1990′s and fluctuated round about 400 Bq kg^{-1} in the early 2000′s in the monitored polluted Austrian and German forests. The time course is a combination of two factors – an exponential ^{137}Cs decay and a peak in August-December due to mushroom feeding. The fall peak decreases with time (Strebl and Tataruch, 2007; Fielitz et al., 2009). In ten Polish regions, mean levels were considerably lower, 6.3 and 8.7 Bq kg^{-1} in meat of roe deer and red deer, respectively, in 2007 (Rachubik, 2008).

During the yearly monitoring between 1990 and 2003, mean ^{137}Cs concentration in wild boar meat fluctuated round about 2,000 Bq kg^{-1} in an Austrian region (Strebl and Tataruch, 2007). In a very comprehensive set of 2,433 wild boars shot in SW Germany during 2001-2003, 21-26 and 1-9.3% of tested meat samples exceeded the limit 600 Bq kg^{-1} during summer and winter, respectively. An underground mushroom *Elaphomyces granulatus* (common name deer truffles) was found in significantly higher proportions in stomachs with maximum ^{137}Cs contamination than in stomachs with low level of the radioisotope. Mean ^{137}Cs activity concentration of 6,030 Bq kg^{-1} fresh matter was determined in the mushroom (Hohmann and Huckschlag, 2005). In another region of south Germany, very wide range of 5 – 8,266 Bq kg^{-1} of ^{137}Cs was observed in meat of 656 wild boars shot between 1998 and 2008 (Semizhon et al., 2009). Only low activity concentrations of ^{137}Cs, in the order of tens Bq per kg, were reported from Poland (Rachubik, 2008) and the Czech Republic (Dvořák et al., 2010).

In monogastric animals, radiocesium is usually excreted via the kidneys. The rate of radiocesium absorption from the intestines into the blood in domestic pigs reached 90-100%. Retention and excretion have an exponential behavior and biological half-time of the absorbed radiocesium is thus short, about 20-40 days (Stara et al. 1971; Voigt et al., 1989).

The activity of ^{137}Cs in wild boar meat can be reduced to about one quarter of the original level by repeated leaching in table salt solutions (Dvořák et al., 2008).

Thus, the steadily decrease of meat contamination with ^{137}Cs is expected in roe deer and red deer, whereas great variability and in some regions the preservation of recent overrun of the limit of 600 Bq kg^{-1} is supposed for a long time, probably a few decades. While the effective dose from game meat is very low in Poland and the Czech Republic except for hunters (and possibly poachers) and their families, the sale of wild boar meat surpassing 600 Bq kg^{-1}, based on a system of hunters self-control, was banned in Germany.

Acknowledgement

Financial support of the project MSM 6007665806 of the Ministry of Education, Youth and Sports of the Czech Republic was highly acknowledged.

References

Aumann, D.C., Clooth, G., Steffan, B., & Steglich, W. (1989). [Complexation of ^{137}Cs by cap pigment of *Xerocomus badius*.] *Angewandte Chemie, 101*, 495-496 (in German).

Baeza,A., & Guillén, F.J. (2004). Dose due to mushroom ingestion in Spain. *Radiation Protection Dosimetry, 111*, 97-100.

Baeza, A., Hernández, S., Guillén, F.J., Moreno, G., Manjón, J.L., & Pascual, R. (2004). Radiocaesium and natural gamma emitters in mushrooms collected in Spain. *Science of the Total Environment, 318*, 59-71.

Baeza, A., Guillén, J., & Bernedo, J.M. (2005). Soil-fungi transfer coefficients: Importance of the location of mycelium in soil and of the differential availability of radionuclides in soil fractions. *Journal of Environmental Radioactivity, 81*, 89-106.

Baeza, A., & Guillén, J. (2006). Influence of the soil bioavailability of radionuclides on the transfer of uranium and thorium to mushrooms. *Applied Radiation and Isotopes, 64*,1020-1026.

Baeza, A., Guillén, J., Mietelski, J.W., & Gaca, P. (2006a). Soil-to-fungi transfer of ^{90}Sr, $^{239+240}$Pu, and ^{241}Am. *Radiochimica Acta, 94* (2), 75-80.

Baeza, A., Guillén, F.J., Salas, A., & Manjon, J.L. (2006b). Distribution of radionuclides in different parts of a mushroom: Influence of the degree of maturity. *Science of the Total Environment, 359*, 255-266.

Bakken, L.R., & Olsen, R.A. (1990). Accumulation of radiocaesium in fungi. *Canadian Journal of Microbiology, 36*, 704-710.

Barnett, C.L., Beresford, N.A., Self, P.L., Howard, B.J., Frankland, J.C., Fulker, M.J., Dodd, B.A., & Marriott, J.V.R. (1999). Radiocaesium activity concentrations in the fruit-bodies of macrofungi in Great Britain and an assessment of dietary intake habits. *Science of the Total Environment, 231*, 67-83.

Bazała, M.A., Bystrzejewska-Piotrowska, G., Čipáková, A. (2005). Bioaccumulation of ^{137}Cs in wild mushrooms collected in Poland and Slovakia. *Nukleonika, 50*, S15-S18.

Butkus, D. & Dimaviciene, D. (2009). [Investigation of Cs-137 transfer in the system "soil-mushrooms-human".] *Journal of Environmental Engineering and Landscape Management, 17*, 44-50 (in Lithuanian).

Calmon, P., Thiry, Y., Zibold, G., Rantavaara, A., & Fesenko, S. (2009). Transfer parameter values in temperate forest ecosystems: a review. *Journal of Environmental Radioactivity, 100*, 757-766.

Dighton, J., Tugay, T., & Zhdanova, N. (2008). Fungi and ionizing radiation from radionuclides. *FEMS Microbiology Letters, 281*, 109-120.

Duff, M.C., & Ramsey, M.L. (2008). Accumulation of radiocesium by mushrooms in the environment: a literature review. *Journal of Environmental Radioactivity*, *99*, 912-932.

Dvořák, P., Kunová, V., Beňová, K., & Ohera, M. (2006). Radiocesium in mushrooms from selected locations in the Czech Republic and the Slovak Republic. *Radiation and Environmental Biophysics, 45*, 145-151.

Dvořák, P., Kunová, V., Kunová, J. & Beňová K. (2008). Radiocesium activity reduction in boar meat by brining. *Radiation and Environmental Biophysics, 47*, 179-182.

Dvořák, P., Snášel, P., & Beňová, K. (2010). Transfer of radiocesium into wild boar meat. *Acta Veterinaria Brno, 79,* S85-S91.

Eckl, P., Hofmann, W., & Türk, R. (1986). Uptake of natural and man-made radionuclides by lichens and mushrooms. *Radiation and Environmental Biophysics, 25*, 43-54.

FAO/WHO (1991). Food Standards Programme. Levels for radionuclides. Codex Alimentarius, Vol. 1, Section 6.1.

Fielitz, U., Klemt, E., Strebl, F., Tataruch, F., & Zibold, G. (2009). Seasonality of ^{137}Cs in roe deer from Austria and Germany. *Journal of Environmental Radioactivity, 100*, 241-249.

Gilett, A.G., & Crout, N.M.J. (2000). A review of ^{137}Cs transfer to fungi and consequences for modelling environmental transfer. *Journal of Environmental Radioactivity, 48*, 95-121.

Grüter, H. (1964). Eine selektive Anreicherung des Spaltprodukts ^{137}Cs in Pilzen. *Naturwissenschaften, 51(7)*, 161-162 (in German).

Guillén, J., Baeza, A., Ontalba, M.A., & Míguez, M.P. (2009). ^{210}Pb and stable lead content in fungi: transfer from soil. *Science of the Total Environment, 407*, 4320-4326.

Heinrich, G. (1993). Distribution of radiocesium in the different parts of mushrooms. *Journal of Environmental Radioactivity, 18*, 229-245.

Hohmann, U., & Huckschlag, D. (2005). Investigations on the radiocaesium contamination of wild boar (*Sus scrofa*) meat in Rhineland-Palatinate: a stomach content analysis. *European Journal of Wildlife Research, 51*, 263-270.

Horyna, J. (1991). Wild mushrooms – the most significant source of internal contamination. *Isotopenpraxis, 27*, 23-24.

IAEA (1996). International basic safety standards for protection against ionizing radiation and for the safety of radiation sources. Safety Series No. 115, IAEA, Vienna.

Kalač, P. (2001). A review of edible mushroom radioactivity. *Food Chemistry, 75*, 29-35.

Karadeniz, Ö., & Yaprak, G. (2007). Dynamic equilibrium of radiocesium with stable cesium within the soil-mushroom system in Turkish pine forest. *Environmental Pollution, 148*, 316-324.

Karadeniz, Ö., & Yaprak, G. (2010). ^{137}Cs, ^{40}K, alkali-alkaline earth element and heavy metal concentrations in wild mushrooms from Turkey. *Journal of Radioanalytical and Nuclear Chemistry, 285*, 611-619.

Klán, J., Řanda, Z., Benada, J., & Horyna, J. (1988). [Investigation of non-radiactive Rb, Cs, and radiocesium in higher fungi.] *Česká Mykologie, 42*, 158-169 (in Czech).

Krolak, E., Kwapulinski, J., & Fischer, A. (2010). ^{137}Cs and ^{40}K isotopes in forest and wasteland soils in a selected region of eastern Poland 20 years after the Chernobyl accident. *Radiation and Environmental Biophysics, 49*, 229-237.

Malinowska, E., Szefer, P., & Bojanowski, R. (2006). Radionuclides content in *Xerocomus badius* and other commercial mushrooms from several regions of Poland. *Food Chemistry, 97*, 19-24.

Mascanzoni, D. (2009). Long-term transfer of ^{137}Cs from soil to mushrooms in a semi-natural environment. *Journal of Radioanalytical and Nuclear Chemistry, 282*, 427-431.

Mietelski, J.W., Jasińska, M., Kubica, B., Kozak, K., & Macharski, P. (1994). Radioactive contamination of Polish mushrooms. *Science of the Total Environment, 157*, 217-226.

Mietelski, J.W., Baeza, A.S., Guillén, J., Buzinny, M., Tsigankov, N., Gaca, P., Jasińska, M., & Tomankiewicz, E. (2002). Plutonium and other alpha emitters in mushrooms from Poland, Spain and Ukraine. *Applied Radiation and Isotopes, 56*, 717-729.

Mietelski, J.W., Dubchak, S., Błażej, S., Anielska, T., & Turnau, K. (2010). ^{137}Cs and ^{40}K in fruiting bodies of different fungal species collected in a single forest in southern Poland. *Journal of Environmental Radioactivity, 101*, 706-711.

Neukom, H.P., & Gisler, E. (1991). Extraction of radioactive caesium from mushrooms with *Xerocomus badius* as an example. *Lebensmittel-Wissenschaft und Technologie, 24*, 442-444.

Popa, K., Pui, A., Tanase, C., & Irimia, R. (2010). Monitoring of ^{226}Ra and ^{137}Cs radioisotopes on Bistrita Valley and their translocation in spontaneous macromycetes. *Revista de Chimie (Bucharest), 61*, 894-896.

Rachubik, J. (2008). Radiocesium in Polish game meat. *Bulletin of the Veterinary Institute in Pulawy, 52*, 399-403.

Semizhon, T., Putyrskaya, V., Zibold, G., & Klemt, E. (2009). Time-dependency of the ^{137}Cs contamination of wild boar from a region in Southern Germany in the years 1998 to 2008. *Journal of Environmental Radioactivity, 100*, 988-992.

Shutov, V.N., Bruk, G.Y., Basalaeva, L.N., Vasilevitskiy, V.A., Ivanova, N.P., & Kaplun, I.S. (1996). The role of mushrooms and berries in the formation of internal exposure doses to the population of Russia after the Chernobyl accident. *Radiation Protection Dosimetry, 67*, 55-64.

Skibniewska, K.A., & Smoczyński, S.S. (1999). [Influence of cooking on radiocesium contamination of edible mushrooms.] *Roczniki Państwowego Zakladu Higieny, 50 (2)*, 157-162 (in Polish).

Skuterud, L., Travnikova, I.G., Balonov, M.I., Strand, P., & Howard, B.J. (1997). Contribution of fungi to radiocaesium intake by rural population in Russia. *Science of the Total Environment, 193*, 237-242.

Skwarzec, B., & Jakusik, A. (2003): ^{210}Po accumulation by mushrooms from Poland. *Journal of Environmental Monitoring, 5*: 791-794.

Stara, J.F., Nelson, N.S., Della Rosa, R.J., & Bustad, L.K. (1971). Comparative metabolismus of radionuclides in mammals – review. *Health Physics, 20*, 113-137.

Steger, U., Burger, A., Ziegler, W., & Wallnöfer, P.R. (1987). [Distribution of Cs-134 and Cs-137 in foodstuffs processed in the kitchen.] *Deutsche Lebensmittel-Rundschau, 83*(3), 85-88 (in German).

Strebl, F., & Tataruch, F. (2007). Time trends (1986-2003) of radiocesium transfer to roe deer and wild boar in two Austrian forest regions. *Journal of Environmental Radioactivity, 98*, 137-152.

Szántó, Zs., Hult, M., Wätjen, U., & Altzigzoglou, T. (2007). Current radioactivity content of wild edible mushrooms: A candidate for an environmental reference material. *Journal of Radioanalytical and Nuclear Chemistry, 273*, 167-170.

Turhan, S., Koese, A., & Varinlioglu, A. (2007). Radioactivity level in some wild edible mushroom species in Turkey. *Isotopes in Environmental and Health Studies, 43*, 249-256.

Vaaramaa, K., Solatie, D., & Aro, L. (2009). Distribution of ^{210}Pb and ^{210}Po concentrations in wild berries and mushrooms in boreal forest ecosystems. *Science of the Total Environment, 408*, 84-91.

Vinichuk, M.M., & Johanson, K.J. (2003). Accumulation of ^{137}Cs by fungal mycelium in forest ecosystems of Ukraine. *Journal of Environmental Radioactivity, 64*, 27-43.

Vinichuk, M., Taylor, A.F.S., Rosen, K., & Johanson, K.J. (2010). Accumulation of potassium, rubidium and caesium (^{133}Cs and ^{137}Cs) in various fractions of soil and fungi in a Swedish forest. *Science of the Total Environment, 408*, 2543-2548.

Voigt, G., Pröhl, G., Müller, H., Bauer, T., Lindner, J.P., Probstmeier, G., & Röhrmoser, G. (1989). Determination of the transfer of cesium and iodine from feed into domestic animals. *Science of the Total Environment, 85*, 329-338.

Yoshida, S., Muramatsu, Y., Dvornik, A.M., Zhuchenko, T.A., & Linkov, I. (2004). Equilibrium of radiocesium with stable cesium within the biological cycle of contaminated forest ecosystems. *Journal of Environmental Radioactivity, 75*, 301-313.

Zarubina, N.E. (2004). [The content of accidental radionuclides in mushrooms from the 30-km zone of Chernobyl Nuclear Power Station.] *Mikologiya i Fitopatologiya, 38* (3), 36-40 (in Russian).

In: Mushrooms: Types, Properties and Nutrition
Editors: S. Andres and N. Baumann

ISBN 978-1-61470-110-1

Chapter 11

WHITE BUTTON MUSHROOMS, QUALITY EVALUATION AND SHELF LIFE PREDICTION

***Masoud Taghizadeh*[1], *Aoife Gowen*[2] *and Colm O'Donnell*[2]**
[1] Ferdowsi University of Mashhad (FUM)
[2] University College Dublin (UCD)

ABSTRACT

White button mushroom (*Agaricus bisporus*) is the most commonly cultivated mushroom species which comprises about 32% of world mushroom production. It is a good source of proteins, amino acids, carbohydrates, vitamins (mainly vitamin D and vitamin B-complex) and minerals, such as iron, calcium, phosphorous and potassium. *Agaricus bisporus* mushroom is valued for its white appearance and browning of its cap is an indicator of poor quality. Browning and bruising of the mushroom surface lead to reduced shelf-life and lower financial returns to producers, therefore there is a need for objective evaluation of mushroom quality to ensure that only high quality produce reaches the market. Traditional methods have been widely used to evaluate mushroom quality and shelf life based on several quality parameters such as visual appearance, size, colour, maturity stage, development stage, microbial growth and weight loss. However, today's mushroom industry demands non-destructive, rapid and accurate on-line sensor technologies to monitor and control mushroom quality and safety. In this chapter, some traditional quality grading methods as well as application of novel techniques for mushroom quality and shelf life evaluation are discussed.

INTRODUCTION

Mushrooms are the edible fleshy fruiting bodies of all fungi with stalks and caps, which may be gathered wild or grown under cultivation. While there are over 14,000 mushroom species, only about 3,000 are edible, about 700 have known medicinal properties, and less than 1% is recognized as poisonous [1]. The most commonly cultivated mushroom species is *Agaricus bisporus* (also known as white button mushrooms). *Agaricus bisporus* comprises about 32% of world mushroom production [2]. It is a good source of proteins, amino acids, carbohydrates, vitamins (mainly vitamin D and vitamin B-complex) and minerals, such as iron, calcium, phosphorous and potassium [3, 4]. Figure 1 shows the main parts of an *Agaricus bisporus* mushroom.

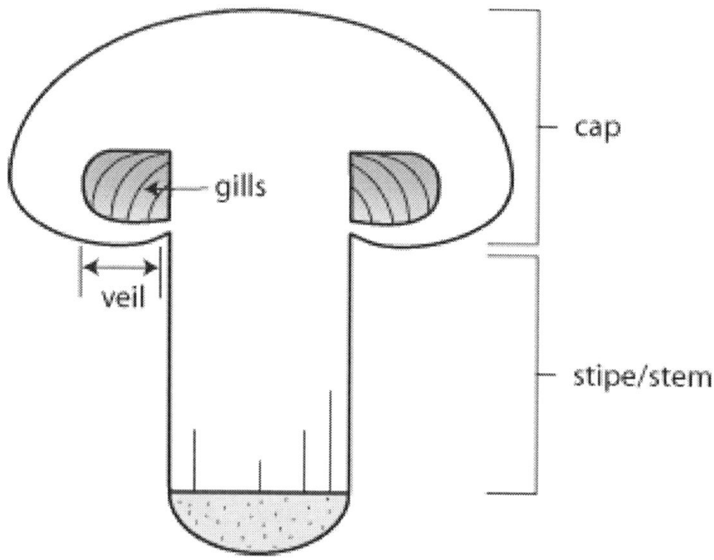

Figure 1. Schematic of *Agaricus bisporus* mushroom.

MUSHROOM SHELF-LIFE AND QUALITY CLASSIFICATION

Cultivated mushrooms (*Agaricus bisporus*) have a short shelf-life of between 3 and 5 days at ambient temperature [5, 6] and from 8 to 10 days under refrigeration conditions [7]. They are highly perishable due to their thin and porous epidermal structure resulting in high respiration rates which induce deterioration after harvest. The short shelf-life of white button mushrooms is an impediment to the distribution and marketing of the fresh produce [8]. Mushrooms' limited shelf-life is mainly due to enzymatic browning during storage [9].

A high quality mushroom is fresh, white, blemish free, with uniform closed caps or buttons and free of compost and casing layer residues [10]. *Agaricus bisporus* mushrooms are valued particularly for their white appearance. Browning of the mushroom cap is an indicator of poor quality [11]. In fact, one of the most important aspects of fresh mushroom quality is whiteness [12]. Quality of cultivated mushrooms is classified as follows:

Table 1. Mushroom (*Agaricus bisporus*) quality classification [13]

Classification	Characteristics in detail
Extra	Superior quality. Well formed. Characteristic of the commercial type in shape, appearance, development and colouring. Only very slight superficial defects allowed if these do not affect appearance, quality or post-harvest life. Practically free of casing material.

Class I	Good quality. Characteristic of the commercial type in shape, appearance, development and colouring. Defects admitted, if they do not affect appearance, quality or post-harvest life: slight defects in shape or colouring slight superficial bruising slight traces of casing material
Class II	Defects admitted if they retain their essential characteristics of quality, the keeping quality and presentation: defects in shape and colouring slight bruising and damage to stalk hollow stalks slight internal moisture of stalks traces of casing material

TRADITIONAL MUSHROOM QUALITY AND SHELF-LIFE EVALUATION METHODS

There are several parameters that determine the quality of mushrooms, such as visual appearance, size, colour, maturity stage, development stage, microbial growth and weight loss [14]. Table 2 demonstrates the most commonly used traditional methods for evaluation of mushroom quality.

Table 2. Traditional methods for quality and shelf-life evaluation of mushrooms

Parameter measured	Description
Colour	L, a and b values, which are associated with lightness, redness and yellowness of mushrooms respectively, are measured
Moisture content	Weight or moisture loss is measured to monitor transpiration rate during the storage period [15]
Maturity index	It is based on the extent of cap opening and is measured to investigate the maturity or development stage of mushroom

The quality of mushrooms declines rapidly after harvest mainly because they have no cuticle to protect them from physical damage, microbial attack or moisture loss [16]. The most important factors affecting mushroom quality and shelf-life are outlined below.

Enzymatic Browning

Mushrooms' high tyrosinase and phenolic content makes them prone to enzymatic browning [17] which is the major cause of quality losses that accounts for reduction in market value [5]. Different factors can induce activation of tyrosinase in mushrooms. Long term or improper storage, mishandling during picking and inappropriate transport may cause discoloration in mushrooms. A succession of biochemical and chemical reactions results in the browning of aged and damaged mushrooms. Therefore, an increase in browning becomes a recognisable signal of aging [18].

Polyphenoloxidase (PPO) is the principal enzyme responsible for browning reactions. This enzyme plays an important role in browning of damaged mushrooms in the presence of oxygen [19]. Therefore, any initial damage can accelerate activation of PPO and the resultant enzymatic reactions.

Physical Damage

Mushrooms are very delicate produce and their quality is highly dependent on manufacturing processes, transport and storage conditions [20]. Improper transport or mishandling during picking can critically affect quality and shelf-life of mushrooms by accelerating physiological and biochemical changes which lead to browning and sliminess [8] in addition to early breaking of the veil, expansion and darkening of the caps and gills and elongation of the stem [21, 22, 23].

More importantly, due to mushrooms' soft percarp, they are prone to tissue damage and bruising which can activate the browning process [24]. In fact, physical and mechanical damage can be regarded as one of the most common cause of defects that can lower the quality of mushrooms after harvest and over the storage period.

Colour

Colour of the mushroom cap is the most important quality factor in *Agaricus bisporus* production. Browning and bruising of the mushroom surface lead to reduced shelf-life and lower financial returns to producers, therefore there is a need for objective evaluation of mushroom quality to ensure that only high quality produce are retailed to the consumer [25]. Conventional mushroom quality grading methods are based on their luminosity or L- value. Gormley and O'Sullivan [12] correlated L- values with sensory analysis in order to develop an objective mushroom grading scale.

Table 4 shows mushroom quality classification according to L-value reported by Gormley and O'Sullivan [12]. Other colour parameters such as a and b values, which are associated with redness and yellowness of mushrooms, have also been measured as quality indicators [16, 26].

Table 3. The most common factors affecting mushroom quality and shelf-life

Factor	Example
Microbial activity	*Pseudomonas* species– virus X
Physical damage	Mishandling during picking- Improper transport, Polyphenol oxidase (PPO)
Moisture loss	High transpiration due to improper storage condition
Colour	Browning due to enzymatic activities
Foreign bodies	Compost particles – pests (Sciarids, Phorids etc.)

Development Stages

Opening of the cap, breaking of the velum, growth and darkening of the gills and stripe elongation occurs in mushrooms after harvest [27]. Guthrie [28] described a maturity index based on the extent of cap opening on a 7 point scale which can be associated with the development stages of mushrooms. It has been reported as one of the indices can be correlated to quality and shelf-life of mushrooms [29, 30].

Due to contact nature of the traditional quality evaluation methods, they are not feasible for on-line use in routine quality measurement. Some of these methods are also time consuming, labour intensive, destructive and sometimes need prolonged sample preparation. Due to these limitations, mushroom processors require the development of new on-line robust and non-destructive evaluation tools for the assessment of shelf-life and quality.

Table 4. L-value classification of mushroom quality [12]

L-value	Classified as:
93 >	Excellent
90 – 93	Very good
86 – 89	Good
80 – 85	Reasonable
80 >	Not acceptable at wholesale level
69 – 79	Poor
69 >	Not acceptable at retail level

NOVEL TECHNIQUES IN MUSHROOM QUALITY AND SHELF LIFE EVALUATION

The food industry demands non-destructive, rapid and accurate on-line sensor technologies to monitor and control product quality and safety. A concept that will meet the aforementioned demands is process analytical technology (PAT) which involves the use of raw material properties, manufacturing parameters, process monitoring, and chemometric techniques to produce finished products of acceptable quality. According to the American Food and Drug Administration (FDA) guideline, PAT can be defined as a system for designing, analysing, and controlling manufacturing through the measurement of critical quality and performance parameters. In the 21st century, the implementation of PAT tools in manufacturing processes will provide opportunities for waste reduction and improved final product quality. Process analytical technology can assist process understanding and facilitate scale-up. PAT in combination with multivariate data analysis must identify the critical factors that should be monitored to guarantee quality, safety and process efficiency in the food industry. There is a need for robust and easy to use analytical tools to be developed and adapted according to the requirements of industrial food processing operations. To date, PAT has not been extensively exploited in the food industry. This is mainly due to a lack of process sensor technologies suitable for food manufacturing.

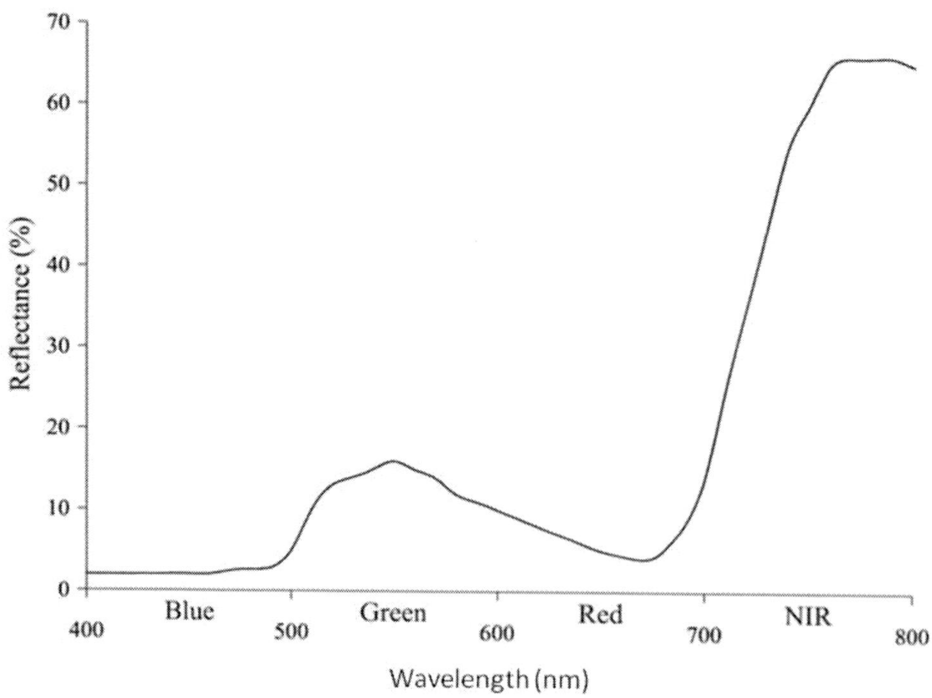

Figure 2. A typical reflectance spectrum in Vis-NIR wavelength range of maize [33].

Application of Vis- NIR for Mushroom Quality Evaluation

Vis-NIR spectroscopy is based on the absorption of electromagnetic radiation in the wavelength range of 400 - 2500 nm. Spectra of food samples comprise broad bands arising from overlapping absorptions corresponding mainly to overtones and combinations of vibrational modes involving C–H, O–H and N–H chemical bonds [31]. The concentrations of constituents such as water, protein, fat and carbohydrate can in principle be determined using classical absorption Vis-NIR spectroscopy. However, for most food samples, this chemical information is obscured by changes in the spectra caused by physical properties such as particle size. Thus, Vis-NIR spectroscopy is a secondary method requiring calibration against a reference method for the constituent of interest [32]. A typical reflectance spectrum is illustrated in Figure 2.

Application of Vis-NIR spectroscopy in mushroom quality control has been reported in several research studies (Table 5). Roy et al. [15] employed Vis-NIR spectroscopy in the wavelength range of 600 to 2200 nm for non-destructive prediction of the moisture content of fresh mushrooms. They concluded that to develop an accurate prediction model using Vis-NIR spectroscopy, evaluation of many spectral and data processing technique, establishment of a library of spectra for the product and expansion of the calibration with selected important new spectra is necessary.

Moreover, the use of NIR spectroscopy to predict yield of fruiting bodies of Agaricus bisporus from compost was evaluated [34]. Rajarathnam et al. [35] studied the biochemical changes associated with mushroom browning in *Agaricus bisporus* using Fourier transform infrared spectroscopy. The resultant spectral showed a characteristic decrease in phenolic hydroxyls and first derivative spectra was used to assess relative peak intensities. Esquerre et al. [40] studied the potential of Vis-NIR spectroscopy for identification of bruise damage on mushroom samples. They suggested the possibility of developing an automated system for grading mushrooms based on reflectance in the visible and near infrared wavelength ranges. It was also found that the most important spectral changes associated with bruise damage, occur in the visible part of the spectrum, indicating that this region would be useful for quality evaluation of mushrooms.

Table 5. Reported application of Vis-NIR spectroscopy for mushroom quality assessment

Parameter	Measurement mode	Wavelength Range	Classification technique	Reference
Moisture content	Reflectance	600 - 2200 nm	PLSR	Roy et al., 1993 [15]
Compost quality	Reflectance	400 - 2498 nm	PCA	Sharma et al., 2000 [34]
Biochemical changes	Transmittance	400 - 500 nm	First derivative	Rajarathnam et al., 2003 [35]
Mushrooms species	Reflectance	700 - 3050 nm	PCA	Liu et al., 2004 [36]

Table 5. (Continued)

Parameter	Measurement mode	Wavelength Range	Classification technique	Reference
Mushroom yield production	Reflectance	400 - 2498 nm	PCA	Sharma et al., 2005 [37]
Different parts of fruiting body	Reflectance	700 - 3050 nm	PCA	Liu et al., 2005 [38]
Casing soil quality	Reflectance	400 - 2498 nm	PLSR	Baxter et al., 2008 [39]
Bruise damage	Reflectance	400 - 2498 nm	PLS	Esquerre et al., 2009 [40]

Application of RGB Imaging for Mushroom Quality Evaluation

RGB imaging systems or red-green-blue (RGB) colour vision systems consist of 4 basic steps, which are capture of a digital image, storage, processing and features extraction. Image processing did not become a practical technique suitable for broad use in industry until the early 1980s. Since this period several advances were made in computer and imaging techniques. Colour image processing was employed in the food industry in the 1990s. Robotic vision became popular in the food industry for determining size, shape, colour and texture. This technique was employed for inspection of raw materials and packing systems and was a significant advancement over manual inspection. In recent years, image processing technology is one of the most attractive on-line techniques for evaluating and quantifying several parameters such as size, shape, structure and colour in different types of food. It is considered a rapid, economical, consistent and non-destructive tool for determining the quality of food products [41].

In Figure 3, pixels which are the smallest addressable element of a digital image have been illustrated.

The image processing analysis for an acquired image basically consists of converting the image to digital form, obtaining an improved image by applying pre-processing operations, segmentation of the image to partition it into non-overlapping regions, measuring the characteristics of objects such as size, shape, colour, etc., and identifying the object by classifying them into different groups. The captured image is subject to various types of noises, which may degrade the image quality [42]. In order to improve the quality of an image, operations need to be performed on it to remove or decrease degradations suffered by the image during its acquisition.

The purpose of pre-processing is to suppress unwilling distortions or enhance some image features that are important for further processing, and creates a more suitable image than the original one for a specific application. Two types of image pre-processing approaches can be identified for food quality evaluation: pixel pre-processing and local pre-processing. Pixel pre-processing is a simple but important image processing technique, which converts an input image into an output image in such a way that each output pixel corresponds directly to the input pixel having the same coordinates. However, local pre-processing methods use a small neighbourhood of a pixel in an input image to produce a new brightness value in the output image, which are also called filtration [43].

Conventional RGB imaging systems are useful for many food sorting and grading operations, but they tend to be poor identifiers of surface features sensitive to wavebands other than RGB, such as low but potentially harmful concentrations of contamination on foods.

RGB imaging systems have been investigated for monitoring quality in mushrooms (Table 6).

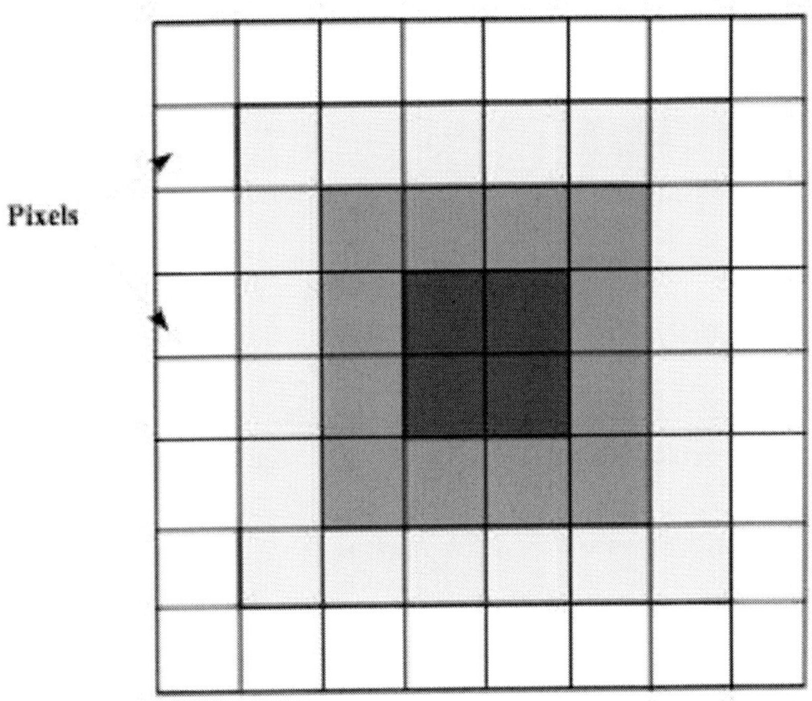

Figure 3. An illustration of pixels on a digital image.

Van de Vooren et al. [45] applied various image analysis techniques to obtain morphological parameters from greyscale images of different mushrooms cultivars, using just 4 parameters which enabled classification of 80% of the cultivars studied. Heinemann et al. [46] investigated the utility of a monochrome camera for mushroom grading (in terms of size, shape, colour, veil opening and stem cut), reporting an average misclassification rate of 20% which compared favourably with the ability of human inspectors. More recently, Vizhanyo and Felfoldi [48] reported a technique to distinguish between diseased mushrooms and those that had experienced natural browning by transforming colour image into CIELAB a* and b* colour axes, with 81% of the diseased region on a test material being correctly classified. The imaging and spectroscopic methods outlined above were shown to perform well for mushroom quality prediction. In addition, Aguirre et al. [51] used greyscale images to examine browning and brown spotting in mushrooms. They suggested that the kinetics of colour degradation and spotting follows a logistic pattern, and the best storage condition to delay the onset of browning and spotting could be found at high relative humidity (< 90%) and temperatures as high as 11 °C.

Table 6. Reported application of RGB imaging for mushroom quality assessment

Application	Reference
Variety testing	Van de Vooren et al., 1991 [44]
Cultivar identification	Van de Vooren et al., 1992 [45]
Grading	Heinemann et al., 1994 [46]
Harvesting	Reed and Tillett, 1995 [47]
Defect detection	Vizhanyo and Felfoldi, 2000 [48]
Grading	Chen and Ting, 2004 [49]
Identification	Yu et al., 2005 [50]
Modelling browning in mushroom	Aguirre et al., 2009 [51]

Application of Hyperspectral Imaging (HSI) for Mushroom Quality Evaluation

In recent years, HSI has emerged as a powerful PAT tool. HSI is a novel technique that combines imaging and spectroscopy to acquire both spatial and spectral information from an object. It enables images at a large number of wavebands (typically >100) with greater resolution to be examined. In this way, HSI combines the advantages of imaging and spectroscopy. Wavelength ranges typically employed in HSI for food control range from the visible through to near infrared regions (~ 400 - 2500 nm).

HSI involves the acquisition of images across a large, usually contiguous series of narrow spectral bands comparable to single-point spectroscopic techniques, thus integrating the complete spatial and chemical information of the sample. The acquired data can be considered as an image cube which is called a *hypercube* which has two spatial and one spectral dimensions (Figure 5.). The *hypercube* can be treated as a series of spatially resolved spectra (pixels) or as a series of spectrally resolved images (Neil et al., 2004). HSI measurement systems typically comprise a light source, lens, spectrograph, detector, and data acquisition system (see Figure 5).

There are two main methods for *hypercube* acquisition. The first method is known as the "staring imager" configuration, in which sample and camera are kept stationary and single images are recorded for each wavelength. Tuneable filters are usually applied for this configuration, which is typically used in pharmaceutical analysis. Images recorded at different wavelengths are combined and the spectra are then calculated. The second approach, which is applicable to conveyor belt survey, is known as the "push-broom" scanning configuration. This method requires a relative movement between the camera and sample to scan over the surface. In this arrangement, holographic prisms are applied. The imaging system records the spatial information line wise and provides spectral information for each pixel along the line by projection along the second axis of the two-dimensional camera chip.

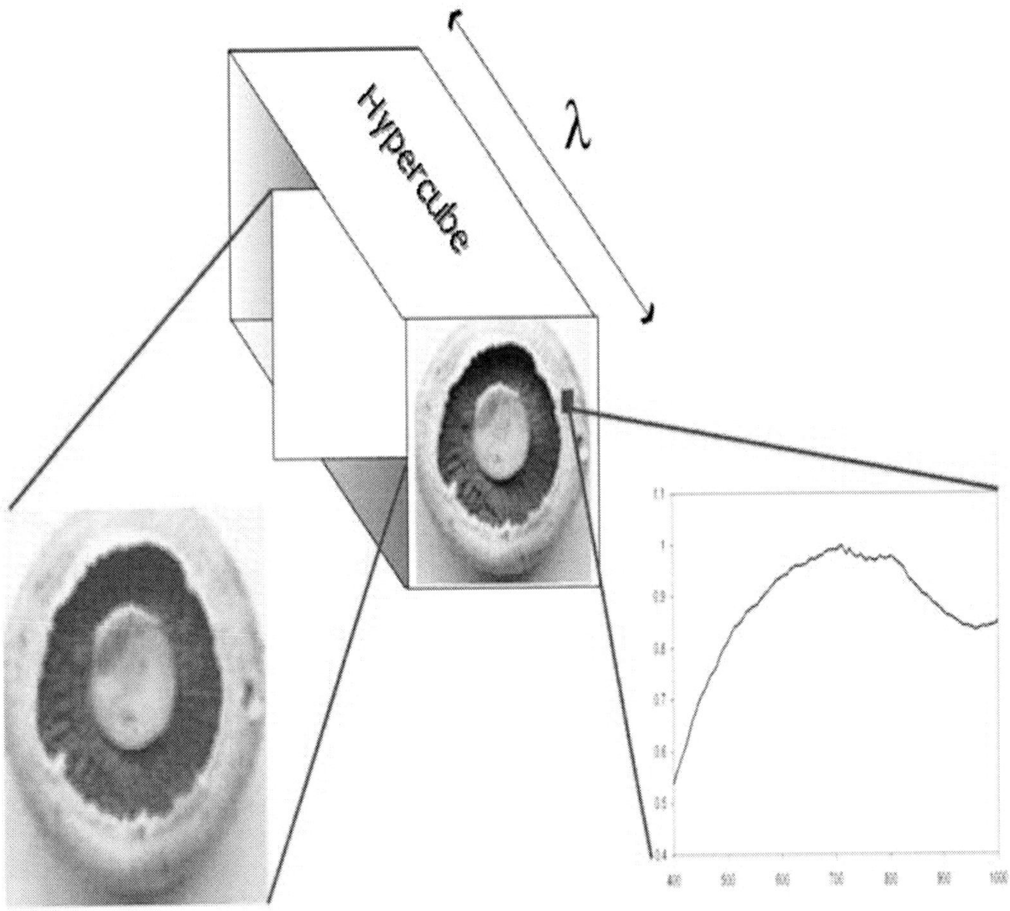

Figure 5. A *hypercube* which contains 2 spatial and 1 spectral dimension.

In an HSI system, image contrast is based on chemical differences between the various components of the sample. After the desired chemical contrast is derived, images can be further processed and converted to values that may be used to reliably classify the samples. Chemometric tools, including PCA and PLS are regularly employed to process the data and transform qualitative information contained within an image into quantitative metrics. The true value of HSI is exploited with the use of appropriate statistical and image analysis techniques.

With the aim of decreasing spectral variability introduced by sample morphology, which is one of the most important challenges in HSI data analysis, it is desirable to apply spectral or spatial pre-processing to the HSI data. Pixel spectra obtained from each region is subjected to chemometric pre-treatments such as multiplicative scatter correction (MSC) and standard normal variation (SNV) pre-processing [52]. Application of HSI in food quality and safety control has been reported by several researchers. Table 7 lists a summary of selected key published works in which use of hyperspectral imaging in food quality control have been reported.

O'Donnell et al. (2007 to 2010) were first to investigate HSI for estimation of different parameters affecting quality and shelf-life of white button mushrooms (Table 8). Gowen et al. [68] used an HSI system operating in the wavelength range of 400 - 1000 nm to detect bruise damage on white button mushrooms. They investigated 2 data reduction methods. In the first method, they applied PCA to the *hypercube* of each sample and the second PC (PC 2) was used for identification of bruise damages on mushrooms' surfaces. In the second method, a data set comprising of average spectra from bruise damaged and undamaged regions were obtained and PCA was applied to this data set. The undamaged and bruise damaged regions were separable along the resultant first PC (PC 1) axis. Overall, it was found the second method performed better than the first when applied to a set of independent mushroom samples.

Quality deterioration of sliced mushrooms was also investigated by O'Donnell et al. using the same system [69]. Multiple linear regression (MLR) and principal component regression (PCR) models were developed to correlate reflectance and various quality parameters including moisture content, colour and texture. Twenty optimal wavelengths for quality prediction were selected after performing an exhaustive search for the best subsets of predictor variables. PCR applied to the set of optimal wavelengths gave the best performance as compared to MLR and PCR on the entire wavelength range. Application of HSI for early detection of freeze damage in *Agaricus bisporus* mushrooms were also investigated by Gowen et al. [70]. A procedure based on PCA and linear discriminant analysis (LDA) was developed with accuracy of higher than 95 % for classification of freeze-damaged mushrooms after only 45 min thawing at which time freeze-thaw damage was not visibly evident. Water matrix co-ordinated in mushrooms subjected to mechanical vibration was also investigated using an HSI system operating in 1000 – 1700 nm wavelength range [71]. Partial least square regression was used to examine changes in sample spectra arising from perturbation. It was found that mechanical vibration increased the concentration of weakly hydrogen-bonded water and decreased that of strongly hydrogen-bonded water in the mushrooms without causing changes in the bulk moisture content. Gaston et al. [72] studied the ability of HSI to predict the activity of Polyphenol oxidase (PPO) enzyme on mushroom caps. Various modelling approach were investigate and PCR model with MSC pre-treated data were found to be the best. It was found that HSI can be used for rapid identification of mushrooms with a higher likelihood to develop enzymatic browning, hence aiding produce management decision makers in the industry. Gowen et al. [73] studied the effect of polymer packaging films (polyvinyl chloride (PVC) and polyethylene terephthalate (PET)) on spectral and spatial features of HSI data in order to investigate the potential of HSI for quality evaluation of packaged products. They concluded that hyperspectral imaging of food products packaged in polymeric films is feasible, although care should be taken to reduce the variability introduced by the film through the use of appropriate spectral or image processing techniques. It can be seen from Table 8 that high equipment cost has limited the use of HSI in the spectral regions above 1700 nm.

Taghizadeh et al. [74] compared the potential of HSI in predicting quality attributes of mushrooms to the conventional RGB imaging. It was found that an RGB image capture system has a lower ability to monitor the quality of mushrooms than a HSI system. Different model performance indicators demonstrated the reasonably high performance of HSI based models (e.g. R^2 = 0.72) to predict L-value, a common mushroom quality indicator, whereas RGB based models showed very poor performance.

Conclusion

To facilitate industrial adoption of novel technologies especially HSI for on-line mushrooms grading, validation at pilot or commercial-scale is required. While RGB imaging systems are low-cost, easy to use and do not require complicated data analysis; they tend to be poor identifiers of surface features. On the other hand, HSI is a powerful platform technology for food quality control, currently there are two major barriers to its widespread adoption in the food industry. The first is the high capital cost of HSI systems (camera, spectrograph etc.). The second limiting factor arises from the relatively lengthy time necessary for Hypercube image acquisition, processing and classification. In order to facilitate online imaging based mushroom grading on an industrial scale, a multi-spectral approach which is cost effective and rapid would address the main challenges encountered for routine use of HSI in industry. Multispectral usually operates on 10 wavebands or less. However, further work is required to identify the key spectral bands necessary for mushroom quality assessment to facilitate the development of a low-cost high speed imaging system for identification of sub-standard batches of mushrooms. In addition, it can be expected that future developments in optical system components, such as improved cameras and faster hardware will shorten processing and classification time, facilitating increased adoption of HSI monitoring system in the mushroom industry.

Reviewed By

Dr. S.M.A. Razavi (s.razavi@um.ac.ir).

Table 7. Selected key published works on hyperspectral imaging of food products

Product	Parameter	Measurement mode	Wavelength Range	Classification technique	Reference
Laboratory scale					
Tomato	Ripeness	Reflectance	396 - 736 nm	Linear discriminant analysis (LDA)	Polder et al., 2000 [53]
Wheat	Fungal disease	Reflectance	550 - 750 nm	PCA	Muhammad and Larsolle 2003 [54]
Apple	Surface defects	Reflectance	430 - 900 nm	Band difference (BD)	Mehl et al., 2004 [55]
Apple	Bitter pit	Reflectance	954 - 1350 nm	PLS	Nicolai et al., 2006 [56]
Apple	Fecal contamination	Reflectance	450 - 900 nm	Band ratio (BR)	Liu et al., 2007 [57]
Apple	Postharvest quality	Fluorescence	500 - 1040 nm	PC- Artificial neural network (ANN)	Noh and Lu 2007 [58]
Walnut	Walnut shell and pulp	Fluorescence	425 - 775 nm	Vector machine	Jiang et al., 2007 [59]
Codfish	Nematodes	Transmittance	350 - 950 nm	PLS	Heia et al., 2007 [60]
Beef	Tenderness	Reflectance	400 - 1000 nm	PCA	Naganathan et al., 2008 [61]
Apple	Chilling injury	Reflectance	400 - 1000 nm	ANN	ElMasry et al., 2009 [62]
Wheat kernels	Insect damage	Reflectance	1000 - 1600 nm	PCA	Singh et al., 2009 [63]
Citrus	Green fruit	Reflectance	369 - 1042 nm	LDA	Okamoto and Lee 2009 [64]
Pickles	Internal defects and colour surface	Reflectance and Transmittance	400 - 1000 nm	PCA	Ariana and Lu 2010 [65]
Plant scale					
Poultry	Organic residues	Reflectance	425 - 951 nm	BR	Cho et al., 2007 [66]
Poultry	Diseased chicken	Reflectance	385 - 735 nm	Fuzzy logic	Chao et al., 2007 [67]

Table 8. Reported application of HSI for mushroom quality assessment

Parameter	Measurement mode	Wavelength Range	Classification technique	Reference
Bruise damage	Reflectance	400 - 1000 nm	PCA	Gowen et al., 2008a
Quality of sliced mushrooms	Reflectance	400 - 1000 nm	Multiple linear regression (MLR) and principal component regression (PCR)	Gowen et al., 2008b [69]
Freeze damage	Reflectance	400 - 1000 nm	PCA- LDA	Gowen et al., 2009a [70]
Effect of vibration on water matrix	Reflectance	1000 - 1700 nm	PLSR	Gowen et al., 2009b [71]
PPO activity on mushroom cap	Reflectance	400 - 1000 nm	MLR, PCR and PLSR	Gaston et al., 2010 [72]
Influence of packaging film on HSI data	Reflectance and transmission	450 - 950 nm	PCA	Gowen et al., 2010 [73]

REFERENCES

[1] Chang, S.T., Miles, P.G. 2004. *Mushrooms: cultivation, nutritional value, medicinal effect, and environmental impact (*Second edition). Boca raton, Florida, CRC Press.

[2] Chang, S.T. 1999. World production of edible and medicinal mushrooms in 1997 with emphasis on Lentinus edods (Berk). Sing. In China. *International Journal of Medicinal Mushroom* 1, 291–301.

[3] Racz, L., Papp, L., Prokai, B., Kovacz, Z. 1996. Trace element determination in cultivated mushrooms: an investigation of manganese, nickel, and cadmium intake in cultivated mushrooms using ICP atomic emission. *Microchemical Journal* 54, 444–451.

[4] Genccelep, H., Uzun, y., Tuncturk, Y., Demirel, K. 2009. Determination of mineral contents of wild-grown edible mushrooms. *Food Chemistry* 113, 1033–1036.

[5] Mohapatra, D., Frias, J.M., Oliveira, F.A.R., Bira, Z.M., Kerry, J. 2008. Development and validation of a model to predict enzymatic activity during storage of cultivated mushrooms (*Agaricus bisporus* spp.). *Journal of Food Engineering* 86, 39-48.

[6] Martine, B., Gaelle, L. P., Ronan, G. 2000. Post-harvest treatment with citric acid or hydrogen peroxide to extend the shelf-life of fresh sliced mushrooms *LEBENSMITTEL-WISSENSCHAFT UND-TECHNOLOGIE* 33(4), 285–289.

[7] Burton, K.S. 1989. The quality and storage life of *Agaricus bisporus*. Mushroom Science XII. Proceeding of 12th International congress on the Science and Cultivation of edible fungi, Braunschweig, Germany. 683-688.

[8] Wani, A.M., Hussain, P.R., Meena, R.S., Dar, M.A., Mir, M.A. 2009. Effect of gamma irradiation and sulphitation treatments on keeping quality of white button mushroom *Agaricus bisporus* (J. Lge). *International Journal of Food Science and Technology* 44, 967–973.

[9] Eissa, H.A.A. 2007. Effect of chitosan coating on shelf-life and quality of fresh-cut mushroom. *Journal of Food Quality* 30(5), 623-645.

[10] Berendse, H. 1984. Marketing of mushrooms. *Mushroom Journal* 141:301-305

[11] Green, J.M., Grogan, H., Eastwood, D.C., Burton, K.S. 2008. Investigating genetic and environmental control of brown colour development in the cultivated mushroom *Agaricus bisporus* infected with mushroom virus X. *International Society for Mushroom Science* 17, 41.

[12] Gormley, T.R., O'Sullivan, L. 1975. Use of a simple reflectometer to test mushroom quality, *The Mushroom Journal* 34, 344-346.

[13] Freshfel Europe, the European Fresh Produce Association, 2004. Marektting standards – Cultivated mushrooms (*Agaricus spp.*) regulation no. 1863/2004 of 26 October. http://www.freshfel.org

[14] Aguirre, L., Frias, J.M., Ryan, C.B., Grogan, H. 2008. Assessing the effect of product variability on the management of the quality of mushrooms (*Agaricus bisporus*). *Postharvest Biology and Technology* 49, 247-254.

[15] Roy, S., Ramaswamy, C.A., Shenk, J.S., Westerhaus, O.M., Beelman, R.B. 1993. Determination of moisture content of mushrooms by Vis-NIR spectroscopy. *Journal of the Science of Food and Agriculture* 63, 355-360.

[16] Burton, K.S., Noble, R. 1993. The influence of flush number, bruising and storage temperature on mushroom quality. *Postharvest Biology and Technology* 3, 39-47.

[17] Brennan, M., Le Port, G., Gormley, R. 2000. Postharvest treatment with citric acid or hydrogen peroxide to extend the shelf-life of fresh sliced mushrooms. *Lebensmittel-Wissenschaft und-Technologie* 33, 285–289.

[18] Mau, J.L., Beelman, R.B., Ziegler, G.R., Royse, D.J. 1991. Effect of nutrient supplementation on flavour, quality and shelf-life of the cultivated mushroom, *Agaricus bisporus*. *Mycologia* 83(2), 142-149.

[19] Liu, W., Liu, J., Liu, C., Zhong, Y., Liu, W., Wan, J. 2009. Activation and conformational changes of mushroom polyphenol oxidase by high pressure microfluidization treatment. *Innovative Food Science and Emerging Technologies* 10, 142–147.

[20] Gormley, T.R., 1987. Handling, packaging and transportation of fresh mushrooms. In: Proceedings of the 5th National Mushroom Conference Malahide Co., Dublin, Ireland.

[21] Umiecka, L. 1986. Effect of the material quality, treatment, packaging methods and storage conditions on export quality of several mushroom races. *Biul Warzyniczy* 29, 271–292.

[22] Lescano, G. 1994. Extension of mushroom (*Agaricus bisporus*) shelf-life by gamma irradiation. *Postharvest Biology and Technology* 4, 255–260.

[23] Eastwood, D. Burton, K. 2002. Mushrooms – a matter of choice and spoiling oneself. *Microbiology Today* 29, 18–19.

[24] Jolivet, S., Arpin, N., Wicher, H.J., Pellon, G. 1998. *Agaricus bisporus* browning: A review. *Mycological Research* 102(12), 1459–1483.

[25] Gonzalez, F.E., Jimenez, A.S, Pardo, V.T. 2006. Quality and shelf-life of packaged fresh-sliced mushrooms stored at two different temperatures. *Agricultural and Food Science* 5, 414-422.

[26] Taghizadeh, M., Gowen, A., O'Donnell, C.P. 2009. Prediction of white button mushroom (*Agaricus bisporus*) moisture content using hyperspectral imaging. *Sensing and Instrumentation for Food Quality and Safety* 3, 219-226.

[27] Hammond, K.B.W., Nichols, R. 1975. Changes in respiration and soluble carbohydrates during the post-harvest storage of mushrooms (*Agaricus bisporus*). *Journal of the science of Food and Agriculture* 26, 835-842.

[28] Guthrie, B.D. 1984. Studies on the control of bacterial deterioration of fresh, washed mushrooms (*Agaricus bisporus / brunescens*). M.S. thesis, The Pennsylvania State University.

[29] Roy, S., Anantheswaran, R.C., Beelman, R.B. 1995. Fresh mushroom quality as affected by modified atmosphere packaging. *Journal of Food Science* 60(2), 334-340.

[30] Kim, K.M., Ko, J.A., Lee, J.S., Park, H.J., Hanna, M. A. 2006. Effect of modified atmosphere packaging on the shelf-life of coated, whole and sliced mushrooms. *LEBENSMITTEL-WISSENSCHAFT UND-TECHNOLOGIE – Food Science and Technology* 39(4), 365-372.

[31] Blanco, M., Villarroya, I. 2002. NIR spectroscopy: a rapid-response analytical tool. *Trends in Analytical Chemistry* 21(4):240-250.

[32] Rodriguez-Otero, J. L., Hermida, M., Centeno, J. 1997. Analysis of dairy products by near-infrared spectroscopy: A review. *Journal of Agricultural and Food Chemistry* 45 (8), 2815-2819.

[33] Inman, D., Khosla, R., Mayfield, T. 2005. On-the-go active remote sensing for efficient crop nitrogen management. *Sensor Review* 25(3), 209-214.

[34] Sharma, H.S.S., Kilpatrick, M. 2000. Use of near-infrared spectroscopy to predict potential mushroom (*Agaricus bisporus*) yield of phase II compost. *Applied Spectroscopy* 54(1), 44-47.

[35] Rajarathnam, S., Nanjarajurs, M.S., Rashmi, S. 2003. Biochemical changes associated with mushroom browning in *Agaricus bisporus* (Lange) Imbach and *Pleurotus florida* (Block and Tsao): commercial implications. *Journal of the Science of Food and Agriculture* 83(14), 1531-1537.

[36] Liu, G., Liu, J.H., Yang, A.M., Dong, Q., Song, D.S. 2004. Identification of edible mushrooms by Fourier transform infrared spectroscopy. *Spectroscopy and Spectral Analysis* 24(8), 941-945.

[37] Sharma, H.S.S., Kilpatrick, M., Lyong, G. 2005. Prediction of potential mushroom yield by visible and near-infrared spectroscopy using fresh phase II compost. *Applied Spectroscopy* 59(8), 1054-1059.

[38] Liu, G., Liu, J.H., Song, D.S., Dong, Q., Ou, J.M., Yang, A.M. 2005. Fourier transform infrared spectroscopic study of different parts of wild growing edible mushrooms. *Spectroscopy and Spectral Analysis* 25(7), 1053-1056.

[39] Baxter, S.P., Sharma, H.S.S., Kilpatrick, M. 2008. Quality assessment of mushroom casing soil using visible and near infrared spectroscopy. *Journal of Near Infrared Spectroscopy* 16(3), 317-325.

[40] Esquerre, C., Gowen, A.A., O'Donnell, C.P., Downey, G. 2009. Initial studies on the quantitation of bruise damage and freshness in mushrooms using visible-near-infrared spectroscopy. *Journal of Agricultural and Food Chemistry* 57, 1903-1907.

[41] Jelinski, T., Du, C.J., Sun, D.W., Fornal, J. 2007. Inspection of the distribution and amount of ingredients in pasteurized cheese by computer vision. *Journal of Food Engineering* 83, 3-9.

[42] Du, C.J., Sun, D-W. 2004. Recent developments in the applications of image processing techniques for food quality evaluation. *Trends in Food Science & Technology* 15, 230-249.

[43] Du, C.J., Cheng, J.D. 2007. Computer vision. In Process Analytical Technology for the Food Industry. C.P. O'Donnell, C. Fagan, P.J. Cullen (Editors), (Chapter 7) Springer – International Publisher (In press, 2011).

[44] Van de Vooren, J.G., Polder, G., Van der Heijden, G.W.A.M. 1991. Application of image analysis for variety testing of mushroom. *Euphytica* 57(3), 245-250.

[45] Van de Vooren, J.G., Polder, G., Van der Heijden, G.W.A.M. 1992. Identification of mushroom cultivars using image analysis. Transactions of the ASAE 35, 347–350.

[46] Heinemann, P.H., Hughes, R., Morrow, C.T., Sommer, H.J., Beelmam, R.B., Wuest P.J. 1994. Grading of mushrooms using a machine vision system. *Transactions of the ASAE* 37, 1671-1677.

[47] Reed, J.N., Tillett, R.D. 1995. Initial experiments robotic mushroom harvesting. *Mechatronics* 4(3), 265-279.

[48] Vizhanyo, T., Felfoldi, J. 2000. Enhancing colour differences in images of diseased mushrooms. *Computers and Electronics in Agriculture* 26(2), 187–198.

[49] Chen, H.H., Ting, C.H. 2004. The development of a machine vision system for shitake grading. *Journal of Food Quality* 27(5), 352-365.

[50] Yu, G., Zhao, Y., Li, G., Shi, H. 2005. Algorithm for locating individual mushroom description of its contour using machine vision. *Transactions of the Chinese Society of Agriculture Engineering* 21(6), 101-104.

[51] Aguirre, L., Frias, J.M., Ryan, C.B., Grogan, H. 2009. Modelling browning and brown spotting of mushrooms (*Agaricus bisporus*) stored in controlled environmental conditions using image analysis. *Journal of Food Engineering* 91, 280-286.

[52] Burger, J., Geladi, P. 2007. Spectral pre-treatments of hyperspectral near infrared images: analysis of diffuse reflectance scattering. *Journal of Near Infrared Spectroscopy* 15(1), 29-38.

[53] 53. Polder, G., Heijden, G.W.A.M., Young, I.T. 2000. Hyperspectral image analysis for measuring ripeness of tomatoes. ASAE International Meering, Milwaukee, Wisconsin, July 9- 12.

[54] Muhammed, H.H., Larsolle, A. 2003. Feature vector based analysis of hyperspectral crop reflectance data for discrimination and quantification of fungal disease severity in wheat. *Biosystems Engineering* 86(2), 125-134.

[55] Mehl, P.M., Chen, Y.R., Kim, M.S., Chan, D.E. 2004. Development of hyperspectral imaging technique for the detection of apple surface defects and contaminations. *Journal of Food Engineering* 61(1), 67-81.

[56] Nicolai, B., Lotze, E., Peirs, A., Scheerlinck, N., Theron, K. 2006. Non-destructive measurement of bitter pit in apple fruit using NIR hyperspectral imaging. *Postharvest Biology and Technology* 40, 1-6.

[57] Liu, Y., Chen, Y.R., Kim, M.S., Chan, D.E., Lefcourt, A.M. 2007. Development of simple algorithms for the detection of fecal contaminants on apples from visible-near infrared hyperspectral reflectance imaging. *Journal of Food Engineering* 81(2), 412-418.

[58] Noh, H., Lu, R. 2007. Hyperspectral laser-induced fluorescence imaging for assessing apple fruit quality. *Postharvest Biology and Technology* 43(2), 193-201.

[59] Jiang, L., Zhu, B., Rao, X., Berney, G., Tao, Y. 2007. Discrimination of black walnut shell and pulp in hyperspectral fluorescence imagery using Gaussian kernel function approach. *Journal of Food Engineering* 81(1), 108-117.

[60] Heia, K., Sivertsen, A., Stormo, S., Elvevoll, E., Wold, J., Nilsen, H. 2007. Detection of nematodes in Cod (*Gadus morhua*) fillets by imaging spectroscopy. *Journal of Food Science* 72(1), E011-E015.

[61] Naganathan, G.K., Grimes, L.M., Subbiah, J., Calkins, C.R., Samal, A., Meyer, G.E. 2008. Visible/near-infrared hyperspectral imaging for beef tenderness prediction. *Computers and Electronics in Agriculture* 64(2), 225-233.

[62] ElMasry, G., Wang, N., Vigneault, C., Qiao, J., ElSayed, A. 2008. Early detection of apple bruises on different background colour using hyperspectral imaging.

[63] Singh, C.B., Jayas, D.S., Paliwal, J., White, N.D.G. 2009. Detection of insect-damaged wheat kernels using near-infrared hyperspectral imaging. *Journal of Stored Products Research* 45(3), 151-158.

[64] Okamoto, H., Lee, W.S. 2009. Green citrus detection using hyperspectral imaging. *Computers and Electronics in Agriculture* 66(2), 201-208.

[65] Ariana, D., Lu, R. 2010. Evaluation of internal defect and surface color of whole pickles using hyperspectral imaging. *Journal of Food Engineering* 96(4), 583-590.

[66] Cho, B.K., Chen, Y.R., Kim, M.S. 2007. Multispectral detection of organic residues on poultry processing plant equipment based on hyperspectral reflectance imaging technique. *Computers and Electronics in Agriculture* 57(2), 177-189.

[67] Chao K., Yang, C.C., Chen, Y.R., Kim, M.S., Chan, D.E. 2007. Research Note: Hyperspectral-multispectral line-scan imaging system for automated poultry carcasses inspection applications for food safety. *Poultry Science* 86, 2450-2460.

[68] Gowen, A., O'Donnell, C., Taghizadeh, M., Cullen, P.J., Downey, G. 2008. Hyperspectral imaging combined with principal component analysis for surface damage detection on white mushrooms (*Agaricus bisporus*). *Journal of Chemometrics*, 22(3-4), 259-267.

[69] Gowen, A.A., O'Donnell, C.P., Taghizadeh, M., Gaston, E., O'Gorman, A., Cullen, P.J., Frias, J.M., Downey, G. 2008. Hyperspectral imaging for the investigation of quality deterioration in sliced mushrooms (*Agaricus bisporus*) during storage. *Sensing and Instrumentation for Food Quality and Safety* 2(3), 133-143.

[70] Gowen, A., Taghizadeh, M., O'Donnell, C.P. 2009. Identification of mushrooms subjected to freeze damage using hyperspectral imaging. *Journal of Food Engineering* 93, 7–12.

[71] Gowen, A.A., Tsenkova, R., Esquerre, C., Downey, G., O'Donnell, C.P. 2009. Use of near infrared hyperspectral imaging to identify water matrix co- ordinates in mushrooms (*Agaricus bisporus*) subjected to mechanical vibration. *Journal of Near Infrared Spectroscopy* 17(6), 363–371.

[72] Gaston, E., Frias, J.M., Cullen, P.J., O'Donnell, C.P., Gowen, A.A. 2010. Prediction of polyphenol oxidase activity using visible near-infrared hyperspectral imaging on mushroom (*Agaricus bisporus*) caps. *Journal of Agricultural and Food Chemistry* 58(10), 6226-6233.

[73] Gowen, A.A., O'Donnell, C.P., Esquerre, C., Downey, G. 2010. Influence of polymer packaging films on hyperspectral imaging data in the visible-near infrared (450 - 950 nm) wavelength range. *Applied Spectroscopy* 64(3), 304-312.

[74] Taghizadeh, M., Gowen, A., O'Donnell, C.P. 2010. Comparison of hyperspectral imaging with conventional digital imaging for quality evaluation of *Agaricus bisporus* mushrooms. Research Note. *Biosystems Engineering* 108(2), 191-194.

In: Mushrooms: Types, Properties and Nutrition
Editors: S. Andres and N. Baumann
ISBN 978-1-61470-110-1

Chapter 12

BASIDIOMYCETES MUSHROOMS OF BRAZILIAN TROPICAL RAINFORESTS: BIODIVERSITY AND BIOLOGICAL APPLICATIONS

***Luiz H. Rosa*[1], *Susana Johann*[1], *Carlos L. Zani*[2], and *Carlos A. Rosa*[1]**

[1]Departamento de Microbiologia, Instituto de Ciências Biológicas, Universidade Federal de Minas Gerais, Belo Horizonte, MG, Brazil.
[2]Laboratório de Química de Produtos Naturais, Centro de Pesquisas René Rachou, Fundação Oswaldo Cruz, Belo Horizonte, MG, Brazil

SUMMARY

Basidiomycetes mushrooms comprise a diversity of gill fungi that occur in most terrestrial ecosystems worldwide. However, data about their diversity and biological applications in tropical ecosystems remain almost unknown. Some *Basidiomycetes* species from Brazilian tropical rainforests have been described as sources of primary and secondary bioactive compounds with antimicrobial, anti-inflammatory, immunomodulatory, antiparasitic, and antitumoral metabolites and as nutraceutical foods. In addition, some *Basidiomycetes* have demonstrated potential as producers of interesting prototype molecules for the development of drugs useful in medicine. The purpose of the present chapter will be to summarize the current status of the diversity, biotechnological use, and ability to produce different bioactive compounds of *Basidiomycetes* mushrooms from Brazilian tropical rainforests.

1. DIVERSITY OF *BASIDIOMYCETES* LIVING IN THE BRAZILIAN TROPICAL RAINFOREST

Basidiomycetes fungi are defined to include species that are unicellular (yeasts) or, primarily, multi-cellular and capable of producing complex basidioma. Their main characteristic is the production of basidia and basidiospores, which are specialized cells and sexual spores, respectively. They live in all terrestrial ecosystems of the world and decay organic matter. This fungal group includes saprotrophic, mycorrhizal, and pathogenic fungi that occur in various shapes, sizes, and colors (Figure 1). Commonly called mushrooms, bolets, puffballs, earthstars, stinkhorns, bird's nest fungi, jelly fungi, bracket or shelf fungi, the *Basidiomycetes* comprise approximately 30,000 species. These fungi appear during all seasons, but rain stimulates rapid growth when organic matter or their decomposition products are readily available. The *Basidiomycetes*, clade euagarics of Hibbett & Thorn (2001), is the most diverse class in the phylum *Basidiomycota*, with approximately 1,037 genera, 112 families, and 16 orders (Kirk et al., 2008). They are a critical component of the degradation of organic matter, providing habitat and food for many organisms such as bacteria, other fungi, insects, and mammals. According to Hawksworth (2004), there has been some, but not enough, progress in the exploration of fungi in tropical forests. The study of *Basidiomycetes* species richness is relevant for its high biological diversity, their ecological and evolution relationship with other organisms in ecosystems, and biotechnological applications that have become an integral part of human welfare. *Basidiomycetes* are a critical component in different ecosystems because they degrade organic matter and provide habitat and food for many organisms, such as bacteria, other fungi, insects, and mammals.

Figure 1. Different species and morphologies of *Basidiomycetes* fungi found in tropical rainforests in Brazil. (a) Tremellales fungus, (b) Schizophyllum commune, (c) *Leucocoprinus fragilissimus*, (d) *Geastrum* sp., (e) *Ramaria* sp., and (f) *Dictyophora* sp.

Tropical rainforests in South America are characterized by high rainfall and different plant species, which form a rich layer of organic matter over the soil. The organic matter associated with the temperatures range of 21 to 45°C makes this ecosystem amply favorable to the development of fungal communities, especially *Basidiomycetes* taxa. In Brazil, the tropical rainforest occurs in the Amazon region, along the Brazilian coast, and across the country in some preserved natural fragments. Unfortunately, the Brazilian tropical rainforest has been depredated due to human impact, and several species, including *Basidiomycetes* fungi, could be faced with extinction. In addition, the Brazilian tropical rainforest has been considered the second-most threatened tropical forest among 15 regions recognized as hot spots of biodiversity (Myers et al., 2000). The diversity of mycota in these forests is relatively unknown but could potentially supply effective compounds for biotechnological use.

Over the past few decades, few studies have been published on members of *Basidiomycetes* from Brazil. However, in many geographical tropical areas, it is clear that the known number of *Basidiomycetes* species represents only a small fraction of the real number of species. *Basidiomycetes* of Brazilian ecosystems are poorly known, especially those living in the Brazilian tropical rainforest, in which some regions have never been explored. The mycological studies of Brazilian ecosystems began with the contributions of collectors and naturalists from Europe. Various fungal species were deposited in museums and gardens in Europe (mainly in England, France, and Germany). According to Fidalgo (1970), *Pycnoporus sanguineus* (L.: Fr.) Murr. (Figure 2a) was the oldest fungus known to be collected from Brazil and was deposited in the Paris Natural History Museum.

The total number of known species of fungi worldwide is unclear. According to Kirk et al. (2008), there are 99,000 known fungal species in the different ecosystems of the world, and approximately 14,000 taxa (14%) may occur in Brazilian biomes (Lewinsohn and Prado 2006). According to Forzza et al. (2010), in Brazil, the tropical rainforests shelter the 1,664 fungal species that are known, and several species could be endemic. In addition, Forzza et al. (2010) proposed that for all Brazilian ecosystems there 376 genera and 1,730 species (417 are endemics). According to Pegler (1997), the *Basidiomycetes* communities of Brazilian tropical rainforests might be closely compared to those found throughout the Caribbean and regions of tropical America. In addition, terrestrial and forest litter taxa from *Agaricales* and *Aphyllophorales* as well as *Gasteromycetes*, a polyphyletic group, are dominant orders of *Basidiomycetes* fungi in Brazilian tropical rainforest. Some *Basidiomycetes* found in tropical rainforests are rare species (Rosa et al., 2009). Among them is *Agaricus martineziensis* Heinem. (Figure 2b) from Argentina, which was described with the name *Agaricus volvatus*, probably due to the well-developed volva at the base of stipe. Apparently it is a rare species since it was reported once only from the type locality, and afterwards from Mexico (Capelari et al., 2006). Some studies suggest that *Basidiomycetes* fungi represent an important fungal community in tropical rainforests. However, there is a lack of information about their real diversity, ecology, and evolution.

Figure 2. Structures of basidiomes of (a) *Pycnoporus sanguineus* and (b) *Agaricus martineziensis* found in tropical rainforests in Brazil.

2. BIOTECHNOLOGICAL USE OF *BASIDIOMYCETES*

Basidiomycetes fungi are a major source of biologically active natural products that provide a rich variety of active secondary metabolites. These fungi have received special attention from researchers over the last few years due to their potential as primary and secondary bioactive molecules with antimicrobial, cytotoxic, immunomodulatory, and antiparasitic activities. In addition, some *Basidiomycetes* species have been used as source of vitamins, protein, enzymes, and food.

2.1. Bioactive Primary Metabolites Produced by *Basidiomycetes*

The healing properties of mushrooms have been know for thousands of years, with the first recorded report of their medicinal qualities dating from around 300 BC (Brown & Gordon, 2003). The polysaccharides found in *Basidiomycetes* may have important medicinal properties, including acting against cancer, diabetes, arteriosclerosis and, hepatitis (Guterrez et al., 2004). Polysaccharides are polymers of sugars (monosaccharides) joined to each other by glycosidic linkages. These are very complex molecules, as covalent bonds can occur between many pairs of carbon atoms (Daba & Ezeronye, 2003). The polysaccharides of mushrooms occur mostly as glucans (α or β) and heterogalactans (Rosado et al., 2003; Daba & Ezeronye, 2003).

Glucans present in mushrooms are bioactive polysaccharides and are widely studied compounds for their pharmacological properties (Queiroz et al., 2010). In fungi, these polysaccharides are major structural components of the cell wall. The therapeutic properties of glucans appear to depend on their chemical structure and on the spatial conformation of each macromolecule (Brown and Gordon, 2003).

β-Glucans are a heterogeneous group of glucose polymers consisting of a backbone of β(1→3)-linked β-D-glucopyranosyl units with β(1→6) linked side chains varying in distribution and length. These polysaccharides are major structural components of the cell wall in fungi and are also found in plants and some bacteria (Brown & Gordon, 2003). Small differences in the basic structure of β-glucans result in peculiar characteristics with new biotechnological applications (Chakraborty et al., 2004). It is known that the immunomodulatory effects of β-glucans are influenced by their degree of branching, polymer length, and tertiary structure, but there is still no consensus on the basic structural requirements for biological activity (Brown & Gordon, 2003). *In vitro* studies suggest that β-glucans can directly activate leukocytes, stimulating their phagocytic, cytotoxic, and antimicrobial activities, including the production of reactive oxygen and nitrogen intermediates (Brown & Gordon, 2003).

Some works have studied the occurrence of β-glucans in several fruiting bodies of *Basidiomycetes* mushrooms from Brazilian rainforests, including the species *Pleurotus eryngii, P. florida, P. pulmonarius* and *P. ostreatoroseus* (Carbonero et al., 2006; Santo-Neves et al., 2008; Smiderle et al., 2008). β-glucan has also been isolated from the fruiting bodies of *Laetiporus sulphureus* collected from trees grown in Curitiba, State of Parana, Brazil (Alquini et al., 2004). *Pleurotus eryngii, P. ostreatoroseus, P. pulmonaris* and *P. florida* have a branched β-glucan, with a main chain of (1→3)-linked-Glc*p* (glucopyranoside) residues, substituted at O-6 by single-unit β-Glc*p* side chains at, on average, every third or fourth residue of the backbone (Carbonero et al., 2006; Santo-Neves et al., 2008; Smiderle et al., 2008). The fruiting bodies of *L. sulphureus* have a linear β-glucan with a (1→3)-linked main chain, namely, laminaran (Alquini et al., 2004).

The anti-inflammatory and antinociceptive activities of β-glucan from *P. pulmonarius* mushrooms were tested in a rodent model. The acetic acid-induced writhing reaction in mice, a typical model for the assessment of the analgesic or anti-inflammatory properties of new agents, was used. Systemically administered glucan (0.3–3 mg/kg) caused a marked and dose-dependent anti-inflammatory response, demonstrated by the inhibition of leukocyte migration to injured tissues (82 ±6%), at an ID_{50} of 1.19 (0.74–1.92) mg/kg. Animals previously treated with the glucan (3 mg/kg i.p.) showed a reduction of 85 ±5% of writhes after receiving the acetic acid injection. To confirm the analgesic activity of the glucan, the formalin test was performed, and the glucan (3–30 mg/kg, intraperitoneal injection) also caused significant inhibition of both the early (neurogenic pain) and the late (inflammatory pain) phases of formalin-induced licking. These data showed that the glucans had potent anti-inflammatory and analgesic (antinociceptive) activities, which possibly act by the inhibition of pro-inflammatory cytokines (Smiderle et al., 2008).

Queiroz et al. (2010) studied the extract obtained from the mushrooms *Caripia montagnei* (family *Podoscyphaceae*) collected in a private Atlantic forest reserve in the state of Rio Grande do Norte in Northeast Region of Brazil. These authors observed that *C. montagnei* contains polysaccharides with immunomodulatory properties. Chemical analysis, infrared spectroscopy and nuclear magnetic resonance (NMR) showed that the carbohydrate fraction contained β-glucans (63.3 ±4.1%) and proteins (2.2 ±0.3%). To assess the immunomodulatory potential of *C. montagnei* glucans, inflammation was induced using two models: thioglycolate-induced peritonitis and carrageenan-induced plantar edema in Swiss mice and Wistar rats, respectively. Both models showed that *C. montagnei* glucans have anti-inflammatory characteristics at a concentration of 50 mg/kg of body weight. The *C. montagnei* glucans did not exhibit signs of inducing cytotoxicity in the leukocytes obtained from human peripheral blood (Queiroz et al., 2010).

One the most famous *Basidiomycetes* that has been studied for its medicinal properties is the *Agaricus brasiliensis* Wasser *et al.* (*Agaricus blazei* Murrill ss. Heinemann) (Figure 3), which is popularly known as the sun mushroom, god's mushroom, Royal Sun *Agaricus*, Himematsutake, or medicinal mushroom. *Agaricus brasiliensis* occurs naturally in Brazilian tropical rainforests and exhibits different medicinal properties, including antitumoral (Mizuno et al., 1990) and immunostimulatory activities (Ito et al., 1997). The main known bioactive substances of this *Basidiomycetes* fungus are polysaccharides obtained from the basidiomes, and its immunomodulatory activity is attributed mainly to β-glucans. *A. brasiliensis* represents a promising target to obtain bioactive compounds, and it has been studied as source of metabolites with antimutagenic (Souza-Paccola et al., 2004), antidiabetic (Kim et al., 2005), antimicrobial (Bernardshaw et al. 2005) and antioxidant (Soares et al., 2009; Mourão et al., 2011) properties. *Agaricus brasiliensis* is now produced on an industrial scale in some countries such as China, Japan, and Brazil (Kaneno et al., 2004) and has been traditionally used as a health food supplement for preventing cancer, combating physical and emotional stress, stimulating immunity, improving quality of life for diabetics, combating diseases such as osteoporosis, gastric ulcers, and chronic hepatitis, and acting as an effective antioxidant

(Guterrez et al., 2004; Urben, 2007). Glucans, especially (1→6)-β-D-glucan, from the water-soluble fraction of *A. brasiliensis* appear to be the main antitumoral components (Sarcoma 180/mice) (Mizuno et al., 1990). Fujimiya et al. (1998) reported that *A. brasiliensis* fractions containing (1→4)-α-D-glucan and (1→6)-β-D-glucan components show antitumoral activity against fibrosarcoma through the activation of specific tumor-infiltrating lymphocytes. Ellertsen et al. (2006) presumed that the effect of extracts of *A. brasiliensis* on infections and immune-related diseases is to activate certain parts of the immune system. These authors examined the changes of gene expression on a human monocyte cell line (THP-1) caused by one solution of an aqueous extract of *A. brasiliensis* (Gold label, obtained from ACE Co., Ltd, Gifu-ken, Japan) containing 300 μg/ml of β-glucans. Genes related to immune function, particularly proinflammatory genes such as the interleukins IL1B and IL8, were selectively up-regulated, and there was a particular increase in mRNA for the cytokines IL1A, CXCL1 (chemokine), CXCL2 and CXCL3 as well as PTGS2 (cyclooxygenase 2). Therefore, the results offer a demonstration of how *A. brasiliensis* preparations may act on human monocytes.

Figure 3. Basidiomes of medicinal *Basidiomycetes Agaricus brasiliensis* (*Agaricus blazei*), originally described in the Brazilian tropical rainforest.

Other important polysaccharides isolated from mushrooms are heteropolymers such as heterogalactans, which usually have a main chain of (1→6)-linked α-D-galactopyranose that can be substituted at O-2 by L-Fuc*p*, 3-*O*-D-Man*p*-L-Fucp or D-Man*p* groups (Carbonero et al., 2008). This heterogalactans has been isolated from *Pleurotus eryngii, P. ostreatoroseus, P. pulmonaris, P. ostreatus*, and *A. brasiliensis* collected from rainforests in Brazil (Rosado et al., 2003; Carbonero et al., 2008; Smiderle et al., 2008; Komura et al., 2010). Carbonero et al. (2008) described the structural features of an unusual linear, partially 3-*O*-methylated (1→6)-linked α-D-galactan containing Gal and 3-Me-Gal in a 3:1 molar ratio isolated from the fruiting bodies of *P. eryngii* and *P. ostreatoroseus*. Rosado et al. (2003) reported that *P. ostreatoroseus* and *P. ostreatus* 'florida' also have a main chain consisting of (1→6)-linked α-D-galactopyranosyl residues containing 3-*O*-Me-α-D-galactopyranoses, with part of these units being substituted at the position O-2 with β-D-mannopyranose residues. The heteropolysaccharides found were similar, with differences only in the number of methyl groups, which suggests that the amount of methyl groups can vary among species. The presence of partially 3-*O*-methylated mannogalactan appears to be typical of *Pleurotus* species.

Pleurotus pulmonarius, collected at the town of Reserva (State of Paraná, Brazil), presented a mannogalactan with a main chain of (1→6)-linked α-D-galactopyranosyl and 3-O-methyl-α-D-galactopyranosyl units, both of which were partially substituted at O-2 by β-D-mannopyranosyl non-reducing ends. To evaluate the antinociceptive and anti-inflammatory properties of the mannogalactan of *P. pulmonaris*, its effect on the acetic acid-induced writhing reaction in mice was examined. It caused a marked and dose-dependent inhibition of the nociceptive response with an ID_{50} of 16.2 (14.7–17.7) mg/kg and inhibition of 93 ±3% at a dose of 30 mg/kg, while dexamethasone (2 mg/kg, s.c.) and indomethacin (10 mg/kg, i.p.), used as positive controls, inhibited the number of writhes by 30 ±8% and 92 ±5%, respectively. However, mannogalactan (3–30 mg/kg) did not inhibit leukocyte infiltration (total cell migration) induced by acetic acid, nor did it modify either the peritoneal capillary permeability or leukocyte infiltration caused by acetic acid (Smiderle et al., 2008).

The fucogalactans isolated from *A. brasiliensis* have a (1→6)-linked α-D-Gal*p* main chain partially substituted at O-2 by non-reducing end units of α-L-Fuc*p*. In *A. blazei*, fucogalactan has a similar main chain with O-2 substitution, but is partially methylated at HO-3 and has 2.5% non-reducing end units of β-D-Gal. The biological action of the fucogalactans was tested for probable antinociceptive and anti-inflammatory effects in abdominal constriction, peritoneal capillary permeability and leukocyte infiltration caused by an intraperitoneal injection of acetic acid in mice. The fucogalactan from *A. brasiliensis* exhibited 39% antinociceptive inhibition (ID_{50} > 100 mg/kg) and no anti-inflammatory activity. *Agaricus brasiliensis* exhibited an inhibition of 39% at an ID_{50} of 0.33 mg/kg and also inhibited 61% (ID_{50} 5.0 mg/kg) of total cell migration and 32% of peritoneal capillary permeability, which is related to the anti-inflammatory effect. These results show that these heterogalactans have antinociceptive and/or anti-inflammatory activities, which are related to their structures. The influence of their structure on the biological response was found by using similar polymers with small differences in the main chain, such as the presence of O-methyl groups, or in the side chains. As a result, determining the fine structure of the polysaccharides is important to understand the relationship between structure and biological activity (Komura et al., 2010).

2.2. Bioactive Secondary Metabolites Produced by *Basidiomycetes*

The search for new substances of pharmaceutical or agricultural importance can be conducted through screening large collections, employing assays designed to detect compounds of pharmacologically or agriculturally relevant targets. According to Anke (1997), the first researchers that studied the *Basidiomycetes* as source of bioactive secondary metabolites were Anchel, Hervey, Wilkins, and their co-workers in 1941. Furthering these studies, Kavanagh et al. (1950) discovered the diterpene pleuromutilin, the first metabolite of *Pleurotus mutilus* and *P. passeckerianus* that led to the development of a commercial antibiotic, tiamulin, which is used as antimicrobial drug against mycoplasm infections in animals. After this study, several other works characterized different bioactive secondary molecules of *Basidiomycetes* species around the world. However, interest in the metabolites produced by *Basidiomycetes* declined as *Streptomycetes* bacteria were considered to be a more prolific and easier to manipulate source of antibiotics (Anke, 1989). However, with the development of new fermentation and purification technologies, *Basidiomycetes* are again receiving attention as potential sources of new classes of antibiotics (Rosa et al., 2003). The use of the diversity of wild *Basidiomycetes* living in Brazilian tropical rainforests as a source of bioactive secondary metabolites is recent. The first studies to utilize the capability of wild *Basidiomycetes* of Brazil to produce bioactive secondary metabolites focused on the antimicrobial (Rosa et al. 2003), antitumoral, immunosuppressive, trypanocidal, and leishmanicidal (Rosa et al. 2009) activities. In these studies, 84 *Basidiomycetes* species were screened, and several species showed at least one biological activity (Table 1).

Oudemansiella canarii collected in the Brazilian tropical rainforest showed a wide antifungal spectrum, inhibiting the growth of the pathogenic yeasts *Candida albicans*, *C. glabrata*, *C. krusei*, and *C. tropicalis*. *Oudemansiella* species are known to produce several bioactive compounds called oudemansins (1-3) and strobilurins (4) (Figure 4), which are able to inhibit fungal growth at very low concentrations. In further studies using bioautographic assays, *Oudemansiella* also exhibited strong inhibition of the phytopathogenic fungus *Cladosporium sphaerospermum* (Rosa et al., 2005). Chromatographic fractionation guided by this assay allowed for the isolation of oudemansin A (Figure 4), an antifungal compound that showed a minimum inhibitory concentration (MIC) of 1.25 μg/spot in a bioautographic assay.

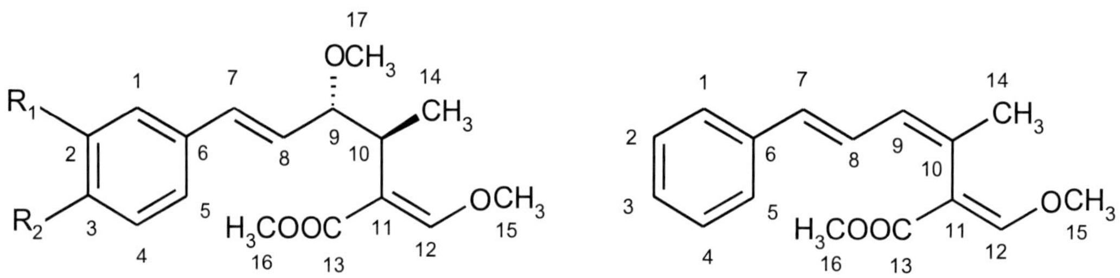

Oudemansin A **(1)** R_1 = H R_2 = H
Oudemansin B **(2)** R_1 = OCH_3 R_2 = Cl
Oudemansin X **(3)** R1 = H R2 = OC_3H

Strobilurin A **(4)**

Figure 4. Structures of oudemansin A, oudemansin B, oudemansin +, and strobilurin A.

The culture extract of *Agrocybe perfecta*, an agaricoid fungus of the family *Bolbitiaceae* that was obtained in the Brazilian tropical rainforest, was investigated for its biological activities (Rosa et al., 2006). The extract showed significant activity against the recombinant enzyme trypanothione reductase (TryR) from *Trypanosoma cruzi*, lymphocyte proliferation in human peripheral mononuclear cells (PBMCs), the human cancer cell lines UACC-62 (melanoma), MCF-7 (mammary), and TK-10 (kidney). The chromatographic fractionation of the extract was monitored by the above bioassays. Agrocybin (Figure 5), a known polyacetylene amide, demonstrated an IC_{50} of 2 μM in the TryR assay but killed only 60% of the trypomastigote form of *T. cruzi* in infected murine blood, even at 680 μM. Agrocybin showed strong cytotoxic activity, with GI_{50} values between 9 to 24.5 μM. In addition to a preliminary investigation to determine if the tumoral cells were dying via necrosis or apoptosis, flow cytometry analysis to detect hypodiploid nuclei showed that 10 μg/mL agrocybin reduced the cellular viability and increased the percentage of hypodiploid cells, suggesting that it killed UACC-62 and TK-10 cells through induction of apoptosis. In addition, other *Basidiomycetes* living in Brazilian tropical rainforests, such as the genera *Agaricus, Climacodon, Gloeoporus, Hexagonia, Irpex, Leucoagaricus, Marasmius, Nothopanus, Pycnoporus, Phellinus*, and *Tyromyces*, were able to produce crude extracts with antimicrobial activity. These can therefore represent a source of bioactive secondary metabolites (Rosa et al., 2003).

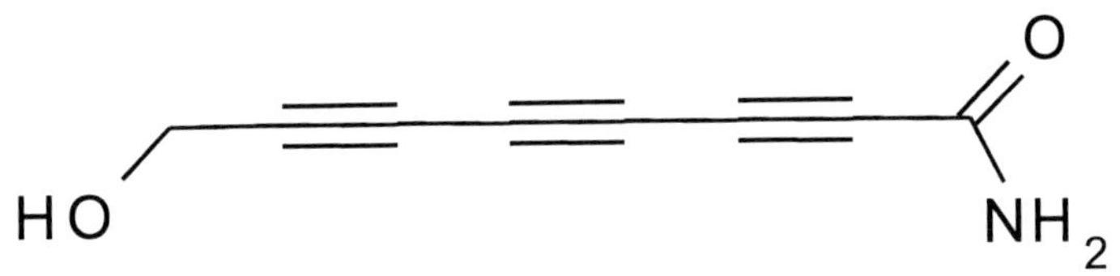

Figure 5. Structure of agrocybin, produced by *Agrocybe perfecta* living in Brazilian tropical rainforests.

New antitumor compounds from *Basidiomycetes* have recently attracted great interest. Crude extracts produced by *Lentinus* species showed cytotoxic activity against human tumor cells and low activity against normal human cells. The genus *Lentinus* Fr. (*Polyporaceae*) (Figure 6) comprises 40 widespread species (Kirk et al., 2008), most of which are saprobes and confined to the tropics. A literature survey showed that only a few *Lentinus* species have been investigated as sources of natural bioactive products. It is not known how many *Lentinus* species occur in Brazilian tropical rainforests.

Figure 6. Basidiomes of *Lentinus strigosus* (Schwein.) Fr. collected in Brazilian tropical rainforests.

Table 1. Biological activities of *Basidiomycetes* fungal taxa collected in Brazilian tropical rainforests

Taxa	Biological activities					
	Antifungal	Antibacterial	Antitumoral	Immunosuppressive	Leishmanicidal	Trypanocidal
Agaricus nigrecentulus Heinem.	-	+	+	-	-	-
Agaricus trinitatensis Baker & Dale	-	-	-	+	-	+
Agrocype *perfecta* (Rick) Sing.	+	-	+	+	-	+
Auricularia fuscosuccinea (Mont.) Farl.	-	-	+	+	-	
Climacodon pulcherrimus (Berk. & Curt.) Nikol.	+	-	-	+	-	+
Coprinus sp.	-	-	-	+	-	+
Cymatoderma dendriticum (Pers.) Reid	-	-	-	-	-	+
Gloeoporus thelephoroides (Hook.) Cunn.	-	+	-	-	-	-
Gloeoporus sp.	-	-	-	+	-	+
Gymnopilus areolatus Murr.	-	-	-	-	+	+
Gymnopilus aureobrunneus (Berk. & Curt.) Murr.	-	-	-	+	-	+
He+agonia hydnoides (Sw.: Fr.) K. Fid.	-	+	-	-	-	-
Irpe+ lacteus (Fr.: Fr.) Cooke	+	+	+	+	+	+
Lentinus bertieri (Fr.) Fr.	-	-	+	+	-	-
Lentinus critinus (L.: Fr.) Fr.	-	-	-	+	-	-
Lentinus strigosus (Schwein.) Fr.	-	-	+	+	+	+
Lentinus villosus Klotzsch	-	-	-	-	-	+
Lentinus zeyheri Berk.	-	-	-	+	-	+
Lepiota sp.	-	-	+	-	-	+
Leucoagaricus cinereus (Que´l.) Bom. & Boiff.	-	+	-	+	-	+
Marasmius bellus Berk.	-	+	-	-	-	
Marasmius cladophyllus Berk.	-	-	+	+	-	+
Marasmius sp.	-	+	+	+	-	+
Merulius corium (Pers.) Fr.	-	-	-	-	-	+
Nothopanus hygrophanus (Mont.) Sing.	-	+	-	-	+	-
Oudemansiella canarii (Jungh.) Hohn	+	+	-	-	-	-

Peniophora cinerea (Fr.) Cook	-	-	-	-	-	+
Phellinus gilvus (Schw.) Pat.	-	-	-	+	-	+
Phellinus sp.	-	+	-	+	-	+
Pleurotus flabellatus (Berk. & Broome) Sacc.	-	-	+	+	+	+
Pluteus cubensis (Murr.) dennis	-	-	-	-	-	+
Pluteus sp.	-	-	+	+	-	+
Pycnoporus sanguineus (L.: Fr.) Murr.	+	+	+	+	-	+
Tyromyces duracinus (Pat.) Murr.	-	+	-	-	-	-
Tyromyces sp.	-	-	-	+	-	+

The crude extract *Lentinus strigosus* (family *Polyporaceae*) found in Brazilian tropical rainforests showed inhibitory activity against *Trypanosoma cruzi* trypanothione reductase (TryR) at 20 μg/mL. This is a key enzyme that helps *Trypanosoma* and *Leishmania* parasites maintain an intracellular reducing environment by reduction of its natural substrate trypanothione disulfide and has been used as a target for the discovery trypanocidal and leishmanicidal drugs. Two panepoxydol derivatives, neopanepoxydol and panepoxydone (Figure 7), were isolated using a bioassay-guided fractionation protocol. Hypnophilin and panepoxydone displayed IC_{50} values of 0.8 and 38.9 μM, respectively, in the trypanothione reductase assay (Cota et al., 2008). The activity of hypnophilin was confirmed in a secondary assay with the intracellular amastigote forms of *T. cruzi*, in which it presented an IC_{50} value of 2.5 μM. Quantitative flow cytometry experiments demonstrated that hypnophilin at 4 μM also reduced the proliferation of human peripheral blood mononuclear cells, without any apparent interference with the viability of lymphocytes and monocytes. The host immune response plays a pivotal role in the adverse events triggered by antigen release during treatment with trypanocidal drugs. The panepoxydone activity was accompanied by a significant inhibitory effect on PHA-induced PBMC proliferation, suggesting a cytotoxic action. Incubation of human PBMCs with panepoxydone reduced the percentage of CD16+ and CD14+ cells and down-regulated CD19+, CD4+ and CD8+ cells, while hypnophilin did not alter any of the phenotypes analyzed. These data indicate that hypnophilin may be considered to be a prototype for the design of drugs for chemotherapy against diseases caused by *Trypanosomatidae* (Souza-Fagundes et al., 2010). These results encourage new studies to search for natural products from *Lentinus* species present in other Brazilian ecosystems.

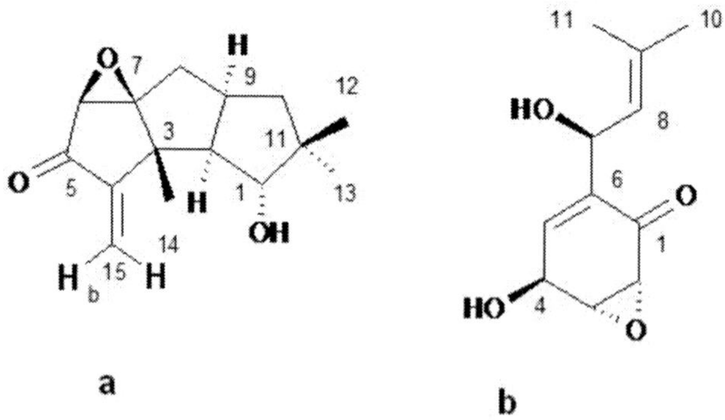

Figure 7. Structure of (a) hypnophilin and (b) panepoxydone isolated from the crude extract of *Lentinus strigosus*.

Basidiomycetes as Source of Nutraceutical Food

Basidiomycetes mushrooms serve as delicacies for human consumption and as nutraceuticals as the "food that also cures". Mushrooms, the fruiting bodies of *Basidiomycetes* fungi, contain substances of various kinds that are highly valued as medicines, flavoring and perfumes (Kues and Liu, 2000). These fungi show a high protein content when compared to beef, containing approximately 14.8% dry weight protein; some *Agaricales* species have 22.5% dry weight protein (Barbisan et al., 2003; Novaes et al., 2007). The fiber has a high molecular weight, which is excreted practically undigested and unabsorbed, and contains chitin, heteropolysaccharides, and glucans (Itoth et al., 1994; Novaes et al., 2007). In addition, some *Basidiomycetes* are sources of the water-soluble vitamins thiamine, riboflavin, niacin, biotin, and ascorbic acid, although the amounts differ between species. They contain large quantities of minerals, especially phosphorus, sodium, calcium, and potassium. Heavy metals like lead, mercury, and copper can also be found in small amounts (Novaes et al., 2007).

Agaricus brasiliensis, known in Brazil as "*cogumelo medicinal*", has been widely cultivated in the country because of its medicinal properties, including immunomodulatory and antitumor activities (Mizuno et al. 1990; Novaes et al., 2007). This mushroom is rich in vitamins, proteins and minerals and is commercialized in several countries as a nutraceutical product, including partially refined extract or dried biomass from the mushroom made into a capsule or tablet (Chang & Buswell 1996). *Agaricus brasiliensis* is harvested in Brazil mostly in the immature stage, when the cap is still closed, to meet exportation standards. Although immature fruiting bodies have not yet achieved their highest biomass at this stage, it is at this point that they reach the highest market value for exportation. Farmers usually discard the mature fruiting bodies (Camelini et al., 2005). The quality of a nutraceutical is dependent on the chemical composition of the fruiting body, particularly in relation to the content of ß-glucans. Camelini et al. (2005) examined the structural evolution of water-soluble polysaccharides, such as the ß-glucans of *A. brasiliensis*, in three different stages of fruiting body maturity and observed that the yield and structural diversity of glucans increased as the fruiting bodies matured. Cap-opened, more fragile mature fruiting bodies of *A. brasiliensis* should be selected over immature ones for the production of nutraceuticals. This strategy will provide the consumer with a higher diversity of glucans, optimizing the bioactivities such as the antitumoral activity and allowing farmers an efficient and profitable use of the mushroom biomass.

Pleurotus spp. is a *Basidiomycetes* distributed worldwide throughout hardwood forests. Some species are found in the Brazilian tropical rainforest. *Pleurotus* spp. are commonly called oyster mushrooms. All known species of *Pleurotus* are edible, and several are commercially cultivated. The popularity of the genus has increased because they are a good source of non-starchy carbohydrates, have a high amount of dietary fiber, and contain moderate quantities of good quality proteins, including most of the essential amino acids, minerals and vitamins (Ragunathan et al., 1996; Santos-Neves et al., 2008). *Pleurotus florida* is an excellent edible and highly nutritious mushroom and is a common species in tropical countries. In addition to having high nutritional value and being easily cultivated, these mushrooms are a good source of compounds, such as polysaccharides, that can act as biological response modifiers (Santos-Neves et al., 2008).

Wild mushrooms that are still not cultivated commercially have great ethnomycological importance because they constitute a much appreciated food (Boa, 2004). The Sanamas group of Yanomami, a indigenous Brazilian group that live in the tropical rainforest of the Amazon, have a long history of mycophilia and use some edible mushrooms, such as *Polyporus tricholoma*, *Polyporus alveolaris, Polyporus aquosus*, *Polyporus indigenus, Lentinus cubensis, L. crinitus, L. strigosus*, *L. velutinus, Pleurotus concavus, Favolus brumeolus, F. tenuiculus*, *Echinochaete brachypora, Coriolopsis daedaleoides*, *Trametes ochraceae*, *Psilocybe plutonia*, *Neoclitocybe byssiseda, Gymnopilus earlei, Leucoprinus cheimonoceps, Collybia subpruinosa*, and *C. pseudocalopus* (Vargas-Isla & Ishikawa, 2008; Cardoso et al., 2010). Logging bans in several countries have renewed interest in non-wood forest products as an alternative source of income and jobs for people previously employed in forestry (Boa, 2004). However, although the potential of the diversity of mushrooms in tropical ecosystems has been discussed extensively, little has been explored scientifically (Vargas-Isla and Ishikawa, 2008).

CONCLUSION

The Brazilian tropical rainforest represents a favorable ecosystem for the development of a rich mycota, in which the *Basidiomycetes* deserve an important role. The study and discovery of novel bioactive metabolites from *Basidiomycetes* mushrooms living in Brazilian tropical rainforests is a challenging endeavor that can bring important rewards when successful. After the screening for biological potential, these tropical *Basidiomycetes* taxa can be submitted to an investigation of their rich arsenal of bioactive secondary metabolites, which can be used as molecular scaffolds for the development of new drugs to control different diseases. The *Basidiomycetes* mushrooms present in tropical rainforests are unexplored as food. Like *Agaricus brasiliensis*, which has been shown to have medicinal and nutritional proprieties, others species are waiting to be discovered and used as nutraceutical food. The diversity, chemical composition, and pharmacological potential of Brazilian *Basidiomycetes* waits to be discovered.

REFERENCES

Anke, T. (1989). Basidiomycetes: a source for new bioactive secondary metabolites. *Progress in the Industrial Microbiology,* 27: 51-66.

Alquini, G.; Carbonero, E.R.; Rosado, F.R.; Cosentino, C.; Iacomini, M. (2004). Polysaccharides from the fruit bodies of the basidiomycete *Laetiporus sulphureus* (Bull.: Fr.) Murr. *FEMS Microbiology Letters*, 230, 47-52.

Barbisan, L.F.; Spinardi-Barbisan, A.L.S.; Moreira, E.L.T.; Salvadori, D.M.F.; Ribeiro, L.R.; Eira, A.F.; Camargo, J.L.V. (2003) *Agaricus blazei* (Himematsutake) does not alter the development of rat diethylnitrosamine-initiated hepatic preneoplastic foci. *Cancer Science*, 94,188-192.

Bernardshaw, S.; Johnson, E.; Hetland, G. (2005). An extract of the mushroom *Agaricus blazei* Murill administered orally protects against systemic *Streptococcus pneumoniae* infection in mice. *Scandinavian Journal of Immunology*, 62, 393-398.

Boa, E. Wild edible fungi. A global overview of their and importance to people. FAO. Rome; 2004.

Brown, G. & Gordon, S. (2003). Fungal β-glucans and mammalian immunity. *Immunity*, 19, 311–315.

Capelari, M.; Rosa, L.H.; Lachance, M.-A. (2006). Description and affinities of *Agaricus martineziensis*, a rare species. *Fungal Diversity*, 21, 11-18.

Carbonero, E.R.; Gracher, A.H.P.; Komura, D.L.; Marcon , R.; Freitas, C.S.; Baggio, C.H.; Santos, A.R.S.; Torri, G.; Gorin, P.A.J.; Iacomini, M. (2008) *Lentinus edodes* heterogalactan: Antinociceptive and anti-inflammatory effects. *Food Chemistry*, 111, 531–537.

Carbonero, E.R.; Gracher, A.H.P.; Rosa, M.C.C.; Torri, G.; Sassaki, G.L.; Gorin, P. A.J.; Iacomini, M. (2008). Unusual partially 3-O-methylated a-galactan from mushrooms of the genus *Pleurotus*. *Phytochemistry*, 69, 252-257.

Carbonero, E.R.; Gracher, A.H.P.; Smiderle, F.R.; Rosado, F.R.; Sassaki, G.L.; Gorin, P.A.J.; Iacomini, M. A. (2006). ß-glucan from the fruit bodies of edible mushrooms *Pleurotus eryngii* and *Pleurotus ostreatoroseus*. *Carbohydrate Research*, 66, 252-257.

Camelini, C.M.; Maraschin, M.; Mendonça, M.M.; Zucco C.; Ferreira, A. G.; Tavares, L.A. (2005). Structural characterization of ß-glucans of *Agaricus brasiliensis* in different stages of fruiting body maturity and their use in nutraceutical products. *Biotechnology Letters*, 27, 1295-1299.

Cardoso, D.B.O.S.; Queiroz, L.P.; Bandeira, F.; Goés-Neto, A. (2010). Correlations between indigenous Brazilian folk classifications fungi and their systematics. *Journal of Ethnobiology*, 30, 255-267.

Chakraborty, I.; Mondal, S.; Pramanik, M.; Rout, D.; Islam. S.S. (2004). Structural investigation of a water-soluble glucan from an edible mushroom *Astraeus hygrometricus*. *Carbohydrate Research*, 339, 2249-2254.

Chang, S.T. & Buswell, J.A. (1996) Mushroom nutriceuticals. *World Journal of Microbiology and Biotechnology*, 12, 473-476.

Cota, B.B.; Rosa, L.H.; Fagundes, E.M.S.; Martins-Filho, O.A., Correa-Oliveira, R.; Romanha, A.J.; Rosa, C.A.; Zani, C.L. (2008). A potent trypanocidal component from the fungus *Lentinus strigosus* inhibits trypanothione reductase and modulates PBMC proliferation. *Memórias do Instituto Oswaldo Cruz*, 103, 263-270.

Daba, A.S. & Ezeronye, O.U. (2003). Anti-cancer effect of polysaccharides isolated from higher basidiomycetes mushrooms. *African Journal of Biotechnology*, 2, 672-678.

Ellertsen, L.K.; Hetland, G.; Johnson, E.; Grinde, B. (2006). Effect of a medicinal extract from *Agaricus blazei* Murill on gene expression in a human monocyte cell line as examined by microarrays and immuno assays. *International Immunopharmacology*, 6, 133-143

Fidalgo, O. (1970). Adições à história da micologia brasileira: I a coleta mais antiga, *Richia*, 5, 1-3.

Forzza, R.C. Catálogo de plantas e fungos do Brasil, volume 1. Rio de Janeiro: Andrea Jakobsson Estúdio: Instituto de Pesquisas Jardim Botânico do Rio de Janeiro, 2010.

Fujimiya, Y.; Suzuki, Y.; Oshiman, K.; Kobori, H.; Moriguchi, K.; Nakashima, H.; Matumoto, Y.; Takahara, S.; Ebina,T.; Katakura, R. (1998). Selective tumoricidal effect of soluble proteoglucan extracted from the basidiomycete, *Agaricus blazei* Murill, mediated via natural killer cell activation and apoptosis. *Cancer Immunology, Immunotherapy*, 46:147-159.

Guterrez, Z.R.; Mantovani, M.S.; Eira, A.F.; Ribeiro, L.R.; Jordão, B.Q. (2004). Variation of the antimutagenicity effects of water extracts of *Agaricus blazei* Murrill *in vitro*. *Toxicol in vitro*, 18, 301-309.

Hawksworth, D.L. (2004). Fungal diversity and its implications for genetic resource collections. *Studies Mycology*, 50, 9-18.

Hibbett, DS; Thorn, RG.: Basidiomycota: Homobasidiomycetes. In: Esser K, Lemke PA. The Mycota: a comprehensive treatise on fungi as experimental systems for basic and applied research. Vol. 7, part B: Systematics and Evolution. Berlin: Springer-Verlag: 2001; 121-168.

Ito, H.; Shimura, K.; Itoh, H.; Kawade, M. (1997). Antitumor effects of a new polysaccharide-protein complex (ATOM) prepared from *Agaricus blazei* (Iwade strain 101) “Himematsutake”; and its mechanisms in tumor-bearing mice. *Anticancer Research*, 17, 277-284.

Itoh, H.; Ito, H.; Amano, H.; Noda, H. (1994). Inhibitory action of a (1-6)-beta-glucan-protein complex (F III-2-b) isolated from *Agaricus blazei* Murill ("himematsutake") on Metha-A fibrosarcoma- bearing mice and its antitumor mechanism. *The Japanese Journal of Pharmacology*, 66, 265-271.

Kaneno, R.; Fontanari, L.M.; Santos, S.A.; Stasic, L.C. D.; Rodrigues-Filho E.; Eira, A.F. (2004). Effects of extracts from Brazilian sun-mushroom (*Agaricus blazei*) on the NK activity and lymphoproliferative responsiveness of Ehrlich tumor-bearing mice. *Food Chemistry Toxicology*, 42, 909-916.

Kavanagh, F.; Hervey, A.; Robbins, W.J. (1950). Antibiotic substances from basidiomycetes. VI. *Agrocybe dura. Procedings os the National Academy of Science of the United States of America*, 36, 102-106.

Kirk, P.M.; Cannon, P.F.; David, J.C.; Stalpers, J.A. Dictionary of the Fungi, 11th ed. Wallingford: CABI Publishing; 2008.

Kim, Y.W.; Kim, K.H.; Choi, H.J.; Lee, D.S. (2005). Anti-diabetic activity of ß-glucans and their enzymatically hydrolyzed oligosaccharides from *Agaricus blazei. Biotechnology Letters*, 27, 483-487.

Komura, D.L.; Carbonero, E.R.; Gracher, A.H.P.; Baggio, C.H.; Freitas, C.S.; Marcon, R.; Santos, A.R.S.; Gorin, P.A.J.; Iacomini, M. (2010). Structure of *Agaricus* spp. fucogalactans and their anti-inflammatory and antinociceptive properties. *Bioresource Technology*, 101, 6192-6199.

Kues, Y. & Liu, U. (2000). Fruiting body production on basidiomycetes. Applied of Microbiology and Biotechnology, 54, 141-152. Barbisan, L.F.; Spinardi-Barbisan, A.L.S.; Moreira, E.L.T.; Salvadori, D.M.F.; Ribeiro, L.R.; Eira, A.F.; Camargo, J.L.V. (2003) *Agaricus blazei* (Himematsutake) does not alter the development of rat diethylnitrosamine-initiated hepatic preneoplastic foci. *Cancer Science*, 94,188-192.

Lewinsohn, T.M.; Prado, P.I.: Síntese do conhecimento atual da biodiversidade brasileira. In: Lewinsohn T.M. (org.). Avaliação do estado do conhecimento da biodiversidade brasileira. *Brasília: Ministério do Meio Ambiente*; 2006; 21-109.

Mizuno, T.; Hagiwara, T.; Nakamura, T.; Ito, H.; Shimura, K.; Sumiya, T.; Asakura, Akihiro. (1990). Antitumor activity and some properties of water-soluble polysaccharides from "Himematsutake," the fruiting body of *Agaricus blazei* Murill. *Agricultural and Biological Chemistry*, 54, 2889-2896.

Mol, P.C. & Wessels, J.G.H. (1990). Differences in wall structure between substrate hyphae and hyphae of fruit-body stipes in *Agaricus bisporus. Mycologycal Research*, 94, 472-479.

Mourão, F.; Umeo, S.H.; Takemura, O.S.; Linde, G.A.; Colauto, N.B. (2011). Antioxidant activity of *Agaricus brasiliensis* basidiocarps on different maturation phases. *Brazilian Journal of Microbiology*, 42, 197-202.

Myers, N.; Mittermeier, R.A.; Mittermeier, C.G.; Fonseca, G.A.B.; Kent, J. (2000) Biodiversity hotspots for conservation priorities. *Nature*, 403, 853–858.

Novaes, M.R.C.G.; Novaes, L.C.G.; Taveira, V.C. (2007). Natural products from *Agaricales* medicinal mushrooms: biology, nutritional properties, and pharmacological effects on Cancer. *Revista Brasileira de Cancerologia*, 53, 411-420.

Pegler, DN. (1997). The agarics of São Paulo, Brazil: an account of the agaricoid fungi (Holobasidiomycetes) of São Paulo State, Brazil. England: Royal Botanic Gardens, Kew; 1997.

Queiroz, L.S.; Nascimento, M.S.; Cruz, A.K.M.; Castro, A.J.G.; Moura, M.F.V.; Baseia, I.G.; Araújo, R.M.; Benevides, N.M.B.; Lima, L.F.A.; Leite, E.L. (2010). Glucans from the *Caripia montagnei* mushroom present anti-inflammatory activity. *International Immunopharmacology*, 10, 34-42.

Ragunathan, R.; Gurusamy, R.; Palaniswamy, M.; Swaminathan, K. (1996). Cultivation of *Pleurotus* spp. on various agro-residues *Food Chemistry*, 55, 139-144.

Rosa, L.H. & Capelari, M. (2009). *Agaricales* fungi from Atlantic Rain forest fragments in Minas Gerais, Brazil. *Brazilian Journal of Microbiology*, 40, 846-851.

Rosa, L.H.; Cota, B.B.; Machado, K.M.G.; Rosa, C.A.; Zani, C.L. (2005). Antifungal and other biological activities from *Oudemansiella canarii* (*Basidiomycota*). *World Journal of Microbiology and Biotechnology*, 21, 983-987.

Rosa, L.H.; Machado, K.M.G.; Jacob, C.C.; Capelari, M.; Rosa, C.A.; Zani, C.L. (2003). Screening of Brazilian *Basidiomycetes* for antimicrobial activity. *Memorias do Instituto Oswaldo Cruz*, 98, 967-974.

Rosa, L.H.; Souza-Fagundes, E.M.; Machado, K.M.G.; Alves,T.M.A., Martins-Filho, O.A.; Romanha, A.J.; Oliveira, R.C.; Rosa, C.A.; Zani, C.L. (2006). Cytotoxic, immunosuppressive and trypanocidal activities of agrocybin, a polyacetylene produced by *Agrocybe perfecta* (*Basidiomycota*). *World Journal of Microbiology and Biotechnology*, 22, 539–545.

Rosado, F.R.; Carbonero, E.R.; Claudino, R.F.; Tischer, C.A.; Kemmelmeier, C.; Iacomini, M. (2003). The presence of partially 3-O-methylated mannogalactan from the fruit bodies of edible basidiomycetes *Pleurotus ostreatus* 'Forida' Berk. and *Pleurotus ostreatoroseus* Sing. *FEMS Microbiology Letters*, 221, 119-124.

Santos-Neves, J.C.; Pereira, M.I.; Carbonero, E.R.; Gracher, A.H.P.; Gorin, P.A.J.; Sassaki, G.L.; Iacomini, M. (2008). A gel-forming ß-glucan isolated from the fruit bodies of the edible mushroom *Pleurotus florida*. *Carbohydrate Research*, 343, 1456-1462.

Smiderle, F.R.; Olsen, L.M.; Carbonero, E.R.; Baggio, C.H.; Freitas, C.S.; Marcon, R.; Santos, A.R.S.; Gorin, P.A.J.; Iacomini, M. (2008). Anti-inflammatory and analgesic properties in a rodent model of a (1→3), (1→6)-linked β-glucan isolated from *Pleurotus pulmonarius*. *European Journal Pharmacology*, 597, 86-91.

Smiderle, F.R.; Olsen, L.M.; Carbonero, E.R.; Marcon, R.; Baggio, C.H.; Freitas, C.S.; Santos, A.R.S.; Torri, G.; Gorin, P.A.J.; Iacomini, M. A. (2008). 3-O-methylated mannogalactan from *Pleurotus pulmonarius*: Structure and antinociceptive effect. *Phytochemistry*, 69, 2731-2736

Soares, A.A.; Souza, C.G.M.; Daniel, F.M.; Ferrari, G.P.; Costa, S.M.; Peralta, R. M. (2009). Antioxidant activity and total phenolic content of *Agaricus brasiliensis* (*Agaricus blazei* Murril) in two stages of maturity. *Food Chemistry*, 112, 775-781

Souza-Paccola, E.A.; Bomfeti, C.A.; Fávaro, L.C.L.; Fonseca, I.C.B.; Paccola-Meirelles, L.D. (2004). Antimutagenic action of *Lentinula edodes* and *Agaricus blazei* on *Aspergillus nidulans* conidia. *Brazilian Journal of Microbiology*, 35, 311-315.

Souza-Fagundes, E.M.; Cota, B.B.; Rosa, L.H.; Romanha, A.J., Corrêa-Oliveira, R.; Rosa, C.A.; Zani, C.L.; Teixeira-Carvalho, A.; Martins-Filho, O.A. (2010). *In vitro* activity of hypnophilin from *Lentinus strigosus*: a potential prototype for Chagas disease and leishmaniasis chemotherapy. *Brazilian Journal of Medical and Biological Research*, 43, 1054-1061.

Urben, A.F. (2007). Caracterização morfológica e fisiológica em acessos de *Agaricus blazei* e *A. sylvaticus*. *Biotecnologia Ciência e Desenvolvimento*, 37, 76-79.

Vargas-Isla, R. & Ishikawa, N.K. (2008). Optimal conditions of *in vitro* mycelial growth of *Lentinus strigosus*, and edible mushroom isolated in the Brazilian Amazon. *Mycoscience*, 49, 215-219.

In: Mushrooms: Types, Properties and Nutrition
Editors: S. Andres and N. Baumann

ISBN 978-1-61470-110-1

Chapter 13

CO-CULTIVATION OF *PLEUROTUS* SPECIES WITH YEASTS

D. N. Novoselova and O. V. Kamzolkina*
* Mycology and algology department, Biological faculty,
Lomonosov MSU, Moscow. Russia

ABSTRACT

Fungi of the genus *Pleurotus* are one of the most popular edible fungi in the world. Most of the cultivaion substrates and wood too is very poor in nitrogen, so one of the greatest problem for all wood decay fungi including *Pleurotus* species is to provide themselves nitrogen nutrition. To satisfy their nitrogen requirement *Pleurotus* fungi can parasitize on the bacteria, yeasts, and algae cells in nature and during the cultivation process.

Some of the wide cultivated *Pleurotus* species (*P. citrinopileatus*, *P. djamor*, *P. eryngii*, *P. ostreatus*, and *P. pulmonarius*) were observed in mixed cultures with 8 species (10 strains) of yeasts from different taxonomic groups (ascomycetous, i.e. *Debaryomyces hansenii, Hanseniaspora uvarum*, *Kluyveromyces marxianus*, *Metschnikowia pulcherrima*, *Saccharomyces cerevisiae*, and basidiomycetous, i.e. *Cryptococcus albidus, Cystofilobasidium capitatum, Rhodotorula minuta*) on agar media. To provide co-cultivation on agar media and on substrate were used water suspension of alive or dead yeast cells, or the suspension supernatant.

The morphological and physiological peculiarities of *Pleurotus* mycelium were described on agar medium in mixed cultures with the suspension of alive/dead yeast cells. Some specialized mycelial structures were found only in the mixed cultures with yeasts, and weren't found in the pure culture, e.g. short nipple-like appendages and coralloid hyphae, which provide contacts between mycelium and yeast cells.

The trophic preferendum was defined for each of five *Pleurotus* species, and than *Pleurotus* fungi were cultivated on substrate in association with the yeast species from the range of trophic preferendum. On substrate all five *Pleurotus* species demonstrated the same reaction on the yeast cells presence. Mycelium growth speed was higher on substrate with yeast suspension. Biomass of the fruiting bodies was higher for different

* Email: o-kamzolkina@yandex.ru

species, and the highest increase was shown for *P. eryngii*, twofold in comparison with the biomass of the fruiting bodies which grew on the substrate without yeasts. Analyses of amino acids and volatile compounds of *P. citrinopileatus* fruiting bodies, cultivated with and without yeast *S. cerevisiae,* demonstated high food value and wide spectrum of odor.

INTRODUCTION

Fungi of the genus *Pleurotus* are one of the most popular edible fungi in the world. *Pleurotus* fungi, in particular the Oyster Mushroom, are widely cultivated in many countries. Some *Pleurotus* species can be used not only in food industry, but in pharmaceutics, perfume industry, and medicine because of producing biologically active substances in mycelium and fruiting bodies (Wasser, Weis, 1999; Wang et al., 2005; Hu et al., 2006). *Pleurotus* spp. as well as many other wood decay fungi grew rapidly on different cellulose-containing substrates in the wide range of environmental conditions. *Pleurotus* species are used in the process of biological degrading of agricultural and woodworking industry waste, such as wheat straw, sunflower seed's peel, cotton waste, saw dust etc. (Philippoussis et al., 2000).

Pleurotus fungi mostly inhabit dead broadleaf trees. There is very low level of nitrogen in wood but the C : N ratio in wood is very high, 300 : 1 to 1000 : 1 (Barron, 2003). Accordingly nitrogen is the limiting factor for fungal growth on wood. To satisfy their nitrogen requirement wood decay fungi can use bacteria, yeasts, algae which live on wood (Blanchette, Shaw, 1978; Hutchison, Barron, 1997), and also the basidiospores of other fungi as a nutrient source (Fries, Swedjemark, 1985). This phenomenon (named parasitism) was shown for many wood decay Basidiomycetes including *Pleurotus* fungi (Barron, 2003).

The mechanism of the wood decay fungal parasitism described by Barron (Barron, 1988; Barron, 2003) consists of the following steps: the production of specialized appendages, the penetration the host cells, the digestion of the contents of the host cells. The specialized appendages, also described as nipple-like appendages (Kamzolkina et al., 2006), form on hyphae to provide the contact between mycelium and microorganism's cell and to absorb nutrients from microorganism's cell. To penetrate the microorganism's colony also the coralloid hyphae is formed (Hutchison, Barron, 1996).

Yeasts are wide spread component of dead wood microflora, and they were found in association with lignicolous fungi in all stages of the wood degradation (Blanchette, Shaw, 1978). Yeasts are lack hydrolytic enzymes so they don't play a direct role in wood decay process but they can enhance this process as a nutrient source for lignicolous fungi (Hutchison, Barron, 1996). Many of basidiomycetous fungi were tested in mixed culture with the yeast cells, 100 out of 198 species of fungi demonstrated positive reaction, i.e. they could parasites to yeasts (Hutchison, Barron, 1996). Oyster mushroom could be parasite to bacteria (Barron, 1988) and also to algae and yeasts both (Hutchison, Barron, 1996; Hutchison, Barron, 1997). This research deals with the cytophysiology of the interactions between *Pleurotus* species and yeasts of different taxonomic groups on agar media and during cultivation process.

Material and Methods

Five *Pleurotus* species were observed: *P. citrinopileatus* Singer (Santana 304P), *P. djamor* (Rumph. ex Fr.) Boedijn (Santana RP), *P. eryngii* (DC.) Gillet (Santana CP), *P. ostreatus* (Jacq.) P. Kumm. (HK-35), *P. pulmonarius* (Fr.) Quel. (BF-32). Eight species (10 strains) of yeasts from different taxonomic groups (ascomycetous and basidiomycetous) were used to create the mixed cultures with *Pleurotus* fungi. Ascomycetous species: *Debaryomyces hansenii* (Zopf) Lodder & Kreger-van Rij, *Hanseniaspora uvarum* (Niehaus) Shehata, Mrak & Phaff, *Kluyveromyces marxianus* (E.C. Hansen) Van der Walt, *Metschnikowia pulcherrima* Pitt & M.W. Mill., *Saccharomyces cerevisiae* Meyen ex E.C. Hansen (3785, 3809, Moment). Basidiomycetous species: *Cryptococcus albidus* (Saito) C.E. Skinner, *Cystofilobasidium capitatum* (Fell, I.L. Hunter & Tallman) Oberw. & Bandoni, *Rhodotorula minuta* (Saito) F.C. Harrison.

Cultivation on Agar Media

Mixed cultures were cultivated on agar media in Petri dish. Two media were used: 2,4 % malt extract agar (MEA) and water agar (WA). Fungal mycelium was inoculated in the center of the media in Petri dish and incubated alone for 7–10 days in the dark at 24 °C. When the fungi colonized one-half of the medium, a suspension of yeasts in sterile water was added near the margin of the fungal colony in the 1,5 – 2 cm distance by three drops 50 µm each. Mixed cultures of fungus and yeasts were incubated in the dark at 24 °C and microscopic observations were carried after 3 – 4 days.

Two types of yeasts suspension were used to create mixed cultures: suspension of alive yeast cells and suspension of dead yeast cells. Suspension of alive yeast cells was made by dissolving the yeast cells from the edge of the sowing needle in 10 ml of sterile water. Suspension of dead yeast cells was made by the same method and was boiled for 30 minutes after the dissolution.

Yeast cells vitality was defined by the rhodamine dying method (Ludovico et al., 2001). For *S. cerevisiae* it was shown that dead cells displayed a diffuse fluorescence in the cytoplasm whereas live cells displayed mitochondrial staining together with some slight non-specific fluorescence at the cell envelope (Ludovico et al., 2001). Microscope Axioskop 40 FL with the digital camera AxioCam MRc was used for all microscopic observations.

Cultivation of *Pleurotus* Fruiting Bodies

Mycelium of *Pleurotus* fungi was grown on wheat grain, prepared according to modified Lemke method (Lemke, 1972), and was incubated in the dark at 24 °C. Cultivation was carried out in 450 ml glass jars. Two types of substrate were used for mushroom cultivation: sunflower seed's peel (for *P. citrinopileatus*, *P. djamor*, *P. ostreatus*, *P. pulmonarius*) and sunflower seed's peel with *Umbeliferae* plants straw (5 : 1, for *P. eryngii*). Substrates were prepared by modified Stamets method (Stamets, 1993). After substrate inoculation 10 ml of yeast cells suspension or 10 ml of supernatant of yeast cells suspension were added to the experimental jars, and 10 ml of sterile water (K № 2) or nothing (K № 1) were added to control jars. Supernatants were created by double centrifuging of the *S. cerevisiae* Moment cells suspension: 1) the suspension was centrifuged immediately after the dissolving of the yeast cells in water (Supernatant № 1); 2) the suspension was centrifuged 2 hours later the dissolving of the yeast cells in water (Supernatant № 2). Jars were incubated in the dark at 24 °C and transferred in the climatic room for mushroom cultivation (17 °C – 19 °C, air humidity 85 – 90%) when mycelium overgrew all the substrate.

Biochemical Analysis of *P. Citrinopileatus* Fruiting Bodies

Amino acid composition of *P. citrinopileatus* fruiting bodies, cultivated with and without yeast *S. cerevisiae,* was analyzed with High-Performance Liquid Chromatography (HPLC) method on chromatograph Hitachi L-8800 (Japan). Samples were prepared by acidic proteins hydrolysis method (Tsugita, Scheffler, 2004). Volatile compounds of *P. citrinopileatus* fruiting bodies, cultivated with and without yeast *S. cerevisiae,* were analysed by gas chromatography mass spectrometry (GC-MS) method on chromatomassspectrometr Pegasus 4D LECO. Volatile compounds were extracted with CH_2Cl_2 (Wood et al., 2004). The compounds were identified by comparison of mass spectra in the NIST computerised mass spectral library. All the experiments were carried out three times.

Results and Discussion

I. Cultivation on Agar Media

On agar the mycelium of all five *Pleurotus* species overgrew the yeast microcolonies. There wasn't shown any features of the chemotaxis because there weren't any differences between the growth speed of the mycelium in pure fungal culture and in the mixed culture with the yeasts. Yeast cells penetrated with the fungal hyphae and free in the contact zone both were dead.

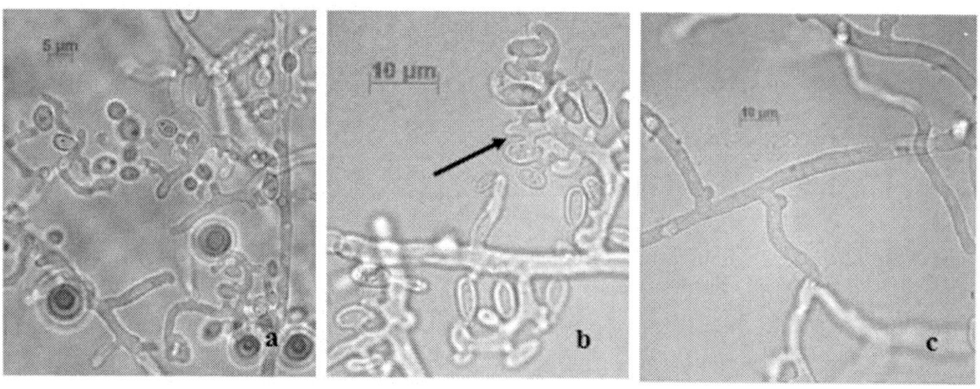

Figure 1. Micromorphology of *P. djamor* mycelium: a, b – mixed culture with yeast; c – pure culture. A) mycelium of *P. djamor* in mixed culture with *M. pulcherrima.* B) coralloid hyphae (arrow) of *P. djamor* in mixed culture with *R. minuta.* C) mycelium of *P. djamor.*

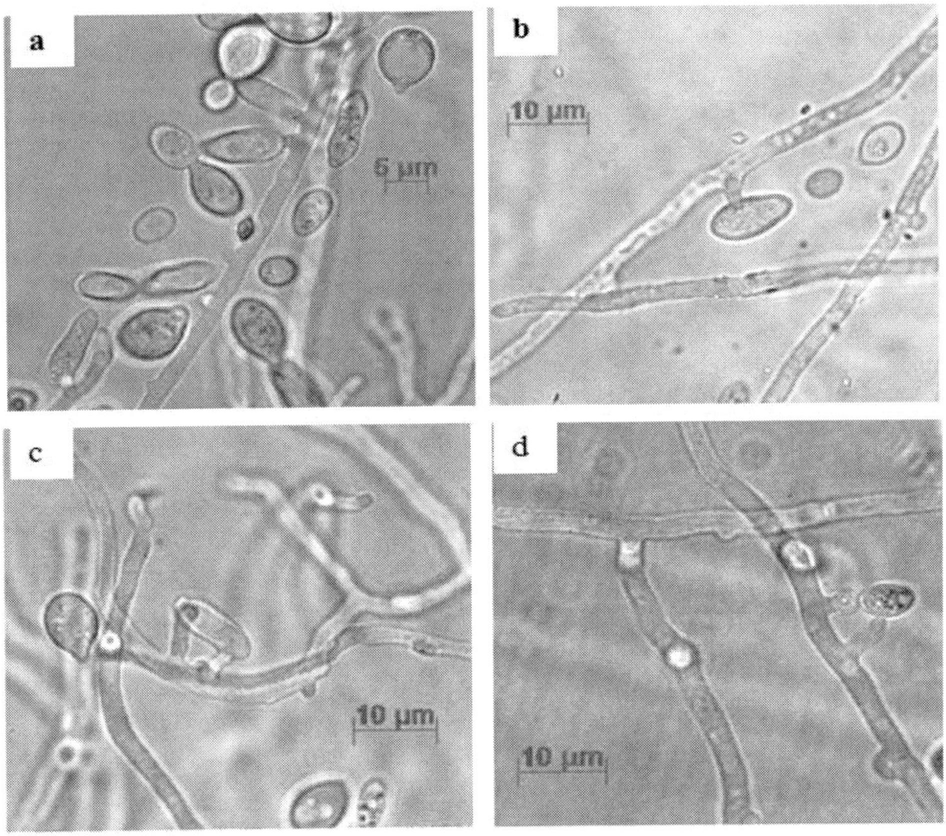

Figure 2. Morphology of the contacts between fungal mycelium and yeast cells: a – mixed culture of *P. citrinopileatus* and *Cr. albidus*, b – mixed culture of *P. eryngii* and *H. uvarum*, c – mixed culture of *P. citrinopileatus* and *Cr. albidus*, d – mixed culture of *P. djamor* and *S. cerevisiae* 3809.

The branching extent of mycelium was higher in the mixed cultures with yeasts, the number of branches in the mixed cultures was 2–8 times higher in 100 μm distance to hypha (Figure 1 a). These changes are supposed to be linked to the requirement of fungus to increase the number of the contacts with yeast cells.

Some specialized mycelial structures were found only in the mixed cultures with yeasts, and weren't found in the pure cultures. Short nipple-like appendages were found in all mixed cultures on water agar. Hyphae formed them to contact with the yeast cells (Figure 1 a, b; 2 a, b). They can be single and in some cultures there were many appendages which form the coralloid hyphae (Figure 1 b). And 2 – 3 appendages might contact with one yeast cell (Figure 2 c, d).

Contacts between hyphae and yeast cells were not found in all mixed cultures. The contacts presence was the main feature which was used to select trophic preferendum for each *Pleurotus* species. Yeast species were called as «trophic preferendum» («+» in table 1) when:

1) there were multiple contacts between the mycelium and yeast cells;
2) branching extent was higher in 2 – 8 times in mixed culture;
3) there were multiple nipple-like appendages;
4) coralloid hyphae formed.

Three species out of eight was defined as trophic preferendum for *P. citrinopileatus* (*Cr. albidus, M. pulcherrima, S. cerevisiae*), six species out of eight was defined as trophic preferendum for *P. djamor* (*Cr. albidus, C. capitatum, K. marxianus, M. pulcherrima, R. minuta, S. cerevisiae*), 1 species out of 8 was defined as trophic preferendum for *P. eryngii* (*S. cerevisiae*), 3 species out of 8 was defined as trophic preferendum for *P. ostreatus* (*H. uvarum, R. minuta, S. cerevisiae*), and 2 species out of 8 was defined as trophic preferendum for *P. pulmonarius* (*H. uvarum, S. cerevisiae*) (table 1). Thus *P. djamor* demonstrated the widest range of trophic preferences, and *P. eryngii* showed the narrowest range. *Saccharomyces cerevisiae* was general for all *Pleurotus* species.

Micromorphology of mycelium differed on various media. Contacts between mycelium and yeast cells were found only on WA. On MEA *P. pulmonarius* hyphae formed few nipple-like appendages, but no coralloid hyphae and contacts between mycelium and yeast cells were found. We suggested that *Pleurotus* mycelium uses the medial sources as nutrition, and only when there are not any accessible nutrient sources in the medium fungus begin to be parasite to yeasts (table 2).

Contacts between hyphae and yeast cells were found only when suspension of alive yeast cells was added to fungal culture on water agar. When we added the suspension of dead (killed by boiling) yeast cells there were few nipple-like appendages (in mixed cultures of *P. citrinopileatus*, *P. djamor*, *P. pulmonarius* with *S. cerevisiae*) or there weren't nipple-like appendages at all (in mixed cultures of *P. eryngii*, *P. ostreatus* with *S. cerevisiae*). No coralloid hyphae and contacts between mycelium and yeast cells were found in this case (table 3).

We supposed that the presence of alive yeast cells by chemical influence on hyphae with yeast metabolites induce the nipple-like appendages formation for mechanical contact between yeast cells and hyphae.

Table 1. Trophic preferendum of five *Pleurotus* species in co-cultivation with yeasts on water agar

Combinations	*P. citrinopileatus*	*P. djamor*	*P. eryngii*	*P. ostreatus*	*P. pulmonarius*
Cr. albidus	+	+	–	–	–
C. capitatum	–	+	–	–	–
D. hansenii	–	–	–	–	–
H. uvarum	–	–	–	+	+
K. marxianus	–	+	–	–	–
M. pulcherrima	+	+	–	–	–
R. minuta	–	+	–	+	–
S. cerevisiae 3785	–	+	–	–	–
S. cerevisiae 3809	+	–	+	+	+
S. cerevisiae Moment	+	+	+	–	+

Table 2. Micromorphological features of *P. pulmonarius* mycelium in pure culture and in mixed culture with *S. cerevisiae* Moment on WA and MEA

Parameters	WA		MEA	
	Pure culture	**Mixed culture with *S. cerevisiae* 3809**	**Pure culture**	**Mixed culture with *S. cerevisiae* 3809**
Distance between the two closest branches on one hypha (μm)	30 ± 6	5 ± 1	50 ± 7	5 ± 1
Nipple-like appendages	–	+	–	+
Coralloid hyphae	–	+	–	–
Contacts between hyphae and yeast cells	–	+	–	–

Table 3. Micromorphological features of *Pleurotus* species mycelium in mixed culture with *S. cerevisiae* Moment on water agar

Combinations *		**Nipple-like appendages**	**Coralloid hyphae**	**Contacts between hyphae and yeast cells**
P. citrinopileatus	Mixed culture with *S. cerevisiae* Moment (alive)	++	+	++
	Mixed culture with *S. cerevisiae* Moment (dead)	+	–	–
P. djamor	Mixed culture with *S. cerevisiae* Moment (alive)	+++	+	+++
	Mixed culture with *S. cerevisiae* Moment (dead)			
P. eryngii	Mixed culture with *S. cerevisiae* Moment (alive)	++	+	+
	Mixed culture with *S. cerevisiae* Moment (dead)	–	–	–
P. ostreatus	Mixed culture with *S. cerevisiae* Moment (alive)	++	–	+
	Mixed culture with *S. cerevisiae* Moment (dead)	–	–	–
P. pulmonarius	Mixed culture with *S. cerevisiae* Moment (alive)	++	+	++
	Mixed culture with *S. cerevisiae* Moment (dead)	+	–	–

* Number of nipple-like appendages and contacts between hyphae and yeast cells in the microscopical field of vision: + single, ++ from one to three, +++ three and more.

II. Cultivation of *Pleurotus* Fruiting Bodies

A. Cultivation of P. Citrinopileatus with Yeast Suspension and Yeast Suspension Supernatant

The highest growth speed was observed in the suspension presence, and the lowest was observed in control samples. *P. citrinopileatus* mycelium overgrew the substrate for 14 days with suspension, for 16 days with supernatant, and for 18 days with sterile water and in control samples (Figure 3).

The fruiting bodies' biomass of *P. citrinopileatus* was different in various samples (table 4). The supernatants № 1–2 addition influenced on the speed of mycelial growth but did not influence on the biomass increase. It is supposed that supernatant consisted of the yeast cell wall components, as water soluble glucanes (Klis et al., 2006), which can be an additional easy accessible nutrient source for mycelium and could speed up it's growth. But for complex effect the alive yeast cells presence is necessary, as they could be a nutrient source for fungus during the growth stage and the fruiting bodies' formation stage, and increased the growth speed and the yield biomass accordingly.

Table 4. Parameters of *P. citrinopileatus* fruiting body formation on sunflower seed's peel in *S. cerevisiae* yeast cells suspension and the suspension supernatant addition

Combinations	Fruiting body formation time (days)	Maximal fruiting body's height (± 5 mm)	Fruiting body's biomass (g)
Control samples	8 – 11	50	7,2 ± 1,0
+ sterile water (10 ml)	8 – 11	50	7,6 ± 0,5
+ *S. cerevisiae* suspension (10 ml)	7	55	11,1 ± 2,0
+ *S. cerevisiae* suspension supernatant № 1 (10 ml)	9 – 11	55	8,5 ± 1,1
+ *S. cerevisiae* suspension supernatant № 2 (10 ml)	11	50	8,0 ± 1,0

B. Co-Cultivation of Pleurotus Species with the Trophic Preferendum Yeasts Suspension

Cultivation of P. citrinopileatus

For *P. citrinopileatus* cultivation *Cr. albidus, M. pulcherrima,* and *S. cerevisiae* 3809, Moment were used. The growth speed of mycelium was higher in the yeast suspension presence (table 5). The mushroom yield increased on 43.6% with *Cr. albidus* suspension addition, and on 39.2 – 44.9% with *S. cerevisiae* 3809, Moment.

Cultivation of *P. djamor*

For P. djamor cultivation Cr. albidus, C. capitatum, K. marxianus, M. pulcherrima, R. minuta, and S. cerevisiae Moment were used. The growth speed of mycelium was higher when the Cr. albidus, K. marxianus, M. pulcherrima, R. minuta, and S. cerevisiae suspension were added (table 6). The mushroom yield increased in the presence of suspension of Cr. albidus (on 24,6%), C. capitatum (on 22,8%), S. cerevisiae Moment (on 49,1%).

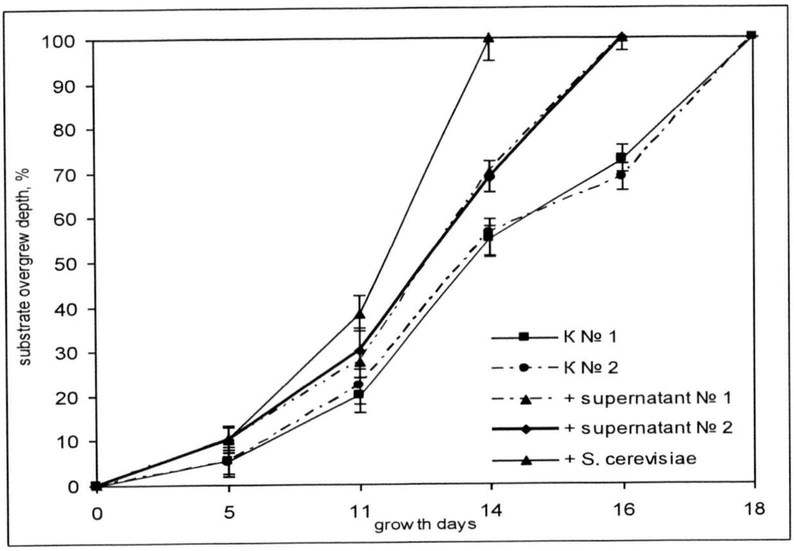

Figure 3. *P. citrinopileatus* mycelium speed of growth on sunflower seed's peel in *S. cerevisiae* yeast cells suspension as well as with supernatant addition.

Table 5. Parameters of *P. citrinopileatus* fruiting body formation on sunflower seed's peel with yeast suspension addition

Combinations	Time of substrate over-growing (days)	Fruiting body formation time (days)	Maximal fruiting body's height (± 5 mm)	Fruiting body's biomass (g)
Control samples	18	6 – 12	75	8,7 ± 1,0
+ *Cr. albidus*	16	6 – 8	80	12,5 ± 2,0
+ *M. pulcherrima*	14	16	65	10,0 ± 0,5
+ *S. cerevisiae* 3809	16	6 – 8	90	15,5 ± 2,0
+ *S. cerevisiae* Moment	16	6	85	14,9 ± 2,5

Table 6. Parameters of *P. djamor* fruiting body formation on sunflower seed's peel with yeast suspension addition

Combinations	Time of substrate over-growing (days)	Fruiting body formation time (days)	Maximal fruiting body's height (± 5 mm)	Fruiting body's biomass (g)
Control samples	18	9	50	5,7 ± 0,9
+ *Cr. albidus*	16	9	70	7,1 ± 0,2
+ *C. capitatum*	18	9	50	7,0 ± 0,1
+ *K. marxianus*	16	7 – 9	65	7,0 ± 0,9
+ *M. pulcherrima*	14	9	45	6,2 ± 0,3
+ *R. minuta*	16	10	60	6,0 ± 0,3
+ *S. cerevisiae* Moment	16	9	65	8,5 ± 1,0

Pleurotus fruiting bodies were the same in control samples and in yeast suspension presence (Figure 4).

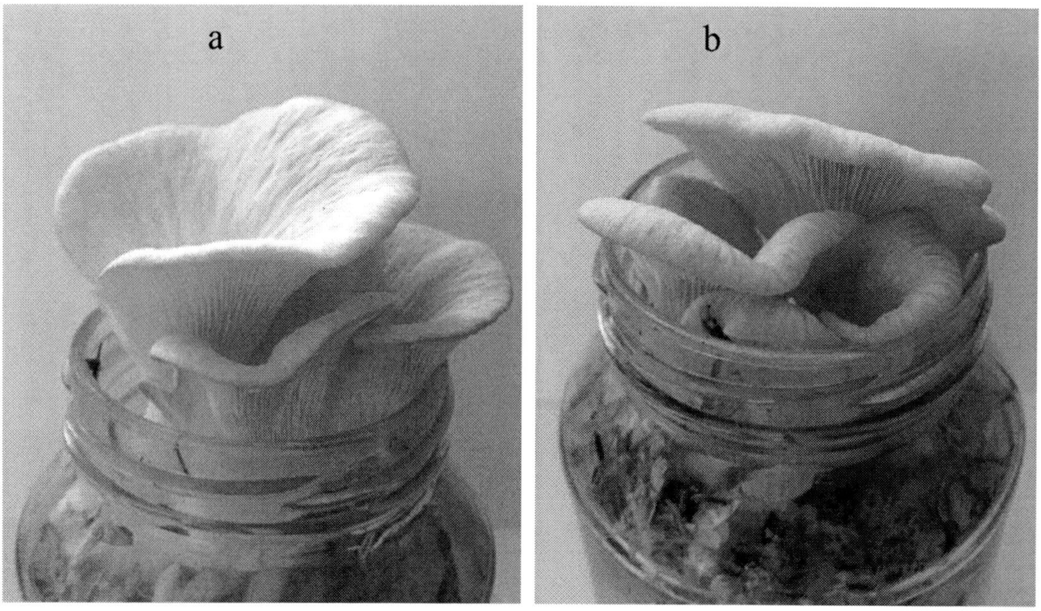

Figure 4. *P. djamor* fruiting bodies on sunflower seed's peel: a – control samples; b – in addition with *S. cerevisiae* Moment yeast cells suspension.

Cultivation of *P. eryngii*

For *P. eryngii* co-cultivation *S. cerevisiae* Moment were used. The growth speed of mycelium was higher in the *S. cerevisiae* suspension presence (table 7) on 2 days. The mushroom yield increased in 2,1 times in the yeast suspension presence.

Cultivation of *P. ostreatus*

For *P. ostreatus* cultivation *H. uvarum, R. minuta,* and *S. cerevisiae* 3809 were used. The growth speed of mycelium was higher in the yeast suspension presence on 4 days (table 8). The mushroom yield increased in the presence of suspension of *H. uvarum* (on 75,7%) and *S. cerevisiae* (on 52,8%).

Cultivation of P. pulmonarius

For *P. pulmonarius* cultivation *H. uvarum* and *S. cerevisiae* Moment were used. The growth speed of mycelium was higher in the yeast suspension presence on 3 days (table 9). The fruiting bodies formed on 4 days earlier when the *S. cerevisiae* Moment suspension was added in the substrate. The mushroom yield increased only in the presence of suspension *S. cerevisiae* on 43,8%.

Table 7. Parameters of *P. eryngii* fruiting body formation on sunflower seed's peel and *Umbeliferae* plants straw with yeast suspension addition

Combinations	Time of substrate overgrowing (days)	Fruiting body formation time (days)	Maximal fruiting body's height (± 5 mm)	Fruiting body's biomass (g)
Control samples	14	16 – 22	40	4,7 ± 2,5
+ *S. cerevisiae* Moment	12	16 – 22	55	9,8 ± 2,5

Table 8. Parameters of *P. ostreatus* fruiting body formation on sunflower seed's peel.

Combinations	Time of substrate overgrowing (days)	Fruiting body formation time (days)	Maximal fruiting body's height (± 5 mm)	Fruiting body's biomass (g)
Control samples	18	11	72,5	7,0 ± 1,0
+ *H. uvarum*	14	11 – 13	52,5	12,3 ± 1,8
+ *R. minuta*	14	13	70	9,0 ± 1,0
+ *S. cerevisiae* 3809	14	11 – 13	55	10,7 ± 0,6

Table 9. Parameters of *P. pulmonarius* fruiting body formation on sunflower seed's peel

Combinations	Time of substrate overgrowing (days)	Fruiting body formation time (days)	Maximal fruiting body's height (± 5 mm)	Fruiting body's biomass (g)
Control samples	18	20	75	8,0 ± 0,5
+ *H. uvarum*	15	16 – 20	65	8,3 ± 0,3
+ *S. cerevisiae* Moment	15	16	70	11,5 ± 1,0

III. Biochemical Analysis of *P. citrinopileatus* Fruiting Bodies

Amino acid concentrations were the same in fruiting bodies, cultivated with and without yeasts (table 10). Total protein concentration was the same too.

Table 10. Amino acid concentration (mg/g dry weight) in *P. citrinopileatus* fruiting bodies, cultivated with and without yeasts *S. cerevisiae*

Amino acid	Control samples	+ *S. cerevisiae*
Asp	21,51 ± 0,35	20,48 ± 1,52
Thr	10,52 ± 0,05	10,33 ± 0,76
Ser	10,99 ± 0,31	10,51 ± 0,68
Glu	40,23 ± 4,16	33,62 ± 4,30
Pro	9,88 ± 0,04	10,00 ± 0,88
Gly	11,10 ± 0,37	10,37 ± 0,80
Ala	16,53 ± 1,44	15,22 ± 1,47
Cys	0,04 ± 0,00	0,04 ± 0,01
Val	13,86 ± 0,26	13,93 ± 1,15
Met	2,88 ± 0,02	3,07 ± 0,24
allo-Ile	0,30 ± 0,01	0,22 ± 0,06
Ile	9,85 ± 0,21	9,92 ± 0,87
Leu	16,86 ± 0,18	16,74 ± 1,58
Tyr	4,90 ± 0,72	4,32 ± 0,17
Phe	10,21 ± 0,30	10,33 ± 0,96
OH-Lys	2,30 ± 0,92	1,22 ± 0,87
Lys	13,90 ± 0,03	14,28 ± 1,25
His	5,80 ± 0,46	5,43 ± 0,61
Arg	11,98 ± 0,83	10,94 ± 2,60
Total protein (%)	21,5 ± 0,5	21,0 ± 1,0

Organoleptic analysis of fruiting bodies showed that the *P. citrinopileatus* flavor differed in the samples, cultivated with and without yeasts *S. cerevisiae*. In the yeast cells presence the flavor was stronger and more sweet and «flowery». GC-MS analysis demonstrated the differences in the composition and in the concentrations of volatile organic compounds of *P. citrinopileatus* fruiting bodies (table 11).

Mushroom flavor depends on the substrate composition and cultivating parameters (Jong, Birmingham, 1993). It is suggested that the yeast cells presence in the substrate results in the changes of *P. citrinopileatus* fruiting bodies flavor. Tributyl acetylcitrate and 2-pentylfuran, which were found only in fruiting bodies cultivated with yeasts, are used in the flavor conservation, so they may affect on the flavor strength's increase. Octanoic acid and indole derivative compounds have sweety and fruity smell in low concentrations, so they could give sweet flavor to mushrooms, cultivated with yeasts.

Table 11. Composition and concentration (μg/g dry weight) of volatile organic compounds of *P. citrinopileatus* fruiting bodies, cultivated with and without yeasts *S. cerevisiae*

Parameters	Control samples	+ *S. cerevisiae*	Flavor *
Benzenesulfonamide, N-ethylmethyl-	0,81	0,76	unknown
n-Hexadecanoic acid	5,23	6,49	unknown
Cyclohexaneethanol, 2-methylene-	0,05	0,05	unknown
2-Methyl-3-methoxy-4H-pyran-4-one	0,11	0,09	caramel
2-Nonenal	–	0,11	fatty
Hexadecanoic acid, 2-oxo-, methyl ester	0,26	0,19	unknown
Furan, 2-pentyl-	0,13	0,45	unknown
3-Octanol	0,07	0,08	fruity, sweet
3-Octanone	0,71	0,69	fruity, sweet
Benzaldehyde, 3,4-dimethoxy-	0,24	0,25	milky, earthy
Indole, 5-methyl-2-(4-pyridyl)-	–	0,45	unknown
9,12-Octadecadienoic acid	3,88	5,81	unknown
Alkylamide	–	0,43	fishy
Benzeneacetaldehyde	0,24	0,26	sweet honey-like
Benzyl Benzoate	0,10	0,12	flowery
Hexanal	0,44	0,45	green
Decene	–	0,11	unknown
Dimethyl sulfone	0,32	0,34	unknown
Dodecanoic acid	0,01	0,03	unknown
Indole	0,13	0,35	unknown
Indole-3-acetaldehyde	0,49	0,57	unknown
Nonanoic acid	0,04	0,05	unknown
Nonenal	0,15	0,17	fatty
Octanoic acid	–	0,15	sweet
Pentadecanoic acid	0,60	0,64	unknown
Tetradecanoic acid	0,24	0,19	unknown
Tributyl acetylcitrate	–	0,69	unknown

* Classification Jong, Birmingham, 1993, Lizarraga-Guerra et al., 1997; Wood et al., 2004; Hanson, 2008.

CONCLUDING REMARKS

Research of mixed cultures of basidiomycetes fungi and microorganisms takes place from middle of XX century, however large majority of investigations have ecological and theoretical significance. Knowledge of practice application of the mixed cultures for the increase of edible mushroom fruit bodies crop are few, have commercial character and did not definition device of fungal-microorganism interaction. Our complex research of the *Pleurotus* mycelium micromorphology revealed special structure for the contacts with the yeast cells: nipple-like appendages, which formed with the alive and dead yeast cells, and coralloid hyphae which formed only in present of the alive yeast cells. The detailed description of co-cultures of five *Pleurotus* species with yeasts made for the first time in present work, the *Pleurotus* species ability to parasite yeast cells was demonstrated on water agar, and the trophic preferendum was identified for each *Pleurotus* species. Co-cultivation of *Pleurotus* species with trophic preferendum yeasts could be useful in mushroom cultivating industry. Yeast suspension addition influences on the mycelium growth speed, thus the total time of fruiting bodies cultivation could be decreased for 2 – 4 days for different *Pleurotus* species. And the main benefit of co-cultivation is the yield increase, the biomass increases on 22,8 – 108% when the suspension of yeast cells, which were defined as trophic preferendum, was added in the substrate. Fruiting bodies of *Pleurotus* which were grown with yeasts have high food value and wide spectrum of odor.

Acknowledgements

We wish to thank Dr. Ivan Chernov (Moscow State University) for the cultures of yeasts. It is a special pleasure to scientific group of the mycelium production company «Santana» (Saratov, Russia) for the cultures of some *Pleurotus* species. We are grateful to the protein analytic group of the A.N. Belozersky Institute of Physico-Chemical Biology for the amino acid analysis carrying, and also to the analytic group of Chemistry faculty (Moscow State University) for GC-MS analysis realization.

References

[1] Barron G.L. Microcolonies of bacteria as a nutrient source for lignicolous and other fungi. (1988). *Canadian Journal of Botany, vol. 66*, p. 2505–2510.

[2] Barron G.L. Predatory fungi, wood decay, and the carbon cycle. (2003). *Biodiversity, vol. 4,* p. 3–9.

[3] Blanchette R.A.; Shaw C.G. Associations among bacteria, yeasts, and Basidiomycetes during wood decay. (1978). *Phytopathology, vol. 68,* p. 631–637.

[4] Fries N.; Swedjemark G. Sporophagy in hymenomycetes (1985). *Experimental mycology, vol. 9,* p. 74–79.

[5] Hanson J.R. *The chemistry of fungi*. UK, Cambridge, Royal Society of Chemistry, 221 p. 2008.

[6] Hu S.H.; Liang Z.C.; Chia Y.C.; Lien J.L.; Chen K.S.; Lee M.Y.; Wang J.C. Antihyperlipidemic and antioxidant effects of extracts from *Pleurotus citrinopileatus*.(2006). *Journal of Agricultural and Food Chemistry, vol. 54,* p. 2103–2110.

[7] Hutchison L.J.; Barron G.L. Parasitism of algae by lignicolous *Basidiomycota* and other fungi. (1997). *Canadian Journal of Botany, vol. 75,* p. 1006–1111.

[8] Hutchison L.J.; Barron G.L. Parasitism of yeasts by lignicolous *Basidiomycota* and other fungi. (1996). *Canadian Journal of Botany. vol. 74,* p. 735–742.

[9] Jong S.C.; Birmingham J.M. *Mushrooms as a source of natural flavor and aroma compounds*. Mushroom Biology and Mushroom Products. Chinese University Press: Hong Kong, p. 345–366. (1993).

[10] Kamzolkina O.V.; Grishanina A.N.; Pancheva E.V.; Volkova V.N.; Kozlova M.V. Micromorphological features of *Pleurotus pulmonarius* (Fr.) Quel. and *P. ostreatus* (Jacq.) P. Kumm. Strains in pure and binary culture with yeasts. (2006). *Cell and Tissue biology, vol. 48,* № 2, p. 153–160.

[11] Klis F.M; Boorsma A.; De Groot P.W.J. Cell wall construction in *Saccharomyces cerevisiae*. (2006). *Yeast, vol. 23*, p. 185–202.

[12] Lemke G. Myzelwachstumsteste mit vier champignonstammen. (1972). *Champignon*, № 128, p.1–5.

[13] Lizarraga-Guerra R.; Guth H.; Lopez M.G. Identification of the Most Potent Odorants in Huitlacoche (*Ustilago maydis*) and Austern Pilzen (*Pleurotus* sp.) by Aroma Extract Dilution Analysis and Static Head-Space Samples. (1997). *J. Agric. Food Chem. vol. 45*, p. 1329-1332.

[14] Ludovico P.; Sansonetty F.; Corte-Real M. Assessment of mitochondrial membrane potential in yeast cell populations by flow cytometry. (2001). *Microbiology*. vol. *147*. p. 3335–3343.

[15] Philippoussis A.; Diamantopoulou P.; Zervakis G.; Ioannidou S. Potential for the cultivation of exotic mushroom species by exploitation of Mediterranean agricultural wastes. In: Van Griensven (ed.). *Science and cultivation of edible fungi*. Balkema, Rotterdam; 2000; p. 523–530.

[16] Stamets P. *Growing Gourmet and Medicinal Mushrooms*. Third Edition. Ten Speed Press. Berceley. Toronto. 574 p. 2000.

[17] Tsugita A.; Scheffler J.-J. A rapid method for acid hydrolysis of protein with a mixture of trifluoroacetic acid and hydrochloric acid. (2004). *European Journal of Biochemistry*, *vol. 124*, № 3, p. 585–588.

[18] Wang J.C.; Hu S.H.; Liang Z.C.; Yeh C.J. Optimization for the production of water-soluble polysaccharide from *Pleurotus citrinopileatus* in submerged culture and its antitumor effect. (2005). *Applied Microbiology and Biotechnology, vol. 67,* p. 759–766.

[19] Wasser S.P.; Weis A.L. Medicinal properties of substances occurring in higher *Basidiomycetes* mushrooms: current perspectives. (1999). *International Journal of Medicinal Mushrooms,* № 1, p. 31–62.

[20] Wood W.F.; Largent D.L.; Henkel T.W. Headspace analysis identifies indole and 1-octen-3-ol as the "coal tar" odor of *Tricholoma inamoenum*. (2004). *Mycological Progress, vol. 3,* № 4, p. 325–328.

In: Mushrooms: Types, Properties and Nutrition
Editors: S. Andres and N. Baumann
ISBN 978-1-61470-110-1

Chapter 14

Antioxidant Properties of the Most Common Edible Mushrooms Consumed in Brazil

Gisele Pezente Ferrari, Andréia Assunção Soares, Gisele Cristina dos Santos Bazanella, Adelar Bracht, Cristina Giatti Marques de Souza, Cinthia Gandolfi Bôer, and Rosane Marina Peralta*

Departamento de Bioquímica, Universidade Estadual de Maringá, 87020900, Maringá, PR, Brazil

Abstract

Mushrooms are functional foods largely consumed in Asia and Europe. The consumption of mushroom in Brazil is very low, in part due to the lack of information about the nutritional value of the edible mushrooms cultivated in the country. The objective of this chapter was to investigate the antioxidant properties and to quantify the phenolic contents of hydroalcoholic extracts of the most commonly mushrooms consumed in Brazil: *Agaricus bisporus* (white button mushroom), *Lentinula edodes* (shiitake), *Pleurotus ostreatus* (shimeji) and *Pleurotus eryngii* (eryngii). Among four edible mushrooms studied, *A. bisporus* presented the highest phenolic contents (19.97±2.01 mg/g dry extract), followed by *P. ostreatus* (16.05±1.10 mg/g), *P. eryngii* (12.74±0.98 mg/g) and *L. edodes* (10.00±0.76 mg/g). The antioxidant activities were evaluated using four methods. A strong correlation was found between the phenolic contents and the antioxidant activity determined by the scavenging activity of DPPH radicals, reducing power and lipid peroxidation inhibition. On the other hand, a weak correlation was found between the phenolic content and the chelating activity against Fe^{2+}. The results obtained allow to conclude that the mushrooms produced in Brazil present antioxidant properties, and that the most popular of them, *A. bisporus*, appears as an excellent option to enrich the diet with antioxidant phenolic compounds.

Keywords: antioxidant activity, commercial mushrooms, edible mushrooms, phenolic compounds, natural products, radical scavenging, reducing power.

* E-mail: rmperalta@uem.br, rosanemperalta@gmail.com
Phone: 554430114715

INTRODUCTION

Many fruits, vegetables, herbs, cereals, sprouts, seeds and edible mushrooms have been investigated for their antioxidant activities in the last years [7,15,17,25,26,29,41,42]. Antioxidants are believed to play a very important role in the body defense system against reactive oxygen species (ROS), which result in oxidative damage to DNA, proteins and lipids and are associated to the development of many chronic and degenerative diseases. Molecules termed antioxidants can influence the oxidation process through simple or complex mechanisms including prevention of chain initiation, binding of transitional metal ion catalysts, decomposition of peroxides, prevention of continued hydrogen abstraction, and radical scavenging [1]. The search of food with antioxidant properties is a very active research domain.

Mushrooms are world wide appreciated for their taste and flavour and are consumed both in fresh and processed form. They are poor in calories and rich in proteins, fibers and carbohydrates [22]. Studies have demonstrated that the regular consumption of mushrooms or consumption of isolated bioactive constituents present in mushrooms is beneficial to health. Mushrooms may thus be considered as functional foods or nutraceutical products [18]. They accumulate a variety of secondary metabolites such as phenolic compounds, polyketides, terpenes and steroids that are possibly involved in their medicinal effects and functional values [37]. It is important to note that the accumulation of these compounds depends on management, processing and maturity at the time of harvest [2,6].

In Brazil, the main cultivated species is *Agaricus bisporus* (white button mushroom), followed by *Lentinula edodes* (shiitake), *Pleurotus ostreatus* (shimeji) and *Pleurotus eryngii* (eryngii) [10,30]. For climatic reasons, the mushroom cultivation is done in the South and Southeast of Brazil, particularly in the regions where the annual mean temperature is around 20°C. The cultivation is still carried out in a rudimentary way, mostly by Japanese descendants who inherited the techniques from many generations of ancestors, but without deeper scientific knowledge. The mushroom consumption in Brazil is very low when compared to other countries in Europe and Asia. However, it is increasing due to the intense advertising of its benefits and cost [10].

Nutritional information on foods is becoming more and more important for both professionals in the food and health areas and for consumers, who show increasing concern about the nutritional quality of the food which makes up or could be introduced into their diets. However, little is known about the nutritional, medicinal and functional values of the edible mushrooms cultivated in Brazil. In order to fill at least partly this gap, the aim of the present chapter was to evaluate the antioxidant potentials of the mushrooms more commonly produced and consumed in Brazil. Four different methods were used and attempts were done to correlate the corresponding antioxidant parameters indicated by each method with the total phenolic contents of the mushrooms.

MATERIAL AND METHODS

Mushrooms and Preparation of Extracts

The mushroom species were commercial samples of *Lentinula edodes* Berk (shiitake), *Pleurotus ostreatus* Jacq Quélet (shimeji), *Agaricus bisporus* Lange (white button mushroom) and *Pleurotus eryngii* DC: Fr Quélet (eryngii mushroom). Samples were kept under refrigeration until the moment of analysis. After the mushroom species were dried to constant weight at 40° C, they were weighed and ground into a fine powder (20 mesh). Dried mushroom samples (5.0 g) were extracted by stirring them with 100 ml of one of following solvents: methanol, ethanol, acetone, ethanol-water (2:1), methanol-water (2:1) and acetone-water (2:1) at 28°C at 150 rpm for 3 h and filtering through filter paper. The residues were then extracted with two additional 100 ml of each solvent as described above. The combined extracts were then rotary evaporated at 40° C to dryness and kept in the dark at 4°C until use.

Determination of Total Reducing Power

The reducing power was assayed as described previously with some modifications [17]. Different concentrations of mushroom extracts (1.0 ml) were mixed with 2.5 ml of phosphate buffer (50 mM, pH 7.0) and 2.5 ml of 1% potassium ferricyanide. The mixture was then incubated at 50° C for 20 min. After, 2.5 ml of trichloroacetic acid (10%) was added to the mixture, which was then centrifuged at 3,000 rpm for 10 min. Finally, 1.25 ml from the supernatant was mixed with 1.25 ml of distilled water and 0.25 ml $FeCl_3$ solution (0.1%, w/v). The absorbance was measured at 700 nm. The assays were carried out in triplicate and the results were expressed as mean values±standard deviations. Higher absorbance values indicate higher reducing power. The extract concentration providing 0.5 of absorbance (EC_{50}) was calculated from the graph of absorbance at 700 nm against the extract concentration. BHT (2,6-di-tert-butyl 4-methyl phenol), a synthetic antioxidant used generally for food, cosmetics and pharmaceuticals, was used as standard.

Lipid Peroxidation Inhibition Assay (β-Carotene-Linoleic Acid Assay)

The lipid peroxidation inhibition wasas evaluated by the β-carotene-linoleate model system as previously described with some modification [34]. Firstly, β-carotene (0.2 mg) was dissolved in 1.0 ml of chloroform. After, 0.02 ml of linoleic acid plus 0.2 ml of Tween 80 was added and the mixture was left standing at room temperature for 15 min. After evaporation of chloroform, 50 ml of oxygenated distilled water was added and the mixture was shaken to form an emulsion (β-carotene-linoleic acid emulsion). Aliquots of 3.0 ml of this emulsion were transferred into test tubes containing 0.2 ml of different concentrations of extracts. The tubes were shaken and incubated at 50° C in a water bath. As soon as the emulsion was added to each tube, the zero time absorbance (A_0) was measured at 470 nm using a spectrophotometer. A second absorbance (A_1) was measured after 120 min. A blank, without β-carotene was prepared for back-ground subtraction. Lipid peroxidation (LPO) inhibition was calculated using the following equation:

$$LPO\,\text{inhibition}(\%) = \frac{A_0 - A_1}{A_0} \times 100 \quad (1)$$

The assays were carried out in triplicate and the results expressed as mean values ± standard deviations. The extract concentration possessing 50% antioxidant activity (EC_{50}) was calculated from the graph of antioxidant activity percentage against extract concentration. BHT was used as standard.

Determination of Chelating Effects on Ferrous Ions

The ferrous ion chelating ability of the mushroom extracts was determined as previously described [35]. A sample (0.7 ml) of each extract was diluted in 0.7 ml of distilled water and mixed with 0.175 ml of $FeCl_2$ (0.5 mM) and the absorbance (A_0) was measured at 550 nm. After, the reaction was initiated by the addition of 0.175 ml ferrozine (0.5 mM). The mixture was shaken vigorously for 1 min and left standing at room temperature for 20 min when the absorbance (A_1) was measured at 550 nm. The percentage of inhibition of the ferrozine-Fe^{2+} complex formation was calculated as follows:

$$\text{Chelating ability } (\%) = \frac{A_0 - A_1}{A_0} \times 100 \quad (2)$$

A lower absorbance indicates higher chelating ability. The extract concentration providing 50% chelating ability (EC_{50}) was calculated from the graph of antioxidant activity percentage against extract concentration. EDTA (ethylenediaminetetraacetic acid) at 0.5 mM was used as the positive control.

Determination of the Scavenging Effect on DPPH$^{\bullet}$ Radicals

The free radical scavenging activities of mushroom extracts were measured by using 1,1 diphenyl-2-picryl-hydrazyl (DPPH$^{\bullet}$) as previously described in slight modifications [35]. Briefly, 150 μl of each extract at various concentrations was added to 2.850 ml of DPPH solution (0.1 mM), vigorously shaken and maintained for 2 h at room temperature in the dark. Methanol was used instead of mushroom extract as a control. Then the absorbance was measured at 515 nm. The capability to scavenge the DPPH radical was calculated using the following equation:

$$\text{DPPH scavenging effect } (\%) = \frac{A_0 - A_1}{A_0} \times 100 \quad (3)$$

In equation [3] A_0 is the absorbance of the control reaction and A_1 the absorbance in the presence of the sample. The extract concentration providing 50% inhibition (EC_{50}) was calculated from the graph of the DPPH scavenging effect against extract concentration. BHT was used as standard.

Determination of Total Phenolic Contents

Total soluble phenolic compounds in the mushroom extracts were measured using the Folin-Ciocalteu's phenol reagent and expressed as gallic acid equivalents [34]. A sample of the methanolic extract was added to distilled water for a final volume of 2 ml. After, it was mixed with 0.3 ml of a saturated sodium carbonate (Na_2CO_3) solution and 0.1 ml of 1 N Folin-Ciocalteu's phenol reagent. The mixture was placed for 1 h at room temperature in the dark. The absorbance was measured at 725 nm against the blank.

Statistical Analysis

All determinations were done in triplicate. The results were reported as means ± standard deviation (SD). The significance of differences among treatment means was determined by one-way analysis of variance (ANOVA) and Tukey test using the program Graph Pad ®, Prism, 3.0. The Pearson coefficients for determination of the relationships between the variables of antioxidant activity and phenolic contents of the fungi extracts were calculated using the same program.

Chemicals

DPPH, ABTS, β-carotene, linoleic acid, Folin-Ciocalteu reagent, gallic acid, and BHT were acquired from Sigma Co®. All other reagents were analytical grade.

Results and Discussion

Yield of Methanolic and Hydroalcoholic Extracts

The extraction yields of the mushroom extracts are shown in Table 1. Six different solvents were used to extract the soluble solid compounds from mushrooms: methanol, ethanol and acetone and mixtures (2:1) of ethanol-water, methanol-water and acetone-water. Methanol is the most common solvent used in the extraction of antioxidant substances from edible mushrooms because high yields are obtained with its use [9,24,43]. The yields obtained using methanol as solvent were similar to those found for several edible mushrooms with yields varying from 15.9 to 43.9% [42]. In all cases, a mixture of organic solvent with water increased significantly the extraction yield. Different conventional solvent extraction of antioxidant molecules of different foods (fruits, vegetables, herbs, cereals, sprouts, seeds and edible mushrooms) has been reported in a laboratory scale using acetone, acetone-water, methanol, ethanol or ethanol-water. Mixtures of acetone and water are good solvents for polar antioxidants, but may lead to an unacceptable level of acetone residue in the extracts [20]. The manipulation of methanol can be dangerous, considering that it is easily absorbed through the respiratory system and skin. The toxicity of methanol is mainly due to the fact that in mammals it is metabolized to formic acid *via* formaldehyde [36]. Among the 6 solvents tested, the mixture of ethanol-water showed to be the most suitable solvent system due this high extraction yield and its acceptability for human consumption.

Antioxidant Properties of Hydroalcoholic Extracts

The hydroalcoholic mushroom extracts were subjected to screening for their antioxidant activities. Four complementary test systems, namely β-carotene-linoleic acid assay, DPPH radical scavenging activities, reducing power and ferrous ion chelating ability were used.

Figure 1A shows the reducing power of the mushroom extracts as a function of their concentration. BHT (0.1 mg/ml), the positive control used in this test, had a reducing power value of 0.771. For all mushroom extracts, the reducing power increased with their concentrations. The hydroalcoholic extracts from white button mushroom presented the highest reducing power: at 5 mg/ml, reducing power was higher than 1.0. Shimeji and shiitake extracts presented a reducing power of 0.5 at 5 mg/ml and the eryngii extract was the less efficient extract, with a reducing power value minor of 0.5. With regard to the reducing powers, the hydroalcoholic and methanolic extracts obtained from Brazilian commercial mushrooms were good and comparable to those of edible and medicinal mushrooms in general [5, 42].

Figure 1B shows the antioxidant activity of the mushroom extracts as measured by the bleaching of the β-carotene-linoleate system and calculated according to equation [1]. BHT, the positive control used in this test, had 92% of antioxidant activity at the concentration 0.01 mg/ml. The lipid peroxidation (LPO) inhibition caused by the mushroom extracts increased when their concentrations were increased. At the concentration of 0.3 mg/ml, all mushroom extracts presented an antioxidant activity around 75%. The white button extract was slightly more efficient than the others.

The chelating abilities of mushroom hydroalcoholic extracts, calculated according to equation [2], are shown in Figure 1C. All mushroom extracts presented chelating effects on ferrous ions and the activities increased with their concentrations. The highest chelating effects were observed in shiitake and white button extracts. EDTA (positive control) presented a high chelating ability of 99.3% at the concentration of 0.03 mg/ml. Tsai et al. [37] reported that the ethanolic extracts of *A. blazei* have a chelating ability of 58.8% at the concentration of 20 mg/ml. These authors suggest that moderate to high ferrous-ions chelating abilities showed by mushrooms could be beneficial to health. Iron can stimulate lipid peroxidation by the Fenton reaction, and also accelerates peroxidation by decomposing lipid hydroperoxides into peroxyl and alkoxyl radicals that can themselves abstract hydrogen and perpetuate the chain reaction of lipid peroxidation [12].

Finally, the antioxidant activities of mushroom extracts were assayed by using the stable free radical DPPH and calculated according to equation [3]. The scavenging activities of the hydroalcoholic and methanolic extracts from all mushroom species on the DPPH radicals increased when the extract concentrations were increased (Figure 1D). The potency of extracts obeyed the sequence *A. bisporus* > *P. ostreatus* > *P. eryngii* > *L. edodes*. The free radical scavenging capacities of these extracts were superior to those found for several other methanolic extracts of edible mushrooms [5,9]. The standard BTH presented a scavenging effect of 92% at the concentration of 0.0.01 mg/ml.

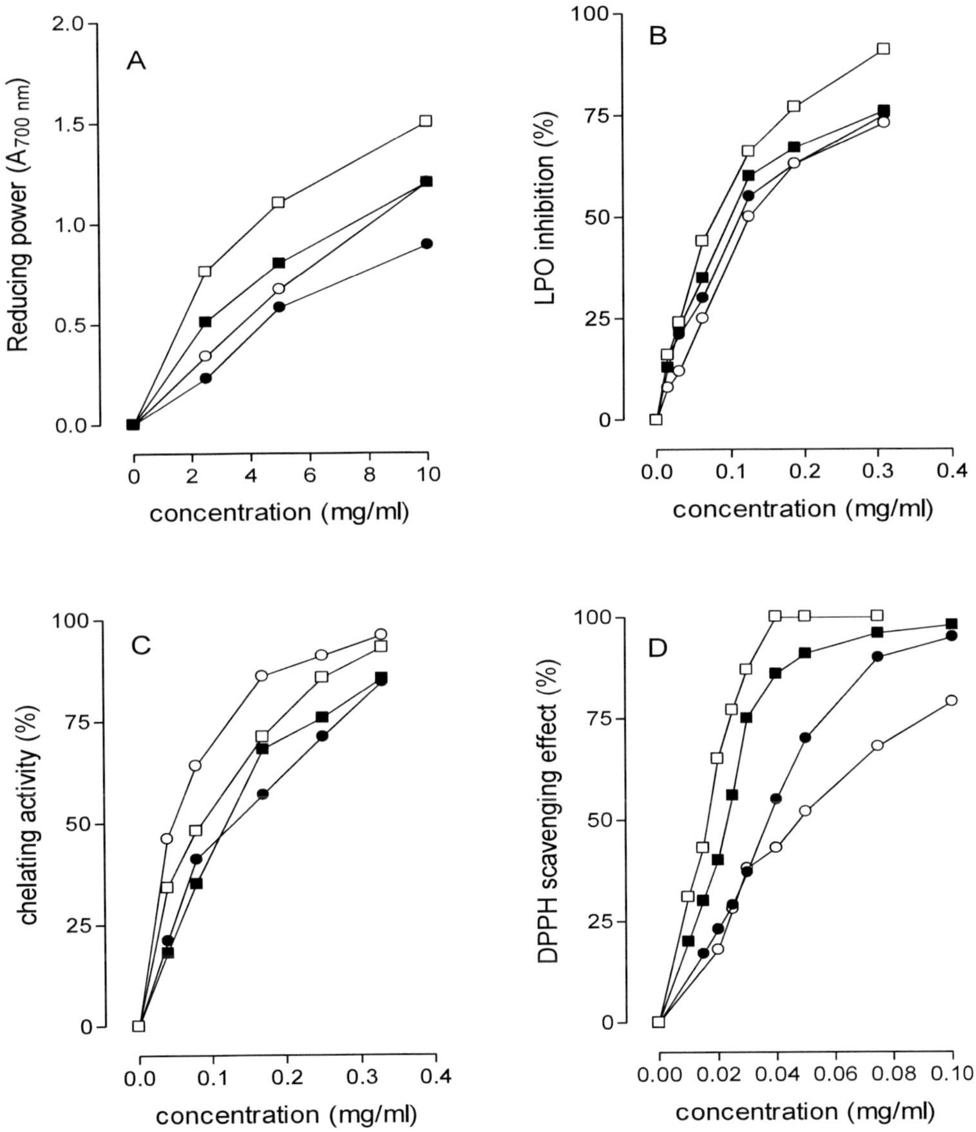

Figure 1. Antioxidant activities of mushroom extracts determined using four complementary methods. A: Reducing power; B: LPO inhibition; C: Metal chelating ability; D: DPPH scavenging activity. Each value is expressed as mean of three independent results. A. bisporus (□); P. eryngii (●); L. edodes (○); P. ostreatus (■).

Table 1. Extraction yields of soluble solid from commercial mushrooms produced in Brazil using different solvents

	Extraction yield (%)					
	EtOH	**MeOH**	**acetone**	**EtOH-H_2O**	**MeOH-H_20**	**Acetone-H_2O**
A. bisporus	18.0 ± 2.1^a	22.4 ± 3.1^a	22.0 ± 2.5^a	42.4 ± 5.1^b	44.0 ± 5.4^b	45.0 ± 5.1^b
P. eryngii	13.4 ± 1.7^a	15.2 ± 2.6^a	15.9 ± 2.0^a	27.7 ± 3.3^b	29.7 ± 1.8^b	28.4 ± 3.3^b
L. edodes	11.0 ± 1.2^a	13.6 ± 1.9^a	14.0 ± 2.0^a	27.2 ± 3.0^b	32.7 ± 4.0^b	33.8 ± 3.9^b
P. ostreatus	24.7 ± 2.9^a	26.6 ± 2.5^a	23.0 ± 1.5^a	39.4 ± 4.8^b	42.3 ± 4.7^b	39.0 ± 5.1^b

Dried mushroom samples (5.0 g) were extracted by stirring them with 100 ml of methanol (MeOH), 100 ml acetone or 100 ml ethanol (EtOH) or 100 ml of a mixture (2:1) of ethanol-water (EtOH-H_2O), methanol-water (MeOH-H_2O) or acetone-water at 28°C at 150 rpm for 4 h and filtering through filter paper. The residues were then extracted with two additional 100 ml of solvent as described above. The combined volumes were then rotary evaporated at 40° C to dryness and kept in the dark at 4°C until use. Means with different capital letters within a line are significantly different ($P<0.05$).

Correlation of Antioxidant Activities and Phenolic Contents

Antioxidant properties of mushrooms are usually related to low-molecular-weight compounds, particularly to the phenolic fractions [11]. For this reason, it is important to analyze the phenolic content in the mushroom extracts. The phenolic content in the extracts decreased according to the sequence white *A. bisporus* > *P. ostreatus* > *P. eryngii* > *L. edodes* (Table 2). To evaluate a possible correlation between the phenolic contents of the extracts and their antioxidant properties, the values of EC_{50} (the mg/ml value of each extract bearing 50% of antioxidant activity) were calculated (Table 2) and plotted against the corresponding phenolic content (Table 3).

A strong correlation was found between the phenolic contents and the antioxidant activity determined by the scavenging activity of DPPH radicals, the reducing power and the beta-carotene bleaching methods. A weak correlation was found, however, between the phenolic content and chelating activity against Fe^{2+} (r=0.2993). In this respect it has been reported that compounds with structures containing two or more of the following functional groups –OH, -SH, -COOH, $-PO_3H_2$, -C=O, $-NR_2$, -S- and –O- in a favorable structure-function configuration can show metal chelation activity [19]. This includes, for example, ubiquitous compounds such as citric, malic, tartaric, oxalic, succinic, lipoic and phytic acids.

The role of the phenolic compounds as antioxidants is amply known. Phenolics are one of the major groups of non-essential dietary components that have been associated with the inhibition of atherosclerosis, cancer, diabetes, and hypertension and to act as antidepressants and general antioxidants [4,8,21,23,27]. The bioactivity of phenolics may be related to their ability to chelate metals, inhibit lipoxygenase and scavenge free radicals [29]. Some common edible mushrooms have currently been found to possess antioxidant activity, which is well correlated with their total phenolic contents [5,38,39]. However, this correlation is not a general one. While high correlation has been described in some reports [5], very weak correlation or even virtual absence of correlation have also been reported [28].

The cause for these apparently discrepant results could be related to the fact that different species may contain different types of polyphenolics with quite different antioxidant activities. For example, in wine, there are many polyphenolics, some of them possessing very high antioxidant activities and others with only moderate antioxidant activities [40]. There are over 5,000 different phenolic compounds described in plants [31].

The identification of phenolic compounds in mushrooms is still in its beginnings and only a few dozen structures have already been described [16]. It would thus not be surprising if it turns out that the phenolic compounds of fungi species showing no correlation or poor correlation between phenolic contents and antioxidant activity are precisely those in which phenolics with low antioxidant activity predominates over those with high antioxidant activity.

Table 2. Phenolic contents and antioxidant activities of mushroom hydroalcoholic extracts determined using four complementary methods

Mushroom	Phenolics (mg/g)[e]	Antioxidant activities - EC_{50} values (mg/ml)			
		Reducing power[a]	Chelating ability[b]	LPO inhibition[c]	DPPH test[d]
A. bisporus	19.97±2.01[A]	1.65±0.07[A]	0.09±0.01[A]	0.08±0.01[A]	0.017±0.003[A]
P. eryngii	12.74±0.98[B,C]	4.10±0.34[B]	0.13±0.01[B]	0.10±0.01[A,B]	0.039±0.004[B]
L. edodes	10.00±0.76[B]	3.75±0.28[B]	0.05±0.01[C]	0.12±0.01[B]	0.044±0.004[B]
P. ostreatus	16.05±1.10[C]	2.50±0.10[C]	0.12±0.02[A,B]	0.11±0.01[B]	0.024±0.002[A]

EC_{50} values: the effective concentration where the absorbance was 0.5 for reducing power[(a)]; the effective concentration where the ferrous ions were chelated by 50%[(b)]; the effective concentration at which the antioxidant activity using the β-carotene–linoleic acid assay (LPO inhibition) was inhibited by 50%[(c)]; the effective concentration where DPPH radicals was scavenged by 50%[(d)]; EC_{50} values were obtained by interpolation from linear regression analysis. Each EC_{50} value is expressed as mean ± standard deviation ($n = 3$). Means with different capital letters within a row are significantly different ($P < 0.05$). Concentration of total phenolics as equivalents of gallic acid was expressed as mg/g dry extract[(e)]

Table 3. Correlation between the content of total phenolic compounds and the antioxidant activities of mushroom extracts using different methods

Method	*equation*	*r value*	*correlation*
Reducing power	y= -(0.2426±0.0719)x + (6.564± 1.089)	0.9223	Very strong
LPO inhibition	y= -(0.003343±0.002)x + (0.1516±0.023)	0.8425	Very strong
Chelating activity	y=(0.002511±0.0056)x + (0.06012±0.086)	0.2993	weak
DPPH activity	y= -(0.002881±0.0003884)x + (0.007332±0.005886)	0.9823	Very strong

CONCLUSION

The results presented here indicate that the mushrooms produced in Brazil constitute a rich dietary source of antioxidant molecules and phenolic coumponds. The *A. bisporus* is the main cultivated edible mushroom not only in Brazil but also worldwide. It is usually considered to be of lesser nutritional and medicinal value compared with other cultivated mushrooms. However, recent evidence suggests that *A. bisporus* contains high levels of molecules with nutritional and medicinal importance and this species can thus be considered as a functional food [3,13,14,28,32,33] The data obtained with respect to its antioxidant properties, fully corroborate this new point of view.

ACKNOWLEDGEMENTS

The authors wish to thank to Conselho Nacional de Desenvolvimento Científico e Tecnológico (CNPq), Fundação Araucária and Fundo Paraná for the financial support and the technical assistance of M.A.F. Costa and A. Chaves.

REFERENCES

[1] Ames, BN; Shigenaga, MK; Hagen, TM Oxidants, antioxidants, and the degenerative diseases of aging. *Proc Natl Acad Sci USA*, 1993, 90, 7915–7922.

[2] Barros, L; Baptista, P; Ferreira, ICFR. Effect of *Lactarius piperatus* fruiting body maturity stage on antioxidant activity measured by several biochemical assays. Food Chem., 2007, 45, 1731-1737.

[3] Beelman, RB; Royse, DJ; Chikthimmah, N. Bioactive components in button mushroom *Agaricus bisporus* (J. Lge) Imbach (Agaricomycetideae) of nutritional, medicinal or biological importance. *Int. J. Med. Mushrooms*, 2003, 5, 321-337.

[4] Bobek, P; Nosalova, V; Cerna, S. Effect of pleuran (beta glucan from *Pleurotus ostreatus*) in diet or drinking fluid on colits rats. *Nahrung*, 2001, 45, 360-363.

[5] Cheung, LM; Cheung, PCK; Ooi VEC Antioxidant activity and total phenolics of edible mushroom extracts. *Food Chem.*, 2003, 81:249-255.

[6] Choi, Y; Lee, SM; Chun, J; Lee, HB; Lee, J. Influence of heat treatment on the antioxidant activities and polyphenolic compounds of shiitake (*Lentinus edodes*) mushroom. *Food Chem.*, 2006, 99, 381-387

[7] Dimitrios, B. Sources of natural phenolic antioxidants. *Trends Food Sci. Tech.*, 2006, 17:505-512

[8] Duthie, G; Crozier, A. Plant-derived phenolic antioxidants *Curr. Opin. Lipidol.,* 2000, **11,** 43–47.

[9] Ferreira, ICFR; Baptista, P; Vilas-Boas, M; Barros, L. Free-radical scavenging capacity and reducing power of wild edible mushrooms from northeast Portugal: Individual cap and stipe activity. *Food Chem.*, 2007, 100, 1511-1516.

[10] Furlani, RPZ; Godoy, HT. Nutritional value of edible mushrooms. *Cien. Tecn. Alim.*, 2007, 27 154-157.

[11] Gursoy, N; Sarikurkcu, C; Cengiz, M; Solak, MH. Antioxidant activities, metal contents, total phenolics and flavonoids of seven *Morchella* species. *Food Chem. Toxicol.*, 2009, 47, 2381-2388.

[12] Halliwell, B. (1999). The biological toxicity of free radicals and other reactive oxygen species. In: Aruoma, O.I., Halliwell, B (eds). *Free radicals and food additives* Taylor and Francis Ltd. London, pp 37-57

[13] Jagadish, L; Krishnan, VV; Shenbhagaraman, R; Kaviyarasan, V. Comparative study on the antioxidant, anticancer and antimicrobial property of *Agaricus bisporus* (J.E. Lange) Imbach before and after boiling. *Afr. J. Biotechnol.*, 2009, 8, 654-661.

[14] Kent, D; Sheridan, C; Tomkinson, HA; White, S; Hiscott, P; Grierson, I. Edible mushroom (*Agaricus bisporus*) lectin modulates human retinal pigment epithelial cell behaviour *in vitro*. *Exp. Eye Res.*, 2003, 76, 213-219.

[15] Kris-Etherton, PM; Hecker, KD; Bonanome, A; Coval, SM; Binkoski, AE; Hilpert, KF. Bioactive compounds in foods: their role in the prevention of cardiovascular disease and cancer. *Am. J. Med.*, 2002, 30, 71-88

[16] Kim, M-Y; Seguin, P; Ahn, J-K; Kim, J-J; Chun, S-C; Kim, E-H; Seo, S-H; Kang, E-Y; Kim, S-L; Park, Y-J; Ro, H-M; Chung, III-M. Phenolic compounds concentration and antioxidant activities of edible and medicinal mushrooms from Korea. *J. Agric. Food Chem.*, 2008, 56, 7265-7270

[17] Kuda, T; Tsunekawa, M; Goto, H; Araki, Y. Antioxidant properties of four edible algae harvested in the Noto Peninsula, Japan. *J. Food Comp. Anal.*, 2005, 18, 625-633.

[18] Lakhanpal, T.N; Rana, M. Medicinal and nutraceutical genetic resources of mushrooms. *Plant Gen. Res.*, 2005, 3, 288-303

[19] Lindsay, R.C. (1996). Food additives. In: Fenemma, O.R. (ed). *Food Chemistry*. Marcel Dekker Inc. pp.778–780.

[20] Lu, Y; Foo, LY. Rosmarinic acid derivatives from *Salvia officinalis*. *Phytochemistry*, 1999, **51**, 91–94

[21] Manzi, P; Aguzzi, A; Pizzoferrato, L. Nutritional value of mushrooms widely consumed in Italy. *Food Chem.*, 2001, 73, 321–325.

[22] Manzi, P; Gambelli, L; Marconi, S; Vivanti, V; Pizzoferrato, L. Nutrients in edible mushrooms: an inter-species comparative study. *Food Chem.*, 1999, 65, 477-482.

[23] Mattila, P; Suonpãã, K; Piironen, V. Functional properties of edible mushrooms. *Nutrition*, 2000, 16, 694-696.

[24] Mau, J-L; Tsai, S-Y; Tseng, Y-H; Huang, S-J. Antioxidant properties of methanolic extracts from *Ganoderma tsuge*. *Food Chem.*, 2005, 93, 641-649.

[25] Moure, A; Cruz, JM; Franco, D; Dominguez, JM; Sineiro, J; Dominguez, H; Nunez, MJ; Parajo, JC. Natural antioxidants from resource bioresources. *Food Chem.*, 2001, 72, 145-171

[26] Pichichero, E; Canuti, L; Canini, A. Characterization of the phenolic and flavonoid fractions and antioxidant power of Italian honeys of different botanical origin. *J. Sci. Food Agric.*, 2009, 89, 609-616.

[27] Quiles, JL; Huertas, JR; Battino, M; Mataix, J; Ramirez-Tortosa, MC. Antioxidant nutrients and adriamycin toxicity. *Toxicology*, 2002, 180, 79-95

[28] Ramirez-Anguiano, AC; Santoyo, S, Reglero, G; Soler-Rivas, C. Radical scavenging activities, endogenous oxidative enzymes and total phenols in edible mushrooms commonly consumed in Europe. *J. Sci. Food Agric.*, 2007, 87, 2272-2278

[29] Rice-Evans, C; Miller, NJ; Paganga, G. Antioxidant properties of phenolic compounds. *Trends Plant Sci.*, 1997, 2, 152-159

[30] Ruegger, MJS; Tornisielo, SMT; Bononi, VLR; Capelari, M. Cultivation of the edible mushroom *Oudemansiella canarii* (Jungh) Hohn in lignocellulosic substrates. *Braz. J. Microbiol.*, 2001, 32, 211-214

[31] Sahidi, F; Naczk, M. (1995). *Food phenolics: sources, chemistry, effects, applications.* Technomic Publishing Co. Inc, Lancaster, PA, USA.

[32] Savoie J-M; Minvielle, N; Largeteau, ML Radical-scavenging properties of extracts from the white button mushroom, *Agaricus bisporus. J. Sci. Food Agric.*, 2008, 88, 970-975.

[33] Singla, R; Ghosh, M; Ganguli, A. Phenolics and antioxidant activity of a ready-to-eat snack food prepared from the edible mushroom (*Agaricus bisporus*). *Nut. Food Sci.*, 2009, 39, 227-234.

[34] Singleton, VL; Rossi, JA Colorimetry of total phenolics with phosphomolybdic-phosphotungstic acid reagents. *Am. J. Enol. Viticult.*, 1965, 16, 144-158.

[35] Soares, AA; Souza, CGM; Daniel, FM; Ferrari, GP; Costa, SMG; Peralta, RM. Antioxidant activity and total phenolic content of *Agaricus brasiliensis* (*Agaricus blazei* Murril) in two stages of maturity. *Food Chem.*, 2009, 112, 775-781.

[36] Tephly, TR, The toxicity of methanol. *Life Sci.*, 1991, 48, 1031-1041.

[37] Tsai, S-Y; Tsai, H-U; Mau, JL. Antioxidant properties of *Agaricus blazei*, *Agrocybe cylindracea* and *Boletus edulis*. *LWT Food Sci. Tech.*, 2007, 40, 1392-1402

[38] Turkoglu, A; Duru, ME; Mercan, N; Kivrak, I; Gezer, K. Antioxidant and antimicrobial activities of *Laetiporus sulphureus* (Bull.) Murrrill. Food Chem., 2007, 101, 267-273.

[39] Velioglu, YS; Mazza, G; Gao, L; Oomah, BD. Antioxidant Activity and Total Phenolics in Selected Fruits, Vegetables, and Grain Products. *J. Agric.Food Chem.*, 1998, 46, 4113-4117.

[40] Villano, D; Fernandez-Pachon, S; Troncoso, AM; Garcia-Parrilla, MC. Comparison of antioxidant activity of wine phenolic compounds and metabolites *in vitro*. *Anal. Chim. Acta*, 2005, 538, 391- 398.

[41] Wong, JY; Chye, FY. Antioxidant properties of selected tropical wild edible mushrooms. *J. Food Comp. Anal.*, 2009, 22, 269-277.

[42] Yang, J-H; Lin, H-C; Mau, J-L. Antioxidant properties of several commercial mushrooms. *Food Chem.*, 2002, 77, 229–235.

In: Mushrooms: Types, Properties and Nutrition
Editors: S. Andres and N. Baumann
ISBN 978-1-61470-110-1

Chapter 15

BIOLOGICALLY ACTIVE MUSHROOM PROTEINS

T. B. Ng* and Xiu Juan Ye
School of Biomedical Sciences, Faculty of Medicine,
The Chinese University of Hong Kong, Shatin
New Territories, Hong Kong, China

ABSTRACT

The objective of this article is to review mushroom proteins with biologuical activity. They include antifungal proteins, antibacterial protcins, ribosome inactivating proteins, ribonucleases, ubiquitin- like proteins, lectins, proteases, hemolysins, laccases, xylanases, acid phosphatases, L-amino acid oxidases, α -mannosidases, deoxyribonucleases, carboxylesterases, polysaccharopeptides and polysaccharoproteins, amatoxins, phallotoxins and virotoxins;

Keywords: mushroom, proteins, biological activity.

INTRODUCTION

Mushrooms have been officially designated as healthy foods by the Food and Drug Administration. They have a high content of protein, high-molecular-weight polysaccharides, antioxidants, vitamin B compex, provitamin D2, dietary fiber, selenium, potassium and copper. Mushrooms are also delicious and included in the diet of many people.

Great strides have been made in the isolation and characterization of mushroom proteins. The different types of mushrooms and their biological activities are reviewed below:

* Corresponding author: T.B. Ng
Email: b021770@mailserv.cuhk.edu.hk;
Fax: +852-2603-5123; Phone: +852-2609-6872

Antifungal Proteins

Since pathogennic fungic cause great damage to crops many investigators have dedicated their effects to finding antifungal proteins with a broad spectrum of action and high potency. Antifungal proteins are produced by a variety of animals, and plants. Grenier et al. (2000) reported thaumatin-like proteins from the mushroom *Irpex lacteus* and *Lentinus edodes* and the fungus *Rhizoctonia solani*. The proteins suppress the growth of the yeast *Saccharomyces cerevisiae*. Galan et al (1999) described the enzymatic characteristics of an endo-1,3 β - glucanase from *Agaricus bisporus* without any report of antifungal activity. Lyophyllin is a ribosome inactivating protein with antifungal activity from *Lyophyllum shimeiji*. From the same mashroom, Lyophyllum antifungal protein with higher antifungal activity than lyophyllin has been isolated (Lam and Ng, 2001a). Both lyophyllin and Lyophyllum antifungal protein inhibit cell-free translation in a rabbit reticulocyte system and the activity of HIV-1 reverse transcriptase. The 10-kDa antifungal peptide eryngin from *Pleurotus eryngii* exhibit antifungal activity toward *Fusarium oxysporum* and *Mycosphaerella arachidicola* (Wang and Ng 2004).

Antibacterial Protein

A 44-kDa dimeric protein with a novel N-terminal sequence and potent antibacterial activity toward *Agrobacterium rhizogenes, A. tumefaciens, A. vitis, Xanthomonas oryzae* and *X. malvacearum* was isolated from *Clitocybe sinopica*. However, it was devoid of activity toward a number of other bacteria (Zheng et al., 2010) (20198215).

Ribosome Inactivating Proteins (RIPs)

RIPs have been isolated from flowering plants and fungi. RIPs inhibit protein synthesis by attacking ribosomes and catalyze the breaking of the N-glycosidic bond at A4324 in rat 28s RNA with their N-glycosidase activity. Marmorin and hypsin are RIPs isolated from *Hypsizigus marmoreus* (Wong and Ng; Lam and Ng 2001 b). Lyophyllin is an RIP from the closed related *Lyohyllum shimeiji* (Lam and Ng 2001 a). They potently inhibit cell-free translation in rabbit reticulocyte lysate system and the activity of HIV-1 reverse transcriptase. The N-terminal sequences of lyophyllin and hypsin are strikingly similar.

Calcaelin from *Calvatia caelata* exhibits an N-terminal amino acid sequence with similanty to panaxagin and quinqueginsin which are proteins with cell-free translation inhibitory activity from Chinese ginseng and American ginseng, respectively. Like other RIPs, it inhibits proliferation of cancer cells and mitogenic activity of spleen cells *in vitro* (Ng et al., 2003).

Several RIPs have been isolated from *Flammulina velutipes*, including 40-kDa flammulin, 13.8-kDa velutin, 30-kDa flammin and 19-kDa velin (Wang and Ng, 2000a, 2001a, 2005a). They inhibit cell-free translatioon with an IC_{50} of 0.25 nM, 0.29 nM, 1.4 nM and 2.5 nM, respectively. Velutin inhibits HIV-1 reverse transcriptase and this activity is potentiated by succinylation. The wild mushroom *Polyporus adusta* produces a 16.5-kDa translation inhibiting peptide designated as adustin (Ng and Wang, 2005b). Pluturegin is a 38-kDa RIP from *Pleurotus tuber-regium*. It inhibits cell-free translation with an IC_{50} of 0.5 nM (Wang and Ng, 2001b).

Proteases

Metalloproteases with a molecular mass in the range 18-20-kDa have been isolated from *Armillariella mellea, Grifola frondosa, Pleurotus ostreatus* and *Tricholoma saponaceum* (Nonaka et al., 1995). Prolyl endopeptidases from *Agaricus bisporus* and *Lyophyllum cinerasceus* have a much higher molecular mass (78 kDa) (Sattar et al., 1990; Yoshimoto et al., 1990) while the aspartic protease pleureryn from *P. eryngii* has a lower molecular mass (11.5 kDa). Pleureryn has relatively high pH and tenperature atability.

Ubiquitin-Like Proteisn

Proteins and peptides with N-terminal sequence similarity to ubiquitin have been purified from *C. caelata* (Lam et al., 2001), *Pleurotus ostreatus* (Ng et al., 2002b), *Agrocybe cylindracea* (Ngai et al., 2003) and *Cantharellus cibarius* (Wang et al., 2003b). Some of them manifest antiproliferative activity toward cancer cells (Lam et al., 2001; Nagai et al., 2003). They exhibit ribonuclease activity and cell-free translation inhibitory activity.

Laccases

Laccases, together with manganase peroxidase and lignin peroxidase, are important in lignin degradation. Laccases have been purified from submerged cultures of a variety of mushroom species. They display a blue color. Laccases from solid state cultures, on the other hand ,exhibit a yellowish brown color and different spectra. Most laccases are monomeric have a molecular mass close to 60 kDa. However, *Cantharellus cibarius* laccase is dimeric (Ng and Wang, 2004). *Tricholoma giganteum* laccase has a smaller molecular mass (Wang and Ng, 2004 c).

A 71-kDa laccase with fair thermostability and pH stability was purified from culture medium of *Grifola frondosa* (Nitheranont et al., 2011) (21389619)*Tricholoma mongolicum* laccase exhibited antoproleferative activity toward hepatoma and brease cancer cells (Li et al., 2010) *P.cornucopiae* laccase exerted antiprolifrative activity against hepatoma and leukemia cells (Wong et al., 2010)(19807674). Oyster mushroom laccase inhibited the entry of hepatoma cells (El-Fakharany et al., 2010).

Xylanases

Xylans, the major hemicelluloses in differentiated hardwood are a large reservoir of fixed carbon in nature. A 58-kDa thermostable xylanase from *Fomitopsis pinicola* with an optimal pH of 4.5 and a temperature optimum of 70 ºC exhibits intemal sequence homology to hydrolases from glucose hydrolase family 10 (Shin et al., 2010) (21030826). A 12-kDa xylanase with an optimal pH of 5 and an optimal temperature at 50 ºC with sequence homology to *Neurospora crassa* was isolated from *Termitomyces clypeatus* (Soren et al., 2010) (19128166).

Lentinus edodes xylanase has a molecular mass of 35 kDa and sequence homology to *Aspergillus terreus* xylanase (Lee et al., 2007) (18092465).

Tyrosinases

Tyrosinases are bifunctional copper-containing enzymes. Tyrosinases use molecular oxygen to catalyze oxidation of phenoles to o-diphenols (monophenolase activity) and their subsequent oxidation to o-quinones (catecholase activity), which then auto-polymerase to produce brown polyphenolic catechol melanins. Mushroom tyrosinases are potentially useful in biothchnological applications. Tyrosinases have been reported from *Agaricus bisporus*, *Lentinula edodes*, and *Pycnoporus sanguineus* and *Pholiota nameko* (Kawamura-Konishi et al., 2007)

A. bisporus tyrosinase has a molecular mass of 70 kDa and its isoforms have an isoelectric points between 4.9 and 5.2 (Fan and Fluskey, 2004) (15016563). *P. nameko* tyrosinase exhibits a molecular mass of 42 kDa and contains 1.9 Cu atoms/molecule.

Hemolysins

A 40 kDa thermolabile hemolysin which induced diarrhea and death in mice was isolated from the tosic mushroom *Rhodophyllus rhodopolius*. Its activity was partially destroyed by proteolytic enzyme. Bovine and ovine enythrocytes were not affected. Murine enythrocytes were the most sensitive, followed in order by chicken, rat, horse and human red blood cells (Suzuki et al., 1990).

POLYSACCHAROPEPTIDES AND POLYSACCHAROPROTEINS

Wang et al., (1995) reported a 17-kDa polysacchride-peptide (PSP) complex a mycelial culture of a *Tricholoma* species.which stimulatd T cell proliferation, activated macrophages and inhibited sarcoma growth in mice. The effects were more potent than those of *Coriolus versicolor* PSP. A 15.5-kDa PSP with similar activities was reported from *T. mongolicum* (Wang et al., 1996) Wang et al(1996) showed that PSP from culture medium activated murine macrophages and lymphocytes. The polysaccharoprotein from culture filtrate of *T.lobayense* exhibited antitumor and immunoenhancing activities in mice (Liu et al., 1996).

PSP from *C.versicolor* has a molecular mass of about 100 kDa. It induces cytokine production in experimental animals and exerts antiproliferative activity against tumor cells and antitumor activity *in vivo.* It alleviates the symptoms of cancer patients undergoing radiation therapy or chemotherapy (Ng, 1998). It induces apoptosis in human promyelocytic leukemic HL-60 cells (Yang et al., 2005) (15870943), and exhances the cell cycle dependent activity and apoptotic all death of doxorubin and etoposide in HL-60 cells (Hui et al., 2005) (15944782) and ZR-75-30 breast cancer cells (Wan et al., 2008) (18292947).

Peptide with HIV-1 Reverse Transcriptase Inhibitory Activity

A 4.5-kDa thermostable peptide with the N-terminal sequence KREHGQHCEF with inhibitory activity against HIV-1 reverse transcriptase (IC_{50}=11μM) was isolated from *Russula paludosa* (Wang et al., 2007).

Deoxyribonuclease

A 60-kDa DNase that preferantially digested single-stranded DNA was reported from *Coprinus meiocytes* (Kitamura et al., 1997) (9392529)

Acid Phosphatase

A 145-kDa homotetrameric acid phophatase, with an optimun pH at 4.75 and an optimal tepperature at 65°C, was isolated from *Agaricus bisporus* (Wannet et al., 2000) (15188886). It exhibited broad substrate specificity toward various phosphates. It was inhibited by tartrate and inorganic phosphate.

β -Mannosidase

The 64k-Daβ-mannosidase from fruiting bodies of *Polyporus sulfureus* exhibits a pH optimum between pH 2.4 and pH 3.4 and a Km of 1.6 mM. It releases β-D mannopyranosyl unit from various natural substrates including ovalbumin, human 21-acid-glycoprotein, 6-0-β-D-mannopyranosyl-D-galactose, and 4-0-β-D-mannopyranosyl-N-acetylglucosamine (Wan et al., 1976) (6478).

Carboxylesterase

A 430-kDa homooctameric extracellar carboxylesterase with sequence homology to *Candida rugosa* lipase was reported from *Pleurotus sapidus*. It manifested an optimum temperature at 40 °C, an optimum pH at pH 5.8, and an isoelectric point of 4.5 . It catalyzes the hydrolysis of xanthophyll esters (Zom et al., 2005) (15927887).

Ribonucleases

Ribonucleases have been isolated form many mushroom species including *Pleurotus djamor* (Wu et al., 2010), *Lyophyllum shimeiji* (Zhang et al., 2010), Russula delica (Zhao et al., 2010), *Hypsizigus marmoreus* (Guan et al., 2007), *Agaricus bisporus* (Wang and Ng, 2006), brown oyster mushroom (Xia et al., 2005), *Thelephora ganbajun* (Wang and Ng, 2004), *Pleurotus eryngii* (Ng and Wang , 2004), *Clitocybe maxima (*Wang and Ng, 2004), *Pleurotus sajor –caju* (Ngai and Ng ,2004), *Dictyophora indusiata* (Wang and Ng, 2003), *Russula virescens* (Wang and Ng, 2003), *Termitomyces globulus* (Wang and Ng 2003), *Pleurotus pulmonarius* (Ye and Ng 2002), *Pleurotus tuber-regium* (Wang and Ng, 2001), *Volvariella volvacea* (Wang and Ng, 1999), *Lentinus edodes* (Kobayashi et al., 1992), *Irpex lacteus* (Kobayashi et al., 2003).

Substantial amino acid sequence is detectable only between *L.edodes* and *I. lacteus* RNases polyhomoribonucleotide specificity varies from one RNase to another. Poly C specificity, poly G specificity, poly U specificity, cospecificity toward poly A and poly G, and cospecificity for polyA and polyC, have all been reported. The range of molecular masses runs from 9 kDa to 42 kDa. The optimum temperature for most mushroom RNases lies between 60 °C and 70 °C. Antiproliferative activity toward tumor cells was demonstrated by RNases from *L. shimeiji* (Zhang et al., 2010), *R.delica* (Zhao et al., 2010), *H. marmoreus* (Guan et al., 2007) and *P. sajor-caju*(Ngai and Ng, 2004). *P.sajor-caju* RNase in addition manifests antifungal, antibacterial and antimitogenic activities (Ngai and Ng, 2004).

Lectins

Mushroom lectins have recently been reviewed by Singh et al. (2010) (20105049). Thus only recent publications will be reviewed here.

A 15.9-kDa homodimeric ricin β -like lectin with antiproliferative activity against human leukemic T cells was reported by Pohleven et al (2009) from *Clitocybe nebularis* (19100814). A 17 -kDa lectin with a unique N-terminal inhibitory activity and antiproliferature activity toward breast cancer and hepatoma cells was isolated from the toxic wild mushroom *Inocybe umbrinella* (Zhao et al., 2009) (19111567). A 56-kDa tetrameric N-acetylgalactosamine-specific lectin was reported from *Pleurocybella porrigens* Suzuki et al. (2009).

Zhang et al (2009) reported a dimeric 16-kDa lectin with HIV-1 reverse transcriptase inhibitory and antiproliferative activities from *Pholiota adiposa*. Toxic isolectins were isolated from *Boletus revenatus* (Horibe et al., 2010) (20096904)

Another lectin with a molecular mass of 60-kDa and similar activity was isolated from *Russula delica* (Zhao et al., 2010) A lectin with relatively high thermostability (up to 70 °C), antiproliferative activity toward hepatoma and breast cancer cells and antitumor activity in sarcoma bearing mice was isolated from *Russula lepida* (Zhang et al., 2010) (20378319) A *Hericium erinaceum* lectin with motogenic activity toward spleen cells, antiproliferative activity toward tumor cells, and inhibitory activity HIV-1 reverse transcriptase was reported by Li et al., (2010). A thermostable 30.4-kDa lectin with antiproliferative activity but no HIV-1 reverse transcriptase inhibitory activity was isolated from *Agarius arvensis* by Zhao et al (2011).

A 36-kD dimeric hemolytic lectin was purified from *Amanita virosa* (Antonyuk et al., 2010).

L-Amino Acid Oxidase

Toxophallin is a highly toxic protein from the death cap *Amanita phalloides* that induces apoptosis with a caspase-independent pathway. It is an L-amino acid oxidase (Stasyk et al., 2010) (20121947). Toxovirin from *Amanita virosa* is a thernolabile toxic L-amino acid oxidase (Antonyuk et al., 2010) (20153765).

Amatoxins, Phallotoxins and Virotoxins

Toxic mushrooms including *Amanita*, *Galerina* and *Lepiota* species produce proteinaceous toxins like phallolysin, and peptidic toxins including amatoxins, phallotoxins, and virotoxins (Wieland 1983) (6354951).

CONCLUSION

From the foregoing account an impression can be gathered that a variety of proteins with different biological activities are produced by mushrooms. Some of them are highly toxic and produced by toxic mushrooms. Others have potentially exploitable activities including antoproliferative activity toward cancer cells, antitumor activity toward tumor bearing animals, inhibitory activity toward HIV-1 reverse transcriptase, antifungal activity, and antibacterial activity. These activities are demonstrated by ribosome inactivating proteins, ribonucleases, lectins, uibiquitin-like proteins and peptides, and polysaccharide-peptide and polysaccharide-protein complexes. Hence these molecules can be regarded as antipathogenic. Some of the mushroom enzymes find application e.g. laccases, tyrosinases, xylanases and mannosidases Laccases and xylanases are related to the saprophytic mode of existence of mushrooms. Antifungal proteins, antibacterial proteins, proteases, deoxyribonucleases and ribonucleases may play a defensive function in mushrooms. The last three enzymes break down foreign DNA, RNA and proteins. It is interesting to note that ribonucleases from different mushroom species may have distinctly different N-terminal sequences. The same phenomenon is observed in mushroom antifungal proteins and lectins. On the other hand, sequence homology is detected among some of the mushroom laccases and hemolysins.

The mechanism of action of many of the aforementioned mushroom antifungal proteins, lectins, ribonucleases and ribosome inactivating proteins remain to be elucidated. Efforts should also be dedicated to investigate the medicinal application of some of the mushroom proteins like ribonucleases, ribosome inactivating proteins and antifungal proteins.

In view of the large number of mushroom species, hopefully more mushroom proteins with therapeutic value will be uncovered in the future.

REFERENCES

Antonyuk VO, Klyuchivska OY, Stoika RS.Cytotoxic proteins of Amanita virosa Secr. mushroom: purification, characteristics and action towards mammalian cells. *Toxicon.* 2010, 55:1297-305.

Chan WY, Ng TB, Lam JS, Wong JH, Chu KT, Ngai PH, Lam SK, Wang HX.The mushroom ribosome-inactivating protein lyophyllin exerts deleterious effects on mouse embryonic development in vitro. *Appl Microbiol Biotechnol.* 2010, 85:985-993.

Cui L, Liu QH, Wang HX, Ng TB.An alkaline protease from fresh fruiting bodies of the edible mushroom Pleurotus citrinopileatus. *Appl Microbiol Biotechnol.* 2007 75:81-85.

El-Fakharany EM, Haroun BM, Ng TB, Redwan EROyster mushroom laccase inhibits hepatitis C virus entry into peripheral blood cells and hepatoma cells. *Protein Pept Lett.* 2010,17:1031-9.

Galán B, García Mendoza C, Calonje M, Novaes-Ledieu M.News & notes: production, purification, and properties of an endo-1, 3-beta-glucanase from the basidiomycete Agaricus bisporus. *Curr Microbiol.* 1999,38:190-319.

Guan GP, Wang HX, Ng TB.A novel ribonuclease with antiproliferative activity from fresh fruiting bodies of the edible mushroom Hypsizigus marmoreus. *Biochim Biophys Acta.* 2007 , 1770:1593-1597.

Guan GP, Zhang GQ, Wu YY, Wang HX, Ng TB.Purification and characterization of a novel serine protease from the mushroom Pholiota nameko. *J Biosci Bioeng.* 2011.

Guan GP, Zhang GQ, Wu YY, Wang HX, Ng TB.Purification and characterization of a novel serine protease from the mushroom Pholiota nameko. *J Biosci Bioeng.* 2011 Mar 7. [Epub ahead of print]

Guo Y, Wang H, Ng TB.Isolation of trichogin, an antifungal protein from fresh fruiting bodies of the edible mushroom Tricholoma giganteum. *Peptides.* 2005, 26:575-580.

Han CH, Zhang GQ, Wang HX, Ng TB.Schizolysin, a hemolysin from the split gill mushroom Schizophyllum commune. *FEMS Microbiol Lett.* 2010, 309:115-121.

Han CH, Zhang GQ, Wang HX, Ng TB.Schizolysin, a hemolysin from the split gill mushroom Schizophyllum commune. *FEMS Microbiol Lett.* 2010, 309:115-121.

Hu DD, Zhang RY, Zhang GQ, Wang HX, Ng TB.A laccase with antiproliferative activity against tumor cells from an edible mushroom, white common Agrocybe cylindracea.Phytomedicine. 2011,18:374-379.

Hui KP, Sit WH, Wan JMInduction of S phase cell arrest and caspase activation by polysaccharide peptide isolated from Coriolus versicolor enhanced the cell cycle dependent activity and apoptotic cell death of doxorubicin and etoposide, but not cytarabine in HL-60 cells. *Oncol Rep.* 2005,14:145-155.

Li Y, Zhang G, Ng TB, Wang H.A novel lectin with antiproliferative and HIV-1 reverse transcriptase inhibitory activities from dried fruiting bodies of the monkey head mushroom Hericium erinaceum. *J Biomed Biotechnol.* 2010;2010:716515. Epub 2010 Jun 13.

Li YR, Liu QH, Wang HX, Ng TB.A novel lectin with potent antitumor, mitogenic and HIV-1 reverse transcriptase inhibitory activities from the edible mushroom Pleurotus citrinopileatus. *Biochim Biophys Acta.* 2008, 1780:51-57.

Lu CL, Wang JP, Chen SN.Protease purified from Schizophyllum commune culture broth digests fibrins without activating plasminogen. *Am J Chin Med.* 2010,38:1223-1231.

Ng TB, Lam JS, Wong JH, Lam SK, Ngai PH, Wang HX, Chu KT, Chan WY.Differential abilities of the mushroom ribosome-inactivating proteins hypsin and velutin to perturb normal development of cultured mouse embryos. *Toxicol In Vitro.* 2010, 24:1250-1257.

Ng TB, Lam YW, Wang H.Calcaelin, a new protein with translation-inhibiting, antiproliferative and antimitogenic activities from the mosaic puffball mushroom Calvatia caelata. *Planta Med.* 2003, 69:212-217.

Ng TB, Wang H.Adustin, a small translation-inhibiting polypeptide from fruiting bodies of the wild mushroom Polyporus adusta. *Peptides.* 2004, 25:689-692.

Ng TB, Wang HX.A homodimeric laccase with unique characteristics from the yellow mushroom Cantharellus cibarius. *Biochem Biophys Res Commun.* 2004, 313:37-41.

Ng TB, Wang HX.A novel ribonuclease from fruiting bodies of the common edible mushroom Pleurotus eryngii. *Peptides.* 2004, 25:1365-1368.

Ng TBA review of research on the protein-bound polysaccharide (polysaccharopeptide, PSP) from the mushroom Coriolus versicolor (Basidiomycetes: Polyporaceae). *Gen Pharmacol.* 1998, 30:1-4.

Ngai PH, Wang HX, Ng TB.Purification and characterization of a ubiquitin-like peptide with macrophage stimulating, antiproliferative and ribonuclease activities from the mushroom Agrocybe cylindracea. *Peptides.* 2003, 24:639-645.

Nitheranont T, Watanabe A, Asada Y.Extracellular Laccase Produced by an Edible Basidiomycetous Mushroom, Grifola frondosa: Purification and Characterization. *Biosci Biotechnol Biochem.* 2011. [Epub ahead of print]

Pohleven J, Obermajer N, Sabotic J, Anzlovar S, Sepcić K, Kos J, Kralj B, Strukelj B, Brzin J.Purification, characterization and cloning of a ricin B-like lectin from mushroom Clitocybe nebularis with antiproliferative activity against human leukemic T cells. *Biochim Biophys Acta.* 2009, 1790:173-181.

Shen MH, Kim JS, Sapkota K, Park SE, Choi BS, Kim S, Lee HH, Kim CS, Chun HS, Ryoo CI, Kim Purification, characterization, and cloning of fibrinolytic metalloprotease from Pleurotus ostreatus mycelia. *J Microbiol Biotechnol.* 2007, 17:1271-1283.

Singh RS, Bhari R, Kaur HP.Mushroom lectins: current status and future perspectives. *Crit Rev Biotechnol.* 2010 , 30:99-126.

Stasyk T, Lutsik-Kordovsky M, Wernstedt C, Antonyuk V, Klyuchivska O, Souchelnytskyi S, Hellman U, StoikaA new highly toxic protein isolated from the death cap Amanita phalloides is an L-amino acid oxidase. *FEBS J.* 2010, 277:1260-1269.

Suzuki T, Amano Y, Fujita M, Kobayashi Y, Dohra H, Hirai H, Murata T, Usui T, Morita T, Kawagishi Purification, characterization, and cDNA cloning of a lectin from the mushroom Pleurocybella porrigens. *Biosci Biotechnol Biochem.* 2009, 73:702-709.

Wan CC, Muldrey JE, Li SC, Li YT.beta-Mannosidase from the mushroom Polyporus sulfureus. *J Biol Chem.* 1976, 251:4384-4388.

Wan JM, Sit WH, Louie JC.Polysaccharopeptide enhances the anticancer activity of doxorubicin and etoposide on human breast cancer cells ZR-75-30. *Int J Oncol.* 2008, 32:689-699.

Wang H, Ng TB, Liu Q.Alveolarin, a novel antifungal polypeptide from the wild mushroom Polyporus alveolaris. *Peptides.* 2004,25:693-696.

Wang H, Ng TB.A ribonuclease from the wild mushroom Boletus griseus. *Appl Microbiol Biotechnol.* 2006, 72:912-916.

Wang H, Ng TB.A ribonuclease with distinctive features from the wild green-headed mushroom Russulus virescens. *Biochem Biophys Res Commun.* 2003, 312:965-968.

Wang H, Ng TB.Eryngin, a novel antifungal peptide from fruiting bodies of the edible mushroom Pleurotus eryngii. *Peptides.* 2004, 25:1-5.

Wang H, Ng TB.Ganodermin, an antifungal protein from fruiting bodies of the medicinal mushroom Ganoderma lucidum. *Peptides.* 2006, 27:27-30.

Wang H, Ng TB.Isolation and characterization of velutin, a novel low-molecular-weight ribosome-inactivating protein from winter mushroom (Flammulina velutipes) fruiting bodies. *Life Sci.* 2001, 68:2151-2158.

Wang H, Ng TB.Isolation of a new ribonuclease from fresh fruiting bodies of the straw mushroom. *Biochem Biophys Res Commun.* 1999, 264:714-718.

Wang H, Ng TB.Isolation of a new ribonuclease from fruiting bodies of the silver plate mushroom Clitocybe maxima. *Peptides.* 2004 , 25:935-939.

Wang H, Ng TB.Isolation of a ribonuclease from fruiting bodies of the wild mushroom Termitomyces globulus. *Peptides.* 2003, 24:973-977.

Wang H, Ng TB.Pleureryn, a novel protease from fresh fruiting bodies of the edible mushroom Pleurotus eryngii. *Biochem Biophys Res Commun.* 2001, 289:750-755.

Wang HX, Liu WK, Ng TB, Ooi VE, Chang ST.Immunomodulatory and antitumor activities of a polysaccharide-peptide complex from a mycelial culture of Tricholoma sp., a local edible mushroom. *Life Sci.* 1995, 57:269-281.

Wang HX, Ng TB, Chiu SW.A distinctive ribonuclease from fresh fruiting bodies of the medicinal mushroom Ganoderma lucidum. *Biochem Biophys Res Commun.* 2004, 314:519-522.

Wang HX, NG TB, Liu WK, Ooi VE, Chang ST.Polysaccharide-peptide complexes from the cultured mycelia of the mushroom Coriolus versicolor and their culture medium activate mouse lymphocytes and macrophages. *Int J Biochem Cell Biol.* 1996, 28:601-607.

Wang HX, Ng TB, Ooi VE, Liu WK, Chang ST.A polysaccharide-peptide complex from cultured mycelia of the mushroom Tricholoma mongolicum with immunoenhancing and antitumor activities. *Biochem Cell Biol.* 1996, 74:95-100.

Wang HX, Ng TB.\A new ribonuclease from the black oyster mushroom Pleurotus ostreatus. *Peptides.* 2004, 25:685-687.

Wang HX, Ng TB.A laccase from the medicinal mushroom Ganoderma lucidum. *Appl Microbiol Biotechnol.* 2006, 72:508-513.

Wang HX, Ng TB.A new laccase from dried fruiting bodies of the monkey head mushroom Hericium erinaceum. *Biochem Biophys Res Commun.* 2004, 322:17-21.

Wang HX, Ng TB.A novel laccase with fair thermostability from the edible wild mushroom (Albatrella dispansus). *Biochem Biophys Res Commun.* 2004, 319:381-385.

Wang HX, Ng TB.A novel ribonuclease from fresh fruiting bodies of the portabella mushroom Agaricus bisporus. *Biochem Cell Biol.* 2006, 84:178-183.

Wang HX, Ng TB.Flammulin: a novel ribosome-inactivating protein from fruiting bodies of the winter mushroom Flammulina velutipes. *Biochem Cell Biol.* 2000,78:699-702.

Wang HX, Ng TB.Isolation of a novel ubiquitin-like protein from Pleurotus ostreatus mushroom with anti-human immunodeficiency virus, translation-inhibitory, and ribonuclease activities. *Biochem Biophys Res Commun.* 2000, 276:587-593.

Wang HX, Ng TB.Isolation of pleuturegin, a novel ribosome-inactivating protein from fresh sclerotia of the edible mushroom Pleurotus tuber-regium. *Biochem Biophys Res Commun.* 2001, 288:718-721.

Wang HX, Ng TB.Purification and characterization of a potent homodimeric guanine-specific ribonuclease from fresh mushroom (Pleurotus tuber-regium) sclerotia. *Int J Biochem Cell Biol.* 2001, 33:483-490.

Wang HX, Ng TB.Purification of a laccase from fruiting bodies of the mushroom Pleurotus eryngii. *Appl Microbiol Biotechnol.* 2006 , 69:521-525.

Wang HX, Ng TB.Purification of a novel low-molecular-mass laccase with HIV-1 reverse transcriptase inhibitory activity from the mushroom Tricholoma giganteum. *Biochem Biophys Res Commun.* 2004, 315:450-454.

Wang HX, Ng TB.Purification of a novel ribonuclease from dried fruiting bodies of the edible wild mushroom Thelephora ganbajun. *Biochem Biophys Res Commun.* 2004, 324:855-859.

Wang HX, Ngai HK, Ng TB A ubiquitin-like peptide with ribonuclease activity against various polyhomoribonucleotides from the yellow mushroom Cantharellus cibarius. *Peptides.* 2003, 24:509-513.

Wang J, Wang HX, Ng TB.A peptide with HIV-1 reverse transcriptase inhibitory activity from the medicinal mushroom Russula paludosa. *Peptides.* 2007, 28:560-565.

Wieland T.The toxic peptides from Amanita mushrooms. *Int J Pept Protein Res.* 1983, 22:257-76.

Wong JH, Ng TB, Jiang Y, Liu F, Sze SC, Zhang KY.Purification and characterization of a Laccase with inhibitory activity toward HIV-1 reverse transcriptase and tumor cells from an edible mushroom (Pleurotus cornucopiae). *Protein Pept Lett.* 2010,17(8):1040-7.

Wong JH, Wang HX, Ng TB.Marmorin, a new ribosome inactivating protein with antiproliferative and HIV-1 reverse transcriptase inhibitory activities from the mushroom Hypsizigus marmoreus. *Appl Microbiol Biotechnol.* 2008, 81:669-674.

Wu X, Zheng S, Cui L, Wang H, Ng TB.Isolation and characterization of a novel ribonuclease from the pink oyster mushroom Pleurotus djamor. *J Gen Appl Microbiol.* 2010, 56:231-9.

Yang X, Sit WH, Chan DK, Wan JMThe cell death process of the anticancer agent polysaccharide-peptide (PSP) in human promyelocytic leukemic HL-60 cells. *Oncol Rep.* 2005,13:1201-1210.

Zhang G, Sun J, Wang H, Ng TB.First isolation and characterization of a novel lectin with potent antitumor activity from a Russula mushroom. *Phytomedicine.* 2010, 17:775-781.

Zhang GQ, Sun J, Wang HX, Ng TB.A novel lectin with antiproliferative activity from the medicinal mushroom Pholiota adiposa. *Acta Biochim Pol.* 2009, 56:415-421.

Zhang RY, Zhang GQ, Hu DD, Wang HX, Ng TB.A novel ribonuclease with antiproliferative activity from fresh fruiting bodies of the edible mushroom Lyophyllum shimeiji. *Biochem Genet.* 2010 , 48:658-68.

Zhao JK, Wang HX, Ng TB.Purification and characterization of a novel lectin from the toxic wild mushroom Inocybe umbrinella. *Toxicon.* 2009 , 53:360-366.

Zhao JK, Zhao YC, Li SH, Wang HX, Ng TB.Isolation and characterization of a novel thermostable lectin from the wild edible mushroom Agaricus arvensis. *J Basic Microbiol.* 2011

Zhao S, Zhao Y, Li S, Zhang G, Wang H, Ng TB.An antiproliferative ribonuclease from fruiting bodies of the wild mushroom Russula delica. *J Microbiol Biotechnol.* 2010, 20:693-699.

Zhao S, Zhao Y, Li S, Zhao J, Zhang G, Wang H, Ng TB.A novel lectin with highly potent antiproliferative and HIV-1 reverse transcriptase inhibitory activities from the edible wild mushroom Russula delica. *Glycoconj J.* 2010, 27:259-265.

Zhao S, Zhao Y, Li S, Zhao J, Zhang G, Wang H, Ng TB.A novel lectin with highly potent antiproliferative and HIV-1 reverse transcriptase inhibitory activities from the edible wild mushroom Russula delica. *Glycoconj J.* 2010, Jan 12. [Epub ahead of print]

Zheng S, Liu Q, Zhang G, Wang H, Ng TB.Purification and characterization of an antibacterial protein from dried fruiting bodies of the wild mushroom Clitocybe sinopica. *Acta Biochim Pol.* 2010, 57:43-48.

Zorn H, Bouws H, Takenberg M, Nimtz M, Getzlaff R, Breithaupt DE, Berger RG.An extracellular carboxylesterase from the basidiomycete Pleurotus sapidus hydrolyses xanthophyll esters. *Biol Chem.* 2005, 386:435-440.

In: Mushrooms: Types, Properties and Nutrition
Editors: S. Andres and N. Baumann
ISBN 978-1-61470-110-1

Chapter 16

ACTIVATION AND STABILIZATION OF MUSHROOM TYROSINASE BY ADDITION OF POLYETHYLENE GLYCOLS

Zhen Yang*[*], *Ya-Jun Yue and Ting-Wei Chen
College of Life Sciences, Shenzhen University, Shenzhen 518060, China

ABSTRACT

Effect of polyethylene glycol (PEG) and its molecular weight on the activity and stability of mushroom tyrosinase was studied systematically. The kinetic parameters (such as K_m, V_{max}, optimal pH and temperature, and activation energy) and the thermostability of the enzyme at different temperatures were determined and compared in the aqueous solution with and without addition of PEGs. Polyethylene glycol showed the ability of both activating and stabilizing the enzyme (e.g., a 1.6-fold increase in V_{max}/K_m and a 1.8-fold increase in $t_{1/2}$ were achieved in the presence of PEG 10000), and such PEG-induced activation and stabilization effect was molecular weight dependent. Both our activity and stability data suggest that PEG can affect the enzyme activity and stability basically via its interaction with the enzyme so as to modify the enzyme's active site and conformation or steric structure.

Keywords: Mushroom tyrosinase, polyphenol oxidase, polyethylene glycol, enzyme activation, enzyme stabilization.

INTRODUCTION

Tyrosinase is a copper-containing enzyme that catalyzes the hydroxylation of monophenols to *o*-diphenols with oxygen and their subsequent oxidation to *o*-quinones

[*] Corresponding author: Dr. Zhen Yang
College of Life Sciences, Shenzhen University, Shenzhen 518060, Chin
Email: zyang@szu.edu.cn; a

(Robb, 1984). It is widely distributed in microorganisms, animals, and plants. As a key enzyme in melanin biosynthesis, tyrosinase is involved in browning of plants and in determining the color of mammalian skin and hair. Tyrosinase applications have been well recognized, especially in food, cosmetic, and pharmaceutical industries. Tyrosinase extracted from mushrooms has been found to be catalytically active not only in aqueous solution, but also in nonaqueous environments such as organic solvents (Yang & Robb, 1993) and ionic liquids (Yang et al., 2008). It is of great interest in organic synthesis because of its specific regioselectivity (Kazandjian & Klibanov, 1985).

Catalytic performance of an enzyme can be modified by means of protein engineering, chemical modification, and addition of some additives such as polyols and sugars (Polizzi et al., 2007). Polyethylene glycol (PEG) is one of the major additives that has been extensively employed in biomedical and biotechnological applications due to its unique properties and its general compatibility with biological systems (Harris, 1992). Although PEGs have been often reported to confer stabilization (Simon et al., 2004; Spreti et al., 2004) and activation (Mukerjea et al., 2006; Pancera et al., 2002) to a protein in solution, the mechanism for such effects is still unclear yet.

One of our major research interests is in optimization of the catalytic performance of mushroom tyrosinase. Our recent work has demonstrated that different from its cationic and anionic partners, the nonionic surfactant, Brij 52 (polyethylene glycol hexadecyl ether, $C_{16}H_{33}(OCH_2CH_2)_nOH$, n~2), can both activate and stabilize mushroom tyrosinase (Yang et al., 2007). The mechanism for such effect is worth a further investigation. We speculated that the Brij 52-induced activation and stabilization was related to the presence of the polyethylene glycol fragment in the surfactant. Therefore in this paper, we tried to investigate how PEGs affect both activity and stability of mushroom tyrosinase. It is worth mentioning that our aim is to study the effect of PEGs as additives, rather than as chemical modifiers. By examining the impact of PEGs with different molecular weights, this study will provide a better understanding on the effect of PEGs on enzyme performance and on the general mechanism of the PEG-enzyme interaction.

EXPERIMENTAL

Materials

Fresh white mushrooms were purchased from a local supermarket in Shenzhen, China. L-dopa (3,4-dihydroxy-L-phenylalanine) was obtained from Sigma-Aldrich. Polyethylene glycols with different molecular weights (200, 1000, 2000, 4000, 6000, 10000) were supplied from ABCR GmbH & Co. KG (Karlsruhe, Germany).

Enzyme Preparation

The enzyme solution was obtained as the supernatant after centrifugation of an extract of fresh mushrooms (50 g) in 100 ml Na phosphate buffer (50 mM, pH 6.0) and was used directly for kinetic assays (Yang & Robb, 1993). For practical storage purpose, the enzyme used for thermal inactivation studies went through additional steps. The fresh enzyme solution was subjected to addition of $(NH_4)_2SO_4$, and the precipitate obtained after centrifugation was mixed with celite 545 (Sigma) and dried in a vacuum oven as indicated in (Yang & Robb, 1993). The enzyme solution used for stability tests was obtained as the supernatant after mixing 20 mg of such enzyme powders thoroughly with 1 ml of 50 mM phosphate buffer (pH 6.0). The enzyme was prepared exactly following the above procedures, and experimental results obtained with different enzyme preparations were proved to be reproducible.

Activity Studies

The activity of the enzyme solution was determined by following the formation of dopachrome spectrophotometrically at 30°C as indicated in (Yang et al., 2007). After addition of 50 μl enzyme extract to a cuvette containing 1.2 ml phosphate buffer (50 mM, pH 6.0) and 0.8 ml 10 mM L-dopa (prepared in the same phosphate buffer), the solution was immediately mixed and the increase in absorbance at 475 nm (indicating the formation of dopachrome) was recorded with the Pharmacia Biotech Ultraspec 2000 UV/Vis spectrophotometer equipped with a thermostated cell. The initial rate was calculated from the linear part of the progress curve (i.e. over the first 1-2 min). Both phosphate buffer and the substrate (L-dopa) solution were prepared containing a fixed concentration of a PEG: 100 mM for both PEGs 200 (0.02 g/ml) and 2000 (0.2 g/ml), and 30 mM for PEG 10000 (0.3 g/ml). A lower concentration for PEG 10000 was used due to its lower solubility in water. For determination of the K_m and V_{max} values, reactions were carried out at 30°C in a cuvette where 50 μl of enzyme solution was added to 2 ml 50 mM phosphate buffer containing 0-4.0 mM L-dopa (7 substrate concentrations were selected as significantly below and above the observed K_m: 0.25, 0.5, 0.75, 1.0, 2.0, 3.0, and 4.0 mM). The K_m and V_{max} values of the enzyme were obtained from the Lineweaver-Burk plots. Optimum pH was determined by measuring tyrosinase activity at 30°C over a pH range of 5.0-8.0 in 50 mM phosphate buffer. It has to be noted that the pH values reported here were the apparent pH's as they were recorded by the pH electrode in the fairly concentrated PEG solutions instead of in dilute aqueous media. Optimum temperature was obtained by measuring the activity of tyrosinase in 50 mM phosphate buffer (pH 6.0) at various reaction temperatures (20-65°C), and the data were used to calculate the activation energy according to the Arrhenius equation. The PEG concentrations in the enzyme assay solution were 50 mM for PEGs 200, 2000, 4000 and 10 mM for PEGs 6000 and 10000 (equivalent to 0.01, 0.1, 0.2, 0.06, and 0.1 g/ml, respectively). Experiments have confirmed that the presence of either the PEGs or the substrate did not affect the pH of the phosphate buffer significantly. All the experiments throughout this study have been repeated at least twice, with the normal experimental errors falling in the range of ±5%.

Stability Studies

The enzyme solution in 50 mM phosphate buffer (pH 6.0) containing 0.1 g/ml of each PEG was incubated in a water bath controlled at different temperatures (40°C, 50°C, 60°C, 70°C). Periodically 50 μl of each enzyme solution was taken for activity assay at 30°C in 2 ml of 50 mM phosphate buffer (pH 6.0) containing 4.0 mM L-dopa. The half-life ($t_{1/2}$) for each enzyme solution was measured from the time course of the enzyme deactivation.

RESULTS AND DISCUSSION

Activity Studies

In order to study whether PEGs have special effects on enzyme activity, kinetic parameters of mushroom tyrosinase, such as K_m, V_{max}, optimal pH and temperature, and activation energy, in the absence and presence of different PEGs, were determined. Figure 1 shows the Lineweaver-Burk plots for the enzyme in the PEG-free buffer solution (as the control) and in the buffer containing PEG 200, 2000, and 10000. All the four plots show good linearity, with an R^2 value of 0.9957, 0.9876, 0.9928, and 0.9979, respectively.

The K_m, V_{max} values of the enzyme did vary upon the addition of different PEGs, and the variation seemed to be more serious with an increase in the molecular weight of the PEG (Figure 2). For all the PEGs that were tested, addition of PEG led to a simultaneous decrease in both K_m and V_{max}. As the reduction in K_m was more pronounced, a higher catalytic efficiency (V_{max}/K_m) of the enzyme was achieved. Therefore in the buffer containing PEG 10000, the enzyme obtained a 1.6-fold increase in its catalytic efficiency as a result of 2.1- and 1.3-fold reductions in both K_m and V_{max}, respectively.

pH dependence profiles of mushroom tyrosinase in the absence and presence of the three different PEGs were depicted in Figure 3. The optimal pH for the native enzyme was 6.5, which was not shifted by addition of PEG 200. A right shift was observed when a PEG with a higher molecular weight was present.

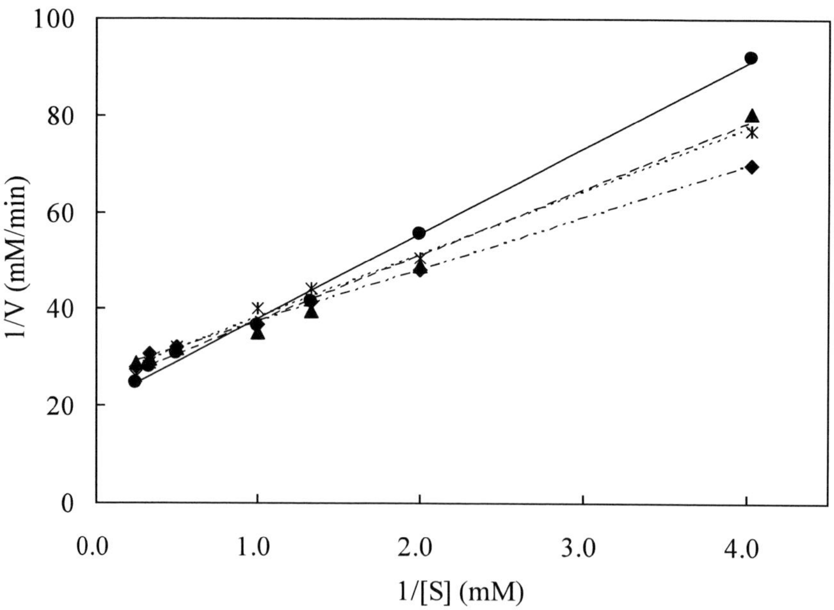

Figure 1. Lineweaver-Burk plots for the enzyme in the PEG-free buffer solution (•) and in the buffer containing PEG 200 (▲), 2000 (✱), and 10000 (♦). V represents the initial reaction rate.

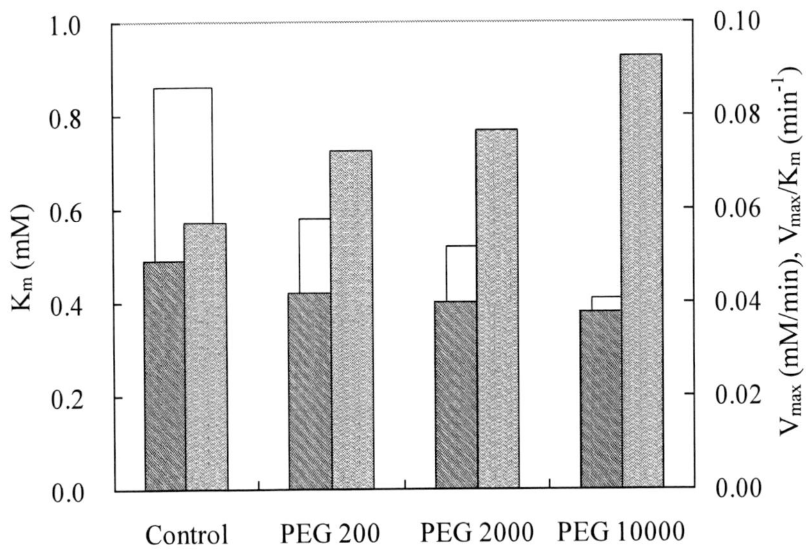

Figure 2. Kinetic parameters obtained by the enzyme in pure phosphate buffer (as the control) and in the buffer solution containing PEG 200, 1000, and 10000: K_m (empty bars), V_{max} (striped bars), and V_{max}/K_m (waved bars).

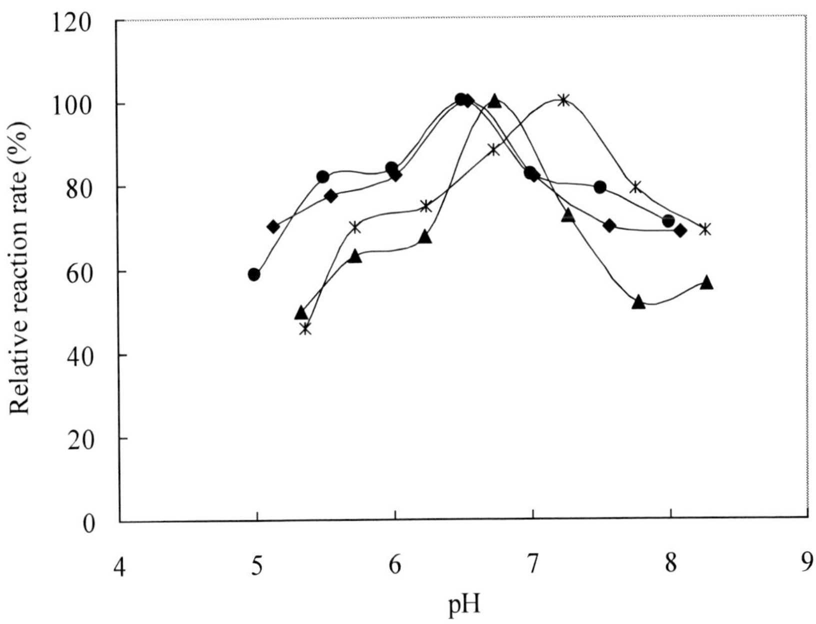

Figure 3. pH dependence of mushroom tyrosinase in pure phosphate buffer (as the control) (•) and in the buffer solution containing PEG 200 (♦), 1000 (▲), and 10000 (✱).

Figure 4 shows the variation of enzyme activity upon the change in reaction temperature whether or not PEG was involved. The optimal reaction temperature for the enzyme in phosphate buffer was 50°C. This seemed to be slightly lowered by addition of PEG 200. PEG 2000 did not affect the optimal temperature of the enzyme, but a 5°C increase was observed when PEG 4000 or PEG 10000 was present. This seemed to suggest that the presence of a PEG with a high enough molecular weight might trigger an enhancement in the optimal reaction temperature of the enzyme. Above the optimum temperature the enzyme showed a progressively lower reaction rate derived from a linear slope. The data in Figure 4 were used to calculate the activation energy of the enzyme according to the Arrhenius equation, and the results are 39.8, 39.9, 42.3, 43.5, 42.3, and 33.1 kJ/mol for the enzyme in pure buffer and in the buffer containing PEG 200, 2000, 4000, 6000, and 10000, respectively.

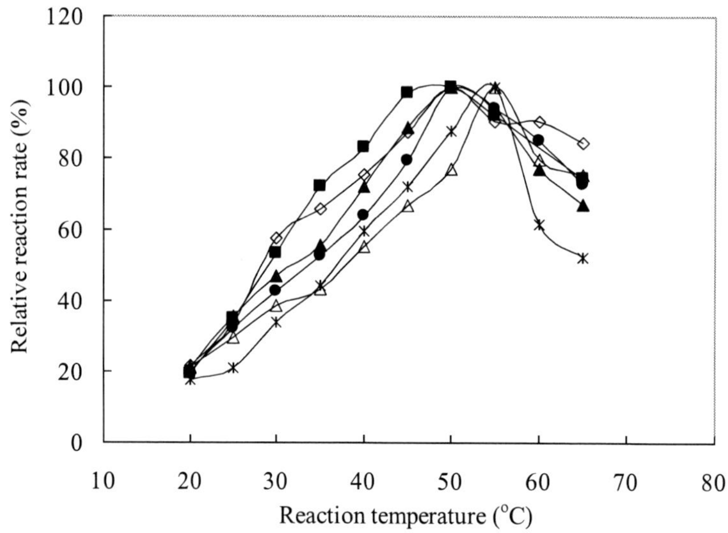

Figure 4. Variation of enzyme activity in PEG-free phosphate buffer (as the control) (•) and in solutions containing PEG 200 (■), 2000 (▲), 4000 (✱), 6000 (◇), and 10000 (Δ) upon the change in reaction temperature.

Stability Studies

Thermostability of the enzyme in the presence of different PEGs was determined by following the loss of enzyme activity at different temperatures (40-70°C) as a function of the incubation time. The time-dependent loss in activity was used to calculate the half life of the enzyme. Enzyme activity usually declined gradually with incubation time. Interestingly, when tyrosinase was incubated in phosphate buffer containing a PEG with a high molecular weight such as PEG 4000, 6000, or 10000, there was usually a short initial period where the enzyme's residual activity increased to up to 147% of its original value before dropping gradually. An example was shown in Figure 5. The thermoinactivation data did not follow the classical first-order kinetics, suggesting a more complex deactivation mechanism that the enzyme might follow.

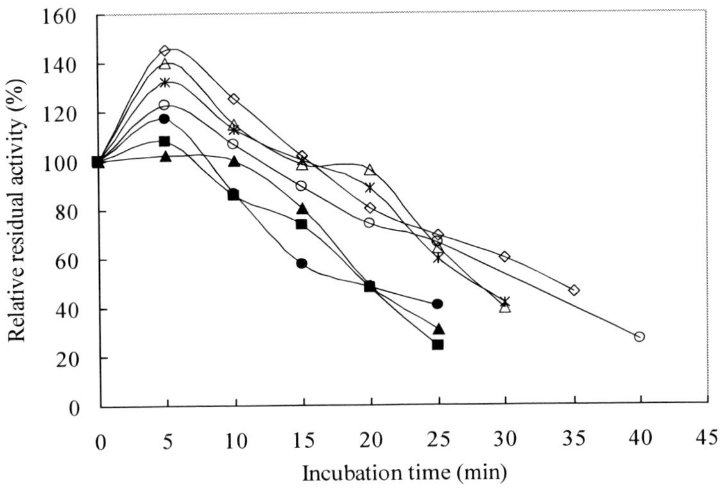

Figure 5. Time course of enzyme inactivation at 60°C in PEG-free phosphate buffer (as the control) (•) and in the buffer containing PEG 200 (■), 1000 (▲), 2000 (✱), 4000 (Δ), 6000 (o), and 10000 (◊).

The half lives of the enzyme in the presence of various PEGs were illustrated in Figure 6. PEG did present a stabilization effect on the enzyme, and this effect seemed to be associated with the molecular weight of the PEG, generally with a higher half life obtained by the enzyme in the presence of a PEG with a higher molecular weight. For instance, at 40°C the half life for the enzyme incubated in phosphate buffer was 31.4 hrs. This value was raised to 45.9 hrs when 0.1 g/ml of PEG 4000 was present. The half lives for the enzyme when incubated in phosphate buffer containing PEG 6000 and PEG 10000 were expected to be much higher, as under such condition the enzyme remained almost 100% active within the incubation period set for the experiment. At 60°C, a 1.8-fold increase in the half life was obtained by the enzyme when incubated in the presence of 0.1 g/ml PEG 10000.

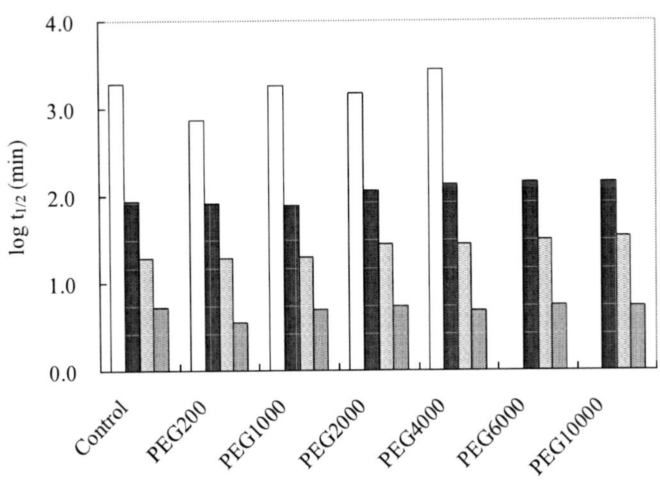

Figure 6. Comparison of the half-lives ($t_{1/2}$, min) obtained by the enzyme at four different incubation temperatures in the absence (as the control) and presence of different PEGs: 40°C (empty bars), 50°C (striped bars), 60°C (dotted bars), and 70°C (crossed bars).

Discussion

By comparing the above experimental data with the results obtained for the same enzyme in the presence of Brij 52 (Yang et al., 2007), one can find that both additives (PEGs and Brij 52) shared the same impact on the catalytic performance of mushroom tyrosinase:

1) The catalytic efficiency of the enzyme (V_{max}/K_m) was enhanced, and such enhancement was basically associated with the decrease in the enzyme's K_m value;
2) No significant shift in both pH and temperature optima was observed when either Brij 52 or low molecular weight PEGs (e.g., PEG 200) was present;
3) The thermostability of the enzyme was increased in the presence of either Brij 52 or PEGs with high molecular weights.

Therefore, these results demonstrate that PEGs can both activate and stabilize mushroom tyrosinase, hence reinforcing our previous speculation that achievement of activation and stabilization by mushroom tyrosinase in the presence of Brij 52 was largely attributed to the PEG fragment in the surfactant.

Although not very common, there have been a few reports in literature regarding the PEG-induced activation effect on enzymes. Addition of 0.4% (w/v) PEG 10000 or PEG 15000 respectively to α-amylases from two sources resulted in an enhancement (up to 150% of original) in the enzyme activity, and this activation effect has been demonstrated to be due to the interaction of PEG with the enzyme rather than with the substrate, leading to the formation of highly active species from a series of less active forms (Mukerjea et al., 2006). A 20% increase in activity of glucose-6-phosphate dehydrogenase was also obtained in aqueous buffer solution containing 20% (w/v) PEG 400 or PEG 4000 (Pancera et al., 2002). In the case of pyruvate kinase, addition of 5% (w/v) PEG 8000 caused a 2.6-fold increase in V_{max} and 12.5- and 2.0-fold reductions in K_m values for the two substrates, phosphoenolpyruvate and ADP, respectively (Podestá & Plaxton, 1993).

In our situation, mushroom tyrosinase obtained a reduction in both K_m and V_{max} values with addition of PEGs, and the reduction was more serious when the PEG with a higher molecular weight was used. The lower K_m values indicate that the PEGs play an important role in modifying the active site of the enzyme, triggering an active conformation which facilitates its affinity for the substrate. On the other hand, the active site of the enzyme may also be partially blocked by the PEG surrounding the enzyme molecule, with a more serious effect in the presence of a PEG with an increasing molecular mass or a longer chain length. Also, the flexibility of the enzyme molecule may be limited by the surrounding of the PEG molecules around. As a result, a less effective enzyme concentration is available for catalysis, thus triggering a lower V_{max}. This dual function of the PEG may be used to account for our experimental result that the activation energy for the enzyme did not go down with an increase in the molecular weight of the PEG to match the enzyme activity data. Fortunately, the effect of PEG was more pronounced on the K_m than on the V_{max} values, thus offering a higher catalytic efficiency.

PEG has been popular for its role as a stabilizer for proteins. PEG-induced stabilization has been achieved by alkaline protease against heat (Beg and Gupta 2003), by catalase against pH (Costa et al., 2002), by chloroperoxidase against H_2O_2 (Spreti et al., 2004), and by α-chymotrypsin against organic solvents such as ethanol, dioxane and acetonitrile (Simon et al., 2002, 2004; László et al., 2001). Our previous work has also revealed that covalent attachment of a PEG 5000 on subtilisin Carlsberg resulted in enhanced enzyme stability against both pH and temperature (Yang et al., 1996).

This study offered another example of enzyme stabilization induced by PEG. An enzyme goes through deactivation under thermal stress, basically due to the disruption of the hydrophobic interactions, hydrogen bonds, and other interactions that are responsible for maintenance of the tertiary protein structure. PEG has been taken as one of the most useful protein salting-out agents, and this effectiveness is largely contributed to the PEG-induced preferential hydration of the protein, which is increased with an increase in the molecular weight of the PEG mainly due to its large excluded volume effect (Arakawa & Timasheff, 1985). This can be used to explain the relationship between the enzyme's thermostability and the molecular weight of the PEG as shown in Figure 6. The viewpoint that a protein preserves a more compact and more stable tertiary structure in the presence of a PEG with a higher molecular weight (up to 20000) was supported by the UV CD measurements carried out by Simon et al. (2004). The same study has also confirmed that α-chymotrypsin was stabilized in 60% ethanol with addition of PEG 6000, 12000, and 20000 (rather than in the presence of PEG 400), but this PEG-induced stabilization was actually not molecular weight-dependent. A previous study conducted by the same authors (Simon et al., 2002) has also revealed that the loss of enzyme stability was not necessarily accompanied by a loss in the enzyme's steric structure. Therefore, the effect of PEG on the enzyme stability-structure relationship is worth a further investigation.

In fact, a PEG is a polyether with repeating units of [-CH_2CH_2O-] and two hydroxyl end groups. On the one hand, it becomes more hydrophobic as its molecular weight (and hence its chain length) increases (Forciniti et al., 1991). This would strengthen the hydrophobic interactions of the enzyme molecule, thus offering higher enzyme stability. On the other hand, it is necessary to consider the contribution of the hydroxyl end group (Pancera *et al.* 2002). Both ethanol and glycols (Herskovits et al., 1970) have been reported to be denaturants to proteins. Knoll and Herman (1981) have pointed out that the presence of PEG in aqueous solution could enhance the protein's denaturation by lowering the polarity of the solvent, and could stabilize the compact native conformation of a protein by favoring the unfolded structure which has a larger excluded volume relative to the folded conformation; the net effect would depend on the structure of the PEG and the protein. When a PEG has a short chain length, its hydroxyl end groups may be more polar and stronger in competing with H-bonds in the enzyme molecule, thus disrupting the enzyme conformation and even denaturing the enzyme. This can be used to explain the unusual behavior that mushroom tyrosinase exhibited in the presence of PEG 200 as compared to that in solutions containing PEGs with higher molecular weights: Unequivocally low half-lives were obtained by the enzyme in the buffer with addition of PEG 200 (and sometimes PEG 1000) at all the four incubation temperatures tested (Figure 6); unlike other PEGs that had an ability to slightly raise both the optimal pH and temperature for the enzyme, PEG 200 did not shift the pH optimum (Figure 3), and instead of increasing the enzyme's optimal temperature it has even slightly lowered it (Figure 4). Indeed, with the aid of small angle X-ray scattering measurements, Pancera *et al.* (2002) found out that in PEG 4000-containing solution glucose-6-phosphate dehydrogenase assumed a conformation close to its crystalline structure, whereas in PEG 400 solution different aggregates of the enzyme were formed as in pure buffer. Simon *et al.* (2004) have also reported that opposite to its higher molecular weight partners, PEG 400 had no effect on stabilizing α-chymotrypsin in 60% ethanol solution.

Overall, PEG can affect the enzyme activity and stability via its interaction with the enzyme to modify the enzyme's active site and conformation or steric structure. Such effect is directly related to the molecular weight of the PEG. A deeper investigation of the basic mechanism involved by utilizing various biophysical techniques is being conducted in our laboratory.

REFERENCES

Arakawa, T., & Timasheff, S. N. (1985). Mechanism of poly(ethylene glycol) interactions with proteins. *Biochem., 24,* 6756-6762.

Beg, Q. K., & Gupta, R. (2003). Purification and characterization of an oxidation-stable, thiol-dependent serine alkaline protease from *Bacillus mojavensis*. *Enz. Microb. Technol., 32,* 294-304.

Costa, S. A., Tzanov, T., Carneiro, A. F., Paar, A., Gübitz, G. M., & Cavaco-Paulo, A. (2002). Studies of stabilization of native catalase using additives. *Enz. Microb. Technol.,* 30, 387-391.

Forciniti, D., Hall, C. K., & Kula, M. R. (1991). Protein partitioning at the isoelectric point: Influence of polymer molecular weight and concentration and protein size. *Biotechnol. Bioeng., 38,* 986-994.

Harris, J. M. (1992). *Poly(ethylene Glycol) Chemistry; Biotechnical and Biomedical Applications.* New York: Plenum Press.

Herskovits, T. T., Gadegbeku, B., & Jaillet, H. (1970). On the structural stability and solvent denaturation of proteins. I. Denaturation by the alcohols and glycols. *J. Biol. Chem., 245,* 2588-2598.

Kazandjian, R. Z., & Klibanov, A. M. (1985). Regioselective oxidation of phenols catalyzed by polyphenol oxidase in chloroform. *J. Am. Chem. Soc., 107,* 5448-5450.

Knoll, D. A., & Hermans, J. (1981). Effect of poly(ethylene glyol) on protein denaturation and model compound pKa's. *Biopolymers, 20,* 1747-1750.

László, K., Szava, A., & Simon, L. M. (2001). Stabilization of various α-chymotrypsin forms in aqueous-organic media by additives. *J. Mol. Catal. B: Enzym., 16,* 141-146.

Mukerjea, R., Slocum, G., Mukerjea, R., & Robyt, J. F. (2006). Significant differences in the activities of α-amylases in the absence and presence of polyethylene glycol assayed on eight starches solubilized by two methods. *Carbohydr. Res., 341,* 2049-2054.

Pancera, S. M., da Silva, L. H. M., Loh, W., Itri, R., Pessoa Jr, A., & Petri, D. F. S. (2002). The effect of poly(ethylene glycol) on the activity and structure of glucose-6-phosphate dehydrogenase in solution. *Colloids Surf. B: Biointerfaces, 26,* 291-300.

Podestá, F. E., & Plaxton, W. C. (1993). Activation of cytosolic pyruvate kinase by polyethylene glycol. *Plant Physiol., 103,* 285-288.

Polizzi, K. M., Bommarius, A. S., Broering, J. M., & Chaparro-Riggers, J. F. (2007). Stability of biocatalysts. *Curr. Opin. Chem. Biol., 11,* 220-225.

Robb, D. A. (1984). Tyrosinase. In Lontie R (Ed), *Copper proteins and copper enzymes, vol.II.* (pp 207-141). Boca Raton: CRC Press.

Simon, L. M., Kotormán, M., Garab, G., & Laczkó, I. (2002). Effects of polyhydroxy compounds on the structure and activity of α-chymotrypsin. *Biochem. Biophys. Res. Commun., 293,* 416-420.

Simon, L. M., Kotormán, M., Szabó, A., Garab, G., & Laczkó, I. (2004). Effects of polyethylene glycol on stability of α-chymotrypsin in aqueous ethanol solvent. *Biochem. Biophys. Res. Commun.,, 317,* 610-613.

Spreti, N., Germani, R., Incani, A., & Savelli, G. (2004). Stabilization of chloroperoxidase by polyethylene glycols in aqueous media: kinetic studies and synthetic applications. *Biotechnol. Prog., 20,* 96-101.

Yang, Z., Yue, Y.-J., & Xing, M. (2008). Tyrosinase activity in ionic liquids. *Biotechnol. Lett., 30,* 153-158.

Yang, Z., Deng, J., & Chen, L.-F. (2007). Effect of ionic and nonionic surfactants on the activity and stability of mushroom tyrosinase. *J. Mol. Catal. B: Enzym., 47,* 79-85.

Yang, Z., & Robb, D. A. (1993). Comparison of tyrosinase activity and stability in aqueous and nearly nonaqueous environments. *Enz. Microb. Technol., 15,* 1030-1036.

Yang, Z., Domach, M., Auger, R., Yang, F. X., & Russell, A. J. (1996). Polyethylene glycol-induced stabilization of subtilisin. *Enz. Microb. Technol., 18,* 82-89.

In: Mushrooms: Types, Properties and Nutrition
Editors: S. Andres and N. Baumann
ISBN 978-1-61470-110-1

Chapter 17

AGARICUS BM MUSHROOM MAY PROTECT ATHLETES AGAINST INFLAMMATION, INFECTION AND ASTHMA

G. Hetland*
Dept of Cellular Therapy, Oslo University Hospital, Oslo, Norway

ABSTRACT

Athletes are known to be prone to illnesses such as infection and asthma, which is due to exercise-related skewness of the immune response, and injury-related inflammation in the muscle-skeleton system. The Brazilian medicinal mushroom, *Agaricus blazei* Murill (AbM), has strong immunomodulatory properties and is shown to protect against infections, allergy and asthma in animal models. An AbM-based extract (AndoSan^{TM}) is also found to have anti-inflammatory effect both in healthy individuals and in patients with inflammatory bowel diseases. Hence, it is suggested that AbM mushroom extract may be beneficial for athletes during strenuous exercise.

Keywords: *Agaricus blazei* M., medicinal mushroom, allergy.

INTRODUCTION

Strenuous exercise suppresses cellular immunity by lowering type-1/type-2 cytokine ratio and increases susceptibility to infections in addition to inducing a systemic inflammatory response [21]. In elite swimmers also the frequency of allergy, asthma and airway hyper-responsiveness is increased [4]. A recent report confirmed that bronchial hyper responsiveness (BMR) is frequently found among adolescent competitive swimmers [20]. This is explained by lower type-1/type-2 cytokine ratio in athletes [21] because type-1

* Correspondence: Geir Hetland, MD, PhD, Dept. of Cellular Therapy, Oslo University Hospital, Ullernchausseen 70, 0310 Oslo, Norway
E-mail: geir.hetland@rikshospitalet.no

cytokines such as IFNγ protect against infections and cancer, whereas type-2 cytokines like IL-4 promote allergy and asthma. Other elite endurance athletes at risk are cross-country skiers [14]. A previous report from Osterud et al. [16] concluded that endurance ski racing causes the mobilization of activated leukocytes that may in fact induce activation of the coagulation system. This is a finding closer associated with that in patients (e.g. disseminated intravascular coagulation, DIC, in sepsis) than in healthy individuals.

Agaricus bM (AbM) is an edible medicinal *Basidiomycetes* mushroom of Brazilian origin related to the champignon (*Agaricus bisporus*), which is used in traditional medicine and cultivated commercially for its positive health effects. AbM is a known immunomodulator [15] and has antitumor and other effects in mouse models [11]. Recently, we have examined the effect of the AbM-based extract, AndoSan™, on bacterial sepsis [3,2] and allergy [7] in mouse models, and inflammation in humans [12, 9]. There are also anecdotal reports from athletes who have tried AbM extract during strenuous exercise periods, and say that they were less prone to sickness such as colds/other upper respiratory infections and asthma, and recovered faster from injuries.

Here, electronic database PubMed was used as search tool using key words *Agaricus blazei* in combination with inflammation, infection, allergy, or asthma. Examples given are figures from own work with the AbM-based extract, AndoSan™ (www.immunopharma.net), which is a Japanese-produced, GMP-certified mushroom extract containing ca. 82% of AbM, and 15% and 3% of the related *Basidiomycetes* mushrooms, *Hericium erinaceum* and *Grifola frondosa*, respectively.

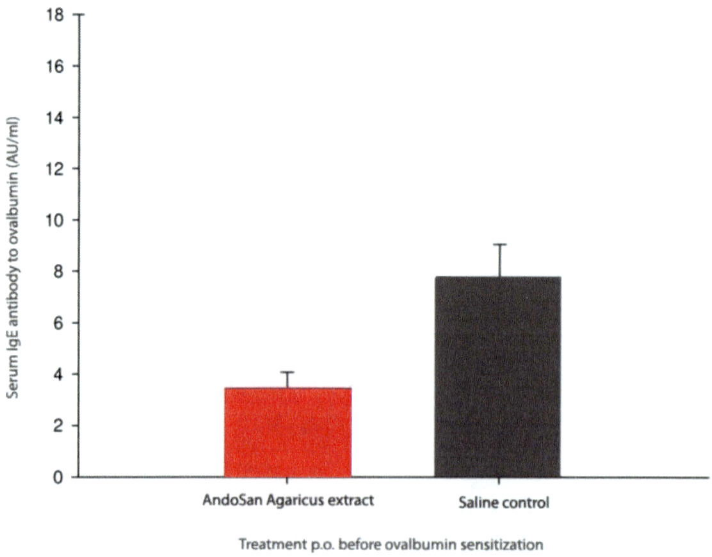

Figure 1. (from [7], previously published by a BMC journal, which allows reuse): Female NIH/Ola mice were given either 200 μl of the AndoSan™ AbM-based extract or PBS intragastrically on day -1 and 10 μg of ovalbumin s.c. on day 0 and again on day 20, before exsanguination for serum on day 26. IgE anti-ovalbumin levels were lower in the AbM- than PBS-treated groups ($p<0.01$). Similar results were found if AbM extract or PBS was given 3 weeks after the allergen immunization (not shown, please see [7]). IgG2a anti-ovalbumin levels (Th1 response) tended to show the opposite result (not shown).

Asthma and Allergy

An *Agaricus blazei* water extract has been shown to inhibit mast cell-mediated anaphylaxis-like reaction in the mouse [6]. Another group has reported that AbM extract ameliorated the skewded Th1/Th2 balance in asthma-induced mice [22]. This was determined by reduced levels of IgE and IgG1 (both due to Th2 response) but not IgG2a (Th1 response) OVA-specific antibodies, in addition to reduced total and eosinophil cell numbers in bronchial alveolar lavage fluids, as well as concentration of IL-5, the chemotactic factor for eosinophils [22]. This has been confirmed by a recent [5] study and our own using AndoSan^{TM} [7] on OVA-sensitized allergy in mice, in which OVA-sensitized allergy in the mice was suppressed by oral AbM administration. This was determined by decreased serum IgE anti-OVA levels (Figure 1), and decreased IL-4 and -5 levels in splenocytes from the sacrificed animals, giving a higher type-1/type-2 cytokine ratio in the cell cultures from the AbM-treated animals [7].

Infection

We have found that the AbM-based extract, AndoSan^{TM}, protected against Gram positive and Gram negative sepsis in mice [3,2]. This was shown by reduced bacteremia in the mushroom group compared with the saline controls and increased survival in the former; 50% vs 13% (Figure 2) and 34% vs 0%, in the two murine infection models, respectively. Also, we have observed a little, but not statistically significant reduction, in viral load after oral 7 days' intake of AndoSan^{TM} in patients with chronic hepatitis C virus infection [10]. Other effects on vira are that of ethanol extracts of an isolated polysaccharide of AbM, which inhibited initial stage of the replication of poliovirus [8]. In line with this, Sorimachi et al. [19] have previously shown that AbM extract inhibited the cytopathic effect of WEE virus in cell culture.

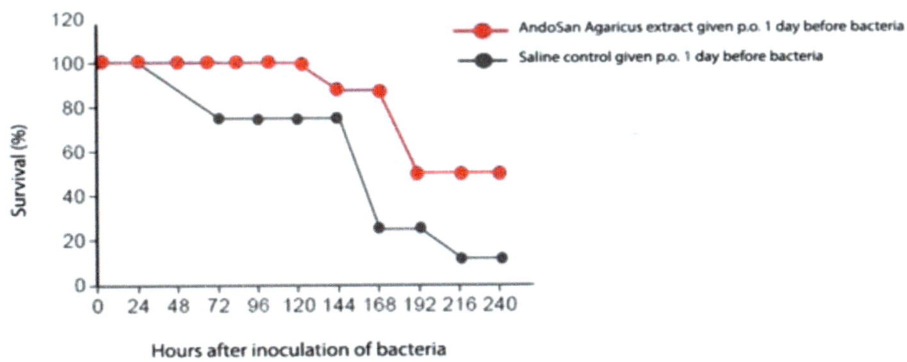

Figure 2. (from [3], permission granted for republication by Scand J Immunol, where the figure was originally published). 5-6 week old female inbred NIH/Ola mice were given either 200 μl of AndoSan^{TM} AbM extract or phosphate-buffered saline (PBS) intragastrically a day before i.p. injection of 1 million colony-forming units of *Streptococcus pneumoniae* serotype 6B. There was a significant difference ($p<0.05$) between survival after treatment with AndoSan^{TM} (red line) and PBS (black line).

INFLAMMATION

In traditional Chinese medicine, AbM is used in the treatment of tissue inflammation as well as cancer, and the mushroom has been shown to suppress phytohemagglutinin-induced PBMC proliferation non cytotoxically by arresting the cells in the G1 cell cycle [13]. Since inflammation is brought about by leukocytes moving to the sites of tissue injury to repair it, also leukocyte proliferation is important for the inflammatory response. In addition to reduced levels of proinflammatory cytokines in the blood of healthy individuals, we have similar observations in patients with the inflammatory bowel diseases, Crohn's disease and ulcerative colitis [9]. In the latter group there was also a significant reduction in the initially high levels of the inflammatory marker, calprotectin, in feces. In healthy volunteers (Figure 3), who are more alike athletes, and IBD patients 12 days'AndoSan™ intake reduced levels of the pro-inflammatory cytokines TNFα, IL-1β and IL-6, and also of IL-17 in plasma. Some of these patients spontaneously reported less joint symptoms (Crohn's disease) and less diarrhea and blood in the stools (ulcerative colitis). IL-6 and TGF-β helps differentiate Th0 cells to Th17 cells [1], which produce IL-17 that can induce autoimmune inflammation in diseases such as multiple sclerosis and rheumatoid arthritis where it also is responsible for joint destruction [18]. In the IBD patients also the important proinflammatory cytokines MIP-1β, MCP-1 and IL-8 were significantly reduced, as well as growth factors G-CSF and GM-CSF for granulocytes and monocytes. Hence, a full double-blinded, placebo-controlled trial is planned with the AbM extract in such patients.

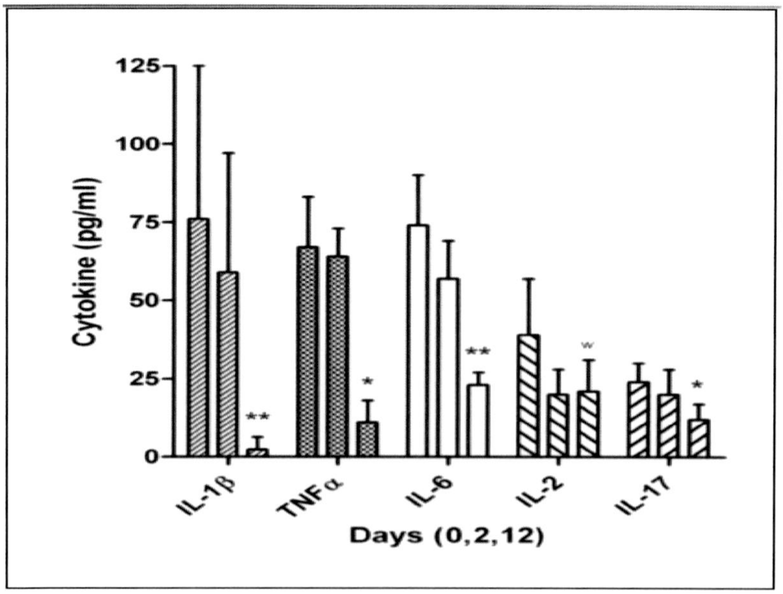

Figure 3. (from [12], permission granted for republication by Scand J Immunol, where the figure was originally published). Significantly (p<0.05) reduced levels (mean ± SEM) of four cytokines in the blood of 8 healthy volunteers after intake of 60 ml a day of AndoSan™ for 12 days. Days 0, 2 and 12 are depicted by the first, second and third bar graph from the left respectively. The *p*-values compare with the cytokine levels at day 0 prior to intake of AndoSan™. $^{w}p = 0.05$, $*p < 0.05$, $**p < 0.01$ respectively. The chemokine MIP-1β that was most reduced by the AbM-based extract in IBD patients [9], was not measured here.

AbM is further shown to abrogate tetrachloride-induced liver injury in rats as determined by reduced serum ALT and AST levels, and reduced hepatic necrosis and fibrosis induced by the agent [23]. One may speculate whether AbM also can abrogate trichloramine-induced bronchial damage and the following asthma in competitive swimmers. ALT and AST are enzymes for liver function that are raised during inflammation of the liver, and AST may also be raised during inflammation of the heart following infarction, and of striped muscle [17]. AbM may in fact counteract the systemic inflammation that is shown to follow exhaustive exercise [21].

CONCLUSION

The positive effects obtained with AndoSan™ on infection, allergy and inflammation indicates that AbM-based mushroom extract may improve athletes' health. This includes training-related inflammation and asthma (inflammation of the bronchi) in athletes such as elite swimmers and cross-country skiers who are prone to developing asthma because of the inhalation of toxic substances and large volumes of freezing air, respectively.

Conflict of interest: G Hetland is shareholder of Immunopharma, which develops AndoSan™ into medicine.

REFERENCES

[1] Afzali B, Lombardi RI, Lechler I, Lord GM. The role of T helper 17 (Th17) and regulatory T regs (Treg) in human organ transplantation and autoimmune disease. *Clin Exp Immunol.* 2007; 148: 32-46.

[2] Bernardshaw S, Hetland G, Grinde B, Johnson E. An extract of the mushroom *Agaricus blazei* Murill protects against lethal septicemia in a mouse model for fecal peritonitis. *Shock,* 2006; 25: 420-425.

[3] Bernardshaw S, Johnson E, Hetland G. An extract of the mushroom *Agaricus blazei* Murill administered orally protects against systemic *Streptococcus pneumoniae* infection in mice. *Scand J Immunol.* 2005; 62: 393-398.

[4] Bougault V, Turmel J, Levesque B, Boulet LP. The respiratory health of swimmers. *Sports Med.* 2009; 39: 295-312, Review.

[5] Bouike G, Nishitani Y, Shiomi H, Yoshida M, Azuma T, Hashimoto T, Kanazawa K, Mizuno M. Oral treatment with extract of *Agaricus blazei* Murill enhanced Th1 response through intestinal epithelial cells and suppressed OVA-sensitised allergy in mice. *Evid Comp Alt Med.* 2011: 2011; 1-11.

[6] Choi YH, Yan GH, Chai OH, Choi YH, Zhang X, Lim JM, Kim JH, Lee MS, Han EH, Kim HT, Song CH. Inhibitory effects of Agaricus blazei on mast cell-mediated anaphylaxis-like reactions. *Biol Pharm Bull.* 2006; 29: 1366-1371.

[7] Ellertsen LK, Hetland G. An extract of the medicinal mushroom Agaricus blazei Murill can protect against allergy. *BMC Clin Mol Allergy*, 2009; Epub 7: 6.

[8] Faccin LC, Benati F, Rincao VP, Mantovani MS, Soares SA, Gonzaga ML, Nozawa C, Carvalho Linhares RE. Antiviral activity of aqueous and ethanol extracts and of an

isolated polysaccharide from Agaricus brasiliensis against poliovirus type 1. *Lett Appl Microbio.l* 2007; 45: 24-28.

[9] Forland DT, Johnson E, Satre L, Lyberg T, Lygren I, Hetland G. Effect of an extract based on the medicinal mushroom *Agaricus blazei* Murill on expression of cytokines and calprotectin in patients with ulcerative colitis and Crohn's disease. *Scand J Immunol.* 2011; 73: 66-75.

[10] Grinde B, Hetland G, Johnson E. Effects on gene expression and viral load of a medicinal extract from *Agaricus blazei* in patients with chronic hepatitis C infection. Int Immunopharm 2006; 6: 1311-1314.

[11] Hetland G, Johnson E, Lyberg T, Bernardshaw S, Tryggestad AMA, Grinde B. Effects of the medicinal mushroom *Agaricus blazei* Murill on immunity, infection and cancer. *Scand J Immunol.* 2008; 69: 363-370, Review.

[12] Johnson E, Forland DT, Satre L, Bernardshaw SV, Lyberg T, Hetland G. Effect of an extract based on the medicinal mushroom *Agaricus blazei* Murill on release of cytokines, chemokines and leukocyte growth factors in human blood *ex vivo* and in vivo. *Scand J Immunol.* 2009; 69: 242-250.

[13] Kuo YC, Huang Y-L, Chen C-C, Lin Y-S, Chuang K-A, Tsai W-J . Cell cycle progression and cytokine gene expression of human peripheral blood mononuclear cells modulated by *Agaricus blazei*. *J Lab Clin Med.* 2002; 140: 176-187.

[14] Larsson K, Ohlsen P, Larsson L, Malmberg P, Rydstrom PO, Ulriksen H. High prevalence of asthma in cross country skiers. *BMJ*, 1993; 307: 1326-1329.

[15] Lima CU, Cordova CO, Nobrega Ode T, Funghetto SS, Karnikowski MG. Does the Agaricus blazei Murill mushroom have properties that affect the immune system? An integrative review. *J Med Food*, 2011;14: 2-8.

[16] Osterud B, Olsen JO, Wilsgård L. Effect of strenuous exercise on blood monocytes and their relation to coagulation. *Med Sci Sports Exerc.* 1989; 21: 374-378.

[17] Panteghini M. Diagnostic applications of nucleic acid probes in the clinical laboratory. Aspartate aminotransferase isoenzymes. *Clin Biochem.* 1990; 23: 311-319.

[18] Reynolds JM, Pornpimon A, Dong C. Th17 Cytokines in host immunity: A balancing act IL-17 family member cytokines: Regulation and function in innate immunity. *Cytokine Growth Fact Rev.* 2010; 21: 413-423.

[19] Sorimachi K, Akimoto K, Ikehara Y, Inafuku K, Okubo A, Yamazaki S. Secretion of TNF- alpha, IL-8 and nitric oxide by macrophages activated with *Agaricus blazei* Murill fractions in vitro. *Cell Struct Func.* 2001; 26: 103-108.

[20] Stadelmann K, Stensrud T, Carlsen K-H. Respiratory Symptoms and Bronchial Responsiveness in Competitive Swimmers. *Med Sci Sports Exerc.* 2011: 43: 375–381.

[21] Suzuki K, Nakaji S, Yamada M, Totsuka M, Sato K, Sugawara K. Systemic inflammatory response to exhaustive exercise. Cytokine kinetics. *Exerc Immunol Rev.* 2002; 8: 6-48.

[22] Takimoto H, Kato H, Kaneko M, Kumazawa Y. Amelioration of skewed Th1/Th2 balance in tumor-bearing and asthma-induced mice by oral administration of *Agaricus blazei* extracts. *Immunopharm Immunotoxicol.* 2008; 30: 747-760.

[23] Wu MF, Hsu YM, Tang MC, Chen HC, Chung JG, Lu HF, Lin JP, Tang NY, Yeh C, Yeh MY. Agaricus blazei Murill extract abrogates CC14-induced liver injury in rats. *In Vivo,* 2011; 25: 35-40.

In: Mushrooms: Types, Properties and Nutrition
Editors: S. Andres and N. Baumann
ISBN 978-1-61470-110-1

Chapter 18

BIOLOGICAL ACTIVITIES OF *PLEUROTUS* MUSHROOMS

D. T. U. Abeytunga
Department of Chemistry, University of Colombo, Sri Lanka

ABSTRACT

Pleurotus mushrooms are popular edible mushrooms all over the world. Studies on this genus have shown that these mushrooms possess anticancer, antimicrobial, antiviral, antioxidant, antinociceptive, antidiabetic activities and cholesterol lowering properties. Proteins, enzymes, polysaccharides such as glucans, and glycoproteins isolated from *Pleurotus* mushrooms are responsible for the said activities. Proteins such as ubiquitins, hemolysins and lectins have shown antiviral activity. It is noteworthy, that anti-HIV activity was observed in all 3 classes of aforementioned proteins. Lignolytic enzymes such as laccases present in *Pleurotus* have helped in converting agricultural waste to animal feed. Glucans are known to possess anticancer activity, antiviral activity and cholesterol lowering properties. Water insoluble glucans present in *Pleurotus* have been converted to water soluble carboxymethylated and sulfated glucans having antitumor and antiviral activities respectively. A few secondary plant metabolites are identified from *Pleurotus* mushrooms such as pleurotin which has promising anticancer properties and lovastain with cholesterol lowering properties. An oxidized form of the abundant fungal steroid ergosterol has shown antifungal activity while ergosterol itself was implicated in bringing about inhibition of lipid peroxidation. The total phenolic content is also positively correlated to the antioxidant activity. Carbon-8 compounds arising from the hydroperoxidation of linoleic acid present in the fruiting body, is identified as a compound responsible for the production of the pleasant aroma in the mushroom. A mixture containing such volatile compounds mimicking the concentration in the fruiting bodies has shown antibacterial activity.

Keywords: *Pleurotus* mushrooms, anticancer, antimicrobial, antiviral, antioxidant, antinociceptive, antidiabetic, cholesterol lowering activity

Introduction

There are a number of *Pleurotus* mushrooms and most of them are edible. In general, they are cultivated on a wide range of plant wastes such as saw dust, waste cotton, rice straw and other agricultural wastes often enclosed in plastic bags. *Pleurotus* mushroom production in the world is about 25% of the total world production of cultivated mushrooms.

Most taxonomic and phylogenetic studies of Basidiomycota have been based on the analysis of morphological characteristics. Same is true for *Pleurotus* mushrooms. Mitochondrial RNA is generally preserved within a species of an organism and can be used as a marker in taxonomy. A comparative study of mitochondrial small-subunit (SSU) rRNA was conducted by Gonzalez & Labarère, 2000 to evaluate the phylogenetic relatedness of 16 *Pleurotus* species. It was confirmed that *P. ostreatus* and *P. florida* represent a single species, as well as *P. pulmonarius* and *P. sajor-caju*. This phylogenetic analysis also made it possible to assess the relative positions of *P. rattenburyi*, *P. lampas*, *P. sapidus*, *P. colombinus* and *P. eryngii* in the phylogenetic tree.

Among the *Pleurotus* species, *P. ostreatus* is the most popular species which is commonly referred to as the "oyster mushroom" (Fig. 1). The name "oyster" stems from its appearance as a white shell. *P. eryngii* is the largest species in the genus *Pleurotus*, and it is called the "king oyster mushroom". *P. tuber-regium* and *P. cystidiosus* are commonly referred to as the tiger milk mushroom and abalone respectively.

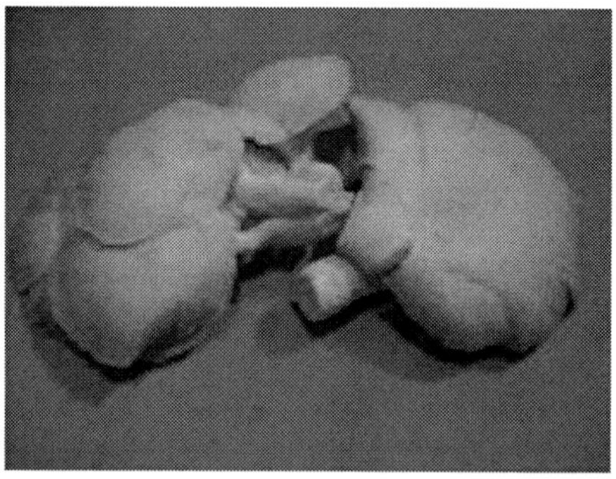

Figure 1. The popolar oyster mushroom; *Pleurotus ostreatus*.

The medicinal value of the genus *Pleurotus* has been reviewed by Gunde-Cimerman in 1999. Hence in this chapter more emphasis will be given to the studies conducted after 1999.

Nutritional Value

The chemical composition and the nutritional value of different species in genus *Pleurotus* have been studied by many. Most fresh mushrooms contain ~ 90% moisture (Crisan and Sands 1978). According to the same report, mushrooms in general contain 44–74% carbohydrates, 19-35% proteins, 1.1-8.3 % lipids of the dry weight of mushrooms. Yang *et al.*, in 2001 compared the composition of *Flammulina velutipes*, *Lentinula edodes*, *Pleurotus cystidiosus* and *Pleurotus ostreatus*. The two *Pleurotus* species were found to contain 61-63% carbohydrates, 15-23% proteins, 5-8% crude fiber and 2-3% lipids.

The aforementioned literature supports mushrooms as a good source of proteins compared to vegetables. However, some studies show varying protein contents (1.7 g/100 g – 2.3 g/100 g) in *P. ostreatus*. In a study conducted by Mattila *et al.*, 2002, it is proven that the net protein content of *P. ostreatus* is 1.97 g/100 g and it contains most of the essential amino acids.

Another study on the nutritional value of *Pleurotus* (Guardia *et al.*, 2005) gives the composition of proteins, carbohydrates, fats, dietary fiber and energy value of several species of *Pleurotus*. According to this study, the dry weight percentage of carbohydrates (including dietary fiber), proteins, lipids in one variety of *P. eryngii* is 63.1%, 23.7% and 8.4% respectively. This study concludes that *Pleurotus* species are a good source of vitamins, minerals and are hypocaloric.

The percentages of vitamins, mineral elements and phenolic compounds in *P. ostreatus* were studied by Mattila *et al.*, 2001. It was found that *P. ostreatus* is a good source of vitamins such as vitamin C, niacin and folates. This mushroom is also a good source of many mineral elements such as Potassium, Phosphorus, Zinc and Iron.

An interesting study reports (Pizzoferrato *et al.*, 2000) the use of solid-state ^{13}C CP MAS NMR technique in monitoring the chemical composition on the solid state of the intact mushroom sample. Three *Pleurotus* species, namely, *P. ostreatus*, *P. eryngii* and *P. pulmunarius*, already characterized by chemical analyses for protein and dietary fiber components were subjected to the above study. The authors conclude that the ratio of protein to polysaccharides determined by NMR or chemical methodologies could form the basis of a taxonomical study.

A detailed study of the volatile flavor compounds and taste components of *P. eryngii* was conducted by Mau *et al.*, 1998. Mushrooms were categorized as large, small mushrooms and the base of the mushrooms. The two categories and the base of the mushrooms were studied separately. The identified volatile compounds were 3-octanone, 1-octen-3-one, 3-octanol, 1-octen-3-ol, benzaldehyde, 1-octanol and 2-octen-1-ol. Carbon-8 volatile compounds such as 3-octanol, 3-octanone, 2-octen-1-ol, 1-octen-3-ol contribute to the aroma in most mushrooms. The major compound in both small and large mushrooms was benzaldehyde and both categories of mushrooms were high in total free amino acids. It was concluded that although the two categories of mushrooms have similar compositions of volatile compounds and non volatile flavor components, the base of the mushrooms has very different flavor profiles.

The carbon-8 volatile compound 1-octen-3-ol which contribute to the aroma in most mushroom species is formed enzymatically by the oxidative cleavage of linoleic acid. Lipooxygenase is believed to be involved in hydroperoxidation of linoleic acid which will then break to give this carbon-8 compound and a carbon-10 compound. Hence, lipooxygenase is a key enzyme in the biosynthesis of 1-octen-3-ol which is a volatile flavor component. Kuribayashi *et al.*, 2002 purified and characterized a lipooxygenase from *P. ostreatus,* a mushroom which gives a pleasant aroma.

ENZYMES AND PROTEINS FROM PLEUROTUS

During fruiting body formation of any mushroom, enzymes play an important role at various stages of its development. These enzymes could be used in bio-pulping, bio-bleaching, bio-transformation and bio-remediation in different industries, if enough quantities can be isolated by the development of unique fermentation procedures. Different enzymes & proteins such as laccases, ribonucleases, ubiquitins, lectins, hemolysins and antifungal proteins have been isolated from *Pleurotus*. Tandem mass spectrometry in combination with database searches has been used (Matis *et al.*, 2005) in the identification of these fungal proteins.

Laccases are copper containing oxidase enzymes which act on molecules like phenols causing one electron oxidations (Fig. 2). Hence, laccases play an important role in lignin formation as well as degradation. During this reaction laccases reduce molecular oxygen to water.

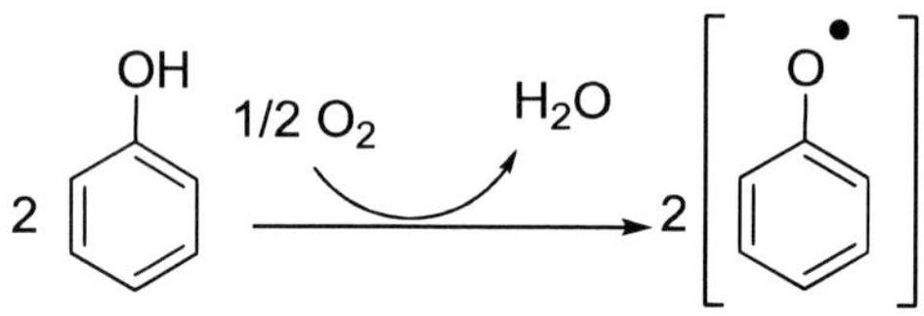

Figure 2. Laccases can catalyze one electron oxidations of phenols.

P. sapidus was identified as the fungus producing considerable amount of laccases during the screening of 46 basidiomycetes species for laccase production by Linke *et al.*, 2005. The laccase was identified using tandem mass spectrometry and this study shows the reusability of the mycelium in laccase production. Prior to this study, laccases have been isolated from *P. eryngii* by Munoz *et al.*, in 1997 and from *P. ostreatus* by Mansur in 2003. Recently, attempts have been made by Liu *et al.*, 2009 and Palyannan & Sathishkumar 2010 to optimize the laccase production by *P. ostreatus* and *P. florida* respectively (As mentioned in the introduction, according to Gonzalez & Labarère, 2000, *P. ostreatus* and *P. florida* represent a single species). Laccases from *Pleurotus* have important applications such as bleaching in the textile & dye industry (Rodriquez *et al.*, 1999, Sathishkumar *et al.*, 2010), delignification of craft pulps (Bourbonnais *et al.*, 1997) and bioremediation (Rodriquez *et al.*, 2004).

Enzymatic activity of *Pleurotus* can be used to convert agricultural waste to animal feed. The use of *P. ostreatus* in processing tomato waste or pomace was studied by Assi & King 2007. Tomato pomace consists of peels, cores, trimming, seeds and unprocessed green tomatoes and this waste is an excellent source of nutrients in poultry feed. However, the high fiber content in tomato pomace is undesirable and this study was undertaken to transform the fibrous ingredients to digestible nutrients using *P. ostreatus.* Such biotechnological applications of *Pleurotus* were reviewed by Cohen *et al.,* 2002.

Ribonuclease (abbreviated as RNase) is an enzyme which can cleave RNA into small fragments. A 12 kD ribonuclease was isolated from the edible mushroom *P. sajor-caju* (Ngai and Ng 2004) which is essentially a poly U-specific. This specificity has not been reported for RNases isolated from mushrooms prior to this study. This ribonuclease possesses antifungal, antibacterial and antiproliferative activities.

Ubiquitin is a small regulatory protein found in almost all tissues. Among other functions its main function is to bind to other proteins in order to label them for destruction. Ubiquitin like proteins have been isolated from mushrooms. The 9.5 kDa peptide isolated from *P. sajor-caju* (Ng *et al.*, 2002) with a yield of 0.25 mg/kg mushroom, has an N-terminal sequence similar to ubiquitin. The isolated ubiquitin from *P. sajor-caju* possesses translation-inhibitory and RNase activities.

Hemolysins are proteins which cause red blood cells to undergo lysis. Lytic activity of hemolysins are important for nutrient acquisition in bacteria and its function in mushrooms is largely unknown. However, it is speculated that the function of hemolysins is the initiation of fruiting bodies in mushrooms. Hemolysins; ostreolysin and eryngeolysin have been isolated from *P. ostreatus* (Berne *et al.*, 2002) and *P. eryngii* (Ngai & Ng 2006) respectively. These studies have indicated that ostreolysin is devoid of insecticidal activity and eryngeolysin lacks antifungal activity towards many fungal species. The lack of antifungal activity may be explained by referring to the study of the interaction of ostreolysin with membrane lipids (Sepcic *et al.*, 2004). This study concludes that a membrane should contain both cholesterol and sphingomyelin for lytic action of ostreolysin. If this observation is extrapolated to eryngeolysin, the lack of antifungal activity of this protein could be attributed to the presence of ergosterol instead of cholesterol in fungal membranes.

Lectins are proteins which show a high specificity towards sugar moieties in glycoproteins and glycolipids and they can precipitate these glycoconjugates as well as other polysaccharides. Lectins have been isolated from popular edible and medicinal mushrooms and the structures and biological activities of these lectins vary widely. Some lectins exhibit potent mitogenic activities while others are non-mitogenic. The lectin isolated from *P. flabellatus* (Ho *et al.*, 2004) did not stimulate cell proliferation.

ANTIMICROBIAL ACTIVITIES

Not much research has been carried out to determine the antimicrobial activity of *Pleurotus* species. The antimicrobial activity of fourteen strains of *P. ostreatus* was determined by Gerasimenya *et al.*, 2002. The extracts of submerged cultures show some antimicrobial activity. However extracts of the fruiting bodies showed no activity. These results are somewhat in agreement with the results from our laboratory on antimicrobial activity.

We have shown that the extracts of *P. cystidiosus* have no activity against *E. coli*, *Staphylococcus* or *Klebsiella* under test conditions. However, the antifungal activity tests against several plant pathogenic fungi; *Colletotrichum gloeosporioides*, *C. acutatum*, *Alternaria*, *Curvularia* and *Rhizoctonia solani* showed that the acetone extract of *P. cystidiosus* was active against *C. gloeosporioides* (Menikpurage *et al.*, 2009). Antifungal assay guided fractionation of the acetone extract lead to the isolation of an oxidized form of the fungal steroid ergosterol; 3β, 5α, 6β-trihydroxyergosta-7,22-diene (Fig. 3). Ergosterol is a common steroid found in fungal cell membranes. However, the function of this oxidized ergosterol in the mushroom is unknown.

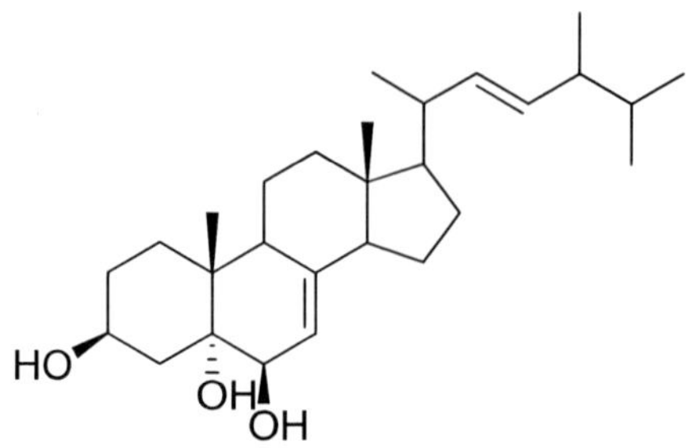

Figure 3. 3beta, 5alpha, 6beta-trihydroxyergosta-7,22-diene.

Literature reveals the use of *Pleurotus* species in bio-control of several fungal diseases. A study conducted by Ruiz-Duenas & Martinez 1996 shows the use of *P. eryngii* in the control of *Fusarium oxysporum*. This study concludes that the glucanases in the mushrooms degrade the hyphal cell walls of *F. oxysporum* and thereby, the fungus loses its cytoplasmic content. It is interesting to note that *Pleurotus* species too contain glucans. However, the glucanases may be acting only upon other fungal species. *Pleurotus* species are also active against *Ralstonia solanacearum* (El-Fallal & Moussa, 2008), the cause of brown rot disease in potatoes.

Submerged culture extracts of *P. tuber-regium* was subjected to further fractionation and tested against *Staphylococcus aureus* by XiJun *et al.*, 2009. Aqueous and ethanol extracts of the mycelium have shown antibacterial activity, but not the nonpolar hexane fraction.

An antifungal peptide with a molecular mass of 10 kDa (Wang & Ng 2004) was isolated from fruiting bodies of the mushroom *P. eryngii*. This peptide is referred to as eryngin. As the literature does not support the presence of secondary metabolites with antimicrobial activity, except the oxidized ergosterol isolated in our laboratory, we can speculate, that in general, the antimicrobial activity of *Pleurotus* species may be due to the enzymes or proteins present in the fungus.

The antibacterial activity of volatile compounds of *P. ostreatus* were determined by Beltran-Garcia *et al.* 1997. Several methods, namely vacuum distillation, nitrogen flow conveyance and extraction by carbon tetrachloride were used to extract the volatile components. Although the first two methods give only the volatiles present in the mushroom, extraction to CCl_4 may not necessarily reflect the volatile constituent profile of the mushroom. The compounds extracted by CCl_4 were identified as 3-octanone, 3-octanol, 1-octen-3-ol, benzaldehyde, 1-octanol, benzaldehyde. A mixture containing these volatile compounds mimicking the concentration in the fruiting body, completely inhibits eight bacterial species used in the antibacterial assay.

ANTICANCER AND ANTITUMOR ACTIVITIES

Studies on antitumor and anticancer activities of *Pleurotus* mushrooms indicate that these activities are mainly due to the polysaccharides present in the fungi. Although the role of polysaccharides as antitumor agents are under debate, the review by Zhang *et al.* 2007 indicates that the polysaccharides are known to be the most potent mushroom derived substances with antitumor and immune-modulating properties. The oral administration of these polysaccharides is known to exert antitumor activity by three mechanisms:

- Cancer preventing activity: can stop the formation of the cancer or oncogenesis
- Immune-enhancing activity: polysaccharides do not directly kill tumor cells, they help the host to adapt to various biological stresses and support major systems
- Direct tumor inhibition activity: induces the apoptosis of tumor cells.

The early studies conducted by various research groups identify *Pleurotus* mushrooms as a source of dietary fibers but do not report any antitumor activity. The study conducted by Cheung in 1996 confirms the presence of dietary fiber both in the mycelium and the fruiting body of *P. sajor-caju* and the analysis of the sugar composition identifies the major dietary fiber as beta glucan. Two years later, Cheung & Lee 1998 compared the dietary fiber content in *P. sajor-caju,* which has a soft and fleshy fruiting body, with that of *P. tuber-regium,* which has a hard and solid sclerotia. Two analytical methods were used in this study, namely, the enzymatic approach based on the Association of Official Analytical Chemists (AOAC) and nonenzymatic method based on the sequential use of solvents. This study indicates that the total dietary fiber obtained by the enzymatic method was higher than by the chemical method. In general, the fiber content in *P. tuber-regium* is higher than in *P. sajor-caju.* This work also indicates that the dominant monosaccharide present in the fiber material of both mushrooms is glucose (78-89%) and glucosamine (7-12%) which attests the presence of beta glucans and chitin as major polysaccharides.

Due to the significance of beta glucans as possible antitumor agents, Cheung & Lee in 2000 have taken further steps to characterize the mushroom dietary fiber (nonstarch polysaccharides) in *P. tuber-regium*. This study shows that the beta glucan in *P. tuber-regium* is highly branched and contains a mixture of beta-1,3, beta-1,6 and beta-1,4 sugar linkages. As it is unusual for beta-1,3 and beta-1,4 linkages to be present in the same polysaccharide, the study concludes that this mushroom has beta-1,3 and beta-1,6 glucans, cellulose and chitin as the major polysaccharides. It is noteworthy that lentinan and schizophyllan (isolated from other mushrooms) which are known antitumor compounds are also beta-glucans. Hence Zhang *et al.*, 2001 have extensively studied the antitumor activity of beta glucans isolated from *P. tuber-regium*. The hot alkali extract of the mushroom dietary fiber with average molecular weight 5.8-17.1 x 10^4 administered at a dose of 20 mg/kg body weight of mice implanted with solid tumor was effective in inhibiting tumor proliferation. Structure of this glucan was proven to be a 1,3 linked beta-glucan with one 1,6 branch at every third glycosidic residue (Fig. 4).

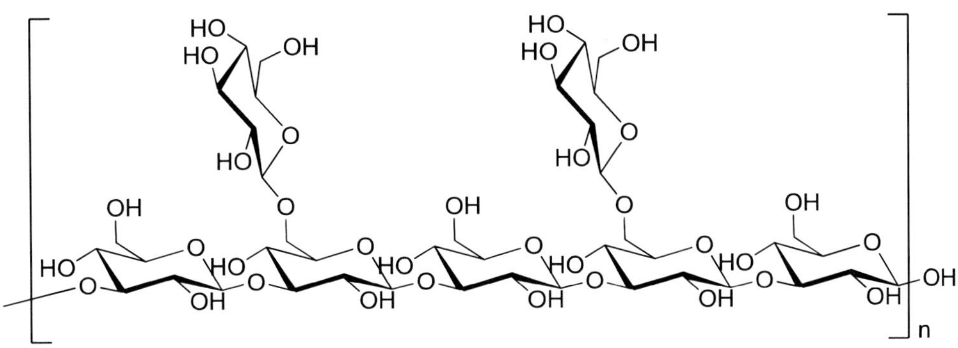

Figure 4. Possible structure for the glucans isolated from *Pleurotus*.

The commonly known antitumor glucans such as lentinan and schizophyllan (molecular weight 40 x 10^4) share the same chemical structure although the molecular weight of the antitumor glucan from *P. tuber-regium* is much lower (5.8-17.1 x 10^4). It is also interesting to note that lentinan and schizophyllan have a triple-helix conformation whereas the water insoluble glucan from *P. tuber-regium* adopts a random-coil chain conformation in DMSO. Zhang *et al.* 2004b successfully converted this water insoluble glucan to a water soluble carboxymethylated beta-glucan having antitumor activity.

It is interesting to note that the same group (Zhang *et al.*, 2006) recently worked on the possible cellular mechanism of this antiproliferative action of the beta-glucan isolated from *P. tuber-regium*. This study was undertaken to find out whether the antitumor activity is due to immunopotentiation activity or to the direct inhibition of tumor cells as discussed at the beginning of this section on anticancer/antitumor activity. Instead of the native water-insoluble beta glucan isolated from *P. tuber-regium*, the water-soluble carboxymethylated glucan with enhanced anti-tumor activity was used in this mechanistic study. The anti-proliferative activity of carboxymethylated glucan on the human breast carcinoma MCF-7 cells *in vitro* was dose dependent with an IC_{50} of 204 μg/mL. This cell proliferation of MCF-7 was inhibited by cell-cycle arrest and apoptosis induction. Further, in another study (Wong *et al.*, 2007) the water-soluble non-starch polysaccharides extracted from the fruiting body of *P. tuber-regium* showed the strongest cytotoxicity (IC_{50}: 25 μg/mL) against human acute promyelocytic leukemia cells (LL-60).

In order to understand the structure anti-tumor activity relationship of mushroom polysaccharides, it is important to isolate polysaccharides within a comparatively small range of molecular weights. Another problem is that the extracted glucans are mixtures with different degrees of polymerization and the product quality and consistency is difficult to maintain. Further, mechanistic studies on polysaccharide-protein complexes are even more challenging as there is lack of information on the association between these two macromolecules. Zhang *et al.* 2011 have developed a gradient precipitation procedure with water and ammonium sulphate to fractionate the hot-water extract of *P. geesteranus*. This method could be used to obtain polysaccharide fractions, free of proteins with high chemical uniformity and narrow molecular weight distribution.

The same group has conducted several studies on mushroom dietary fibers without any mention of antitumor activity in early 2000. Wong *et al.*, 2003 highlights the consumption of dietary fiber in bringing about a protective effect against coronary heart disease, diabetes, colon cancer, constipation and gives the microstructural characteristics of the insoluble and soluble dietary fiber isolated from *P. tuber-regium*. Further, these researchers have developed an enzymatic procedure for preparing dietary fiber from *P. tuber-regium* using industrial food grade glycolytic and proteolytic enzymes and have studied physicochemical and functional properties such as pH, color, water binding capacity, oil holding capacity and emulsifying properties (Wong & Cheung 2005). This paper too does not study the antitumor activity but gives the scale-up enzymatic preparation of beta-glucan type dietary fiber from *P. tuber-regium* for food industry.

A protein-polysaccharide from *P. tuber-regium* has been isolated and characterized by Tao *et al.* 2007 (a). They have also conducted structural studies [Tao *et al.* 2007 (b)] of a water-insoluble polysaccharide isolated from *P. tuber-regium*.

All the aforementioned studies suggest that only the polysaccharides present in *Pleurtous* are significant contributors to the anticancer/antitumor properties. Robbins *et al.*, 1947 isolated a hexocyclic compound, Pleurotin (Fig. 5), from *P. griseus* which was shown to be active against gram positive bacteria. Today, Pleurotin is of great interest as it is a potential antitumor agent.

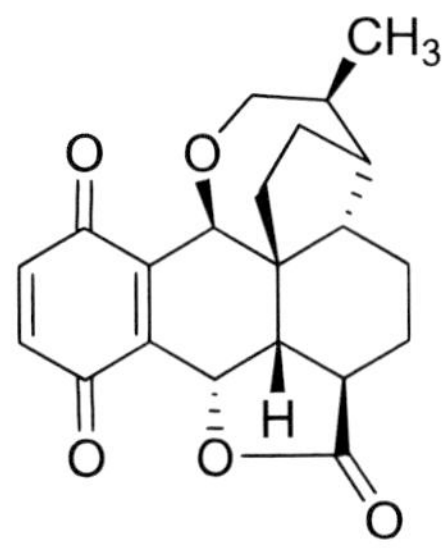

Figure 5. Pleurotin.

Thioredoxin-1 which is over expressed in tumors, is a redox protein involved in regulating cell division. Thioredoxin, thioredoxin reductase and NADPH are referred to as the thioredoxin system. Inhibitors of thioredoxin-thioredoxin reductase are potent antitumor agents and Pleurotin is known to inhibit the thioredoxin system (Wipf *et al.*, 2004). Hence it is considered to be a lead compound in anticancer therapy.

ANTIVIRAL PROPERTIES

Both glucans and proteins isolated from *Pleurotus* species were identified as the constituents responsible for antiviral activity.

Six water-insoluble hot alkali fractions containing beta-glucans from *P. tuber-regium* were sulfated to obtain water-soluble sulfated glucans by Zhang *et al.*, 2004a. Both native beta-glucans and sulfated glucans were tested against herpes simplex virus type 1 (HSV-1), herpes simplex virus type 2 (HSV-2), respiratory syncytial virus and influenza A virus. Some sulfated glucans showed anti-viral activity against HSV-1 and HSV-2 by binding to the viral particles, possibly by the interaction of their negative charges with the positively charged glycoproteins on the viral surface.

A study conducted by Selegean *et al.*, 2009 proves that the polysaccharide containing fractions of *P. ostreatus* have immunomodulatory activity against an infectious bursal disease caused by a virus. Combination of a commercial vaccine and polysaccharides of *P. ostreatus* have enhanced antibody production towards the infectious bursal disease.

Proteins obtained from *P. eryngii* were shown to have very high antiviral activity against tobacco mosaic virus by Fu *et al.*, 2003.

Wang & Ng 2000 isolated an ubiquitin-like protein (molecular weight was 12.5 kDa) with anti-HIV activity from *P. ostreatus*. Isolation involved ion exchange chromatography on DEAE-cellulose, affinity chromatography on Affi-gel blue gel, ion exchange chromatography on SP-Sepharose and Mono Q and gel filtration on Superdex 75.

A lectin isolated from *P. citrinopileatus* (Li *et al.*, 2008) showed 80% inhibition of tumor growth when administered intraperitonealy at 5 mg/kg daily for 20 days in mice bearing sarcoma 180. The same lectin inhibited HIV-1 reverse transcriptase with an IC_{50} of 0.93 μM. The isolation of lectin comprised three ion exchange steps and one gel filtration step, namely, ion exchange chromatography on DEAE-cellulose, CM-celluloses and Q-Sepharose and gel filtration on Superdex 75.

Hemolysin isolated from *P. nebrodensis* by Lv *et al.*, 2009 was named as nebrodeolysin. This nebrodeolysin exhibited some antiviral action on HIV-1. Although this anti HIV-1 activity is not very remarkable it is assumed that the action is brought about by suppressing the infection of the virus through interaction with the cell membrane. Nebrodeolysin showed activity against human cancer cells as well.

CHOLESTEROL LOWERING PROPERTIES

The cholesterol lowering ability of mushrooms has been known to mankind for a long time. Consumption of the famous shiitake (*Lentinus edodes*) mushrooms lowers blood cholesterol levels by as much as 45% and the compound responsible for the activity has been identified as eritadenine.

The hypocholesterolemic ability of *Pleurotus* has also been known (Bobek *et al.*, 1991, Chorvathova *et al.*, 1993) for over two decades. Bobek *et al.*, 1995a indicate that the addition of 5% oyster mushroom to a diet containing 0.3% cholesterol reduces the cholesterol content by 30% in rat serum. The same treatment can reduce the very low density and low density lipoproteins in the serum and the activity of HMG-CoA reductase. This activity is dose dependent (Bobek *et al.*, 1997) as the addition of 1.0%, 2.5%, and 5.0% of oyster mushroom to the diet reduced the serum cholesterol levels by 11%, 31%, and 46% respectively. Time dependency of the hypocholesterolemic activity was also studied (Bobek *et al.*, 1998) by the same group. Hypocholesterolemic activity of the same mushroom on rabbits was reported by Bobek & Galbavy 1999. Studies conducted by groups in Japan (Hossain *et al.*, 2003) and in India (Bajaj *et al.*, 1997) endorse the use of *P. ostreatus* in improving the atherogenic lipid profile in hypercholesterolemic rats.

Attempts were also made to identify the fraction (Bobek *et al.*, 1995b, Bobek *et al.*, 1996) containing hypocholesterolemic activity. Feeding of a diet containing 5% of powdered oyster mushroom or an equal amount of mushroom ethanolic extract along with 0.3% cholesterol for 12 weeks reduced the cholesterol level in serum by 52% and 33% respectively. These results indicate that although the ethanol extract retains the cholesterol lowering ability, the powder of the whole mushroom is better in reducing cholesterol levels. It is also interesting to note that both the whole mushroom and the ethanol extract, decreased the cholesterol in very-low-density lipoproteins and increased the cholesterol in high-density lipoproteins, indicating that *Pleurotus* mushrooms are potential health foods. Further, it is shown that the whole oyster mushroom decreased cholesterol absorption while no such effect was observed with the mushroom extract. The whole mushroom contains the dietary fibers whereas the ethanol extract is devoid of such fibers. One inference that may be drawn from this observation is the involvement of dietary fiber in reduction of cholesterol absorption.

Studies conducted at our laboratory (Abeytunga *et al.*, 2008) indicate that both *P. ostreatus* and *P. cystidiosus* possess cholesterol lowering ability in hypercholesterolemic rats. Better activity was seen with *P. cystidiosus*. However, the polar acetone extract of *P. cystidiosus* did not possess the said activity, once again suggesting that the fibers in mushroom may play an important role in hypocholesterolemic activity. We have strengthened this speculation by analyzing the fecal matter and proving that the cholesterol excretion is high in rats fed with the mushroom.

Statins are inhibitors of 3-hydroxy-3-methylglutaryl-CoA reductase (HMG-CoA reductase), a rate-limiting enzyme of cholesterol biosynthesis. Literature indicates that some fungal genera can produce lovastatin (Fig. 6), a statin with cholesterol lowering activity. Gunde-Cimerman & Cimerman 1995 have detected lovastatin in the fruiting bodies as well as in the mycelium of *P. ostreatus*.

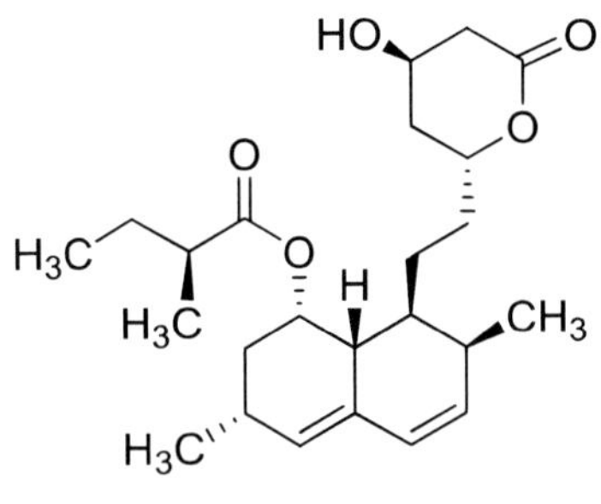

Figure 6. Lovastain.

Antioxidant Activities

Literature reveals that *Pleurotus* species also possess promising antioxidant activity (Yang *et al.* 2002, Cheung *et al.* 2003, Cheung & Cheung 2005). It was shown that *Pleurotus* species contain moderate antioxidant activity in a comparative study (Puttaraju *et al.*, 2006) of the antioxidant potential of 23 species of mushroom extracts. Several assays were used to evaluate the antioxidant potential, such as inhibition of lipid peroxidation, determination of reducing power, free radical scavenging ability and determination of total phenolic constituents. These results are comparable with the results obtained by Kim *et al.*, 2008 where total phenolic content and the antioxidative activity of five edible and five medicinal mushrooms of Korea were compared. It was found that the DPPH radical scavenging activity of *P. ostreatus* and *P. eryngii* were not as high as that of *Ganoderma lucidum*.

The methanol extract of *P. pulmonarius* has shown (Jose *et al.* 2002), significant hydroxyl radical-scavenging and lipid peroxidation-inhibiting activities and the IC_{50} values were 476.7 and 960.0 μg/mL respectively. However, this fraction did not contain any superoxide radical-scavenging activity. The same study shows that the extract has antitumor and antiinflammatory activity comparable to the standard reference drugs, cisplatin and diclofenac respectively.

Pleuran (β-1,3-D glucan) isolated from fibers of *P. ostreatus* has stimulated the activity of antioxidant enzymes such as superoxide dismutase and glutathione peroxidase in rat liver and has reduced the glutathione levels significantly in rat colon (Bobek & Galbavy, 2001).

The antioxidative activity and cytotoxicity of *P. cystidiosus* are not reported. Our studies have shown (unpublished) that the edible mushroom *P. cystidiosus* is a rich source of antioxidants and has cytotoxic activity against *Hep-2* carcinoma cells.

P. florida and *P. pulmonaris* also possess profound antioxidant and antitumor activities (Ajith and Janardhanan, 2007).

We have shown the inhibition of lipid peroxidation by *P. ostreatus* using Thiobarbituric (TBARS) assay. Three solvent extracts of *P. ostreatus,* namely, the acetone, dichloromethane and hexane extracts were used. The antioxidant indexes at 1.25 mg/ml concentration of acetone, dichloromethane and hexane were 38%, 43% and 36% respectively. Ergosterol was isolated and characterized from the dichloromethane extract. The antioxidant index of ergosterol at 1.25 mg/ml was 73% when tested using the same method. There exists a possibility that ergosterol can inhibit the peroxidation of common fatty acids present in egg yolk substrate (which was used for TBARS assay). The relative susceptibilities for peroxidation of ergosterol, linoleic acid and linolenic acid were investigated using computational calculations (Dissanayake *et al.,* 2009). It was found that the addition of an oxygen molecule to ergosterol (Fig. 7) is kinetically more feasible than the addition of the same to linoleic or linolenic acids. The results of the theoretical investigation support the experimental observation that ergosterol is capable of inhibiting lipid peroxidation.

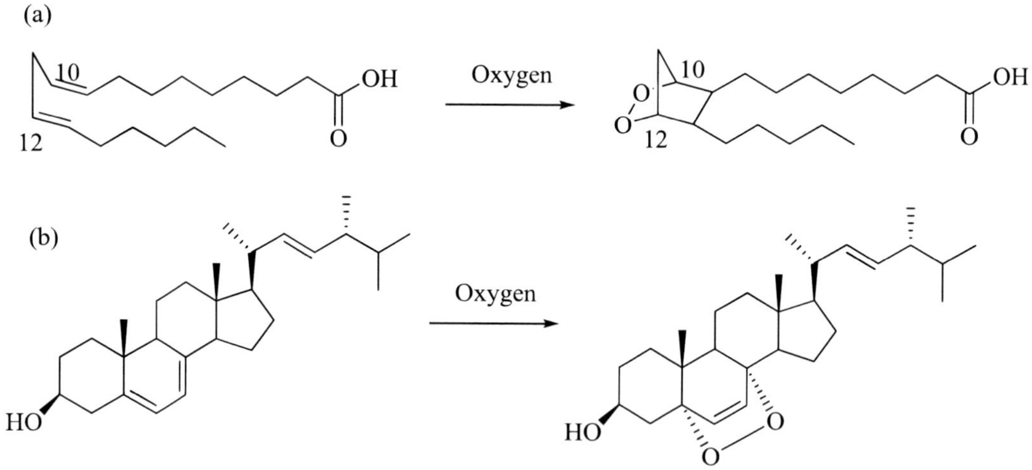

Figure 7. (a) Formation of oxygen adduct with linoleic acid, the adduct breaks down to give malonaldehyde (b) Oxygen addition to ergosterol results in ergosterol peroxide.

It is generally accepted that supplementation of the diet with fruits and vegetables is beneficial in minimizing the oxidative damage in the ageing process of the body. In a study conducted by Jayakumar *et al.*, 2007, it was shown that the extract of *P. ostreatus* can protect the liver, kidney, heart and brain of aged rats against oxidative stress by reducing the intensity of lipid peroxidation and enhancing the activities of antioxidants.

Another recent study by Jayakumar *et al.*, 2009 has very promising results on the antioxidant activities of *P. ostreatus*. This study concludes that the ethanolic extract of this mushroom is rich in antioxidants such as phenols, ascorbic acid, alpha tocopherol, beta-carotene and flavonoid compounds and it can be used as a food supplement or as a pharmaceutical agent.

Antinociceptive and Antiinflammatory Activities

We have shown the antinociceptive potential (Vasudewa *et al.*, 2007) of *P. ostreatus*. It has acute, orally active, moderate to strong pain relieving properties against both neurogenic pain (as shown by the hot plate test and the first phase of formalin test) and continuous inflammatory pain (as shown by the second phase of the formalin test). Antinociceptive activity of *P. ostreatus* was not gender dependent. However, it is of particular interest to note that the antinociceptive action of *P. ostreatus* on females was dependent on the phase of the estrous cycle: Most potent in di-estrous and least potent in pro-estrous and estrous. The antinociceptive activity of *P. ostreatus* had rapid onset (within 1 h) and a short duration of action (1 h in males and 2 h in females). This activity was dose-dependent. Naloxone an opioid receptor antagonist blocked the *P. ostreatus* induced antinociception indicating opioid receptor mediation. *P. cystidiosus* was also shown to have anti-nociceptive activity (Kudahewa et al., 2008).

When *P. ostreatus* was subjected to a bio-assay directed fractionation the antinociceptive activity was retained in the acetone fraction (Vasudewa *et al.*, 2008). In order to investigate the polarity of the compound responsible for the antinociceptive activity, the acetone extract of *P. ostreatus* was subjected to liquid-liquid partitioning using hexane, dichloromethane, ethyl acetate and water. The activity was retained in the water fraction. This fraction was further fractionated on C-18 reverse phase silica gel. A very polar fraction from this purification retained extremely high activity which had a rapid onset and lasted for 3 h. It is reported that Polysccharides such as Pleuran have been isolated from *Pleurotus* species (discussed in the section on anticancer agents). However, we are not certain whether such polysaccharides are responsible for the antinociceptive activity.

Anti-inflammatory activity of the methanol extract of *P. florida* was reported by Jose *et al.*, 2004. This extract at 500 and 1000 mg/kg doses has shown anti-inflammatory activity against inflammation induced by carrageenan and chronic inflammation by formalin. Further, this effect was comparable to the effect shown by the reference drug, diclofenac.

Antidiabetic Activities

People with type 2 diabetes are known to benefit from different mushrooms. Further, mushrooms have no starch and less sugars and are therefore considered suitable as a health food for those who are prone to diabetes. However, limited studies have been conducted on the blood glucose lowering activity of *Pleurotus*. The hypoglycaemic effect of *P. eryngii* on streptozotocin-induced diabetic rats was reported for the first time by Kang *et al.*, 2001. Recently Kim *et al.*, 2010, have fed four weeks old mice with a diet containing 5% *P. eryngii* for 7 weeks. The serum glucose levels and blood glycated hemoglobin were reduced by 12.1 % and 9.7 % respectively by this treatment. Further, this mushroom treatment improved the insulin sensitivity suggesting the use of *P. eryngii* in prevention and management of type 2 diabetes.

The anti-diabetic activity of *P. pulmonarius* was studied by Badole *et al.*, 2006. Oral administration of aqueous extracts of *P. pulmonarius* to alloxan-induced diabetic mice by Badole *et al.*, 2006 reduced the serum glucose level at doses 250, 500 and 1000 mg/kg when fed for 28 days.

Further studies indicate (Badole & Bodhankar, 2007) that single oral administration of an aqueous extract of *P. pulmonarius* (500 mg/kg) to alloxan-induced diabetic mice reduce the serum glucose level at 2, 4, and 6 hrs after administration. In the subacute study, when extract was fed once a day for 28 days, significant reduction in the serum glucose level was observed. A combination of the mushroom extract (500 mg/kg) with acarbose, a alpha-glycosidase inhibitor which competitively inhibits the digestion of oligosaccharide to monosaccharides (50 mg/kg) shows a synergistic antihyperglycemic effect which may be due to increased insulin secretion or an increased glucose threshold. Concomitant administration of metformin and the aqueous extract of *P. pulmonarius* was also studied by the same group (Badole & Bodhankar 2008). It was observed that this combined administration also produced a synergistic antihyperglycemic effect than either treatment alone.

Another *Pleurotus* mushroom was studied for its antihyperglycemic properties by Hu *et al.*, 2006. The water soluble polysaccharides extracted from the edible mushroom *P. citrinopileatus* were fed to rats with streptozotocin induced diabetes. The loss in body weight was less and the fasting blood glucose levels were 44% lower than the rats fed with a diet without the mushroom extracts.

Although the studies conducted on *Pleurotus* do not identify the constituents responsible for this activity, previous studies conducted on maitake mushroom (Manohar *et al.*, 2002) shows the presence of antidiabetic activity in the water soluble fraction containing beta-glucans. The aforementioned studies have also shown activity in the water soluble polysaccharide extracts indicating once again another use of the beta-glucans present in the mushrooms.

CONCLUSION

There is ample evidence in the literature to support the fact that *Pleurotus* mushrooms possess a variety of biological activities. Such biological activities may not be due to the presence of secondary metabolites, as not many natural products have been isolated and characterized from *Pleurotus* mushrooms. The reported activities are mainly due to the polysaccharides, proteins and glycoproteins present in the fungus. However, compounds such as polyphenols and ergosterol are proven to be directly involved in activities such as antioxidative activity.

Pleurotus mushrooms can be conveniently incorporated into the diet as a nutritional supplement. The nutritional value brings many health benefits. On the other hand, the considerable fiber content will help to reduce problems such as constipation. *Pleurotus* mushrooms, being low in caloric content are suitable in diets for obese and diabetic people. Further, production of *Pleurotus* derived nutraceuticals or dietary supplements would definitely have a wide scope in the future market. Attempts have already been taken to develop *Pleurotus* derived food products such as fermented oyster-mushroom sausage (Chockchaisawasdee *et al.*, 2010).

The large scale production of *Pleurotus* mushrooms is essential for this purpose. The major obstacle in using mushroom fruiting bodies on a large scale is the lack of information on the environmental and physiological control over the production of fruiting bodies. Hence studies on *Pleurotus* cultivation techniques (Gregori *et al.*, 2007) will have a great demand in the future. Various agricultural wastes have been successfully used as substrates for the cultivation of *Pleurotus* mushrooms (Melo de Carvalho *et al.*, 2010) and this knowledge would help to develop economically viable methods for large scale cultivation. Further, the fruiting bodies are complex structures and their chemical composition can vary depending on the substrate used for cultivation. Such changes in the biochemical nature will create problems in standardization of the products produced from such mushrooms.

Mushrooms are known to absorb heavy metals; the mycelium channels the heavy metal to the mushroom itself. Hence, care must be taken when choosing substrates for large scale production of mushrooms. In a study where the heavy metal analysis of mushrooms were conducted (Mattila *et al.*, 2001) it was found that *Agaricus bisporous* (brown) contained large amounts of selenium and *Lentinus edodes* contained large amounts of cadmium. However, the heavy metal concentration in *P. ostreatus* was not high as to be alarmed.

As the availability of fruiting bodies grown on solid substrates on a large scale is questionable, submerged mycelium cultures of the *Pleurotus* fungus are used as an alternative method to obtain mushroom extracts. Smith *et al.*, 2002 suggest that submerged mycelium cultures produce more uniform and reproducible biomass for different industrial products. Although the extracts of *Pleurotus* mushrooms have shown various activities, it is our belief that products derived from the whole mushroom may have more health benefits as a dietary supplement. More importantly, isolation of glucans or obtaining glucan enriched fractions from *Pleurotus* and derivatizations such as carboxymethylations and sulfonation of glucans would lead to the development of pharmaceutically useful products in the future.

References

Abeytunga, D.T.U., Menikpurage, I.P., Ratnasooriya, W.D., Pathirana, W. (2008). *Cholesterol lowering edible mushroom tablets and powder satches*. Patent No: 14471, National Intellectual Property Office of Sri Lanka.

Ajith, T.A., Janardhanan, K.K. (2007). Indian medicinal mushrooms as a source of antioxidant and antitumor agents. *Journal of Clinical Biochemistry and Nutrition*, 40(03), 157-162.

Assi, J.A., King, A.J. (2007). Assessment of selected antioxidants in tomato pomace subsequent to treatment with the edible oyster mushroom, *Pleurotus ostreatus*, under solid-state fermentation. *Journal of Agricultural and Food Chemistry*, 55, 9095-9098.

Badole, S.L., Bodhankar, S.L. (2007). Interaction of aqueous extract of *Pleurotus pulmonarius* (Fr.) Quel.-Champ with acarbose in alloxan induced diabetic mice. *Journal of Applied Biomedicine*, 5, 157-166.

Badole, S.L., Bodhankar S.L. (2008). Effect of concomitant administration of aqueous extract of *Pleurotus pulmonarius* and metformin in diabetic mice. *Journal of Complementary and Integrative Medicine,* 5, 27.

Badole, S.L., Shah, S.N., Patel, N.M., Thakurdesai, P.A., Bodhankar, S.L. (2006). Hypoglycemic activity of aqueous extract of *Pleurotus pulmonarius* in Alloxan induced diabetic mice. *Pharmaceutical Biology*, 44, 421-425.

Bajaj, M., Vadhera, S., Brar, A.P., Soni, G.L. (1997). Role of oyster mushroom (*Pleurotus florida*) as hypocholesterolemic/antiatherogenic agent. *Indian Journal of Experimental Biology*, 35(10), 1070-1075.

Beltran-Garcia, M.J., Estrarron-Espinosa, M., Ogura T., (1997). Volatile compounds secreted by the oyster mushroom (*Pleurotus ostreatus*) and their antibacterial activities. *Journal of Agricultural and Food Chemistry,* 45, 4049-4052.

Berne, S., Krizaj, I., Pohleven F., Tuck, T., Macek, P., Sepcic, K. (2002). *Pleurotus* and *Agrocybe* hemolysins, new proteins hypothetically improved in fungal fruiting. *Biochimica et Biophysica Acta.*, 1570, 153-159.

Bobek, P., Galbavy, S. (1999). Hypocholesterolemic and antiatherogenic effect of oyster mushroom (*Pleurotus ostreatus*) in rabbits. *Nahrung.,* 43 (5), 339-342.

Bobek, P., Galbavy, S. (2001). Effect of pleuran (beta-glucan from *Pleurotus ostreatus*) on the antioxidant status of the organism and on dimethylhydrazine-induced precancerous lesions in rat colon. *British Journal of Biomedical Science*, 58(03), 164-168.

Bobek, P., Ginter, E., Jurcovicova, M., Kuniak, L. (1991). Cholesterol-lowering effect of the mushroom *Pleurotus ostreatus* in hereditary hypercholesterolemic rats. *Annals of Nutrition and Metabolism*, 35 (4), 191-195.

Bobek, P., Hromadova, M., Ozdin, L. (1995a). Oyster mushroom (*Pleurotus ostreatus*) reduces the activity of 3-hydroxy-3-methylglutaryl CoA reductase in rat liver microsomes. *Experientia*, 51 (6), 589-591.

Bobek, P., Ozdin, L., Galbavy, S. (1998). Dose and time dependent hypocholesterolemic effect of oyster mushroom (*Pleurotus ostreatus*) in rats. *Nutrition*, 14(3), 282-286.

Bobek, P., Ozdin, L., Kajaba, I. (1997). Dose dependent hypocholesterolemic effect of oyster mushroom (*Pleurotus ostreatus*) in rats. *Physiological Research*, 46, 327-329.

Bobek, P., Ozdin, L., Kuniak, L. (1995b). The effect of oyster mushroom (*Pleurotus ostreatus*), its ethanolic extract and extraction residues on cholesterol levels in serum, lipoproteins and liver of rat. *Nahrung*, 39 (1), 98-99.

Bobek, P., Ozdin, L., Kuniak, L. (1996). Effect of oyster mushroom (*Pleurotus ostreatus*) and its ethanolic extract in the diet on absorption and turnover of cholesterol in hypercholesterolemic rats. *Nahrung.,* 40 (4), 222-224.

Bourbonnais, R., Paice, M.G., Freiermuth, B., Bodie, E., Borneman, S. (1997) Reactivities of various mediators and laccasese with kraft pulp and lignin model compounds. *Applied and Environmental Microbiology*, 63 (12), 4627-4632.

Cheung, P.C.K. (1996). Dietary fiber content and composition of some cultivated edible mushroom fruiting bodies and mycelia. *Journal of Agricultural and Food Chemistry*, 44, 468-471.

Cheung, L.M., Cheung, P.C.K. (2005). Mushroom extracts with antioxidant activity against lipid peroxidation. *Food Chemistry*, 89 (3), 403-409.

Cheung, L.M., Cheung, P.C.K., Ooi, V.E.C. (2003). Antioxidant activity and total phenolics of edible mushroom extracts. *Food Chemistry*, 81 (2), 249-255.

Cheung, P.C.K., Lee, M.Y. (1998). Comparative chemical analysis of fiber material prepared by enzymatic and chemical methods from two mushrooms (*Pleurotus sajor-caju and Pleurotus tuber-regium*). *Journal of Agricultural and Food Chemistry*, 46, 4854-4857.

Cheung, P.C.K., Lee, M.Y. (2000). Fractionation and characterization of mushroom dietary fiber (nonstarch polysaccharides) as potential nutraceuticals from sclerotia of *Pleurotus tuber-regium* (Fries) Singer. *Journal of Agricultural and Food Chemistry*, 48, 3148-3151.

Chockchaisawasdee, S., Namjaidee, S., Pochana, S., Stathopoulos, C.E. (2010). Development of fermented oyster-mushroom sausage. *Asian Journal of Food and Agro-Industry*, 3(01), 35-43.

Chorvathova, V., Bobek, P., Ginter, E., Klvanova, J. (1993). Effect of the oyster fungus on glycaemia and cholesterolaemia in rats with insulin-dependent diabetes. *Physiological Research,* 42 (3), 175-179.

Cohen, R., Persky, L., Hadar, Y. (2002). Biotechnological applications and potential of wood-degrading mushrooms of the genus *Pleurotus. Applied Microbiology and Biotechnology,* 58, 582-594.

Crisan, E.V., Sands, A. (1978). Nutritional value. In Chang, S.T. & Hayes. W.A. *The biology and cultivation of edible mushrooms,* New York, Academic Press., pp. 137-165.

Dissanayake, D.P., Abeytunga, D.T.U., Vasudewa, N.S., Ratnasooriya, W.D. (2009). Inhibition of lipid peroxidation by extracts of *Pleurotus ostreatus. Pharmacognosy Magazine*, 4 (19), 266-271.

El-Fallal, A.A., Moussa, Z. (2008). Prospects for biocontrol of brown rot disease of potato *in vitro* and under greenhouse conditions. *Plant Pathology Journal*, 7, 54-64.

Fu, M., Lin, J., Wu, Z., Lin, Q., Xie, L. (2003). Screening of proteins anti-tobacco mosaic virus in *Pleurotus eryngii. Wei sheng Wu Xue Bao*., 43(1), 29-34.

Gerasimenya, V.P., Efremenkova, O.V., Kamzolkina, O. V., Bogush, T.A., Tolstych, I.V., Zenkova, V.A. (2002). Antimicrobial and antitoxical action of edible and medicinal mushroom *Pleurotus ostreatus* (Jacq.:fr.) Kumm. Extracts. *International Journal of Medicinal Mushrooms*, 4, 127-132.

Gonzalez, P., Labarère, J. (2000). Phylogenetic relationships of *Pleurotus* species according to the sequence and secondary structure of the mitochondrial small-subunit rRNA V4, V6 and V9 domains. *Microbiology*, 146, 209-221.

Gregori, A., Svagelj, M., Pohleven, J. (2007). Cultivation techniques and medicinal properties of *Pleurotus* spp., *Food Technology Biotechnology*, 45(3), 236-247.

Guardia, M.L., Venturella, G., Venturella, F. (2005). On the chemical composition and nutritional value of *Pleurotus* Taxa growing on Umbelliferous plants (Apiaceae). *Journal of Agricultural and Food Chemistry*, 53, 5997-6002.

Gunde-Cimerman, N. (1999). Medicinal value of the genus *Pleurotus* (Fr.) P. Karst. (Agaricales, Basidiomycetes). *International Journal of Medicinal Mushrooms*, 1, 69-80.

Gunde-Cimerman, N., Cimerman, A. (1995). *Pleurotus* fruiting bodies contain the inhibitor of 3-hydroxy-3-methylglutaryl-coenzyme A reductase-lovastatin. *Expeimental Mycology*, 19(1), 1-6.

Ho, J.C.K., Sze, S.C.W., Shen, W.Z., Liu, W.K. (2004). Mitogenic activity of edible mushroom lectins. *Biochimica et Biophysica Acta*, 1671, 9-17.

Hossain, S., Hashimoto, M., Choudhury, E.K., Alam, N., Hussain, S., Hasan, M., Choudhury, S.K., Mahmud, I. (2003). Dietery mushroom (*Pleurotus ostreatus*) ameliorates atherogenic lipid in hypocholesterolemic rats. *Clinical and Experimental Pharmacology and Physiology*, *30* (7), 470-475.

Hu, S.H., Wang, J.C., Lien, J.L., Liaw, E.T., Lee, M.Y. (2006). Antihyperglycemic effect of polysaccharide from fermented broth of *Pleurotus citrinopileatus*. *Applied Microbiology and Biotechnology,* 70 (1), 107-113.

Jayakumar, T., Thomas, P.A., Geraldine, P. (2007). Protective effect of an extract of the oyster mushroom, *Pleurotus ostreatus*, on antioxidants of major organs of aged rats. *Experimental Gerontology*, 47, 183-191.

Jayakumar, T., Thomas, P.A., Geraldine, P. (2009). In-vitro antioxidant activities of an ethanolic extract of the oyster mushroom, *Pleurotus ostreatus*. *Innovative Food Science and Emerging Technologies*, 10, 228-234.

Jose, N., Ajith, T.A., Janardhanan, K.K. (2002). Antioxidant, anti-inflammatory, and antitumor activities of culinary-medicinal mushroom *Pleurotus pulmonarius* (Fr.) Quel. (Agaricomycetideae). *International Journal of Medicinal Mushrooms*. 4(4), 112.

Jose, N., Ajith, T.A., Janardhanan, K.K. (2004). Methanol extract of the oyster mushroom, *Pleurotus florida*, inhibits inflammation and platelet aggregation. *Phytotherapy Research,* 18, 43-46.

Kang, T.S., Kang, M.S., Sung, J.M., Kang, A.S., Shon, H.R., Lee, S.Y. (2001). Effect of *Pleurotus eryngii* on the blood glucose and cholesterol in diabetic rats. *Korean Journal of Mycology*, 29(2), 86-90.

Kim, J.I., Kang, M.J., Im, J., Seo, Y.J., Lee, Y.M., Song, J.H., Lee, J.H., Kim, M.E. (2010). Effect of King Oyster Mushroom (*Pleurotus eryngii*) on insulin resistance and dyslipidemia in db/db Mice. *Food Science and Biotechnology*, 19(1), 239-242.

Kim, M., Seguin, P., Ahn, J., Kim, J., Chun, S., Kim, E., Seo, S., Kang, E., Kim, S., Park, Y., Ro, H., Chung, I. (2008). Phenolic compound concentration and antioxidant activities of edible and medicinal mushrooms from Korea. *Journal of Agricultural and Food Chemistry,* 56, 7265-7270.

Kudahewa, D.D., Abeytunga, D.T.U., Ratnasooriya, W.D. (2008). Antinociceptive activity of *Pleurotus cystidiosus*, an edible mushroom in rats. *Pharmacognosy Magazine*, 4(13), 37-41.

Kuribayashi, T., Kaise, H., Uno, C., Hara, T., Hayakawa, T., Joh, T. (2002). Purification and characterization of lipoxygenase from *Pleurotus ostreatus*. *Journal of Agricultural and Food Chemistry*, 50, 1247-1253.

Li, Y.R., Liu, Q.H., Wang, H.X., Ng, T.B. (2008). A novel lectin with potent antitumor, mitogenic and HIV-1 reverse transcriptase inhibitory activities from the edible mushroom *Pleurotus citrinopileatus*. *Biochimica et Biophysica Acta*, 1780, 51-57.

Linke, D., Bouws, H., Peters, T., Nimtz, M., Berger, R.G., Zorn, H. (2005). Laccases of *Pleurotus sapidus*: Characterization and cloning. *Journal of Agricultural and Food Chemistry*, 53, 9498-9505.

Liu, L., Lin, Z., Zheng, T., Lin, L., Zheng, C., Lin, Z., Wang, S., Wang, Z. (2009). Fermentation optimization and characterization of the laccase from *Pleurotus ostreatus* strain 10969. *Enzymes and Microbial Technology*, 44 (6-7), 426-433.

Lv, H., Kong, Y., Yao, Q., Zhang, B., Leng, F., Bian, H., Balzarini, J., Damme, E., Bao, J. (2009). Nebrodeolysin, a novel hemolytic protein form mushroom *Pleurotus nebrodensis* with apoptosis-inducing and anti-HIV-1 effects. *Phytomedicine: International Journal of Phytotherapy & Phytopharmacology,* 16, 198-205.

Manohar, V., Talpur, N., Echard, B.W., Lieberman, S., Preuss, H. G. (2002). Effects of a water soluble extract of maitake mushroom on circulating glucose/insulin concentrations in KK mice. *Diabetes Obesity and Metabolism*, 4, 43-48.

Mansur, M. (2003). The white-rot fungus *Pleurotus ostreatus* secretes laccase isozymes with different substrate specificities. *Mycologia*, 95(6), 1013-1020.

Matis, M., Zakelj-Mavric, M., Peter-Katalinic, J., (2005). Mass spectrometry and database search in the analysis of proteins from the fungus *Pleurotus ostreatus*. *Proteomics*, 5, 67-75.

Mattila, P., Konko, K., Eurola, M., Pihlava, J., Astola, J., Vahteristo, L., Hietaniemi, V., Kumpulainen, J., Valtonen, M., Piironen, V. (2001). Contents of vitamins, mineral elements, and some phenolic compounds in cultivated mushrooms. *Journal of Agricultural and Food Chemistry*, 49, 2343-2348.

Mattila, P., Salo-Vaananen, P., Konko, K., Aro, H., Jalava, T. (2002). Basic composition and amino acid contents of mushrooms cultivated in Finland. *Journal of Agricultural and Food Chemistry,* 50, 6419-6422.

Mau, J., Lin, Y., Chen, P., Wu, Y., Peng, J. (1998). Flavour compounds in king oyster mushrooms *Pleurtous eryngii*. *Journal of Agricultural and Food Chemistry*, 46, 4587-4591.

Melo de Carvalho, C.S., Sales-Campos, C., Nogueira de Andrade, M.C. (2010). Mushrooms of the *Pleurotus* genus: A review of cultivation techniques. *Interciencia*, 35, 177-182.

Menikpurage, I.P., Abeytunga, D.T.U., Jacobsen, N. E., Wijesundara, R.L.C. (2009). An oxidized ergosterol from *Pleurotus cystidiosus* active against anthracnose causing *Colletotrichum gloeosporioides*. *Mycopathologia*, 167(3), 155-162.

Munoz, C., Guillen, F., Martinez, A.T., Martinez, M.J. (1997). Induction and characterization of laccase in the ligninolytic fungus *Pleurotus eryngii*. *Current Microbiology*, 34 (1), 1-5.

Ng, T.B., Lam, S.K., Chan, S.Y. (2002). A ubiquitin-like peptide from the mushroom *Pleurotus sajor-caju* exhibits relatively potent translation-inhibitory and ribonuclease activities. *Peptides*, 23, 1361-1365.

Ngai, P.H.K., Ng, T.B. (2004). A ribonulease with antimicrobial, antimitogenic and antiproliferative activities from the edible mushroom *Pleurotus sajor-caju*. *Peptides*, 25, 11-17.

Ngai, P.H.K., Ng, T.B. (2006). A hemolysin from the mushroom *Pleurotous eryngii*. *Applied Microbiology and Biotechnology,* 72, 1185-1191.

Palyannan, T., Sathishkumar, P. (2010). Production of laccase from *Pleurotus florida* NCIM 1243 using Plackett-Burman design and response surface methodology. *Journal of basic microbiology*, 50(4), 325-335.

Pizzoferrato, L., Manzi, P., Bertocchi, F., Fanelli, C., Rotilio, G., Paci, M. (2000). Solid-state ^{13}C CP MAS NMR spectroscopy of mushrooms gives directly the ratio between proteins and polysaccharides. *Journal of Agricultural and Food Chemistry*, 48, 5484-5488.

Puttaraju, N.G., Venkateshaiah, S.U., Dharmesh, S.M., Urs, S.M.N., Somasundaram, R. (2006). Antioxidant activity of indigenous edible mushrooms. *Journal of Agricultural and Food Chemistry*, 54, 9764-9772.

Robbins, W.J. Kavanagh, F., Hervey, A., (1947). Antibiotic substances from basidiomycetes, *Pleurotus griseus*. *Proceedings of the National Academy of Science,* U.S.A., 33(6), 171-176.

Rodriguez, E., Nuero, O, Guillen, F., Martinez, A.T., Martinez, M.J. (2004). Degradation of phenolic and non-phenolic aromatic pollutants by four *Pleurotus* species: The role of laccase and versatile peroxidase. *Soil Biology & Biochemistry*, 36, 909-916.

Rodriquez, E., Pickard, M.A., Vazquez-Duhalt, R. (1999). Industrial dye decolorization by laccases from ligninolytic fungi. *Current Microbiology*, 38 (1), 27-32.

Ruiz-Duenas, F.J., Martinez, M.J. (1996). Enzymatic activities of *Trametes versicolor* and *Pleurotus eryngii* implicated in biocontrol of *Fusarium oxysporum* f. sp. *Lycopersici. Current Microbiology*, 32(3), 151-155.

Sathishkumar, P., Murugesan, K., Palyannan, T. (2010). Production of laccase from *Pleurotus florida* using agrowastes and efficient decolorization of Reactive blue 198. *Journal of Basic Microbiology,* 50, 360-367.

Selegean, M., Putz, M., Rugea, T. (2009). Effect of the polysaccharide extract from the edible mushroom *Pleurotus ostreatus* against infectious bursal disease virus. *International Journal of Molecular Sciences*, 10, 3616-3634.

Sepcic, K., Berne, S., Rebolj, K., Batista, U., Plemenitas, A., Sentjurc, M., Macek, P. (2004). Ostreolysin, a pore-forming protein from the oyster mushroom, interacts specifically with membrane cholesterol-rich lipid domains. *FEBS Letters*, 575, 81-85.

Smith, J.E., Rowan, N.J., Sullivan, R. (2002). Medicinal mushrooms: a rapidly developing area of biotechnology for cancer therapy and other bioactivities. *Biotechnology Letters,* 24, 1839-1845.

Tao, Y., Yan, F., Zhang, L. (2007a). Fractionation and characterization of a protein-polysaccharide complex from *Pleurotus tuber-regium* sclerotia. *Journal of Polymer Science: Part B: Polymer Physics,* 45, 2546-2554.

Tao, Y., Zhang, L., Yan F., Wu, X. (2007b). Chain conformation of water-insoluble hyperbranched polysaccharide from fungus. *Biomacromolecules*, 8, 2321-2328.

Vasudewa, N.S., Abeytunga, D.T.U., Ratnasooriya, W.D. (2007). Antinociceptive activity of *Pleurotus ostreatus*, an edible mushroom, in rats. *Pharmaceutical Biology*, 45(7), 533-540.

Vasudewa, N.S., Abeytunga, D.T.U., Ratnasooriya, W.D. (2008). Activity-directed fractionation of *Pleurotus ostreatus* in the search for analgesics. *Pharmaceutical Biology*, 46(05), 295-301.

XiJun, F., HongYu, X., ZhenMing, L., ZhengHong, X. (2009). Antibacterial activity of submerged cultures of *Pleurotus tuber-regium. Acta Edulis Fungi*, 16(4), 50-53.

Yang, J.H., Lin, H.C., Mau, J.L. (2001). Non-volatile taste components of several commercial mushrooms. *Food Chemistry*, 72, 465-471.

Yang, J.H., Lin, H.C., Mau, J.L. (2002). Antioxidant properties of several commercial mushrooms. *Food Chemistry*, 77 (2), 229-235.

Wang, H.X., Ng, T.B. (2000). Isolation of a novel ubiquitin-like protein from *Pleurotus ostreatus* mushroom with anti-human immunodeficiency virus, translation-inhibitory, and ribonuclease activities. *Biochemical and Biophysical Research Communications*, 276 (2), 587-593.

Wang, H., Ng, T.B. (2004). Eryngin, a novel antifungal peptide from fruiting bodies of the edible mushroom *Pleurotus eryngii. Peptides*, 25, 1-5.

Wipf, P., Lynch, S.M., Birmingham, A., Tamayo, G., Jimenez, A., Campos, N., Powis, G. (2004). Natural product based inhibitors of the thioredoxin-thioredoxin reductase system. *Organic & Biomolecular Chemistry*, 2, 1651-1658.

Wong, K.H., Cheung, P.C.K (2005). Dietary fibers from mushroom sclerotia: Preparation and physicochemical and functional properties. *Journal of Agricultural and Food Chemistry*, 53, 9395-9400.

Wong, K.H., Cheung, P.C.K., Wu, J.Z. (2003). Biochemical and microstructural characteristics of insoluble and soluble dietary fiber prepared from mushroom sclerotia of *Pleurotus tuber-regium*, *Polyporus rhinoceros* and *Wolfiporia cocos*. *Journal of Agricultural and Food Chemistry*, 51, 7197-7202.

Wong, S.M., Wong, K.K., Chiu, L.C.M., Cheung, P.C.K (2007). Non-starch polysaccharides from different developmental stages of *Pleurotus tuber-regium* inhibited the growth of human acute promyelocytic leukemia HL-60 cells by cell-cycle arrest and/or apoptotic induction. *Carbohydrate Polymers*, 68, 206-217.

Zhang, M., Cheung, P.C.K., Chiu, L.C.M., Wong, E.Y.L., Ooi, V.E.C. (2006). Cell-cycle arrest and apoptosis induction in human breast carcinoma MCF-7 cells by carboxymethylated beta-glucan from the mushroom sclerotia of *Pleurotus tuber-regium*. *Carbohydrate Polymers*, 66, 455-462.

Zhang, M., Cheung, P.C., Ooi, V.E., Zhang, L. (2004a). Evaluation of sulfated fungal beta-glucans from the sclerotium of *Pleurotus tuber-regium* as a potential water-soluble anti-viral agent. *Carbohydrate Research*, 339(13), 2297-2301.

Zhang, M., Cheung, P.C.K., Zhang, L. (2001). Evaluation of mushroom dietary fiber (nonstarch polysaccharides) from sclerotia of *Pleurotus tuber-regium* (Fries) Singer as a potential antitumor agent. *Journal of Agricultural and Food Chemistry*, 49, 5059-5062.

Zhang, M., Cheung, P.C.K., Zhang, L., Chiu, C.M., Ooi, V.E.C. (2004b). Carboxymethylated beta-glucans from mushroom sclerotium of *Pleurotus tuber-regium* as novel water-soluble anti-tumor agent. *Carbohydrate Polymers*, 57(3), 319-325.

Zhang, M., Cui, S.W., Cheung, P.C.K., Wang, Q. (2007). Antitumor polysaccharides from mushrooms: a review on their isolation process, structural characteristics and antitumor activity. *Trends in Food Science & Technology*, 18, 4-10.

Zhang, M., Zhu, L., Cui, S.W., Wang, Q., Zhou, T., Shen, H. (2011). Fractionation, partial characterization and bioactivity of water-soluble polysaccharides and polysaccharide-protein complexes from *Pleurotus geesteranus*. *International journal of Biological macromolecules*, 48, 5-12.

In: Mushrooms: Types, Properties and Nutrition
Editors: S. Andres and N. Baumann
ISBN 978-1-61470-110-1

Chapter 19

MEDICINAL ACTIVITIES OF MUSHROOM POLYSACCHROPEPTIDES

Tzi Bun Ng*, Jack Ho Wong±, Helen Chan, Tak Fu Tse and Timothy Tam
The Chinese University of Hong Kong,Satin, New Territories,Hong Kong,China and Vita Green Pharmaceutical (H.K.) Limited, Hong Kong, China

ABSTRACT

Mushrooms produce a number of molecules with medicinal properties, including polysaccharides, polysaccharide-peptide complexes, lectins, and antifungal proteins have been isolated from several mushroom species, but the one from *Coriolus versicolor* designated as PSP is the most extensively studied.PSP induces apoptosis in a variety of tumor cells by regulating the expression of pro- and anti-apoptotic genes.. It inhibits angiogenesis, regulates the production of cytokines, counteracts the action of immunosuppressive agents, and potentiates the action of antitumor drugs.PSP is devoid of toxicity on female reproductive physiology and embryonic development.. Polysaccharide-peptide complexes from other mushroom species also manifest antitumor and immunoregulatory activities.

INTRODUCTION

Biomolecules from medicinal mushrooms constitute a potentially important new source of anticancer therapeutics. Their assimilation into Western drug discovery programs and clinical trials provides a framework for the study and use of other traditional

* Corresponding author: School of Biomedical Sciences, Faculty of Medicine, The Chinese University of Hong Kong,Satin, New Territories,Hong Kong,China and Vita Green Pharmaceutical (H.K.) Limited, 15th floor, Oceanic Industrial Centre, 2 Lee Lok Street, Ap Lei Chau, Hong Kong, China.
± Corresponding author: School of Biomedical Sciences, Faculty of Medicine, The Chinese University of Hong Kong,Satin, New Territories,Hong Kong,China and Vita Green Pharmaceutical (H.K.) Limited, 15th floor, Oceanic Industrial Centre, 2 Lee Lok Street, Ap Lei Chau, Hong Kong, China.

medicines.(Sullivan et al.,2006).The intent of the present article is to review the medicinal properties of polysaccharide-peptide complexes from mushrooms with an emphasis on that from *Coriolus versicolor*.

POLYSACCHAROPEPTIDE FROM CORIOLUS VERSICOLOR

PSP and PSK

Protein-bound polysaccharides, designated as PSK(Polysaccharide-K) and PSP(Polysaccharide-Peptide), have been isolated from the CM-101 strain of the mushroom *Coriolus versicolor* by Japanese researchers and the COV-1 strain by Chinese investigators, respectively,. PSP is approximately 100 kDa. in molecular weight Glutamic and aspartic acids are abundant in its polypeptide component, whereas its polysaccharide component is composed of monosaccharides with alpha-1,4 and beta-1,3 glucosidic linkages. The presence of fucose in PSK and rhamnose and arabinose in PSP distinguishes the two protein-bound polysaccharides, which are otherwise chemically similar. PSP is classified as a biological response modifier. It induces, in experimental animals, increased gamma-interferon production, interleukin-2 production, and T-cell proliferation. It also counteracts the depressive effect of cyclophosphamide on white blood cell count, interleukin-2 production and delayed-type hypersensitivity reaction. Its antiproliferative activity against tumor cell lines and in vivo antitumor activity have been demonstrated. A small peptide with a molecular weight of 16-18 kDa originating from PSP has been produced with antiproliferative and antitumor activities. PSP administered to patients with esophageal cancer, gastric cancer and lung cancer, and who are undergoing radiotherapy or chemotherapy, helps alleviate symptoms and prevents the decline in immune status(Ng,1998). In double-blind trials, PSP significantly extended five-year survival in esophageal cancer. PSP significantly improved quality of life, provided substantial pain relief, and enhanced immune status in 70-97 percent of patients with cancers of the stomach, esophagus, lung, ovary, and cervix.. PSK significantlyprolonged survival at five years or beyond in cancers of the stomach, colon-rectum, esophagus, nasopharynx, and lung (non-small cell types), and in a HLA B40-positive breast cancer subset. (Kidd,2000).

Preparation of PSP

Mushrooms gathered in the wild, cultivated mushrooms, and the mycelial biomass of submerged culture are used to produce the polysaccharopeptides. Submerged cultures are typically carried out in batches lasting 5-7 days and at 25-27 degrees C. Hot water extraction of the biomass is used to recover the thermostable polysaccharopeptides that are concentrated, purified, and dried into a powder for medicinal use..(Cui and Chisti,2003).

PSP Enhances Apoptosis in Tumor Cells

A CV extract(Vita Green) dose-dependently suppressed the proliferation of Raji (B-cell lymphoma), NB-4, and HL-60 (human promyelocytic leukemia) cells by more than 90%. The extract however did not exert any significant cytotoxic effect on normal liver WRL.cells.Nucleosome production in HL-60, NB-4 and Raji cells but not thaty in WRL cells was elevated (Lau et al.,2004).

PSP pre-treatment dose-dependently increased cell cycle perturbation and the apoptotic activity of doxorubicin (Doxo) and etoposide (VP-16) in human promyelocytic leukemia HL-60 cells. t PSP enhanced the apoptotic machinery of Doxo and VP-16 in a cell cycle-dependent manner (Hui et al.,2006)..

A CV extract(Vita Green) dose-dependently suppressed the proliferation of HL-60 cells with augmented nucleosome production from apoptotic cells. Expression of the pro-apoptotic protein Bax in HL-60 cells was up-regulatede after treatment while that of the anti-apoptotic protein Bcl-2 was down-regulated. The release of cytochrome c from mitochondria to cytosol was enhanced. Studies employing the athymic nude mouse xenograft model corroborated the antitumor activity of the CV extract. The CV extract may be useful for the treatment of some forms of human leukemia.(Ho et al., 2006).The CV extract alsoexhibited antiproliferative activity toward three breast tumor cell lines, T-47D, MCF-7, MDA-MB-231, with a potency comparable to that of the chemotherapeutic drug mitomycin C. Nucleosome production in apoptotic MDA-MB-231, MCF-7 and T-47D cells was enhanced. Expression of p53 protein was significantly upregulated only in T-47D cells but not in MCF-7 and MDA-MB-231 cells. The CV extract downregulated Bcl-2 protein expression in MCF-7 and T-47D cells, but not in MDA-MB-231 cells (Ho et al.,2005).

Suppressive Effect of PSP on Angiogenesis

Angiogenesis plays a pivotal role in tumor growth and metastasis, Quantitative analysis of microcorrosion casting of the tumor tissue showed more angiogenic features such as dense sinusoids and hot spots, in control (untreated) than in PSP-treated S180 tumor-bearing mice. Immunostaining of tumor tissues with antibody against the endothelial cell marker (Factor VIII) demonstrated a positive correlation in that both the vascular density and tumor weight were reduced in PSP -treated mice. Morphometric analysis of corrosion casts revealed that, even though the total amount of new vessel production was reduced, the basic tumor type-specific vascular architecture was retained. However, the expression of vascular endothelial cell growth factor in these tumors was suppressed. In conclusion, PSP suppresses tumor growth by inhibiting angiogenesis (Ho et al.,2004).

PSP Potentiates Anticancer Activity of Doxorubicin and Etoposide on Human Breast Cancer Cells

PSP augmented the cytotoxicity of doxorubicin (Doxo), etoposide (VP-16)) to the human breast cancer cell line ZR-75-30 by creating an S-phase trap . The ratio of protein expression of Bcl-xL/Bax was lowered in the cancer cells. This study supports the hypothesis that the anticancer potentials of PSP is not limited to leukemia but may also be used as an adjuvant therapy for breast cancers(Wan et al.,2008)..

Effect of PSP on Cell-Cycle Progression and Expression of Interleukins-1 Beta,-6, and -8 in Promyelocytic HL-60 Leukemic Cells

Hsieh et al.(2002)tested the cytostatic and cytotoxic activities of extracts derived from the polysaccharopeptide (PSP), I'm-Yunity, prepared from strain Cov-1 of the mushroom *Coriolus versicolor,* on HL-60 cells. After incubation with the extracts for 1-3 days the cells displayed dose-and time-dependent decline in growth and viability. Flow cytometric analysis revealed partial cell arrest in the G(1) phase at low concentrations and induction of apoptosis at high concentrations of ethanol and water extracts, with the latter being more potent. Water extract of I'm-Yunity also significantly increased IL-1 beta and IL-6 but reduced IL-8. (Hsieh et al.,2002).

A standardized ethanol-water extract prepared from CV (Vita Green)exerted proliferative activity toward murine splenic lymphocytes in a time- and dose-dependent manner, maximally by 2.4-fold.The production of two Th1-related cytokines, interleukin (IL)-2 and IL-12,by the CV extract-activated lymphocytes was upregulated at 48h .The levels of two other Th1-related cytokines, interferon (IFN)-gamma and IL-18, was elevated only at 24 h, but not at 48 and 72 h whereas those of two Th2-related cytokines, IL-4 and IL-6, were indiscernible. (Ho et al.,2004).

Effect of PSP on Cyclophosphamide-Induced Immunosuppression

PSP potentiated the cytotoxic effect of cyclophosphamide on HepG2 cells in vitro.In the rat.it decreased the clearance of cyclophosphamide with a simultaneous increase in the area under concentration-time curve, and increased the plasma half-life. Hence PSP may be used as an adjunct to treatment(Chan and Yeung,2006)..PSP was effective in overcoming - cyclophosphamide induced immunosuppression such as reduced lymphocyte proliferation, natural killer cell function, white blood cell production, splenic growth and thymic growth in rats and also in increasing IgG and IL-2 production The data suggest that PSPcould be considered as an adjuvant in conjunction with cyclophosphamide or other chemotherapeutic agent in treatment of cancer patients. (Qian et al.,1997)..

Activation of Mouse Lymphocytes and Macrophages by PSP

Both mouse lymphocytes and macrophages were activated by preparations of polysaccharopeptide from cultured mycelia and culture medium of C. versicolor. However, no direct cytotoxic activity against fibroblasts, hepatoma cells and choriocarcinoma cells could be demonstrated.(Wang et al.,1996).

Lack of an Effect of PSP on Female Reproductive System and Embryonic Development in Mice

PSPdid not affect ovarian steroidogenesis, ovulation and midterm gestation in ICR mice. It did not exert an adverse effect on mouse embryonic development either, as evidenced by the lack of an effect on somite number, axial length and the incidence of abnormalities in heartbeat, yolk sac circulation, optic vesicle, otic vesicle, shape of body axis, forelimb buds, branchial apparatus, cranial neural tube and head size.(Ng and Chan,1997).

Hyperalgesic Effect of PSP

Diphenhydramine inhibited PSP-induced writhing. In the hot-plate test, PSP dose-dependently shortened the hind paw withdrawal latency, indicative of a hyperalgesic effect which was reduced by pretreatment with anti-inflammatory drugs. PSP-induced hyperalgesia was related to the release of inflammatory mediators including prostaglandin E2, tumor necrosis factor-alpha, interleukin-1beta, and histamine in mouse peritoneal macrophages and mast cells (Chan and Yeung,2006)..

Anti-Human Immunodeficiency Virus Activity of PSP

PSP inhibited the interaction between HIV-1 gp 120 and immobilized CD4 receptor (IC50 = 150 microg/ml) Recombinant HIV-1 reverse transcriptase was also inhibited (IC50 = 6.25 microg/ml), and so was a glycohydrolase enzyme associated with viral glycosylation. Collins and Ng,1997)..

Polysaccharopeptide from Grifola Frondosa

Chan et al.(2011)examined the effects of phosphorylation,, acetylation or esterification on the adjuvant and anticancer effects of crude maitake polysaccharide-peptides (MPSP) . The adjuvant effect and growth inhibitory effect were assessed by using C6 cell inoculated rat model with cyclophosphamide (CPA) treatment and in vitro cell viability assay, respectively. All MPSPs s tested demonstrated significant adjuvant effect to CPA treatment on rats inoculated with C6 cancer cells. In addition, growth inhibitory effect was observed in cancer cells but not in normal brain cells.Only phosphorylation could significantly enhance the adjuvant (in vivo) and growth inhibitory effect. A same rank order (phosphorylated MPSP > esterified MPSP ≥ acetylated MPSP ≥ crude MPSP) of efficacy was observed in both in vivo and in vitro assays. This study showed chemical phosphorylation could markedly potentiate the adjuvant and growth inhibitory effects. (Chan et al.,2011).

GFPPS1b, a polysaccharide-peptide from cultured mycelia of *Grifola frondosa* GF9801, exerted antiproliferative activity toward SGC-7901 cells.. The treated cells manifested morphological characteristics typical of apoptosis including loss of villi and appearance of apoptotic bodies on the cell surface, volume reduction, and chromatin condensation,. Thedata from flow cytometry analysis and annexin V-PI assay disclosed that the cell cycle was arrested in the G(2)/M phase, the subdiploid peak of DNA typical of apoptosis was also present.. In agarose gel electrophoresis DNA isolated from treated cells demonstrated DNA 'laddering characteristic of apoptosis. Decline in mitochondrial trans-membrane potential, upregulation of Bax,, downregulation of Bcl-2, and activation of caspase-3 were noted.The results imply that GFPS1b could suppress SGC-7901 cell growth and reduce cell survival(Cui et al.,2007).

Polysaccharopeptide from Lentinus Edodes

A polysaccharide-peptide complex (LE) from culture mycelia of ***Lentinus edodes*** upregulated the expression levels of IL-2 and TNF-alpha genes and increased production of the two cytokines in murine spleen mononuclear cells and human peripheral blood mononuclear cells. The results indicate that LE may trigger Th immune responses.(Liu et al.,1998).

Polysaccharopeptide from Pleurotus Abalonus

The polysaccharide-peptide complex from abalone mushroom (*Pleurotus abalonus*)-fruiting bodies augmented the activities of the antioxidant enzymes superoxide dismutase, catalase, and glutathione peroxidase in the brain, kidney, and liver of senescence-accelerated mice and up-regulated their gene expression levels and suppressed the malondialdehyde content in the three aforementioned organs. The maximal lifespan of the treated mice was prolonged(Li et al.,2007).

Polysaccharopeptide from Tricholoma Mongolicum

A 15.5 -kDa,polysaccharide-peptide complex from the mycelial culture of *Tricholoma mongolicum,* was capable of activating macrophages, stimulating macrophage antigen-presenting activity, which in turn promoted proliferation of T-cells, and suppressed the growth of sarcoma 180 cells implanted in mice(Wang et al.,1996)..

Polysaccharopeptide from Tricholoma Species

A 17 –kDa polysaccharide-peptide complex (PSPC) from a submerged mycelial culture of *Tricholoma* sp., could activate the macrophages, stimulate the proliferation of T-cells, and inhibit the growth of sarcoma 180 in mice. (Wang et al.,1996).

CONCLUSION

Mushroom polysaccharide-peptide complexes generally display immunostimulatory and antitumor activities.Other activities such as antioxidant,and antiviral activities have also been reported.*Coriolus versicolor* PSP is one of the most intensively studied mushroom polysaccharide-peptide complexes and has been shown to produce beneficial effects on cancer patients, especially in counteracting chemotherapy-induced immunosuppression.As more research is conducted, improved strategies involving PSP may evolve to prolong the lifespan of cancer patients or to save their lives.

REFERENCES

Chan JY, Chan E, Chan SW, Sze SY, Chan MF, Tsui SH, Leung KY, Chan RY, Chung IY. Enhancement of in vitro and in vivo anticancer activities of polysaccharide peptide from *Grifola frondosa* by chemical modifications. *Pharm Biol.* 2011 May 20. [Epub ahead of print]

Chan SL, Yeung JH.Effects of polysaccharide peptide (PSP) from *Coriolus versicolor* on the pharmacokinetics of cyclophosphamide in the rat and cytotoxicity in HepG2 cells. *Food Chem Toxicol.* 2006;44(5):689-94.

Chan SL, Yeung JHPolysaccharide peptides from COV-1 strain of *Coriolus versicolor* induce via inflammatory mediator release in the mouse. *Life Sci.* 2006;78(21):2463-70.

Collins RA, Ng TB.Polysaccharopeptide from *Coriolus versicolor* has potential for use against human immunodeficiency virus type 1 infection. Life Sci. 1997;60(25):PL383-7.

Cui J, Chisti Y.Polysaccharopeptides of *Coriolus versicolor:* physiological activity, uses, and production. *Biotechnol Adv.* 2003 21(2):109-22.

Cui FJ, Li Y, Xu YY, Liu ZQ, Huang DM, Zhang ZC, Tao WY. Induction of apoptosis in SGC-7901 cells by polysaccharide-peptide GFPS1b from the cultured mycelia of *Grifola frondosa* GF9801. *Toxicol In Vitro.* 2007 ;21(3):417-27.

Ho CY, Lau CB, Kim CF, Leung KN, Fung KP, Tse TF, Chan HH, Chow MS. Differential effect of *Coriolus versicolor* (Yunzhi) extract on cytokine production by murine lymphocytes in vitro. *Int Immunopharmacol.* 2004 ;4(12):1549-57.

Ho CY, Kim CF, Leung KN, Fung KP, Tse TF, Chan H, Lau CB. Differential anti-tumor activity of *Coriolus versicolor* (Yunzhi) extract through p53- and/or Bcl-2-dependent apoptotic pathway in human breast cancer cells. *Cancer Biol Ther.* 2005;4(6):638-44.

Ho CY, Kim CF, Leung KN, Fung KP, Tse TF, Chan H, Lau CB. *Coriolus versicolor* (Yunzhi) extract attenuates growth of human leukemia xenografts and induces apoptosis through the mitochondrial pathway. *Oncol Rep.* 2006 ;16(3):609-16.

Ho JC, Konerding MA, Gaumann A, Groth M, Liu WK. Fungal polysaccharopeptide inhibits tumor angiogenesis and tumor growth in mice. *Life Sci.* 2004;75(11):1343-56.

Hsieh TC, Kunicki J, Darzynkiewicz Z, Wu JM.Effects of extracts of *Coriolus versicolor* (I'm-Yunity) on cell-cycle progression and expression of interleukins-1 beta,-6, and -8 in promyelocytic HL-60 leukemic cells and mitogenically stimulated and nonstimulated human lymphocytes. *J Altern Complement Med.* 2002;8(5):591-602.

Hui KP, Sit WH, Wan JMInduction of S phase cell arrest and caspase activation by polysaccharide peptide isolated from *Coriolus versicolor* enhanced the cell cycle dependent activity and apoptotic cell death of doxorubicin and etoposide, but not cytarabine in HL-60 cells. *Oncol Rep.* 2005;14(1):145-55.

Kidd PM.The use of mushroom glucans and proteoglycans in cancer treatment. *Altern Med Rev.* 2000;5(1):4-27.

Lau CB, Ho CY, Kim CF, Leung KN, Fung KP, Tse TF, Chan HH, Chow MS. Cytotoxic activities of *Coriolus versicolor* (Yunzhi) extract on human leukemia and lymphoma cells by induction of apoptosis. *Life Sci.* 2004 ;75(7):797-808.

Li L, Ng TB, Song M, Yuan F, Liu ZK, Wang CL, Jiang Y, Fu M, Liu F. A polysaccharide-peptide complex from abalone mushroom (*Pleurotus abalonus)* fruiting bodies increases activities and gene expression of antioxidant enzymes and reduces lipid peroxidation in senescence-accelerated mice. *Appl Microbiol Biotechnol.* 2007;75(4):863-9.

Liu M, Li J, Kong F, Lin J, Gao Y. Induction of immunomodulating cytokines by a new polysaccharide-peptide complex from culture mycelia of Lentinus edodes. *Immunopharmacology*. 1998;40(3):187-98.

Ng TB.A review of research on the protein-bound polysaccharide (polysaccharopeptide, PSP) from the mushroom *Coriolus versicolor* (Basidiomycetes: Polyporaceae).*Gen Pharmacol.* 1998;30(1):1-4.

Ng TB, Chan WYPolysaccharopeptide from the mushroom *Coriolus versicolor* possesses analgesic activity but does not produce adverse effects on female reproductive or embryonic development in mice. *Gen Pharmacol.* 1997;29(2):269-73.

Qian ZM, Xu MF, Tang PLPolysaccharide peptide (PSP) restores immunosuppression induced by cyclophosphamide in rats. *Am J Chin Med.* 1997;25(1):27-35.

Sullivan R, Smith JE, Rowan NJ.Medicinal mushrooms and cancer therapy: translating a traditional practice into Western medicine. *Perspect Biol Med.* 2006 49(2):159-70.

Wan JM, Sit WH, Louie JC. Polysaccharopeptide enhances the anticancer activity of doxorubicin and etoposide on human breast cancer cells ZR-75-30. *.Int J Oncol.* 2008;32(3):689-99.

Wang HX, NG TB, Liu WK, Ooi VE, Chang ST.Polysaccharide-peptide complexes from the cultured mycelia of the mushroom *Coriolus versicolo*r and their culture medium activate mouse lymphocytes and macrophages. *Int J Biochem Cell Biol.* 1996;28(5):601-7.

Wang HX, Ng TB, Ooi VE, Liu WK, Chang ST. A polysaccharide-peptide complex from cultured mycelia of the mushroom *Tricholoma mongolicum* with immunoenhancing and antitumor activities. *Biochem Cell Biol.* 1996;74(1):95-100.

Wang HX, Liu WK, Ng TB, Ooi VE, Chang ST.Immunomodulatory and antitumor activities of a polysaccharide-peptide complex from a mycelial culture of *Tricholoma* sp., a local edible mushroom. *Life Sci.* 1995;57(3):269-81.

Ho CY, Lau CB, Kim CF, Leung KN, Fung KP, Tse TF, Chan HH, Chow MS. Differential effect of *Coriolus versicolor* (Yunzhi) extract on cytokine production by murine lymphocytes *in vitro*. *Int Immunopharmacol.* 2004 ;4(12):1549-57.

Ho CY, Kim CF, Leung KN, Fung KP, Tse TF, Chan H, Lau CB. Differential anti-tumor activity of *Coriolus versicolor* (Yunzhi) extract through p53- and/or Bcl-2-dependent apoptotic pathway in human breast cancer cells. *Cancer Biol Ther.* 2005;4(6):638-44.

Ho CY, Kim CF, Leung KN, Fung KP, Tse TF, Chan H, Lau CB. *Coriolus versicolor* (Yunzhi) extract attenuates growth of human leukemia xenografts and induces apoptosis through the mitochondrial pathway. *Oncol Rep.* 2006 ;16(3):609-16.

INDEX

A

B

D

E

F

G

H

I

J

K

L

M

N

O

P

Q

R

S

T

U

V

W

X

Y

Z